AF352936

Advances in Construction and Project Management—Volume II

Advances in Construction and Project Management—Volume II

Construction and Digitalisation

Editors

Srinath Perera
Albert P. C. Chan
Dilanthi Amaratunga
Makarand Hastak
Patrizia Lombardi
Sepani Senaratne
Xiaohua Jin
Anil Sawhney

MDPI • Basel • Beijing • Wuhan • Barcelona • Belgrade • Manchester • Tokyo • Cluj • Tianjin

Editors

Srinath Perera
Western Sydney University
Australia

Albert P. C. Chan
Shenzhen Research Institute
of the Hong Kong Polytechnic
University
China

Dilanthi Amaratunga
Disaster Risk Management at
the University of
Huddersfield
UK

Makarand Hastak
Purdue University
USA

Patrizia Lombardi
Politecnico di Torino
Italy

Sepani Senaratne
Western Sydney University
Australia

Xiaohua Jin
Western Sydney University
Australia

Anil Sawhney
Royal Institution of Chartered
Surveyors (RICS) and
Adjunct Faculty at Columbia
University
UK

Editorial Office
MDPI
St. Alban-Anlage 66
4052 Basel, Switzerland

This is a reprint of articles from the Topic published online in the open access journals *Buildings* (ISSN 2075-5309), *Sustainability* (ISSN 2071-1050), *Applied Sciences* (ISSN 2076-3417), *Automation* (ISSN 2673-4052), *Informatics* (ISSN 2227-9709), and *Safety* (ISSN 2313-576X) (available at: https://www.mdpi.com/topics/Construction_Management).

For citation purposes, cite each article independently as indicated on the article page online and as indicated below:

LastName, A.A.; LastName, B.B.; LastName, C.C. Article Title. *Journal Name* **Year**, *Volume Number*, Page Range.

Volume II
ISBN 978-3-0365-7636-7 (Hbk)
ISBN 978-3-0365-7637-4 (PDF)

Volume I-III
ISBN 978-3-0365-7632-9 (Hbk)
ISBN 978-3-0365-7633-6 (PDF)

Cover image courtesy of Srinath Perera

Contents

About the Editors

Srinath Perera

Professor Srinath Perera is chair professor of built environment and construction management and the founding Director of Centre for Smart Modern Construction (c4SMC) at Western Sydney University. He joined WSU in June 2016 after serving as professor of construction economics at Northumbria University, Newcastle, in the UK. He is a Board member and the chair of the future leaders committee of the International Council for Research and Innovation in Building and Construction (CIB, www.cibworld.org).

He is a fellow of the Royal Society of New South Wales (FRSN) and also a fellow of the Australian Institute of Building (AIB). He is a chartered surveyor and a member of the Royal Institution of Chartered Surveyors (RICS), the Australian Institute of Quantity Surveyors (AIQS) and Australian Institute of Project Management (AIPM). He has over 30 years' experience in academia and industry and has worked as a consultant quantity surveyor and project manager in the construction industry.

Professor Perera is a pioneer in the field of construction informatics integrating AI technologies to construction and project management. He co-authored a research monograph, "Advances in Construction ICT and e-Business" (2017) and two internationally recognized textbooks, namely *Cost Studies of Buildings* (2015) and *Contractual Procedures in the Construction Industry* (2017) published by Routledge. He is also the author of "Managing Information Technology Projects: Building a Body of Knowledge in IT Project Management" He has authored over 250 peer reviewed publications and his current research leads work in the areas of blockchain and IoT applications in construction, BIM, Digital Twin, offsite construction, construction business models and construction performance leading to Industry 4.0.

He recently published the *Digitalisation of Construction* report, indicating the status and future directions of digitalization of the NSW construction industry.

Albert P. C. Chan

Professor Albert P. C. Chan is currently PolyU's Dean of Students, associate director of Research Institute for Sustainable Urban Development, and chair professor of Construction Engineering and Management. He earned his MSc degree in construction management and economics from the University of Aston in Birmingham, and a PhD degree in project management from the University of South Australia. Before joining the Department of Building and Real Estate of PolyU in 1996, Professor Chan taught at the University of South Australia as a senior lecturer and deputy head of the School of Building and Planning. He was appointed by PolyU as associate head (teaching) of the Department of Building and Real Estate from 2005 to 2011, associate dean from 2011 to 2013, interim dean of the Faculty of Construction and Environment from 2013 to 2014, and head of the Department of Building and Real Estate from 2015 to 2021. He has been an Adjunct Professor in a number of Mainland and overseas universities.

A chartered construction manager, engineer, project manager and surveyor by profession, Professor Chan is devoted to a myriad of research subjects as varied as project management and project success, construction procurement and relational contracting, public–private partnerships, and construction health and safety, as manifested by his prolific research output of over 1,000 refereed journal papers, international refereed conference papers, consultancy reports, and other articles. Besides being an expert member of the Engineering Panel of the Research Grants Council, HKSAR, since 2015, Professor Chan has also served as an expert member in the Built Environment Panel of FORMAS, Swedish Research Grants Council, and the Faculty of Architectural and the Built

Environment, Delft University of Technology in the Netherlands. Professor Chan was ranked among the top 2% of Scientists in the World for three years in a row starting in 2020.

Dilanthi Amaratunga

Professor Dilanthi Amaratunga holds the chair in Disaster Risk Management at the University of Huddersfield, UK, where she leads the Global Disaster Resilience Centre. She is a leading international expert in disaster resilience, with an extensive academic career that has a strong commitment to encouraging colleagues and students to fulfil their full potential. Her research interests include disaster risk reduction in the built environment; understanding disaster risk, preparedness for response; early warning systems; disaster resilience from the perspective of the social/political, economic, and physical sciences; and compound hazards and systemic risks. She has managed the successful completion of a large number of international research projects (over GBP 20 million), generating significant research outputs and outcomes, with the engagement of many significant research collaborations around the world in partnership with key academic and other stakeholders. To date, she has produced over 500 publications, refereed papers, and reports, and has made over 100 keynote speeches in around 40 countries.

Her outstanding contributions, publications, and services to her field of expertise have been recognised with numerous international awards. Between 2016 and 2019, she was winner of the prestigious 2019 Newton Prize, which recognises the best research and innovation projects which create an impact socially and economically, between Indonesia and the United Kingdom. In 2018, she received the "His Excellency the President of Sri Lanka Award" from the President of Sri Lanka, for here contribution to Disaster Resilience in Sri Lanka. In 2018, she won the UALL International Award, which recognises innovative engagement that creates change in an international and transnational context. She is a fellow of the RICS; fellow of the Royal Geographical Society, UK; fellow of the Higher Education Academy, UK; and fellow/chartered manager of the Chartered Management Institute, UK.

Makarand Hastak

Dr. Makarand Hastak is professor and Dernlan Family Head of Construction Engineering and Management as well as professor of civil engineering at Purdue University. Prof. Hastak is recognized around the world as an expert in construction engineering and management, with specific expertise in the profitability of construction companies, disaster risk reduction, infrastructure management, project control, and risk management. He is a licensed professional engineer (PE), a construction risk insurance specialist (CRIS), and a certified cost professional (CCP). Prof. Hastak has worked on numerous projects sponsored by prestigious funding agencies. As a fellow of the American Council on Education (cohort of 2013-14), his work at Cornell University focused on hybrid RCM budgets, engaged institutions, and public–private partnerships in academia.

He is the current president of the International Council for Research and Innovation in Building and Construction (CIB) (https://cibworld.org/) and serves as the academic advisor to the CII Downstream and Chemicals Committee (DCC) as well as the Department of Building and Real Estate, Hong Kong Polytechnic University. Prof. Hastak has authored/co-authored over 200 publications and reports as well as co-authored and edited three widely used books. He served as (Editor-in-Chief of the ASCE *Journal of Management in Engineering* (2009–2016). Prof. Hastak is a founding member and the past chair of the GLF-CEM, the Global Leadership Forum for Construction Engineering and Management programs.

Dr. Hastak received his BE (civil) from Nagpur University, India, MSCE from the University of Cincinnati, and PhD (civil) from Purdue University, USA. In addition, he is a trained university planner through SCUP and has received training at Cornell University in executive leadership, financial management, and advanced strategic human resource management. He has worked at Georgia Tech, NYU, University of Cincinnati, and currently at Purdue University.

Patrizia Lombardi

Professor Patrizia Lombardi, full professor of Projects Appraisal and Planning Evaluation at Politecnico di Torino, she is currently vice-rector for Sustainable Campus Development and Community Inclusion, after having been deputy rector and Urban and Regional (DIST) Department director. Since 2015, she has been coordinating the University Green Team, dedicated to the implementation of Agenda2030, in the context of the university's third mission. She is also president of the Network of Universities for Sustainable Development (RUS), of which she has been a promoter since 2013 and advisor of a number of Industries, NGOs, European Think Thanks, Joint Research Center of the European Commission and the Italian Ministry of Sustainable Infrastructure and Mobility. At international level, she is an established figure in the field of sustainable development evaluation for over 25 years and has coordinated or served as core partner several pan-European projects. Her research concerns issues of sustainable development assessment of the built environment with reference to decision support tools, including interactive Multicriteria Spatial Decision Support Systems. She is the author of more than 240 publications (h-index 32 Google Scholar with more than 4700 citations) and member of the Editorial boards of various international scientific journals. She has received several scientific and career awards.

Sepani Senaratne

Associate professor Sepani Senaratne is currently the director of Academic Program for UG Construction Management at Western Sydney University in Australia. Sepani has more than 20 years of academic experience in the quantity surveying (QS), construction management (CM) and built environment (BE) disciplines, attached to reputable universities in Sri Lanka, the UK and Australia. Her first degree is in BSc (Honours) first-class in QS from Sri Lanka and her PhD is in CM from the University of Salford, UK. Her key research expertise is in knowledge management and project management applications in construction projects and her research interests expand to include sustainable construction, smart modern construction and cost management areas. She has over 150 publications, including several peer-reviewed journal articles, conference papers, books and reports. Sepani's research has benefited various BE industry sectors and professions such as quantity surveyors, project managers, and contractors in solving project management problems. In 2022, as co-coordinator, Sepani launched a task group in CIB (The International Council for Research and Innovation in Building and Construction) on 'TG 124 on Net Zero Carbon' to create a global discussion and research. The sustainability research and activities that she is currently conducting contribute to United Nations Sustainable Development Goals. She has received several best paper awards at international research conferences, an Emerald Award of Excellence for Highly Commended Paper in 2009 and an Outstanding Research Performances Award by the University of Moratuwa consecutively five years. She is actively serving the academic community as a paper reviewer, postgraduate thesis examiner and member of conference committees.

Xiaohua Jin

Associate Professor Dr Xiaohua Jin is an associate professor in project management and the director of Project Management Programs and the Ddirector of Construction Law Programs at Western Sydney University, Australia. He holds a PhD from the University of Melbourne, Australia.

His main research interests include construction economics; project management; risk management; infrastructure procurement; relational contracting; and ICT in construction. Dr Jin has published over 100 peer-reviewed technical articles and been engaged in many industry-funded research projects. He received the Building Research Excellence Award from the Chartered Institute of Building (CIOB). Dr Jin is a member of the International Council for Research and Innovation in Building and Construction (CIB), the Australian Institute of Project Management (AIPM), and the International Centre for Complex Project Management (ICCPM). He is also a joint coordinator of CIB Working Commission W055 Building Economics. Dr Jin has been an expert referee for the Australian government. He is also an editorial panel member for several internationally renowned journals. He was a construction project manager before transiitoning to academia.

Anil Sawhney

Dr. Sawhney is a construction and infrastructure sector expert, an educator, a researcher, and a ConstructionTech enthusiast. In his role at the RICS, he is involved in producing the construction and infrastructure sector's body of knowledge, standards, guidance, practice statements, education, and training. Anil is the current chair of the International Cost Management Standards (ICMS) Standard Setting Committee and the co-chair of the AECO Working Group of the Digital Twin Consortium. He is also an adjunct faculty at Columbia University, visiting professor at Liverpool John Moores University in the UK, and an adjunct faculty at the University of Southern California. Anil has gathered a rich mix of academic, research, industry, and consulting experience in the USA, India, Canada, the UK, and Australia. In 2020, he co-authored a book entitled *Construction 4.0—Innovation Platform for the Built Environment*. He is currently the co-editor of the *Construction Innovation Journal*. Dr. Sawhney serves on the international editorial board of the *ICE Infrastructure Asset Management journal and the Journal of Information Technology in Construction*.

Preface to "Advances in Construction and Project Management—Volume II"

Construction and project management are two critical areas that play significant roles in society's progress and development. Construction projects play crucial roles in shaping the built environment, with an impact ranging from towering skyscrapers to intricate transportation systems. Effective project management is equally vital in this process, ensuring projects are completed on time, within budget, and to the required quality standards.

The field of construction and project management is constantly evolving, with new technologies, processes, and best practices emerging regularly. Keeping up with these advancements is essential for professionals in these fields, allowing them to ensure that they are delivering the best outcomes for their clients and stakeholders.

This book, entitled *Advances in Construction and Project Management*, compiles a collection of chapters from experts in these fields, covering the latest developments and trends. This publication covers a wide range of topics, including sustainable construction, digital technologies, project risk management, and stakeholder engagement, among others.

Written by leading academics and industry professionals from around the world, the individual chapters provide global perspectives on the subject matter. The authors draw on their experience and research to provide practical insights and solutions to the challenges facing construction and project management professionals today.

This book constitutes an essential resource for anyone involved in the construction or project management industries, including architects, engineers, contractors, project managers, and consultants. It is also an excellent reference for students studying in the disciplines of built environment, architecture, engineering, and construction, providing them with the latest information on the subject matter.

We hope to inspire readers to embrace new technologies, processes, and best practices and continue to advance the fields of construction and project management. We would like to express our gratitude to all the authors who contributed to this book and to the readers for their interest in this important topic. We also wish to acknowledge the Centre for Smart Modern Construction (c4SMC) and their industry partners for continued support and collaborations. I would also like to thank the centre researchers, Dr Samudaya Nanayakkara, Thilini Weerasuriya and Prasad Perera, for helping in the compilation of this topics issue.

Srinath Perera, Albert P. C. Chan, Dilanthi Amaratunga, Makarand Hastak, Patrizia Lombardi, Sepani Senaratne, Xiaohua Jin, and Anil Sawhney

Editors

Article

A Framework for Selecting Construction Project Delivery Method Using Design Structure Matrix

Qingping Zhong [1,*]**, Hui Tang** [1] **and Chuan Chen** [2]

[1] Institute for Disaster Management and Reconstruction, Sichuan University, Chengdu 610065, China; 20120032@git.edu.cn
[2] Business School, Sichuan University, Chengdu 610065, China; chenchuan@scu.edu.cn
*   Correspondence: qpzhong@gzu.edu.cn

Abstract: Determining a project delivery method that matches the characteristics of a construction project is a critical step that affects the success or failure of a project. The Project Delivery Method (PDM) should be adapted to the activities and processes of project implementation. However, the traditional selection method does not come from the internal process of the project which may lead to the delivery method not being able to meet the actual project requirements. This research proposes a DSM-based PDM selection framework model that regroups activities and identifies appropriate PDMs by revealing the dependencies and intensities between activities. The research uses a case to demonstrate the feasibility of the framework. After considering specific project requirements and goals, the framework model can be used as a basis for choosing specific project delivery methods, or as a visualization tool to help owners schedule activities.

Keywords: project delivery method; PDM; procurement selection; delivery selection; design structure matrix; DSM

Citation: Zhong, Q.; Tang, H.; Chen, C. A Framework for Selecting Construction Project Delivery Method Using Design Structure Matrix. *Buildings* **2022**, *12*, 443. https://doi.org/10.3390/buildings12040443

Academic Editors: Srinath Perera, Albert P. C. Chan, Dilanthi Amaratunga, Makarand Hastak, Patrizia Lombardi, Sepani Senaratne, Xiaohua Jin and Anil Sawhney

Received: 28 February 2022
Accepted: 2 April 2022
Published: 5 April 2022

Publisher's Note: MDPI stays neutral with regard to jurisdictional claims in published maps and institutional affiliations.

1. Introduction

The selection of a project delivery method (PDM) is a crucial step in impacting project success [1,2]. A PDM describes the relationship and working methods among project participants in the process of transforming the owner's goal into the completed facilities [3]. It directly affects construction performance including schedule, cost, quality, and efficiency [4–6]. The PDM can be viewed as both a contractual structure and compensation arrangement through which project owners obtain a completed facility that meets their needs [7]. The PDMs in practice are design-bid-build (DBB), design-build (DB), construction management at risk (CMR), engineering-procurement-construction (EPC), and integrated project delivery (IPD) [8,9]. However, the most common approaches are the first three, and the limitations of design-bid-build and the complexity of project features and requirements lead to a greater willingness to use design-build and other delivery methods [10].

In order to select the appropriate project delivery method, researchers have developed many methods based on case performance [11,12] and mathematical models [2–4,13–16]. These models and methods rely more on subjective expert opinions which are interfered with by the preferences, expertise, and abilities of the evaluators and are not very adaptable to constantly changing projects.

The project delivery method reflects the task, organizational, and contractual relationship of the project. Under the requirements of specific goals, new organizational and contractual relationships are created immediately after the tasks are rescheduled, and the corresponding delivery methods are also generated. However, few studies consider the feasibility of this delivery method from the perspective of the internal development process and working relationship.

This paper attempts to develop a framework model for selecting project delivery methods. The framework uses the design structure matrix (DSM) method to analyze the relationship between the activities in the system and optimizes activities through merging, deleting, and changing locations. The appropriate implementer is selected according to the closeness between the activities and, finally, the project delivery method is formed. This study expands the path and perspective of project delivery method selection, reveals the connection between project activities and project delivery methods, and reduces the instability of selecting project delivery methods.

2. Literature Review

2.1. Major Delivery Methods in Practice

Design-Bid-Build (DBB) refers to the sequential and phased project delivery method involving three key players: the owner, the designer (architect), and the general contractor (builder). In this contractual structure, the owner contracts with the designer and the contractor, respectively, monitoring the activities of the designer and the contractor to ensure compliance with the contract requirements [13,17]. The owner signs a contract with the designer first and then signs an agreement with the contractor through bidding after the design contract is completed. There is no direct connection between the designer and the contractor and all information needs to be transmitted after the owner's decision.

Construction management at risk (CMR) might be the preferred project delivery method when owners need a defined completion date and price. The CMR manager is responsible for providing consultation on architectural services in evaluating costs, schedule, materials, and the like, and advising on optimizations and design alternatives, playing the role of a general contractor during the construction phase. A CMR manager is also responsible for monitoring and controlling the construction process in terms of costs, time, and other requirements to ensure a guaranteed maximum price (GMP) for the project [7,17]. Similar to DBB, the owner contracts with both the designer and the CMR manager.

In the design-build (DB) method, the design and construction are carried out by one entity. The owner only needs to sign one contract covering architecture, engineering, and construction and contracts with a single enterprise responsible for design and construction [6]. The owner will give priority to the DB method when he cannot bear too much risk and responsibility. Because it is a single entity responsible for design and construction, it avoids the possible opposition in DBB [7]. Since the contractor is liable for all coordination efforts, the owner's contract administration and site representative risks and costs are reduced.

Integrated Project Delivery (IPD) is defined as "a method of project delivery characterized by a contractual arrangement among a minimum of the owner, constructor, and designer that aligns the commercial interests of all participants" [18]. IPD integrates all elements of the system into a single process that synergistically utilizes the talents and abilities of all participants through all stages of design, fabrication, and construction to optimize project outcomes, increase value, reduce waste, and maximize efficiency [19]. IPD method includes some contract principles and behavior principles that promote participants' early cooperation, increase mutual trust, and integrate multiple participants under one contract.

2.2. Selection of Major Delivery Methods

Project delivery methods have evolved from traditional DBB to IPD, but not all projects are suitable for newly developed delivery methods. The same type of project even may be suitable for different delivery methods. Each project should develop a project delivery method adapted to its characteristics. It is not so much that the project delivery is selected, it is better designed [20,21]. Some researchers hope to summarize the experience of selecting project delivery methods through existing project cases. Alleman et al. [21] investigated 291 US highway projects and believed that the alternative contracting methods (DB and CM) have better cost and schedule benefits and are therefore more suitable for highway construction. Demetracopoulou et al. [11] tested 57 lessons learned from Texas highway projects to help clarify the difficulty of choosing PDM. Franz et al. [22] verified the data of 212 projects to compare the cost and schedule performance of different delivery methods. Performance-based research also includes that these provide useful references for encouraging researchers to fully understand PDM and choices [23–25].

Another part of the researchers' hope is to develop a model of delivery method selection based on summarizing the selection criteria. This in turn includes appropriate determination of selection factors or criteria and reasonable methods. The factors for choosing a PDM are constantly enriched [26]. Decision makers focus early on the specific goals of the project [27,28], and as project complexity increases, factors expand to collaboration, integration, sustainability, corruption prevention, etc., [19,29,30]. The corresponding selection methods and models become more and more complex. By calculating the relative importance of different factors in the project goal hierarchy to choose the most appropriate delivery method, the AHP method has become the most commonly used method [4,13]. The artificial neural network method developed by Chen et al. identified similar projects between the target projects in the database and reduced the dependence on an expert's judgment [3]. Many researchers have to work on fuzzy methods in choosing the appropriate PDM to improve the reliability of decision making [2,14,15,31,32]. Additionally, many researchers developed multi-attribute decision making support tools [16,33–37].

However, these efforts may face some difficulties. The complexity of the project makes the choice of delivery method often inconsistent. When researchers try to use project performance indicators (such as cost, schedule, production efficiency, etc.) to select delivery methods, they often draw inconsistent conclusions. Feghaly et al. [38] concluded that DB was statistically superior to DBB in terms of project speed and intensity. Carpenter and Bausman [39] compared the performance of DBB and CM at Risk in public school construction, but the results showed that no one delivery method could meet all performance requirements. Project delivery methods should meet the requirements of the project characteristics. However, the evolution of projects and environmental changes constantly create new features and requirements which weaken the effectiveness of the model.

2.3. Design Structure Matrix

The design structure matrix (DSM), also known as the dependency structure matrix, has become a widely used modeling framework in research and practice. The DSM is a network modeling tool that reflects the interaction of the system's elements, thereby highlighting the system's architecture (or designed structure) [40]. According to the type of system being modeled, DSM can represent various types of architectures. For example, to model a process architecture, the DSM elements would be the activities in the process, and the interactions would be the flow of information and/or materials between them [41]. The DSM approach allows the project or engineering manager to represent meaningful task relationships to determine a reasonable sequence for the modeled activities [42]. The DSM has been identified as a potential tool to simulate interdependent activities, identify suitable assumptions, and formulate and evaluate the result [43].

The activity-based DSM is basically an N-square matrix that contains an activity list of rows and columns arranged in the same order. The order of activities in a row or column indicates the order of execution. In DSM, the relationship between activities is represented by the "X" mark in off-diagonal cells, which reflects the information flow between activities. The "X" mark above the diagonal indicates the information assumption or premise needed to start an activity. DSM is an N-square graph matrix representation of a process that is especially suitable for modeling the sequence and iterative information relationship between activities in the product development process [44–46].

Three possible relationship types between activities and corresponding DSM expressions are shown in Figure 1.

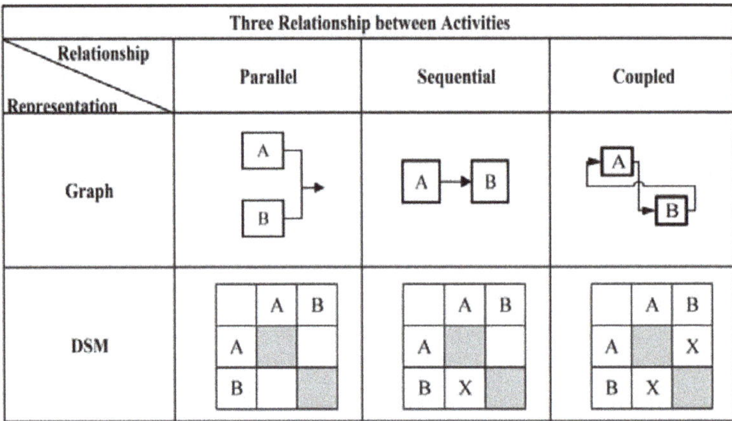

Three Relationship between Activities			
Relationship / Representation	Parallel	Sequential	Coupled
Graph			
DSM			

Figure 1. Three Configurations in DSM Analysis.

The DSM can be divided into four categories: component-based DSM, team-based DSM, activity-based DSM, and parameter-based DSM. The former two DSMs can also be called static DSM, and the last two can be called time-based DSM [47]. They correspond to the four DSM structural directions, as suggested by Yassine, and demonstrate the corresponding analysis methods as shown in Table 1.

Table 1. Four Different Types of Data in DSM (adapted from Yassine 2004).

DSM Types	Representation	Application	Analysis Method
Activity-based DSM	Activities in a process and their input and output	Project scheduling, activity sequencing, and cycle time reduction	Partitioning/Tearing/Banding/ Simulation and Eigenvalue Analysis
Parameter-based DSM	Parameters to determine a design and their relationship	Low level activity sequencing and process construction	
Team-based DSM	Teams in an organization and their relationships	Organizational design, interface management, and team integration	Clustering
Component-based DSM	Components in a product and their relationship	System architecting, Engineering, and design	

Partitioning eliminates or reduces feedback marks [46]. This process reorders activities so that dependencies are below or close to diagonals. When this is completed, we can see which activities are sequential, which can be completed in parallel, and which are coupled or iterative [48].

Tearing is the process of selecting the set of feedback marks that, if removed from the matrix (and then the matrix is re-partitioned), it will make the matrix a lower triangle [48]. Once the hypothesis is made through tearing, the matrix is subdivided to determine the preferred execution sequence [42].

Banding is to add alternating light and dark bands in DSM to show independent (i.e., parallel or concurrent) activities (or system elements) [49]. The collection of bands or levels constitutes the critical path of the system/project [48].

Although DSM is considered an effective tool for planning and sequencing, it is rarely used in construction projects. DSM is mostly used for optimizing activities during the planning and design phases [50–55]. These studies have improved the integration of activities in planning and design, helping engineers and managers to control work more precisely and improve work efficiency. However, research on project delivery method selection based on activity optimization has not been seen.

3. Research Methodology

3.1. The Primary thought of the Study

This study aimed to establish a framework model and method based on the interaction between activities for selecting a project delivery method for construction projects. Therefore, the activity-based DSMs are selected to identify the order and correlation of execution of the main activities and to analyze the likelihood and feasibility of their portfolios to optimize the project delivery process. At the same time, partitioning is chosen as an optimization and analysis tool because it needs to consider the combination of activities.

An example is taken from literature by [48] to explain the basic idea of the framework and then propose the specific steps of the framework in Figure 2. Firstly, the system activities are decomposed, and the spaghetti graph is drawn in Figure 2a. The arrow represents the information relationship between the activities; for example, the arrow B to C means C needs to receive output information from B before it can start. The original DSM is drawn in Figure 2b, and finally, the partitioned DSM is shown in Figure 2c.

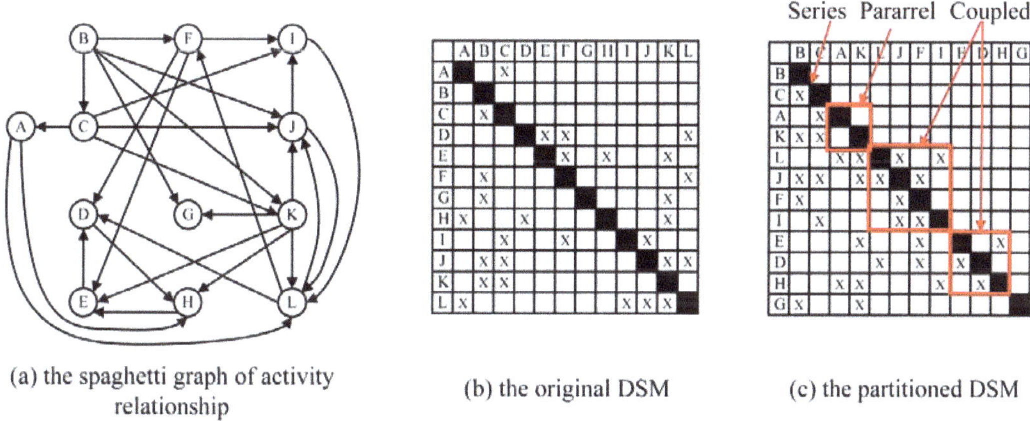

(a) the spaghetti graph of activity relationship

(b) the original DSM

(c) the partitioned DSM

Figure 2. The DSM Method Framework.

Based on the representation of the DSM activity relationship, we can sort out this study's basic ideas and work. The project delivery method can be associated with the representation of three activity relationships in partitioned DSM.

(1) Once the DSM is partitioned, a series of activities are identified and executed in sequence, such as how B and C are sequential in Figure 2. The owner can determine an integrated contract or decentralized contract based on factors such as the closeness between the activities. If the activities are closely linked, they should be managed by one contractor. However, the owner can award contracts to different contractors, and the owner is responsible for the coordination of the activities.

(2) Activity A and K are independent or paralleled, and they can be executed concurrently without information exchange with each other The two activities only need to start after receiving the information of their respective previous work without considering the status of each two of them. So, it is suitable for the owner to entrust them to two contractors independently.

(3) In Figure 2c, a loop is formed in blocks E-D-H: task E first needs to estimate or assume the output of task H, the outcome of E is transmitted to task D, then the output of D flow to task H, and finally, the output of H is fed to task E. At this point, task E starts in a state of uncertainty and incomplete information. Many times can this

uncertainty gradually decrease or converge only after E-D-H iterations occur. If multiple contractors perform separately, this iteration will not be accurately predicted and adequately controlled. There are similar but more complicated relationships among I, L, J, and F, and more upfront planning is required. It is difficult for owners to coordinate different contractors effectively, so they are more suitable for one contractor to conduct integrated management.

3.2. The Procedure of the Selection Model

The selection model for PDM using DSM includes four steps: Identify Requirements or Objectives, Building/Creating the Design Structure Matrix, Project Redesign/Optimization, and Design/Select Project Delivery Method. The model is shown in Figure 3.

Step 1. Identify requirements or objectives. This step consists of three main tasks, including asking the owner and environmental requirements or limitations, identifying project characteristics, and determining project goals. Identifying the environment of the project and clarifying the project owner's needs is the primary task of choosing a suitable project delivery method [4,56,57]. The sequence of activities, the responsibilities of the organization, and the corresponding contract structure are all based on meeting the owner's needs. Project characteristics define the technical nature of the work [58] which will affect the rationality and feasibility of redesigning the process. Project objectives need to be defined broadly in terms of scope, schedule, budget, and project complexity [4].

Step 2. Building/Creating the Design Structure Matrix. Appropriate structural decomposition and accuracy of activity dependencies determine the effectiveness of the DSM approach [48]. This step, therefore, consists of four activities. First, the project manager should fully decompose the project and forms a list of activities whose outputs constitute the entirety of the project entity. The list of activities can be determined by converting existing documents or structured expert interviews [48]. Second, the inputs and outputs of each activity should be determined, which reflect the dependencies between the activities. After the activities and their dependencies are entered into the matrix, an activity-based DSM can be formed. Finally, the marks in the DSM should be checked to confirm whether the relationships between the activities are correct and whether there are activity conflicts. It is worth noting that even if the activities are decomposed the same in different projects, the relationship between activities may still change with the owner's goals and requirements. When schedules are tight, identifying requirements may no longer be an absolute priority activity, but instead needs to be developed gradually through constant feedback during design and construction.

Step 3. Project Redesign/Optimization. After representing the process in the matrix, the project can be redesigned using partitioning, tearing, banding, and clustering. As mentioned previously, this framework focuses on the relationship and regrouping of activities, so partitioning is the main analysis tool.

Step 4. Design/Select Project Delivery Method. In this step, the strength of the relationship and the sequence of activities in the same partition should be checked first from a technical, regulatory, or management perspective. A partition represents the least amount of feedback between activities within it but may be technical, regulatory, or have weak dependencies that are not worth management action. Once it is confirmed that there is no unreasonableness or error, the activities in the partition can be packaged as a basis for assigning responsible persons. Likewise, relationships between activity packages should be examined and combined where feasible. Team activities can be assigned when all activities and activity packages have no relationship conflicts. Finally, a suitable PDM is selected or designed.

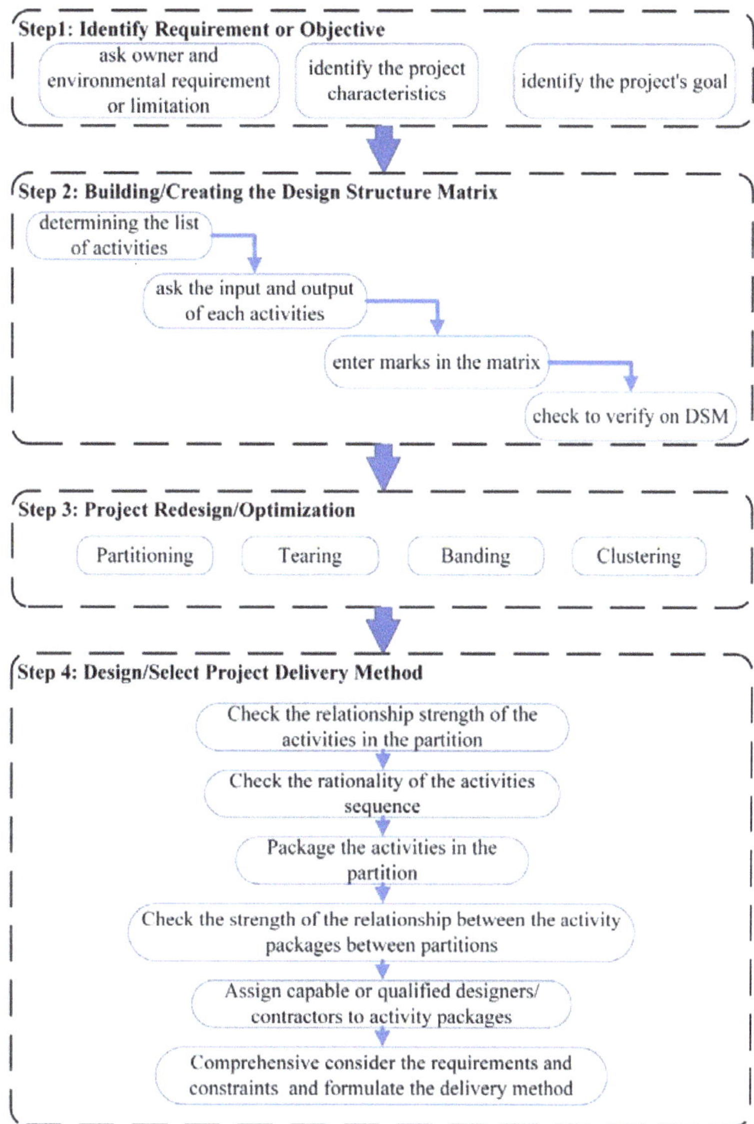

Figure 3. The Framework Model Using DSM to Select the PDM.

4. Case Study

This article uses a case study to describe the feasibility of this method in actual project implementation and uses surveys for verification. The survey asks practitioners their views on the project's activity relationship and inputs the feedback into the model to obtain the simulation results. The feasibility of the method is verified by comparing the simulation results with the actual delivery method.

4.1. Background of the Project

The project is a post-earthquake hospital reconstruction project in China with a total investment of 73.55 million yuan (US$11.37 million) and a total construction area of

13,918 m^2. The project was publicly tendered on 16 January 2018. The winning bidder was determined on 14 February 2018 and the construction of the project began in April 2018 and was finally finished on 7 August 2019. The funding for the project is fiscal funds, which can remain stable and sufficient.

4.2. Identify the Owner's Requirements and Analyze the Working Conditions

According to the central and provincial government's overall plan, the project needed to be delivered before the end of August 2019. The project construction period included design and construction for a total of about 500 days. In order of decreasing importance, the owner put forward the following requirements: to be completed on time or in advance, without quality and safety accidents, reducing environmental damage, improving the ability of the project to resist potential disasters, and improving the local medical level.

The historical weather statistics show that the local area faces regular heavy rains in July, and low temperatures in November, December, January, and February. For about five months of each year, normal construction cannot be carried out and may even be completely shut down. Therefore, the actual available time of the project was about 350 days, which is only 70% of the average time. The owner of this project did not have any management capabilities or experience in similar projects. Moreover, when the project was bidding, the project's detailed design was not completed, and only the plan was made. The final needs of the owner for the project were not precise, and there was the possibility of new requirements midway. The project site was small and challenging to construct. It was close to the river, and the groundwater level was high. The environmental carrying capacity of the project site was fragile, and it was close to a natural heritage protection area, so environmental pollution needed to be minimized as much as possible.

4.3. The Survey and Implementation

The survey was sent out in May 2021. The interviewees were the owner, on-site representative of the owner, designer, contractor enterprise manager, contractor project manager, project production manager, and project supervisor who had participated in the project. The questionnaire asked respondents to review the project implementation process and propose adjustments based on their practical experience.

The basic information of the interviewees and the projects they participated in are shown in Table 2.

Table 2. Respondent Details in Survey.

Respondent Details **By respondent's occupation**		**Quantity**	**Total (%)**
	Project manager	1	11.1
Contractor	Designer	2	22.2
	Production manager	2	22.2
	Enterprise manager	1	11.1
Project Supervisor		1	11.1
Owner		2	22.2
By respondent's working year		**Quantity**	**Total (%)**
≥ 15		2	22.2
$\geq 10, <15$		4	44.4
$\geq 5, <10$		3	33.3
<5		0	0

The contents of the survey mainly include:

(1) Under the circumstance that the constraints cannot change, how can the project activities be adjusted to achieve the owner's goal of 30% ahead of schedule (including deletion, merger, location change, activity association change, etc.)?
(2) If the activity changes, mark the adjusted relationship and location.

(3)   Which current project delivery method is suitable for the adjusted activities?

4.4. Identify Project Activities and Establish the Activity Decomposition Diagram

The project construction process can be decomposed into three major functions or processes: design, preparation, and production. Each can be further divided into more works and activities, and then a node tree can be established. The activities at the bottom of the figure can still be decomposed. For example, 'make detailed design' can still be decomposed into 'design spaces and facades', 'assist in the design of external structures and foundations', 'design frame and roof structures', 'design the complementary structures, surfaces, fittings, and courtyard', and 'prepare a construction specification' [59]. However, these activities are generally completed by different designers within a team. As far as the project delivery method is concerned, the work breakdown below the project work team is no longer necessary. The works and activities are decomposed in Figure 4, and their explanations are below:

(1)   Draw up brief. It is a process to collect the basic information provided by the owner concerning space requirements. The information consists of needs and requirements about the economy, dimension, quality, scheduling, function, etc. Additionally, the possibilities of site situation and availability of resources should be collected. This work is denoted by "A" and can be composed of four activities represented by A1~A4, respectively.

- Identify requirements (A1). The needs of the owner and requirements from outside involve many aspects, including financial requirements, space scale requirements, quality and function requirements, schedule requirements, alternative technical solutions, etc.
- Survey and analyze site information (A2). Analysis of the present situation includes the availability of existing conditions and the possibility of change. Designers and contractors need to analyze the geotechnical condition, city plan, local planning, availability of resources and management systems, etc.
- Establish objectives (A3). This activity formulates and establishes the overall goals of the project. Goals may include establishing the desired attributes and functions developed by the owner, determining regulating requirements, and clarifying the design scope.
- Establish design parameters (A4). Establish design limits, guidelines, and project requirements such as budget, cost, scheduling, quality, constructability, and environmental effects.

(2)   Make conceptual design. Concept design is the forming of abstract concepts using approximate concrete expressions [60]. General concepts such as site use and boundary, architectural consideration, major system types, and materials are explored. Conceptual cost estimates and budgets may also be developed. This work is denoted by "B" and can be composed of four activities represented by B1~B3, respectively.

- Develop preliminary design (B1). This process will determine the project program and terms to define the function. Some drawings, including the basic dimensions of the project, the major architectural components, and structural systems, are developed to illustrate the concept of design and the project scope.
- Coordinate and find compatibility (B2). System schemes between disciplines need to be coordinated for integration. Some checks such as function compatible checks, quality reviews, and standard/code coordination checks should be performed from the macro-level.
- Evaluate and review the preliminary design (B3). The owner reviews the preliminary design from multiple perspectives, including meeting requirements, function, economy, feasibility, legal and government permits, etc., to determine whether the scheme can achieve the expected effect and whether the detailed design can be carried out.

(3) Make detailed design. This process starts with the evaluation of the scheme. The detailed design needs to be elaborated on until the contractor can choose the construction method and purchase materials accordingly. The design process needs to integrate the design process of all disciplines. This work is denoted by "C" and can be composed of three activities represented by C1–C3, respectively.

- Make a detailed design (C1). The detailed design includes activities such as facade design, internal space design, decoration design, structural design, ventilation system design, pipe design, fire protection design, landscape design, etc.
- Check the compatibility of detailed design (C2). The design documents for all disciplines should be checked to ensure compatibility between various professional designs and reduce or eliminate rework due to design conflicts.
- Make the resource checklist (C3). The resource list includes raw materials and equipment. It should list the types, quantities, specifications, models, etc. so that the contractor can purchase resources and arrange the arrival time reasonably.

(4) Acquire contractors. This process includes all activities concerning bidding and tendering. This work is denoted by "D" and can be composed of four activities represented by D1–D4.

- Issue bidding documents (D1). The owner puts forward technical and management capability requirements to the contractor.
- Tendering (D2). The contractor submits documents to the owner to prove that it is suitable for undertaking the project.
- Review and select contractor (D3). The owner reviews the contractor's tender documents, and judges and selects the most suitable contractor.
- Sign contract (D4). The owner and the contractor sign the contract after reaching an agreement through negotiation.

(5) Prepare for construction. The preparation mainly refers to the workforce and material preparation made by the contractor for the construction, including the project team, equipment, materials, etc. This work is denoted by "E" and can be composed of four activities represented by E1–E4, respectively.

- Organize project team (E1). The contractor needs to select a qualified project manager and teams to construct the project.
- Make a construction plan (E2). This plan is about construction scheduling, quality assurance, cost control, and environmental protection.
- Prepare and implement procurement (E3). The contractor needs to make an accurate equipment and material procurement plan and carry out an inquiry, procurement, and storage as planned.
- Prepare site (E4). The construction site must have no legal issues and have the appropriate condition for construction.

(6) Construct project. Implement concrete activities to complete the tasks and objectives specified in the project plan. This work is denoted by "F" and can be composed of four activities represented by F1~F4.

- Plan the daily work (F1). Decompose the overall construction plan to the work to be completed every day according to the schedule, and formulate the personnel and resource allocation plan, quality control measures, and inspection plan.
- Allocate the resources (F2). Allocate sufficient quantity and quality resources to daily work.
- Do the physical work (F3). Arrange appropriate workers and tools to complete daily work and gradually form products.
- Inspect and approve the work (F4). The contractor needs to evaluate the quality and progress of phased products through regular inspection to ensure the project is completed on time and reduce rework.

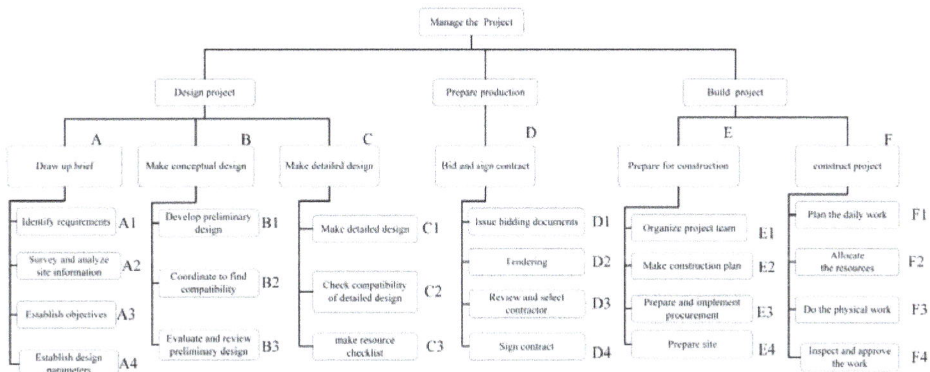

Figure 4. The node tree of the construction project.

4.5. Identify the Relationship between Activities and Establish an Original DSM

It should be noted that there are not only the feedforward and feedback relationships between the internal activities of each stage but also feedforward and feedback relationships between the cross-stage activities. There are three main stages of design, preparation, and construction in the basic model, and there are feedforward and feedback information flows between multiple cross-stage activities. Since the construction stage may encounter different assumptions from the design, the construction activities need feedback information from the initial design to guide the modification, so the possible process cycles appear within the stages and appear between stages. The dependence between activities lists in Table 3. The original DSM is shown in Figure 5.

Activity		#	A1	A2	A3	A4	B1	B2	B3	C1	C2	C3	D1	D2	D3	D4	E1	E2	E3	E4	F1	F2	F3	F4
			1	2	3	4	5	6	7	8	9	10	11	12	13	14	15	16	17	18	19	20	21	22
identify requirements	A1	1	1																					
survey and analyse site information	A2	2		2																				
establish objectives	A3	3	X	X	3		X																	
establish design parameters	A4	4			X	4																		
develop preliminary design	B1	5				X	5	X	X															
coordinate to find compatibilities of	B2	6					X	6																
evaluate and review preliminary design	B3	7						X	7															
make detailed design	C1	8	X							8	X					X	X					X		
check compatibilities of detailed design	C2	9								X	9							X						
make resource checklist	C3	10								X		10												
issue bidding documents	D1	11		X			X						11											
tendering	D2	12											X	12				X						
review and select contractor	D3	13	X												13									
sign contract	D4	14													X	14	X							
organize project team	E1	15														X	15							
make construction plan	E2	16								X								16	X	X				X
prepare and implement procurement	E3	17																X	17					
prepare site	E4	18																X		18				
plan the daily work	F1	19																	X		19	X	X	
allocate the resources	F2	20																		X	X	20	X	
do the physical work	F3	21																			X	X	21	
inspect and approve the work	F4	22																					X	22

Figure 5. The original activity-based DSM of the project.

4.6. The Original DSM Is Manipulated to Eliminate or Reduce the Feedback Marks

For the activity-based DSM, partitioning is the primary method to help a transparent structure emerge. Using this method, sequential, parallel completion, coupled or iterative activities are clearly displayed. The DSM tool is DSM_Program-V2.1 [61]. The partitioned DSM is shown in Figure 6.

Table 3. The Dependence of the activities.

Activity	Depends on
Identify requirements (A1)	—
Survey and analyze site information (A2)	—
Establish objectives (A3)	Identifying requirements (A1), surveying and analyzing site information (A2), and developing a preliminary design (B1)
Establish design parameters (A4)	Establishing objectives (A3)
Develop a preliminary design (B1)	Establishing design parameters (A4), coordinating to find compatibilities of preliminary design (B2), and evaluating and reviewing the preliminary design (B3)
Coordinate to find compatibilities of preliminary design (B2)	Developing a preliminary design (B1)
Evaluate and review the preliminary design (B3)	Coordinating to find compatibilities of preliminary design (B2)
Make the detailed design (C1)	Identifying requirements (A1), checking compatibilities of detailed design (C2), signing the contract (D4), organizing the project team (E1), and doing the physical work (F3)
Check compatibilities of the detailed design (C2)	Making the detailed design (C1) and organizing the project team (E1)
Make a resource checklist (C3)	Making the detailed design (C1)
Issue bidding documents (D1)	Establishing objectives (A3) and evaluating and reviewing the preliminary design (B3)
Tendering (D2)	Issuing bidding documents (D1) and organizing the project team (E1)
Review and select contractor (D3)	Identifying requirements (A1)
Sign contract (D4)	Tendering(D2) and organizing the project team (E1)
Organize project team (E1)	Reviewing and selecting the contractor (D3)
Make a construction plan (E2)	Making the detailed design (C1), organizing the project team (E1), preparing and implementing procurement (E3), preparing the site (E4), and inspecting and approving the work (F4)
Prepare and implement procurement (E3)	Making a resource checklist (C3) and making a construction plan (E2)
Prepare site (E4)	Organizing a project team (E1)
Plan the daily work (F1)	Making a construction plan (E2), allocating the resources (F2), and doing the physical work (F3)
Allocate resources (F2)	Making construction plan (E2), preparing and implementing procurement (E3), and planning the daily work (F1)
Do the physical work (F3)	Planning the daily work (F1) and allocating the resources (F2)
Inspect and approve the work (F4)	Doing the physical work (F3)

4.7. Highlight the Partitioned DSM and Explanations

The partitioned DSM highlights two blocks. The first block includes seven activities such as A3, B1, A4, B2, and B3 which start from the 'establish objective' to the 'evaluate and review preliminary design'. This block represents the main process of the preliminary design. Preliminary design stipulates some design features that cannot be broken through in the detailed design and construction, such as each system's, subsystem's, and component's requirements and functions, a high-level outline of design features that meet each of these requirements, and cost estimates. In many reconstruction projects, determining the project goals clearly and making an acceptable design is not a one-time task. In repeated communication between the designer and the project owner; the project owner can gradually clarify his goals, and the designer can compile satisfactory deliverables.

The second block includes nine activities as E2, C1, C2, E3, C3, F3, F1, F2, and F4 which span broadly from 'make detailed design' to 'inspect and approve the work'. It spans from design to procurement and build and spans from design to procurement and construction. In the reconstruction environment, the interviewees think it is tough for the designer to complete the perfect design alone and deliver it to the purchaser and contractor. More improvement work in practice requires feedback from the contractor during construction. This process must be speedy and smooth. Construction control must be transformed into active control, so it is necessary to monitor the progress every day and revise the plan and resource allocation for the next day in real-time. These feedbacks are all divided into the same block, indicating that they are closely connected and suitable for integration consideration in organizational arrangements.

PARTITIONED DSM			A1	A2	A3	B1	A4	B2	B3	D1	D3	E1	E4	D2	D4	E2	C1	C2	E3	C3	F3	F1	F2	F4
			1	2	3	5	4	6	7	11	13	15	18	12	14	16	8	9	17	10	21	19	20	22
identify requirements	A1	1	1																					
survey and analyse site information	A2	2		2																				
establish objectives	A3	3	X	X	3	X																		
develop preliminary design	B1	5				5	X	X	X															
establish design parameters	A4	4			X		4																	
coordinate to find compatibilities of	B2	6				X		6																
evaluate and review preliminary design	B3	7						X	7															
issue bidding documents	D1	11			X				X	11														
review and select contractor	D3	13	X								13													
organize project team	E1	15									X	15												
prepare site	E4	18										X	18											
tendering	D2	12								X		X		12										
sign contract	D4	14										X		X	14									
make construction plan	E2	16										X	X			16	X		X					X
make detailed design	C1	8	X									X			X		8	X	X					
check compatibilities of detailed design	C2	9										X					X	9						
prepare and implement procurement	E3	17														X			17	X				
make resource checklist	C3	10														X				10				
do the physical work	F3	21																			21	X	X	
plan the daily work	F1	19														X					X	19	X	
allocate the resources	F2	20														X			X			X	20	
inspect and approve the work	F4	22																			X			22

Figure 6. The partitioned DSM of the project.

Some activities after partitioning have changed in order, D3 (evaluating contractor), E1 (building a project team), and E4 (preparing site) before D2 (tendering) and C1 (detailed design). Bidding is a test of the contractor's capability. However, the construction period will be greatly extended if the contractor and bidding are inspected after the detailed design is completed, as in the traditional delivery method. The reconstruction environment is complex and changeable. Contractors need not only sufficient technical force but also strong comprehensive management and coordination capabilities. Therefore, in order to speed up the development of the project, more capable contractors should be evaluated in advance and the contractors should get to know the conditions of the scene earlier in order to make full preparations. The bidding documents submitted by the contractor should show the organization's comprehensive capabilities for the future implementation of the project, such as the organization's ideas and technical arrangements, the handling of emergencies, and the procurement and deployment of resources.

4.8. The Selection of PDM

According to the activity relationship in DSM, we can not only understand the current delivery method but also design a delivery method that is more suitable for project requirements based on the activity relationship. As mentioned previously, selecting PDM also needs to consider requirements and scenarios. Therefore, after considering the owner's requirements, appropriate PDM decisions can be made from the DSM.

The characteristics of the case project include many participants, but the owner was incapable of fully managing and has strong time constraints. The partitioned DSM shows that it could be integrated into one block from detailed design to construction completion which means that it can be implemented by one party. The tasks that needed to be transferred and coordinated by the owner were all completed by the contractor when the contractor carried out multi-stage work. The interaction between activities becomes the internal staff's work with the contractor which will greatly shorten the time and cost of coordination [62]. When the acceleration techniques are adopted, information exchange between personnel and between subcontractors will become more frequent, and the advantages of integrated delivery methods will become more prominent.

Other goals of the owner included controlling investment and improving medical standards. This determines that control cannot be completely abandoned. Early evaluation of the contractor's ability and deep participation in preliminary design could achieve the owner's goal, and at the same time, it could strengthen the owner's control over the main subsystems and avoid large investment deviation and function deviation.

Therefore, according to the partitioned DSM and opinions of interviewees, this project was suitable for delivery with higher integration such as BD or CM or its variation. The project process of partitioned DSM is shown in Figure 7.

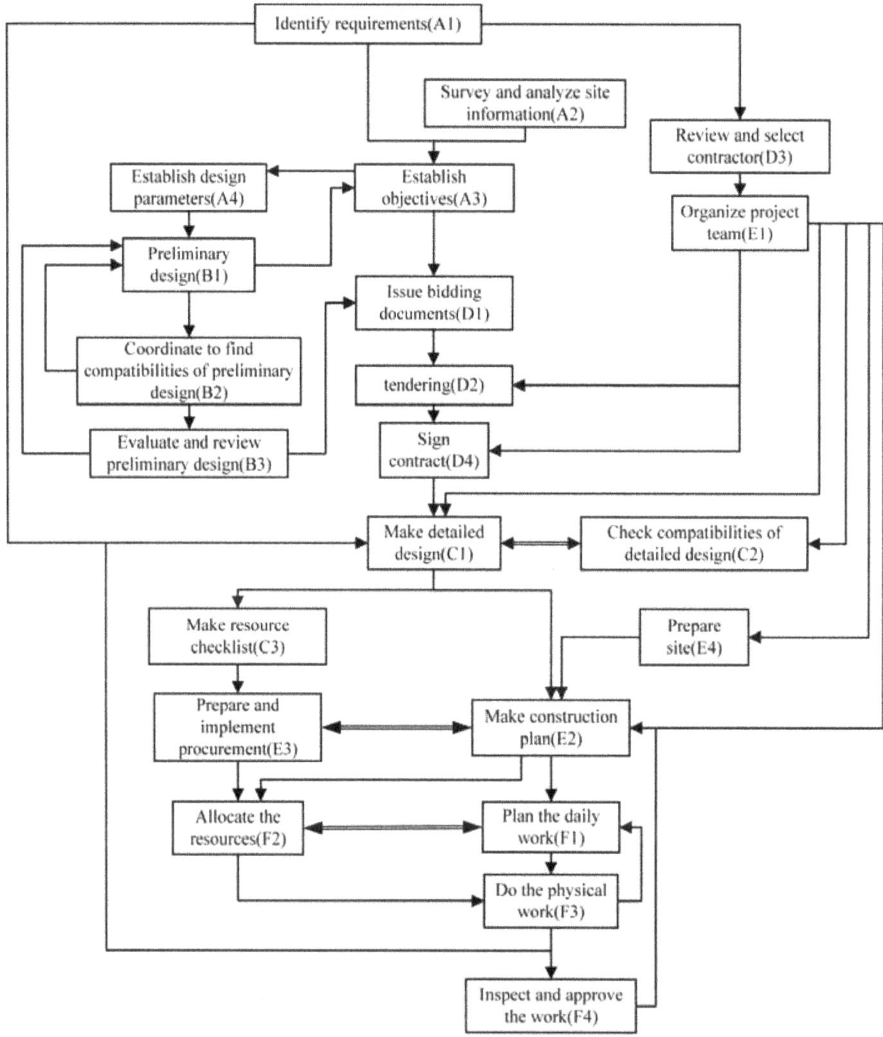

Figure 7. The Project Process based on the partitioned DSM.

This project adopted the EPC delivery method, which is very close to the conclusion obtained by the DSM method. The general contractor was a consortium composed of a design enterprise and a construction enterprise. Both parties participated in the bidding according to the preliminary design completed by the design enterprise entrusted by the owner, and finally won the bid and undertook the detailed design, raw material procurement, and construction. The professionals dispatched by both parties worked together on-site, which reduced the information transmission path, carried out simulation and error correction in advance, and reduced rework.

5. Discussion

5.1. The Type of DSM to Be Used

Successful project implementation needs to understand the project structure and develop and manage a strategy [63]. The selection framework proposed in this study establishes a process model from the relationship between each activity in the construction process and selects the appropriate or optimal delivery method by identifying the relationship between the activities and rearranging the sequence of activities. The framework can also prove the rationality of the selected delivery method or optimize the order of activities in the delivery method.

DSM can be divided into four systems that are interrelated [47]. Which type of DSM to use should depend on the purpose. When only process analysis and optimization are performed, activity-based DSM is used more [64–66]. The parameter-based DSM can better analyze the probability of repetition, the variability of exchanged information, and the impact of iteration [67]. A team-based DSM can clarify how the implementers of various activities communicate and connect with each other. As mentioned earlier, the project delivery method ultimately determines the scope, time, and division of responsibilities of the organization's activities, and limits an organization or group to undertake a single task or multiple related activities. Through appropriate calculations such as partitions and clusters, participants can be combined and divided to achieve the purpose of optimizing the organization. The choice of project delivery method is always a multi-objective optimization problem that needs to be weighed in terms of objectives, management capabilities, and management methods. Constructing a combination of activity-based DSM and team-based DSM under specific target requirements can help decision makers to allocate personnel and responsibilities reasonably.

5.2. Establishing the DSM under the Requirements of a Particular Project

Analysis and examples show that the proposed framework helps to select the most objective delivery method or verify the rationality of the delivery method and optimize activities. However, it should be noted that the needs of the owner always have an important influence on the selection process, and even the relationship between the same type of project activities under the different needs of the owner may be different. Therefore, determining the owner's needs is the fundamental requirement for applying this method.

When the project is under high time pressure, finding connections between activities and their intensity to increase activity overlap and reduce rework becomes the main response method. As time becomes the highest priority goal, the division of design phases will be simplified, design and construction need to be partially paralleled, and the information feedback path of activities needs to be redesigned. Based on the relationship between these activities, one can choose the delivery method that can best achieve time compression. When quality becomes the main goal, due to the uniqueness of construction products, more small-scale coupling activity packages need to appear in the delivery method to ensure that product quality is always controllable and form a final product with satisfactory quality.

5.3. Decomposing the Activities

This case study only carried out a three-level decomposition because the hospital project was only 13,918 m^2 and the scale of the project was not large. In order to ensure that it was completed on time, the owners and contractors were willing to use traditional construction technology rather than innovative technology. The workflow was not much different from regular projects. Project delivery methods vary depending on the scale and technical complexity of the project and will also change during hierarchical decomposition.

Proper decomposition is the key to effectively solving big and difficult problems, which can minimize the interaction between sub-problems [68]. Finding the right level of abstraction to formulate a DSM is not easy since activities can be defined at multiple levels, from very detailed to high-level abstraction [69]. Too much or insufficient task decomposition may lead to management failure. The more decomposition levels, the more specific the underlying activities performed by individuals or small teams. Through continuous decomposition levels, the size of the model increases exponentially, the management organization will increase, and the efficiency will decrease. Too few decomposition levels will result in blurred relationships between activities and unclear division of the management interface. The general rule is to model the process to the level of detail that people want to understand and be able to control the process [47]. The delivery method is expressed as a contract, so it usually only involves the enterprise (sub-enterprise) level and does not need to target individuals or small teams.

5.4. The Expression and Use of Activity Relations

DSM has developed many ways to express the relationship between tasks such as marks, numbers, colors, shadows, etc. This research only uses the most basic markup methods. This expression simply indicates whether there is an interconnection between activities [70]. Other expressions can express more information, such as the probability of overlap or rework between activities [71], interaction strength [72,73], and the duration of the activity [74,75].

In the above information, the connection strength of activities is an important criterion in the choice of project delivery method, especially the delivery method that needs to shorten the project duration. Stronger connections mean that activities can receive more complete information before they can be implemented. Therefore, more effective measures and organizational methods need to be used to ensure the efficiency of information transmission. However, taking additional measures at the same time may bring additional costs. Only when the benefits of relatively strong activity adjustments are greater than the increased costs are adjustments worthwhile, while weak links can be used to diversify risks through contracts and other means. Therefore, follow-up research should deeply analyze the relationship between activity intensity and activity combination from activity intensity.

6. Conclusions

Relationships between activities will become more complicated as building technologies evolve, requiring more flexible delivery methods. A proper PDM will directly affect the effectiveness of the owner's and contractor's organizational arrangements and resource allocation, which will affect the project's success. Therefore, it is necessary to carefully select/design the appropriate PDM at the beginning of the project. Different delivery methods adapt to other activity processes and therefore require different organizational approaches. The delivery method is determined according to the process to maximize the satisfaction of the project requirements. Activity-based DSM can show feedforward and feedback between activities. Operation methods such as partition can optimize and reduce rework caused by information feedback, shorten project duration, and save cost. Depending on the relationship between the activities, the decision maker can choose to delegate certain activities to the appropriate contractor and determine the proper PDM. At the same time, the selected PDM can also be optimized through this framework.

The goal of this paper is to develop an analytical framework that can be used in the early stages of project contracting to demonstrate to participants project activities and the relationships among participants and to support participants' effective allocation and coordination of work. The framework shows the whole process of the project and its activities in a visual way. When decision-makers are faced with specific goals, such as shorter time frames, cost savings, and better organization, this framework can assist decision-makers in comprehensively reviewing the project and making quick judgments to design the appropriate PDM. The difference between the study and previous studies is that the characteristics of the delivery method are designed from the internal process of the project, rather than extracted from the completed project, which is universal and stable and will not be invalid due to the changes in the project.

In summary, the following managerial insights can be helpful for PDM selection.

Process-based PDM selection can reduce decision-making difficulties caused by changes in project characteristics and complexity and improve the pertinence and universality of PDM.

Applying this method to select an appropriate PDM with an organized structure reduces the subjectivity of decision makers.

The framework visualizes the entire process of the project, helping decision makers comprehensively review the project and make quick decisions.

In the absence of experienced decision makers, or the absence of consensus among decision makers, this research will provide good insights to support the final decision.

This research is associated with the following limitations:

- The research did not consider the intensity of the relationship between activities which directly affects the trade-off between the costs and benefits of activity adjustment and then affects the decision results. The empowerment of association strength should be a direction of further research in the future.
- This research only considered the activity-based DSM; the project team staffing and responsibility assignment should be considered in combination with the organization-based DSM in the future.

Author Contributions: Conceptualization, Q.Z. and C.C.; methodology, Q.Z.; software, H.T.; formal analysis, Q.Z.; investigation, H.T.; resources, C.C.; data curation, C.C.; writing—original draft preparation, Q.Z. and H.T.; writing—review and editing, C.C. All authors have read and agreed to the published version of the manuscript.

Funding: The research was supported by the "National Natural Science Foundation of China", the funding number is 71971147.

Institutional Review Board Statement: Not applicable.

Informed Consent Statement: Not applicable.

Conflicts of Interest: The authors declare no conflict of interest.

References

1. Ibbs, W. Alternative methods for choosing an appropriate project delivery system (PDS). *Facilities* **2011**, *29*, 527–541. [CrossRef]
2. Mostafavi, A.; Karamouz, M. Selecting Appropriate Project Delivery System: Fuzzy Approach with Risk Analysis. *J. Constr. Eng.* **2010**, *136*, 923–930. [CrossRef]
3. Chen, Y.Q.; Liu, J.Y.; Li, B.; Lin, B. Project delivery system selection of construction projects in China. *Expert Syst. Appl.* **2011**, *38*, 5456–5462. [CrossRef]
4. Al Khalil, M.I. Selecting the appropriate project delivery method using AHP. *Int. J. Proj. Manag.* **2002**, *20*, 469–474. [CrossRef]
5. Diao, C.Y.; Dong, Y.J.; Cui, Q.B. Project Delivery Selection: Framework and Application in the Utility Industry. In *Construction Research Congress 2018: Infrastructure and Facility Management*; ASCE: New Orleans, LA, USA, 2018; pp. 171–179.
6. Noorzai, E. Performance Analysis of Alternative Contracting Methods for Highway Construction Projects: Case Study for Iran. *J. Infrastruct. Syst.* **2020**, *26*, 04020003. [CrossRef]
7. Mafakheri, F.; Dai, L.; Slezak, D.; Nasiri, F. Project delivery system selection under uncertainty: Multicriteria multilevel decision aid model. *J. Manag. Eng.* **2007**, *23*, 200–206. [CrossRef]
8. Li, H.M.; Qin, K.L.; Li, P. Selection of project delivery approach with unascertained model. *Kybernetes* **2015**, *44*, 238–252. [CrossRef]
9. Qiang, M.; Wen, Q.; Jiang, H.; Yuan, S. Factors governing construction project delivery selection: A content analysis. *Int. J. Proj. Manag.* **2015**, *33*, 1780–1794. [CrossRef]
10. FMI (Fails Management Institute). Design-Build Utilization: Combined Market Study. 2018. Available online: https://dbia.org/wp-content/uploads/2018/06/Design-Build-Market-Research-FMI-2018.pdf (accessed on 20 February 2020).

11. Demetracopoulou, V.; O'Brien, W.J.; Khwaja, N. Lessons Learned from Selection of Project Delivery Methods in Highway Projects: The Texas Experience. *J. Leg. Aff. Disput. Resolut. Eng. Constr.* **2020**, *12*, 04519040. [CrossRef]

12. Alleman, D.; Antoine, A.; Papajohn, D.; Molenaar, K. Desired versus realized benefits of alternative contracting methods on extreme value highway projects. *Proc. Resilient Struct. Sustain. Constr.* **2017**, 1–6. [CrossRef]

13. Mahdi, I.M.; Alreshaid, K. Decision support system for selecting the proper project delivery method using analytical hierarchy process (AHP). *Int. J. Proj. Manag.* **2005**, *23*, 564–572. [CrossRef]

14. An, X.W.; Wang, Z.F.; Li, H.M.; Ding, J.Y. Project Delivery System Selection with Interval-Valued Intuitionistic Fuzzy Set Group Decision-Making Method. *Group Decis. Negot.* **2018**, *27*, 689–707. [CrossRef]

15. Nguyen, P.H.D.; Tran, D.; Lines, B.C. Fuzzy Set Theory Approach to Classify Highway Project Characteristics for Delivery Selection. *J. Constr. Eng.* **2020**, *146*, 04020044. [CrossRef]

16. Xia, B.; Chan, A.P.C.; Yeung, J.F.Y. Developing a Fuzzy Multicriteria Decision-Making Model for Selecting Design-Build Operational Variations. *J. Constr. Eng.* **2011**, *137*, 1176–1184. [CrossRef]

17. Sanvido, V.; Konchar, M. *Selecting Project Delivery Systems*; Project Delivery Institute: State College, PA, USA, 1999.

18. AIA. *AGC Primer on Project Delivery*; AIA: Washington, DC, USA, 2011.

19. Marzouk, M.; Elmesteckawi, L. Analyzing procurement route selection for electric power plants projects using SMART. *J. Civ. Eng. Manag.* **2015**, *21*, 912–922. [CrossRef]

20. Azari, R.; Ballard, G.; Cho, S.; Kim, Y.W. A Dream of Ideal Project Delivery System. *AEI 2011 Build. Integr. Solut.* **2011**, 427–436. [CrossRef]

21. Ding, J.Y.; Wang, N.; Hu, L.C. Framework for Designing Project Delivery and Contract Strategy in Chinese Construction Industry Based on Value-Added Analysis. *Adv. Civ. Eng.* **2018**. [CrossRef]

22. Franz, B.; Molenaar, K.R.; Roberts, B.A.M. Revisiting Project Delivery System Performance from 1998 to 2018. *J. Constr. Eng.* **2020**, *146*, 04020100. [CrossRef]

23. Park, J.; Kwak, Y.H. Design-bid-build (DBB) vs. design-build (DB) in the US public transportation projects: The choice and consequences. *Int. J. Proj. Manag.* **2017**, *35*, 280–295. [CrossRef]

24. Shrestha, P.P.; Fernane, J.D. Performance of design-build and design-bid-build projects for public universities. *J. Constr. Eng.* **2017**, *143*, 04016101. [CrossRef]

25. Tran, D.Q.; Diraviam, G.; Minchin, R.E., Jr. Performance of highway design-bid-build and design-build projects by work types. *J. Constr. Eng.* **2018**, *144*, 04017112. [CrossRef]

26. Ahmed, S.; El-Sayegh, S. Critical review of the evolution of project delivery methods in the construction industry. *Buildings* **2020**, *11*, 11. [CrossRef]

27. Kumaraswamy, M.M.; Dissanayaka, S.M. Developing a decision support system for building project procurement. *Build. Environ.* **2001**, *36*, 337–349. [CrossRef]

28. Ng, S.T.; Luu, D.T.; Chen, S.E.; Lam, K.C. Fuzzy membership functions of procurement selection criteria. *Constr. Manag. Econ.* **2002**, *20*, 285–296. [CrossRef]

29. Deep, S.; Gajendran, T.; Jefferies, M. A systematic review of 'enablers of collaboration' among the participants in construction projects. *Int. J. Constr. Manag.* **2021**, *21*, 919–931. [CrossRef]

30. Roy, V.; Desjardins, D.; Ouellet-Plamondon, C.; Fertel, C. Reflection on integrity management while engaging with third parties in the construction and civil engineering industry. *J. Leg. Aff. Disput. Resolut. Eng. Constr.* **2021**, *13*, 03720005. [CrossRef]

31. Boran, F.E. An integrated intuitionistic fuzzy multi criteria decision making method for facility location selection. *Math. Comput. Appl.* **2011**, *16*, 487–496. [CrossRef]

32. Li, H.M.; Su, L.M.; Cao, Y.C.; Lv, L.L. A pythagorean fuzzy TOPSIS method based on similarity measure and its application to project delivery system selection. *J. Intell. Fuzzy Syst.* **2019**, *37*, 7059–7071. [CrossRef]

33. Khwaja, N.; O'Brien, W.J.; Martinez, M.; Sankaran, B.; O'Connor, J.T.; Bill Hale, W. Innovations in project delivery method selection approach in the Texas Department of Transportation. *J. Manag. Eng.* **2018**, *34*, 05018010. [CrossRef]

34. Feghaly, J.; El Asmar, M.; Ariaratnam, S.; Bearup, W. Selecting project delivery methods for water treatment plants. *Eng. Constr. Archit. Manag.* **2019**, *27*, 936–951. [CrossRef]

35. Zhu, X.; Meng, X.; Chen, Y. A novel decision-making model for selecting a construction project delivery system. *J. Civ. Eng. Manag.* **2020**, *26*, 635–650. [CrossRef]

36. Martin, H.; Lewis, T.M.; Petersen, A. Factors affecting the choice of construction project delivery in developing oil and gas economies. *Archit. Eng. Des. Manag.* **2016**, *12*, 170–188. [CrossRef]

37. Zhu, J.-W.; Zhou, L.-N.; Li, L.; Ali, W. Decision simulation of construction project delivery system under the sustainable construction project management. *Sustainability* **2020**, *12*, 2202. [CrossRef]

38. Feghaly, J.; El Asmar, M.; Ariaratnam, S.T. A comparison of project delivery method performance for water infrastructure capital projects. *Can. J. Civ. Eng.* **2021**, *48*, 691–701. [CrossRef]

39. Carpenter, N.; Bausman, D.C. Project delivery method performance for public school construction: Design-bid-build versus CM at risk. *J. Constr. Eng.* **2016**, *142*, 05016009. [CrossRef]

40. Browning, T.R. Design Structure Matrix Extensions and Innovations: A Survey and New Opportunities. *IEEE Trans. Eng. Manag.* **2016**, *63*, 27–52. [CrossRef]

41. Eppinger, S.D.; Browning, T.R. *Design Structure Matrix Methods and Applications*; MIT Press: Cambridge, MA, USA, 2012; Volume 1.

42. Yassine, A.; Braha, D. Complex concurrent engineering and the design structure matrix method. *Concurr. Eng.* **2003**, *11*, 165–176. [CrossRef]
43. Maheswari, J.U.; Varghese, K. A Structured Approach to Form Dependency Structure Matrix for Construction Projects. In Proceedings of the 22nd International Symposium on Automation and Robotics in Construction, Ferrara, Italy, 11–14 September 2005.
44. Sullivan, J. *Application of the Design Structure Matrix (DSM) to the Real Estate Development Process*; Massachusetts Institute of Technology: Cambridge, MA, USA, 2010.
45. Steward, D. *Systems Analysis and Management: Structure, Strategy, and Design*; Petrocelli Books: New York, NY, USA, 1981.
46. Steward, D.V. The design structure system: A method for managing the design of complex systems. *IEEE Eng. Manag.* **1981**, *28*, 71–74. [CrossRef]
47. Browning, T.R. Applying the design structure matrix to system decomposition and integration problems: A review and new directions. *IEEE Eng. Manag.* **2001**, *48*, 292–306. [CrossRef]
48. Yassine, A. An Introduction to Modeling and Analyzing Complex Product Development Processes Using the Design Structure Matrix (DSM) Method. *Urbana* **2004**, *9*, 1–17.
49. Grose, D. Reengineering the aircraft design process. In Proceedings of the 5th Symposium on Multidisciplinary Analysis and Optimization, Panama City Beach, FL, USA, 7–9 September 1994; p. 4323.
50. Forbes, G.A.; Fleming, D.A.; Duffy, A.; Ball, P. The optimisation of a strategic business process. In Proceedings of the Twentieth International Manufacturing Conference, Cork, Ireland, 3–5 September 2003.
51. Oloufa, A.A.; Hosni, Y.A.; Fayez, M.; Axelsson, P. Using DSM for modeling information flow in construction design projects. *Civ. Eng. Syst.* **2004**, *21*, 105–125. [CrossRef]
52. Liang, L.Y. Grouping decomposition under constraints for design/build life cycle in project delivery system. *Int. J. Technol. Manag.* **2009**, *48*, 168–187. [CrossRef]
53. Hyun, H.; Kim, H.; Lee, H.-S.; Park, M.; Lee, J. Integrated Design Process for Modular Construction Projects to Reduce Rework. *Sustainability* **2020**, *12*, 530. [CrossRef]
54. Ma, Z.; Ma, J. Formulating the application functional requirements of a BIM-based collaboration platform to support IPD projects. *KSCE J. Civ. Eng.* **2017**, *21*, 2011–2026. [CrossRef]
55. Austin, S.; Baldwin, A.; Waskett, P.; Li, B. Analytical design planning technique for programming building design. *Proc. Inst. Civil. Eng. Struct. Build.* **2015**, *134*, 111–118. [CrossRef]
56. Alhazmi, T.; McCaffer, R. Project procurement system selection model. *J. Constr. Eng.* **2000**, *126*, 176–184. [CrossRef]
57. Moon, H.; Cho, K.; Hong, T.; Hyun, C. Selection Model for Delivery Methods for Multifamily-Housing Construction Projects. *J. Manag. Eng.* **2011**, *27*, 106–115. [CrossRef]
58. Molenaar, K.R.; Songer, A.D. Model for public sector design-build project selection. *J. Constr. Eng.* **1998**, *124*, 467–479. [CrossRef]
59. Karhu, V.; Keitilä, M.; Lahdenperä, P. *Construction Process Model: Generic Present-State Systematisation by IDEF0*; VTT Technical Research Centre of Finland: Espoo, Finland, 1997.
60. Takala, T. Design transactions and retrospective planning: Tools for conceptual design. In *Intelligent CAD Systems II: Implementational Issues*; Springer: Berlin/Heidelberg, Germany, 1989.
61. DSM_Program-V2.1. Available online: https://dsmweb.org/excel-macros-for-partitioning-und-simulation/ (accessed on 28 October 2021).
62. Molenaar, K.; Franz, B. *Revisiting Project Delivery Performance. Charles Pankow Foundation*; Construction Industry Institute, University of Colorado, University of Florida: Gainesville, FL, USA, 2018.
63. Brady, T.K. Utilization of Dependency Structure Matrix Analysis to Assess Complex Project Designs. In Proceedings of the ASME 2002 International Design Engineering Technical Conferences and Computers and Information in Engineering Conference, Montreal, QC, Canada, 29 September–2 October 2002; pp. 231–240.
64. Gunawan, I.; Ahsan, K. Project scheduling improvement using design structure matrix. *Int. J. Proj. Organ. Manag.* **2010**, *2*, 311–327. [CrossRef]
65. Zhao, L.; Wang, Z. Process Optimization Calculation Model and Empirical Research of Prefabricated Buildings Based on DSM. In *ICCREM 2019*; American Society of Civil Engineers: Reston, VA, USA, 2019; pp. 614–621.
66. Browning, T.R. Process Integration Using the Design Structure Matrix. *Syst. Eng.* **2002**, *5*, 180–193. [CrossRef]
67. Pektaş, Ş.T.; Pultar, M. Modelling detailed information flows in building design with the parameter-based design structure matrix. *Des. Stud.* **2006**, *27*, 99–122. [CrossRef]
68. Yu, T.-L.; Goldberg, D.E.; Sastry, K.; Lima, C.F.; Pelikan, M. Dependency structure matrix, genetic algorithms, and effective recombination. *Evol. Comput.* **2009**, *17*, 595–626. [CrossRef] [PubMed]
69. Senthilkumar, V.; Varghese, K.; Chandran, A. A web-based system for design interface management of construction projects. *Autom. Constr* **2010**, *19*, 197–212. [CrossRef]
70. Maheswari, J.U.; Varghese, K.; Sridharan, T. Application of Dependency Structure Matrix for Activity Sequencing in Concurrent Engineering Projects. *J. Constr. Eng. Manag.* **2006**, *132*, 482–490. [CrossRef]
71. Ma, G.; Hao, K.; Xiao, Y.; Zhu, T. Critical chain design structure matrix method for construction project scheduling under rework scenarios. *Math. Probl. Eng.* **2019**. [CrossRef]

72.   Yang, Q.; Yao, T.; Lu, T.; Zhang, B. An overlapping-based design structure matrix for measuring interaction strength and clustering analysis in product development project. *IEEE Eng. Manag.* **2013**, *61*, 159–170. [CrossRef]
73.   Zhang, H.; Qiu, W.; Zhang, H. An approach to measuring coupled tasks strength and sequencing of coupled tasks in new product development. *Concurr. Eng.* **2006**, *14*, 305–311. [CrossRef]
74.   Gálvez, E.D.; Ordieres, J.B.; Capuz-Rizo, S.F. Evaluation of Project Duration Uncertainty using the Dependency Structure Matrix and Monte Carlo Simulations. *Rev. De La Construcción. J. Constr.* **2015**, *14*, 72–79. [CrossRef]
75.   Ummer, N.; Maheswari, U.; Matsagar, V.A.; Varghese, K. Factors influencing design iteration with a focus on project duration. *J. Manag. Eng.* **2014**, *30*, 127–130. [CrossRef]

Article

A Multilayer Perception for Estimating the Overall Risk of Residential Projects in the Conceptual Stage

Mohamed Badawy [1], Fahad K. Alqahtani [2,*] and Mohamed Sherif [3]

[1] Structural Engineering Department, Ain Shams University, Cairo 11517, Egypt; mohamed_badawy@eng.asu.edu.eg
[2] Civil Engineering Department, College of Engineering, King Saud University, P.O. Box 800, Riyadh 11421, Saudi Arabia
[3] Department of Civil and Environmental Engineering, College of Engineering, The University of Hawai'i at Mānoa, 2540 Dole Street, Honolulu, HI 96822, USA; msherif@hawaii.edu
*   Correspondence: bfahad@ksu.edu.sa

Abstract: The ability to foresee hazards early plays a critical role in estimating the entire cost of a project. Although several studies have established models to predict the total cost of a project at a conceptual stage, there remains a research vacuum in measuring the overall risk at this stage. Using artificial neural networks, this research provides a strategy for estimating the overall risk in residential projects at the conceptual stage. There are eight important components in the suggested paradigm. The model was created using data from 149 projects. In the first hidden layer in the model, there are five neurons, and in the second hidden layer, there are three neurons. The suggested model's mean absolute error rate was 11.7%. In the conceptual stage of residential projects, the number of floors, the type of interior finishes, and the implementation of risk management processes are the significant aspects that influence the overall risk. The proposed model assists project managers in precisely estimating the project's overall risk, which leads to a more accurate estimation of the contract's entire worth at the conceptual stage, allowing the stakeholders to decide whether or not to proceed with the project.

Keywords: early-stage; overall risk; residential projects; a multilayer perception

Citation: Badawy, M.; Alqahtani, F.K.; Sherif, M. A Multilayer Perception for Estimating the Overall Risk of Residential Projects in the Conceptual Stage. *Buildings* **2022**, *12*, 480. https://doi.org/10.3390/buildings12040480

Academic Editors: Srinath Perera, Albert P. C. Chan, Dilanthi Amaratunga, Makarand Hastak, Patrizia Lombardi, Sepani Senaratne, Xiaohua Jin and Anil Sawhneyand

Received: 17 March 2022
Accepted: 11 April 2022
Published: 13 April 2022

Publisher's Note: MDPI stays neutral with regard to jurisdictional claims in published maps and institutional affiliations.

1. Introduction

A project's cost should be projected with a high degree of precision; however, making a conceptual cost estimate is challenging at this time due to a lack of data [1]. Construction companies require an early budget estimate to assess whether this expenditure is acceptable and, hence, whether the project should be continued or abandoned. To estimate the contract value, the project manager usually calculates the direct cost, indirect cost, profit, and contingencies for project risks. As a result, performing an early risk assessment is critical. Stakeholders can make decisions and choices during the conceptual stage of the projects, which have major impacts on construction duration and costs, but this effect reduces as the project progresses through its life cycle [2]. Negative risks may result in schedule delays and expense overruns [3]. As a result, the project manager should concentrate as much as possible on the major risks [4]. Overall risk estimation suffers many challenges during the conceptual stage of a project due to the limited data provided. A major issue that develops in the early phase is the lack of effective and reliable overall risk estimation approaches. As a result, project decision-makers have begun to focus more on conceptual planning, where a thorough cost analysis is a critical component in achieving the project's objectives [5]. The goal of the estimation process is to make sure that the contract plans and specifications match the cost of completing the project [6].

Developing cost estimation models, both in the planning stage and in the conceptual stages of a project, has been the subject of a lot of research. Despite the importance of overall project risk, there is a gap in the research on assessing the overall risk in the conceptual stage of a project when there is insufficient project information, necessitating the development of a model for predicting the overall risk at the conceptual stage of a project [7]. This research uses artificial neural networks to construct a model for evaluating the overall risk in residential projects, based on a few characteristics that may be easily recognized at a conceptual stage with acceptable accuracy. In other words, this research does not investigate the assessment of individual positive or negative risk variables, but rather the classification of the overall risk of residential projects at a conceptual stage based on the influence on project cost.

As there is a gap in developing a method to estimate the total risk in the conceptual stage, this research aims to propose a model to predict the overall risk of residential buildings at the conceptual stage using an artificial neural network with a multilayer perception.

2. Literature Review

Financial risks are regarded as the most significant risks in construction projects in Egypt and Saudi Arabia, followed by design, political, and construction risks [8]. Lack of money, a tight deadline, design revisions, insufficient information on sustainable design, and a weak definition of sustainable scope were the top hazards to sustainable building projects in the UAE [9]. At the planning stage, risk assessment models have been proposed, using an artificial neural network such as risk assessment in Saudi Arabian building projects [10] or the construction of an expressway [11]. Another model, based on system dynamics and discrete event simulation, was proposed to evaluate the impact of risk factors on project schedules in infrastructure projects [12]. Al-Tabtabai and Alex (2000) provided an ANN-based model for predicting project cost escalation owing to political concerns and the average error was 7% [13]. A model using the Bayesian Belief Network was developed to assess and enhance the implementation of residential construction projects. Improper construction procedures and poor communication were the top risk factors [14]. At a conceptual stage, the most important aspects that determine the overall risk are the use of risk management processes, the entire project duration, contract cost, and contract type [15]. The most essential criteria in the tender process are price, the scope of work, and technical resources [16].

At the conceptual stages, there is a research gap in estimating overall risk. As a result, an essential point to consider is what proportions of errors are acceptable in any model assessing the overall risk. For every equation or model, there is a rate of error, but how can this ratio be judged, meaning how one can determine if the model is accepted or not. It is not fair to judge the error rate of a model in the conceptual stage, where there is not enough information, to the error rate of a model in the design stage where there is sufficient information. It is expected that the error rate is less in the design stage than at the conceptual stage. Therefore, the error rate of any model must be compared with the extent of errors in the same stage. Since there is no research that deal with estimating the total risk of the project in the conceptual stage, the error rate in the proposed model for calculating the total risk of the project in the conceptual stage was compared with the acceptable range of the rates of estimation models for cost estimation models in the conceptual phase. Hence, to estimate the allowable range of percentage errors, the authors relied on a review of past studies on cost estimation at a conceptual stage.

To estimate the cost of school buildings in Korea, ten factors were identified. Three models were developed to calculate the cost of the school buildings, based on 217 projects. The first model was developed using neural network techniques, while the regression analysis was used in the second model and the third model was presented using the support vector machine. The results of the neural network model showed a more accurate estimate than the results of regression analysis or the supporting vector machine models [17]. Two studies were conducted in Gaza to estimate the cost of buildings at an early stage. The first research was based on seven variables and a model was proposed based on information derived from 71 construction projects using artificial neural networks [18]. While the second research developed a model for assessing the cost of construction projects with a high degree of accuracy and without the need for a lot of information, through the use of artificial neural networks. A database of 169 projects was collected from relevant institutions in the Gaza Strip has been adopted. The artificial neural network model has eleven factors as independent inputs [19]. A study to predict the cost of construction projects at the conceptual stage in Taiwan using ten parameters. The research suggested the utilization of the evolutionary fuzzy neural inference model to enhance cost assessment accuracy. The proposed model was relied on eleven factors [20]. In Egypt, a model to assess the cost of a residential building at an early stage using the artificial neural network and data obtained from 174 residential projects. The proposed model depended on four parameters: number of floors, the area of the floor, type of external finishing, and type of internal finishing [21]. The costs of 136 executed projects were utilized to propose an artificial neural network model to predict the preliminary cost of construction projects in Yemen. The suggested model contained 17 factors [22]. In the United States, research was conducted on the difference in the computation of construction costs utilizing artificial neural networks by comparing nineteen variables in 20 projects [23]. In Taiwan, a study has presented a prototype for the rapid assessment of a proposal integrating a probabilistic cost sub-model and a multi-factor assessment sub-model. The cost-based sub-model concentrates on the cost divisions. While the multi-factor assessment sub-model captures the specific elements influencing the cost division. That research is based on 21 variables [24]. The eight previous studies mostly agreed on nine primary factors that can be used for cost estimates at the conceptual stage of a project. These nine parameters are floor area, number of floors, type of foundation, number of elevators, type of slab, type of exterior finishing, interior finishes, type of electromechanical works, and number of basements. Table 1 shows the different sources for each parameter.

Table 1. Sources of parameters influencing the cost estimation.

Factor	[17]	[18]	[19]	[20]	[21]	[22]	[23]	[24]
Floor area	√	√	√	√	√	√	√	√
Number of floors	√	√	√	√	√	√	√	√
Slab type	√		√			√	√	√
Internal finishes	√		√	√	√	√	√	√
Number of elevators		√	√			√	√	√
External finishes			√		√	√	√	√
Foundation type		√	√			√	√	√
Basement	√		√			√	√	
Electromechanical type			√	√		√	√	√

The symbol "√" means the corresponding research determined the corresponding factor as a key factor for identifying the cost estimate at the conceptual stage.

Traditional cost estimation strategies in construction projects are the most often used. They rely on time-consuming manual project cost estimation or Excel spreadsheets, rather than using computerized tools to estimate building costs. Soft computing strategies for conceptual-stage software development were compared by Bhatnagar and Ghose (2012). The feed-forward back propagation neural network model had a mean absolute percentage error (MAPE) of 13%, a cascaded feed-forward back propagation neural network model had a MAPE of 13.6 percent, a layer recurrent neural network model had a MAPE of 11.5 percent, and a fuzzy logic model had a MAPE of 3.9 percent. This means they accepted models up to MAPE with a 13.5% acceptance rate [25].

There is a lot of research that investigates the cost estimates at the conceptual stage. Each research proposed a model with a mean absolute percentage error. Table 2 shows the mean absolute percentage error for some of these prior studies. Which illustrated that the errors in the proposed models were ranged from 4–28.2%. This means that the maximum acceptable mean percentage error in the proposed model at the conceptual stage is 28.2%.

Table 2. The minimum absolute percentage errors for the previous studies.

Reference	Research	MAPE
[26]	Data modelling and the application of a neural network approach to the prediction of total construction costs	16.6
[27]	A neural network approach for early cost estimation of structural systems of buildings	7
[28]	Conceptual cost estimates using evolutionary fuzzy hybrid neural network for projects in the construction industry	10.4
[18]	Early-stage cost estimation of buildings construction projects using artificial neural networks	4
[19]	Cost estimation for building construction projects in the Gaza Strip using an artificial neural network (ANN)	6
[17]	Comparison of school buildings' construction costs' estimation methods using regression analysis, neural network, and support vector machine	5.27
[29]	Estimating water treatment plants' costs using factor analysis and artificial neural networks	21.2
[30]	Conceptual cost estimation model for engineering services in public construction projects	28.2
[31]	Cost estimation of civil construction projects using machine learning paradigm	6.2
[32]	Comparison of artificial intelligence techniques for project conceptual cost prediction	26.3
[21]	A hybrid approach for a cost estimate of residential buildings at the early stage	13.2

Limited studies in the conceptual stage of risk estimation identified four criteria: use of risk management processes, duration of the entire project, total cost, and type of contract. As the total cost of the project can be estimated through nine criteria: floor area, number of floors, type of foundation, number of elevators, type of slab, type of external finishing, internal finishes, type of electromechanical works, and number of basements. Hence, the cost of the project can be replaced by these nine factors. The type of contract was not included due to research limitations, as the research is related to estimating the cost of housing projects based on a fixed price contract only. Hence, the initial list of criteria used to derive the overall project risk at the conceptual stage contained eleven criteria: floor area, number of floors, type of foundation, number of elevators, type of slab, type of exterior finish, interior finishes, type of electromechanical works, number of basements plus the use of risk management processes, and the entire project duration.

3. Methodology

This study's approach is divided into three sections. The first step in achieving the study's goal is to identify the essential components influencing the overall risk assessment by evaluating past research studies that focus on construction cost and risk estimation at a conceptual stage. As a result, eleven construction elements (or processes) were presented in the primary list. Five experts with at least 15 years of experience in the construction of residential projects were randomly selected. The primary questionnaire was presented to the experts using the Delphi technique to determine the parameters in the final questionnaire in three stages. In the first stage, experts were asked to add any missing criteria, if any, that could affect the overall risk and could be discovered in the conceptual stage. Data are collected from experts, revised, and re-sent back to the experts where they are asked to rank each criterion on a five-point Likert scale. After collecting the data from the second round, the averages are calculated and any factor that has a very low impact on estimating the overall risk of the project is removed from the final list. In the third round, the experts are asked to assess whether or not they agree with the final list. The third step is to develop the model. The model is simulated using artificial neural networks using Statistical Package for the Social Sciences (SPSS) software. The critical parameters that affect the estimation of the overall risks at the conceptual stages are considered as inputs to the model while the output is the overall risk of the project. The model may contain one or two hidden layers. The number of neurons in the one-hidden-layer model can be three, four, or five. In models with two hidden layers, the number of neurons can be four in the first layer and three in the second, or five in the first layer and three in the second, or five in the first layer and four in the second. Thus, there are three different groups in terms of the number of neurons in each hidden layer. Hence, six models can be developed. The hyperbolic tangent function was used as an activation function for the hidden layers in six models, and the Sigmoid function was tested as an activation function for the hidden layers in six other models. Hence, twelve Multilayer Perceptron models have been identified and tested. To evaluate the performance of the model, the available data were also randomly divided 5-fold. The first fold contains 29 cases, while the subsequent folds contain 30 cases. Four folds were used to train the network in each model, while the fifth fold was used to evaluate the model. The final proposed model for estimating the overall risk in the conceptual stages is the model with the lowest mean absolute error rate.

4. Identifying the Critical Parameters Affecting the Estimation of Overall Risk at the Conceptual Stages

Literature analysis and earlier research yielded eleven criteria that can be utilized to forecast overall risk in residential projects. Floor space, number of floors, slab type, interior finishes, number of elevators, external finishes, electromechanical type, number of basements, foundation type, risk management implementation, and overall project duration are some of the elements to consider. Five experts with at least 15 years of experience in residential project management used the Delphi technique to select the final parameters. Table 3 shows the demographic information about the experts. The experts were requested to add missing parameters, if any, that could affect the overall risk and could be discovered at a conceptual stage, in the first round. There is no missing factor according to the experts' responses. On a five-point Likert scale, the experts were asked to evaluate the weight of each parameter in the second round. "1" indicates that this element is inconsequential; "2" suggests low importance; "3" indicates moderate significance; "4" indicates high significance; and "5" indicates extremely significant. Equation (1) was used to calculate the relative relevance index based on the responses received. Table 4 displays the relative importance indexes. The lowest number on the Likert scale is "1," and the highest is "5", resulting in a range of four, which will be graded according to the five categories. The zone for each category is 0.8. The very low category has a range of 1 to 1.8. The low category has a range from 1.8 to 2.6, while the range of the medium category is from 2.6 to 3.4. The high category is from 3.4 to 4.2, whereas the very high category is from 4.3 to 5.0.

$$RII = \frac{\sum W}{AN} = \frac{(5n_5 + 4n_4 + 3n_3 + 2n_2 + 1n_1)}{5N} \tag{1}$$

Table 3. Demographic data regarding the experts.

Expert I.D.	Education Level	Experience	Job Title	Company
Expert (1)	Bachelor of Civil Engineering	18	Project manager	Private
Expert (2)	Bachelor of Civil Engineering	15	Risk manager	Private
Expert (3)	Ph.D. in Civil Engineering	22	Project manager	Private
Expert (4)	Bachelor of Civil Engineering	17	Project manager	Public
Expert (5)	Bachelor of Civil Engineering	16	Project manager	Private

Table 4. The relative importance indices of critical parameters.

Expert I.D.	Floor Area	Number of Floors	Slab Type	Internal Finishes	Elevator	External Finishes	Foundation Type	Basement	Electromechanical	Risk Management Application	Total Project Duration
Expert (1)	4	4	1	3	1	3	1	2	2	3	4
Expert (2)	4	4	2	2	1	2	2	3	3	3	5
Expert (3)	3	4	1	1	1	1	1	2	3	4	4
Expert (4)	4	5	1	2	1	2	2	3	4	3	5
Expert (5)	3	4	2	2	2	2	2	2	2	3	4
RII	3.60	4.20	1.40	2.00	1.20	2.00	1.60	2.40	2.80	3.20	4.40
Grade	H	VH	VL	L	VL	L	VL	L	M	M	VH

The five experts agreed that the total duration of the project and the number of floors are the most important factors, with relative importance indices of 4.4 and 4.2, respectively, followed by the floor area, which has a relative importance index of 3.6. Experts agreed that the risk management application and the type of electromechanical factors are considered to have a medium effect on the cost estimation. While the factors of interior finishes, exterior finishes, and the number of basements were considered to have a low impact on cost estimation by experts. Whereas slab type, elevator number, and foundation type had very low impacts on cost estimation. As a result, any factor with an RII of less than 1.8 was eliminated from the final list. As a result, the slab type, elevator number, and foundation type were left off the final list of parameters influencing the total risk prediction in the conceptual stages. The final set of criteria consisted of the remaining eight parameters. Experts were asked to assess whether or not they agreed with the finalist list in the third round. Regarding the final list of criteria, which includes the remaining eight criteria, the experts agreed unanimously.

5. Data Collection

There were eight input parameters and one output variable in the data collected. Less than 200 square meters, 200 to 400 square meters, 400 to 600 square meters, and more than 600 square meters were the four categories for the floor space factor. The authors divided the factors of the number of floors into four categories: one or two floors, three to five stories, six to eight stories, and more than eight stories. The interior finishes variant is categorized into four groups: no interior finishes, basic, semi-finished interior finishes, and luxurious interior finishes. The choice is of the type of semi-finished interior finishes in the case of normal plaster for walls only and there are no paintworks, whereas for the type of basic interior finishes, it is in the case of the presence of paint works for the walls and ceramics for the floors. The type of luxurious interior finishes is chosen in the case of the presence of paint works for the walls and porcelain or marble works for the floors. The external finishing aspect was simply divided into two categories: basic and luxurious. The type of basic external finishing is chosen if the facades of the building have been painted only without any works of marble, Hashemite, or Pharaonic stone, while the external finishing is considered the luxurious type if the facades of the building have been done with any works of marble, Hashemite, or Pharaonic stone. There were two groups for the number of basements parameter: no basement and one basement. The overall project duration parameter was divided into four categories by the authors: less than six months, six months to a year, one year to two years, and more than two years. The risk management process application parameter was split into two categories: no risk management processes were performed on the project and risk management procedures were performed on the project. Electromechanical can be divided into two categories: basic and luxurious. The type of the electromechanical parameter is considered with the basic standards if the scope of work includes the main works of water, electricity, and sewage outside the apartment, but if it includes the internal works of the apartment, the type of the electromechanical parameter is considered a luxury type.

Based on the review of planned cost and actual cost data and using Equation (2), the authors assessed the overall percentage of risk for the completed projects. A project is excluded from the analysis if there is insufficient information about its planned cost or its actual cost. According to Table 5, the overall percentage of risk which is the major outcome variable was divided into three levels: low, medium, and high-risk scores.

$$\%OR = \frac{|AC - PC|}{PC} \times 100 \tag{2}$$

"$\%OR$" represents the overall percentage of risk, "PC" represents planned cost and "AC" represents an actual cost.

Table 5. The Classifications of outputs.

Category	Low	Medium	High
Impact on cost	Less than 10%	10–20%	More than 20%

The authors examined 250 projects and discovered that some data for the eight input variables or the result variable were missing. As a result, only the full data of 149 actual residential projects were accessed. For example, out of the 149 projects analyzed, "case no. the 26" project consisted of a 12-story building with 500 square meters per level, exquisite interior and exterior finishes, and luxurious electromechanical work. This building has one basement and was built in 20 months using risk management procedures, with an overall risk of roughly 12%. Due to the enormous population, it was assumed that the population's size was unlimited, so the sample size could be calculated using Equation (3). Table 6 shows demographic information about the respondents, whereas Table 7 shows demographic information about the inputs gathered from 149 projects.

$$SS = \frac{Z^2 \times p \times (1 - p)}{C^2} \tag{3}$$

where SS stands for sample size, Z stands for 1.96 with a 95% confidence level, p stands for the probability of selection, and C stands for the confidence interval. The sample size in this study was 149 projects, and the p-value was 0.5, hence the confidence interval was 0.08.

Table 6. The demographic data regarding the respondents.

		Work Experience in the Construction Industry		
		From 5 to 10 Years	From 10 to 15 Years	More Than 15 Years
	Site engineer	42	29	0
Job title	Project manager	7	52	13
	senior manager	0	0	6

Table 7. Demographic data of inputs.

Floor Area (A)		Number of Floors (N)		Internal Finishes (IF)		Total Project Duration (D)	
Less than 200	28	one or two	36	N.A	20	Up to 6	32
200 to 400	45	three to five	41	half-finished	49	6 to 12	40
400 to 600	46	six to eight	36	Basic	48	12 to 24	40
more than 600	30	more than 8	36	luxury	32	more than 24	37
External finishes (EF)		**Number of basements (B)**		**Risk management processes (RM)**		**Type of electromechanical (E)**	
Basic	72	No	118	No	132	Basic	77
luxury	77	One	31	Yes	17	luxury	72

Using Equation (4), the authors estimated Cronbach's Alpha for items. Cronbach's Alpha has a threshold of 0.7 [33]. Cronbach's alpha in this study was 0.757, which is higher than 0.7. It indicates that the scale is consistent and does not contradict itself, implying that it will produce the same findings when applied to the same sample again. Validity refers to how accurate a measurement is. The validity of this study was 0.87.

$$\propto = \frac{n}{n-1} \times \left(1 - \frac{\sum_1^n V_i}{V_t}\right) \tag{4}$$

where "n" represents the number of items, V_i represents the variance of item i, and V_t represents the variance of the test score.

6. Model Specification

The model was simulated using artificial neural networks. Due to its ease of use, IBM SPSS software was chosen to construct the model. It has a simple user interface and can be quickly imported and exported from Excel. The model has eight input parameters and just one output. Floor space, number of floors, interior, and exterior finishes, number of basements, total project time, risk management process application, and electromechanical type are all inputs. The output, on the other hand, is the overall risk factor. To evaluate the model's performance, the acquired data were randomly divided into 5-fold cross-validation. The first fold has 29 cases, whereas the subsequent folds have 30. Four folds were utilized to train the network in each model, while the fifth fold was used to evaluate the model. One hidden layer or two hidden layers might be present in a model. As a result, there are two different sorts of hidden layer groups. The number of neurons in the model with one hidden layer can be three, four, or five. In models with two hidden layers, the number of neurons can be four in the first layer and three in the second, or five in the first layer and three in the second, or five in the first layer and four in the second. Thus, there are three different groups in terms of the number of neurons in each hidden layer. The hyperbolic tangent function or the sigmoid function was employed as an activation function for the hidden layers, and both were investigated. Equation (5) can be used to estimate the number of models that can be tested. Twelve Multilayer Perceptron models were identified and tested as a result. The examined models and their mean absolute errors (MAE) in each k-fold are shown in Table 8. Equation (6) can be used to calculate the mean absolute error [34].

$$N_m = N_l \times N_a \times N_g \tag{5}$$

$$MAE = \frac{\left(\sum_{i=1}^N (ER - RS)\right)}{N} \tag{6}$$

where "N_m" stands for the number of models, "N_l" for the number of hidden layers, "N_a" for the number of hidden layer activation functions, and "N_g" for the number of neuron groups. "ER" stands for the model's estimated risk, "RS" for the risk score, and "N" for the number of case studies.

The mean absolute error of any model is equal to the mean error in its k-fold. Hence, the proposed model should have the minimum percentage of MAE. In this study, the MAE was equal to 11.7%, as shown in Table 8. The proposed model consists of two hidden layers: five neurons in the first hidden layer, and three neurons in the second hidden layer. The activation function of the hidden layer was the Hyperbolic Tangent function in the proposed model. Figure 1 illustrates the structure of the proposed model. The real and estimated overall risks are presented in Table 9.

Table 8. Mean absolute error of the models.

Model	H3-0	H4-0	H5-0	H4-3	H5-3	H5-4	S3-0	S4-0	S5-0	S4-3	S5-3	S5-4
No. of hidden layer	1	1	1	2	2	2	1	1	1	2	2	2
No. of neurons in the first layer	3	4	5	4	5	5	3	4	5	4	5	5
No. of neurons in the second layer	-	-	-	3	3	4	-	-	-	3	3	4
Activation Function	H	H	H	H	H	H	S	S	S	S	S	S
K-1	11.4%	10.7%	13.4%	10.7%	13.4%	13.4%	14.8%	15.4%	18.1%	18.1%	15.4%	13.4%
K-2	10.7%	10.7%	13.4%	13.4%	9.4%	10.7%	12.8%	9.4%	14.1%	13.4%	14.1%	10.7%
K-3	12.8%	16.1%	16.8%	16.8%	14.8%	16.1%	20.8%	15.4%	16.8%	15.4%	12.1%	17.4%
K-4	16.1%	15.4%	16.1%	16.8%	10.1%	14.8%	19.5%	12.8%	16.8%	22.1%	18.1%	22.8%
K-5	10.7%	14.8%	16.1%	14.8%	10.7%	16.1%	15.4%	16.8%	18.8%	14.8%	19.5%	20.1%
MAE	12.3%	13.6%	15.2%	14.5%	11.7%	14.2%	16.6%	14.0%	16.9%	16.8%	15.8%	16.9%

"H" stands for the Hidden Layers' Hyperbolic Tangent activation function and "S" stands for the Hidden Layers' Sigmoid activation function.

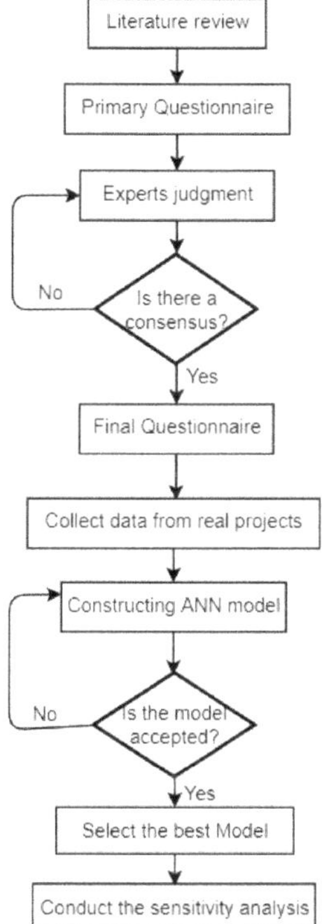

Figure 1. The Methodology of research.

Table 9. Classification of overall risk.

Sample	Classification	Predicted			
		Low	**Medium**	**High**	**Percent Correct**
	Low	73	5	0	95.5%
Training	Medium	5	26	1	85.1%
	High	0	4	5	57.7%
	Low	14	2	0	92.6%
Testing	Medium	0	9	1	85.7%
	High	0	4	0	23.1%

7. Discussion

There is very little research on total risk assessment at the conceptual stage. For example, Oad et al. (2021) determined that price, the scope of work, and technical resources are the most important criteria in the bidding process at the conceptual stage [16]. Another study identified project cost, total project time, contract type, and use of risk management techniques as the primary criteria that can be used to assess overall risk in apartment buildings at the conceptual stage [15]. In this research, the main criteria used to estimate the total risks at the conceptual stage in residential projects are the number of floors, the building area, the interior finishes, the exterior finishes, the number of basements, the total duration of the project, the type of electromechanical, and the application of risk management processes. The scope of work that was identified as a critical factor in the study by Oad et al. (2021) was expressed in the current study by the number of floors, building area, internal and external finishes, and the number of basements. The duration of any activity can be estimated based on the required quantity and the production rate of the available resources. Hence, the technical resources identified by Oad et al. (2021) as critical factors in the conceptual stage were expressed in the quantities that can be inferred from the scope of work and the total duration of the project in the current study. Whereas the project cost component in the conceptual stage, which was identified by Badawy et al. (2022) can be estimated through many previous studies in the conceptual stage, which indicated that the cost can be deduced from the number of floors, building area, interior, and exterior finishes, and the number of basements, which was applied in the current study. Therefore, the eight input variables in the current study are in agreement with previous studies.

In the training phase, the proposed model predicted an average of 104 cases correctly and accurately with a ratio of 87.4% and predicted 15 cases incorrectly. The MAE for the low overall risk classification was 4.5%, and for the medium overall risk, the MAE was 14.9%. Unfortunately, the prediction of the overall risk in the case of the high-risk classification was 42.3%, which is considered a high ratio. In the testing phase, the proposed model predicted an average of 23 cases correctly and accurately with a ratio of 76.7% and predicted 7 cases incorrectly. The MAE for the low overall risk classification was 7.4%, and for the medium overall risk, the MAE was 14.3%. Unfortunately, the prediction of the overall risk in the case of the high-risk classification was 76.9%, which is considered a high ratio. Hence, the results indicated that this model is excellent in predicting the low and medium overall risk at the conceptual stage.

The mean absolute percentage error was 16.6% in an ANN model for estimating the total construction costs [26], while the MAPE was 13.2% in a hybrid technique for a cost assessment of residential projects at the early phase [21]. The MAPE was 26.3% an ANN approaches for cost forecast at the conceptual stage [32], while to estimate the cost of water treatment plants, the model has an error of 21.2% [29]. A model to predict the conceptual cost for engineering services in public construction projects was developed with a MAPE of 28.2 [30]. As a result of reviewing past research on conceptual-stage cost models, it was discovered that a mean absolute error of more than 13% was permitted, implying that the accepted model should have an error of less than 13%. The suggested strategy correctly classified 149 projects with a mean absolute error of 11.7%. Hence, this model can be accepted. The suggested model's acceptability implies that the eight input factors can be utilized to predict the overall risk of residential projects at a conceptual stage. The results of the study agreed with the viewpoint of the five experts who were interviewed to determine the most important criteria in the final list that can be used to predict the overall risk in the conceptual stage of residential construction projects. The most important of these factors was the number of floors, which represents 28.5%. The second top criterion was the interior finishes with 16.3 percent. The execution of the risk management process component ranked third, with 14.4 percent, while the floor area element came fourth, with 11.7 percent. The total project time was the fifth component that had a 10.8% impact on the overall risk forecasting in the conceptual stages, followed by the exterior finishes, which had a 10.2% impact. Finally, the electromechanical type had a weight of 6.2%, and the lowest parameter was the number of basements with a relevance of 1.7 percent. The importance of each component in determining the overall risk in the conceptual stages of residential projects is depicted in Figure 2.

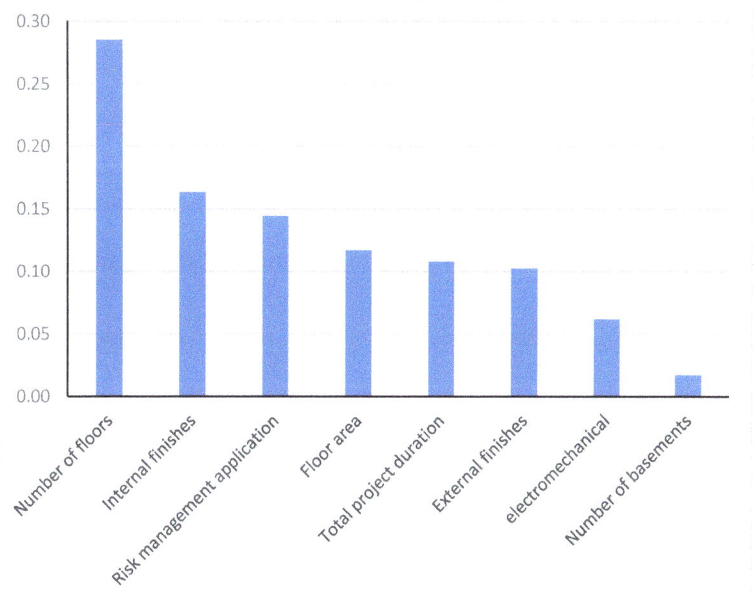

Figure 2. Importance of parameters in estimating the overall risk.

8. Conclusions

Decision-makers aim to predict the estimated value of the project budget in the conceptual stage to assess whether this investment is acceptable or not. The value of the reserve that covers the project's overall risk is included in the project budget. As a result, early on, a comprehensive risk assessment is required. There has been a great deal of research in developing cost estimation models, both in the planning phase and in the conceptual stages of a project. Unfortunately, there is a research vacuum in estimating risk in the conceptual stages of a project due to a lack of knowledge, so this study offers a model to forecast the overall risk at the conceptual stages of a project. A provisional list of essential characteristics, consisting of eleven parameters, was generated through a review of existing research and can be used to anticipate the overall risk in residential buildings at a conceptual stage. After three rounds of Delphi, the experts reached an agreement on the most critical parameters. The parameters for slab type, elevator number, and foundation type were omitted from the final list. Floor area, number of floors, interior finishes, external finishes, number of basements, kind of electromechanical, risk management process implementation, and overall project duration were all included in the final list. Four groups were created for the floor space, the number of floors, interior finishes, and total project duration, while the internal finishes, the number of basements, the risk management method execution, and electromechanical kinds were all divided into two categories. Three levels were assigned to the output variable. Data were gathered from 149 actual residential projects. As a result, the confidence interval was 0.08 at the 95% confidence level. The model was simulated using artificial neural networks. The data were divided into five groups at random. There were twelve Multilayer Perceptron models identified and tested, each with a distinct number of hidden layers and activation functions. The proposed model has two hidden layers, the first of which has five neurons and the second of which has three neurons. In the suggested model, the Hyperbolic Tangent function was used to activate the hidden layer. The MAE was equal to 11.7% in this investigation. The number of floors is the most critical factor in determining the overall risk in the conceptual stages of residential projects, followed by interior finishing, and the risk management procedure. The electromechanical type and the number of basements were the least critical elements. The project manager can use the proposed model to identify residential projects in the conceptual stages as low-risk, medium-risk, or high-risk. As a result, the proposed model can assist stakeholders in deciding whether or not to continue with the project.

9. Limitations of Research

The overall risk and the influence of the important parameters were solely calculated based on the impact on the cost of the residential construction in this study. This study only looked at projects with fixed-price contracts. As a result, projects with cost-reimbursable contracts, for example, will require a re-estimation of the input parameter weights. The eight input criteria can be used in any country to obtain the overall risk at the conceptual phase. The data were obtained from 149 projects in Egypt, which means that the ranking of importance of each criterion may differ from one country to another. Hence, they should be double-checked the ranking of the importance of the criteria before being used in any other country. The user needs to alter the weights of the variables to adapt the model to subsequent times because the data used to produce it came from residential buildings in Egypt built between 2018 and 2020.

Author Contributions: Conceptualization, M.B. and F.K.A.; methodology, M.B.; software, M.B. and M.S.; validation, F.K.A.; formal analysis, M.S.; investigation, M.S.; resources, F.K.A.; data curation, M.B.; writing—original draft preparation, M.B.; writing—review and editing, F.K.A.; visualization, M.S.; supervision, M.B.; project administration, F.K.A.; funding acquisition, F.K.A. All authors have read and agreed to the published version of the manuscript."

Funding: This research was funded by the Research Supporting Project number (RSP-2021/264), King Saud University, Riyadh, Saudi Arabia.

Informed Consent Statement: Not applicable.

Data Availability Statement: The data collected and used for analysis will be available from the corresponding author upon request.

Acknowledgments: The authors extend their appreciation to the Research Supporting Project number (RSP-2021/264), King Saud University, Riyadh, Saudi Arabia for funding this work.

Conflicts of Interest: The authors declare that there are no conflicts of interest regarding the publication of this paper.

References

1. Verlinden, B.; Duflou, J.R.; Collin, P.; Cattrysse, D. Cost Estimation for Sheet Metal Parts Using Multiple Regression and Artificial Neural Networks: A Case Study. *Int. J. Prod. Econ.* **2008**, *111*, 484–492. [CrossRef]
2. Feng, G.L.; Li, L. Application of Genetic Algorithm and Neural Network in Construction Cost Estimate. In Proceedings of the Advanced Materials Research; 2013; Volume 756–759, pp. 3194–3198.
3. Smith, N.J.; Merna, T.; Jobling, P. *Managing Risk in Construction Projects*; John Wiley & Sons, Ltd.: Hoboken, NJ, USA, 2014; ISBN 978-1-118-34723-2.

4. Barber, R.B. Understanding Internally Generated Risks in Projects. *Int. J. Proj. Manag.* **2005**, *23*, 584–590. [CrossRef]
5. Wang, Y.R.; Yu, C.Y.; Chan, H.H. Predicting Construction Cost and Schedule Success Using Artificial Neural Networks Ensemble and Support Vector Machines Classification Models. *Int. J. Proj. Manag.* **2012**, *30*, 470–478. [CrossRef]
6. Antohie, E. Classes of Construction Cost Estimates. *Bul. Institutului Politeh. din Iasi. Sect. Constr. Arhit.* **2009**, *55*, 21.
7. Al-Shanti, Y. A Cost Estimate System for Gaza Strip Construction Contractors, Palastine. Ph.D. Thesis, The Islamic University of Gaza, Gaza Strip, Israel, 2003.
8. Eskander, R.F.A. Risk Assessment Influencing Factors for Arabian Construction Projects Using Analytic Hierarchy Process. *Alexandria Eng. J.* **2018**, *57*, 4207–4218. [CrossRef]
9. El-Sayegh, S.M.; Manjikian, S.; Ibrahim, A.; Abouelyousr, A.; Jabbour, R. Risk Identification and Assessment in Sustainable Construction Projects in the UAE. *Int. J. Constr. Manag.* **2018**, *21*, 327–336. [CrossRef]
10. Al-Sobiei, O.S.; Arditi, D.; Polat, G. Predicting the Risk of Contractor Default in Saudi Arabia Utilizing Artificial Neural Network (ANN) and Genetic Algorithm (GA) Techniques. *Constr. Manag. Econ.* **2005**, *23*, 423–430. [CrossRef]
11. Zichun, Y. The BP Artificial Neural Network Model on Expressway Construction Phase Risk. *Syst. Eng. Procedia* **2012**, *4*, 409–415. [CrossRef]
12. Xu, X.; Wang, J.; Li, C.Z.; Huang, W.; Xia, N. Schedule Risk Analysis of Infrastructure Projects: A Hybrid Dynamic Approach. *Autom. Constr.* **2018**, *95*, 20–34. [CrossRef]
13. Al-Tabtabai, H.; Alex, A.P. Modeling the Cost of Political Risk in International Construction Projects. *Proj. Manag. J.* **2000**, *31*, 4–13. [CrossRef]
14. Odimabo, O.O.; Oduoza, C.F.; Suresh, S. Methodology for Project Risk Assessment of Building Construction Projects Using Bayesian Belief Networks. *Int. J. Constr. Eng. Manag.* **2017**, *6*, 221–234.
15. Badawy, M.; Alqahtani, F.; Hafez, H. Identifying the Risk Factors Affecting the Overall Cost Risk in Residential Projects at the Early Stage. *Ain Shams Eng. J.* **2022**, *13*, 101586. [CrossRef]
16. Oad, P.K.; Kajewski, S.; Kumar, A.; Xia, B. Bid Evaluation and Assessment of Innovation in Road Construction Industry: A Systematic Literature Review. *Civ. Eng. J.* **2021**, *7*, 179–196. [CrossRef]
17. Kim, G.-H.; Shin, J.-M.; Kim, S.; Shin, Y. Comparison of School Building Construction Costs Estimation Methods Using Regression Analysis, Neural Network, and Support Vector Machine. *J. Build. Constr. Plan. Res.* **2013**, *1*, 1–7. [CrossRef]
18. Arafa, M.; Alqedra, M. Early Stage Cost Estimation of Buildings Construction Projects Using Artificial Neural Networks. *J. Artif. Intell.* **2011**, *4*, 63–75. [CrossRef]
19. Shehatto, O. *Cost Estimation for Building Construction Projects in Gaza Strip Using Artificial Neural Network (ANN)*; Islamic University: Gaza, Israel, 2013.
20. Cheng, M.Y.; Tsai, H.C.; Hsieh, W.S. Web-Based Conceptual Cost Estimates for Construction Projects Using Evolutionary Fuzzy Neural Inference Model. *Autom. Constr.* **2009**, *18*, 164–172. [CrossRef]
21. Badawy, M. A Hybrid Approach for a Cost Estimate of Residential Buildings in Egypt at the Early Stage. *Asian J. Civ. Eng.* **2020**, *21*, 763–774. [CrossRef]
22. Hakami, W.; Hassan, A. Preliminary Construction Cost Estimate in Yemen by Artificial Neural Network. *Balt. J. Real Estate Econ. Constr. Manag.* **2019**, *7*, 110–122. [CrossRef]
23. Sonmez, R. Range Estimation of Construction Costs Using Neural Networks with Bootstrap Prediction Intervals. *Expert Syst. Appl.* **2011**, *38*, 9913–9917. [CrossRef]
24. Wang, W.C.; Wang, S.H.; Tsui, Y.K.; Hsu, C.H. A Factor-Based Probabilistic Cost Model to Support Bid-Price Estimation. *Expert Syst. Appl.* **2012**, *39*, 5358–5366. [CrossRef]
25. Bhatnagar, R.; Ghose, M.K. Comparing Soft Computing Techniques For Early Stage Software Development Effort Estimations. *Int. J. Softw. Eng. Appl.* **2012**, *3*, 119–127. [CrossRef]
26. Emsley, M.W.; Lowe, D.J.; Duff, A.R.; Harding, A.; Hickson, A. Data Modelling and the Application of a Neural Network Approach to the Prediction of Total Construction Costs. *Constr. Manag. Econ.* **2002**, *20*, 465–472. [CrossRef]
27. Günaydin, H.M.; Doğan, S.Z. A Neural Network Approach for Early Cost Estimation of Structural Systems of Buildings. *Int. J. Proj. Manag.* **2004**, *22*, 595–602. [CrossRef]
28. Cheng, M.Y.; Tsai, H.C.; Sudjono, E. Conceptual Cost Estimates Using Evolutionary Fuzzy Hybrid Neural Network for Projects in Construction Industry. *Expert Syst. Appl.* **2010**, *37*, 4224–4231. [CrossRef]
29. Marzouk, M.; Elkadi, M. Estimating Water Treatment Plants Costs Using Factor Analysis and Artificial Neural Networks. *J. Clean. Prod.* **2016**, *112*, 4540–4549. [CrossRef]
30. Hyari, K.H.; Al-Daraiseh, A.; El-Mashaleh, M. Conceptual Cost Estimation Model for Engineering Services in Public Construction Projects. *J. Manag. Eng.* **2016**, *32*, 04015021. [CrossRef]
31. Arage, S.S.; Dharwadkar, N.V. Cost Estimation of Civil Construction Projects Using Machine Learning Paradigm. In Proceedings of the International Conference on IoT in Social, Mobile, Analytics and Cloud, I-SMAC 2017, Institute of Electrical and Electronics Engineers Inc., Palladam, India, 10–11 February 2017; pp. 594–599.
32. Elmousalami, H.H. Comparison of Artificial Intelligence Techniques for Project Conceptual Cost Prediction: A Case Study and Comparative Analysis. *IEEE Trans. Eng. Manag.* **2021**, *68*, 183–196. [CrossRef]

33. Cortina, J.M. What Is Coefficient Alpha? An Examination of Theory and Applications. *J. Appl. Psychol.* **1993**, *78*, 98–104. [CrossRef]
34. Urquhart, B.; Ghonima, M.; Nguyen, D.; Kurtz, B.; Chow, C.W.; Kleissl, J. Sky-Imaging Systems for Short-Term Forecasting. In *Solar Energy Forecasting and Resource Assessment*; Elsevier: Amsterdam, The Netherlands, 2013; pp. 195–232. ISBN 978-0-12-397177-7.

 *buildings*

Article

Influence of the Construction Risks on the Cost and Duration of a Project

Azariy Lapidus, Dmitriy Topchiy, Tatyana Kuzmina * and Otari Chapidze

Department of Technology and Organization of Construction Production, Moscow State University of Civil Engineering (National Research University) (MGSU), 26 Yaroslavskoe Shosse, 129337 Moscow, Russia; lapidusaa@mgsu.ru (A.L.); topchiydv@mgsu.ru (D.T.); cod95@inbox.ru (O.C.)
* Correspondence: kuzminatk@mgsu.ru

Abstract: Recent years have witnessed active construction of multi-storey residential buildings. The scale of construction, its timing and limitations in financing contribute to the emergence of risk factors affecting the key parameters of cost and duration of projects. The purpose of this research is to develop the most effective mathematical model to reveal, study and estimate in a timely manner the influence of risk factors on stable implementation of a construction project during its life cycle. The mathematical model of the study is based on the theory of fuzzy sets, including 25 rules used to estimate the influence of a risk factor. An expert survey of leading specialists in the construction industry was performed and risk factors distributed over the stages of the life cycle were listed. Risk factors affecting the sustainability of the life cycle of a multi-storey residential building were identified and ranked. The result of the study shows that the application of the mathematical model will significantly increase the success of construction projects by identifying the critical risk factors in the phases of their life cycle. Since the proposed model is relatively new in Russia, it should be considered as a starting point for a new assessment of the impact of risk factors on projects. The methodology can be improved, and many aspects are still to be analyzed.

Keywords: life cycle; risk factor; risk; project; fuzzy logic

Citation: Lapidus, A.; Topchiy, D.; Kuzmina, T.; Chapidze, O. Influence of the Construction Risks on the Cost and Duration of a Project. *Buildings* **2022**, *12*, 484. https://doi.org/10.3390/buildings12040484

Academic Editor: Srinath Perera

Received: 2 March 2022
Accepted: 11 April 2022
Published: 14 April 2022

Publisher's Note: MDPI stays neutral with regard to jurisdictional claims in published maps and institutional affiliations.

1. Introduction

One of the priorities of each state is to provide affordable, comfortable and safe housing for its citizens. Currently, the purchase of residential real estate requires large financial investments and, in most cases, is associated with multi-year credits (mortgages) and interest-bearing loans. When buying houses for years, possibly for life, people pay special attention to the aesthetics of the building, the comfortable layout and the view from the window, the availability of parking spaces, infrastructure, and the environmental friendliness and safety of the residential area.

In recent years the volume of construction of multi-storey residence buildings with individual architectural and constructive solutions has been growing rapidly. For example, in Russia about 90 million square meters of residential space have been commissioned annually in the last 5 years. [1] The uniqueness of the adopted volume-planning and constructive solutions, the use of new technologies, the scale of construction, the large number of parties involved, tight deadlines and limited funding contribute to the risks affecting the implementation of such projects [2].

The study of project risk factors in recent years has had a vital role in the construction industry, and hundreds of factors have been identified [3,4], affecting the parameters of the cost and duration of a project. Works by Fahimeh Allahi, Lucia Cassettari and Muhammad Saiful Islam note that the cost of a project due to the influence of a factor can increase by up to 20%, and the duration of large construction projects can grow by up to 30% [5,6] Many authors have determined the project risk concept in different ways. Risks may have both positive and negative impacts, affecting the life cycle of a project [7–10]. Risk is closely related to the state of uncertainty, but it is fair to say that in most cases it has a negative impact on the project parameters [11]. There are many risk factors at all stages of the life cycle of a project: organizational, technological, technical, and economic [12,13]. In addition to organizational and technical complexities, managers take into account a growing number of additional parameters, including environmental and social ones. In such circumstances, it is important to understand the real practice of risk analysis and, above all, to assess the risk-driving factors at each stage of the life cycle of the construction project. Nowadays the theme of risks in the construction industry is a relevant and priority issue, and the vector of research in this area should be aimed at developing methods to reduce factors affecting the occurrence of risks in the life cycle of a project and achieve the required technical and economic performance.

However, the Russian Federation still lacks any model of the life cycle of a construction project with the technical risk factors included at each stage of construction, whereas complete and up-to-date documentation may contribute to mitigation of technical, industrial and natural risks.

Taking this into account, this research was focused on selected stages of the life cycle of a construction project exposed to risk factors in the construction of multi-storey residential buildings. As a result of an extensive study of scientific and technical literature, the authors identified, grouped and systematized life cycle risk factors by stages. The description of studies of the stages of the life cycle of a construction project in which the risks occurred, and the systematization of risk factors by project cycle, are presented in more detail in the articles [12,14–17]. Each stage was assigned a risk factor, which in turn enabled the assignment of an identification number to the risk factor and subsequent assessment and control. In addition, the authors chose the expert assessment method to determine the weights of the assessed parameters of risk factors in the absence of statistical data on the above topic of the study. Experts were required to assess risk factors on a five-point scale of probability of risk occurrence, as well as the impact of risks on the cost and duration of the construction project.

The authors applied fuzzy set theory, fuzzy logic and the Dempster-Shafer theory (DS), which allowed establishing the relationship between the results and obtaining a model of risk factors at the considered stages. To assess the impact of risk factors on the construction project it was necessary not only to allocate risk to the right stage and conduct an expert evaluation of the risk factor, but also to process mathematically the results in order to determine the degree of influence and the probability of the impact of factors on such parameters as cost and duration.

As a result, a model showing the factors of risk occurrence at the life cycle stages of a multi-storey residential building construction project was developed, which allows timely assessment of the level of possible risks and their impact on the cost and duration. The results of the study make it possible to model and assess risks in an attempt to investigate real, favorable ways of development at each stage of the life cycle of the construction project, including by mitigating or timely eliminating the risk factors.

The contribution of this article to the construction industry can be described at several levels. The first level is a literature review in which the main direction is residential buildings, risk factors, the life cycle of buildings and structures, mathematical models for analyzing research data are studied. At the second level, an improved life cycle model for a multi-storey residential building in a cramped building is presented and factors are considered and analyzed through mathematical models. At the third level, based on the analysis of the selected mathematical models, critical risks are determined in real time, and the dependences of time and cost are shown.

This approach allows the reader to form a comprehensive structure of the object and quickly predict the integrity of the object's life cycle in the time interval.

2. Literature Review

Understanding the risk scenarios of complex projects is an important step towards achieving the expected level of accuracy in contingencies. This section briefly discusses some relevant research related to understanding the risk scenarios of complex projects in different parts of the world.

The cost of building a residential building is very dynamic and changeable [18]. Price fluctuations may be related to the prices of building materials, human resources and other costs used in construction. This economic uncertainty can have a serious impact on business, especially on long-term projects [18]. To minimize the risk of uncertainty of investment costs for the construction of a residential building, it is necessary to predict the cost of building a residential building.

Compared to classical risk assessment methods, the modified Fuzzy Bayesizan Belief Network (FBBN) system has certain advantages for risk assessment in an uncertain and complex project environment because it shows risk cause-and-effect networks more efficiently. This helps understanding of the root causes of cost overrun risks and requires significantly less probabilistic data to obtain information from experts, which not only saves time and effort for data collection, but also reduces the computational load on the model compared to the widely used FBBN models [19].

Project management plays a big role, project management is now appearing in many organizations, and this trend is constantly growing. However, in today's dynamic environment, the success of such projects is influenced not only by the level of the project management method and the quality of the management team, but the success of the project can also be supported by effective risk management [20,21]. Risk and uncertainty are an integral part of project management [22]. If risk management can be integrated into an organization and used effectively, certain benefits and resource savings can be obtained [23–25]. Risk management also plays a key role in terms of the sustainability of a construction company [26].

It is also worth noting that the choice of the appropriate method of project implementation is one of the most important management decisions, since it has a direct impact on the success of the project and affects key performance indicators such as cost, quality, schedule and safety.

3. Methods

3.1. Data Sourcing

The first stage of the research method involved an extensive review of the scientific literature, focusing on risks in the design and construction of residence buildings, as well as an analysis of already built facilities with identified factors that impact the parameters of duration and cost.

The first stage was not limited only to collecting data on risk factors, since the purpose of the study is related to the residential building life cycle. More importantly, the aim here is to identify risk factors and assign them to each stage of the multi-storey residence building life cycle.

The selected risk factors were analyzed and divided into risk groups, as well as assigned to the stages of the project's life cycle. Figure 1 presents the construction project's life cycle, taking into account risk factors [12]. This model contains all the stages of the life cycle of an object. The essence of this model is that it contains all the risk factors considered in this work. The risks are divided into groups and correlated to the stages. The selection of these factors was carried out by analyzing the scientific literature and studying the objects of analogues, in the documentation of which the quality department recorded the risks that arose at the stage of work.

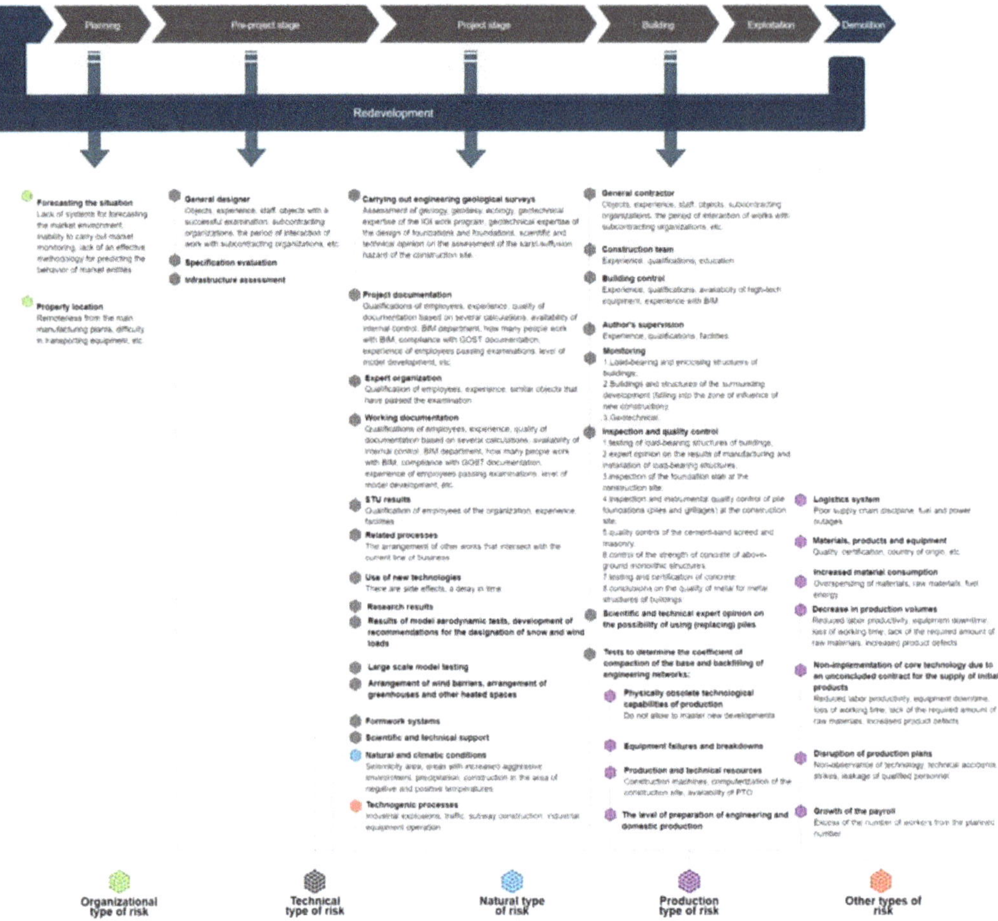

Figure 1. Object life cycle model with risk factors included.

Moreover, this model was complemented by the method of expert assessment, given that it allows determining which risk factor parameters have the most significant impact on the multi-storey residential building life cycle.

The method utilized here consisted of an expert assessment of the proposed risk factors for a 20-storey residence building in Moscow. The experts were required to assess the risk factors on a five-point scale Likert scale of risk occurrence probability, impact on cost, and impact on duration.

The most dangerous stages in the life cycle of a building object are:

1. Planning
2. Pre-project stage
3. Project stage

Expert examination was conducted among the specialists of the construction industry; 60 experts took part in the selection in accordance with the requirements of competencies for an expert, and the number of experts was determined by the proposed method of Ruposov V.L. [27,28].

Basic requirements:

- Academic degree or academic qualification.

- Participation in international scientific and technical cooperation.
- At least 10 years of professional experience.
- Member of NOPRIZ (National Association of Designers and Surveyors) and (or) NOSTROY (National Association of Builders).

The expert's questionnaire is shown in Supplementary Materials Expert questionnaire. As a result of the analysis of the expert assessment, weight indicators of the estimated risk factor parameters were determined.

3.2. Mathematical Model of Data Analysis

To assess the impact of risk factors on the construction project's life cycle, it is required not only to allocate the risk to the desired stage and conduct an expert assessment of the risk factor, but the data of the expert survey should be calculated mathematically to determine not only the degree of impact, but also the likelihood of the factor impacting such parameters as cost and duration.

The mathematical model for the analysis of expert evaluation is based on two theories:

- Fuzzy set theory, fuzzy logic.
- Dempster-Schafer theory (DS).

The fuzzy set theory is a method of experiment planning that is widely used in quantitative analysis of a machine process, especially for quality and risk assessment in engineering [29]. The main limitation of the method is related to the use of statistical mathematics and probability theory in the analysis. A probabilistic attempt is insufficient when the data are scant, as knowledge of their values becomes inaccurate or incomplete [30].

One of the possible solutions for cases where the data is scant is a non-parametric maximum likelihood estimate [31,32]. At the end of the twentieth century, a method based on the idea of fuzzy logic associated with L.A. Zadeh [33] was developed taking into account the possibility of describing the so-called linguistic variable. An example of applying the idea to the perceived risk assessment in the project is presented in the articles [34,35].

Fuzzy logic was first introduced by Professor L.A. Zadeh in 1965 and began to be applied in the 1970s [33,34]. Fuzzy logic is a successful application in the context of fuzzy sets in which the variables are linguistic rather than numeric. Since its development in 1965, it has become the optimal choice for handling data-related inaccuracies and uncertainties in risk assessment tasks [36].

Fuzzy logic is different from binary or Aristotelian logic, which sees everything as binary: yes or no, black or white, zero or one. The values in this logic vary from zero to one [37]. Figure 2 shows the architecture of the fuzzy inference system.

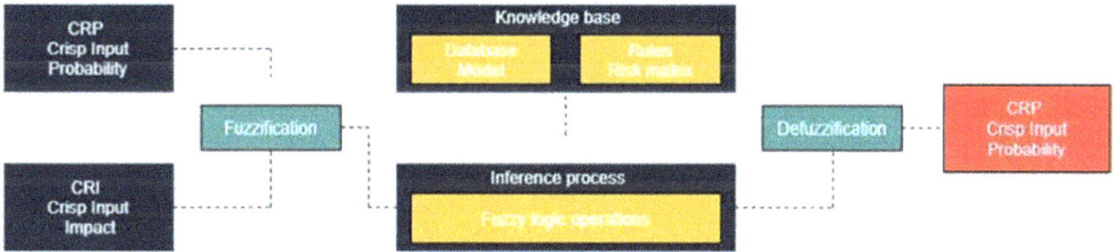

Figure 2. Fuzzy inference system [17].

A fuzzy inference system [17] usually consists of the following components:

- Fuzzificator
- Risk matrix
- Fuzzy inference mechanism
- Defuzzificator

The components of the fuzzy inference system for risk assessment are described below [17,37]. The process of converting explicit variables into linguistic variables is called fuzzification.

Fuzzification is the establishment of a correspondence between the numerical value of the input variable of the fuzzy inference system and the value of the membership function of the corresponding term of the linguistic variable [38].

The input and output data of the fuzzy inference system must first be fuzzy in a fuzzy inference system. The probability of occurrence and the severity of the impact of the risk are considered as two inputs, and the level of risk is considered as the output of the system of fuzzy inferences.

The linguistic expressions and fuzzy sets used for defining the input and output data of a fuzzy inference system are presented in Table 1 [17,38,39].

Table 1. Linguistic terms.

Input and Output Values	Linguistic Term	Definition	Rank
Probability levels Input 1	IM: Improbable	Extremely rare, almost no chance of occurrence.	1
	R: Remote	Chance of manifestation is small.	2
	O: Occasional	Probability to occur is 30–50%.	3
	P: Probable	Probability to occur is very high.	4
	F: Frequent	Probability to occur is almost certain and and inevitable.	5
Levels of impact Input 2	N: Negligible	There is no real negative consequences or a significant threat to the organization or project.	1
	M: Minor	There is little potential for negative consequences, and there is no significant impact on overall success.	2
	MA: Major	Can lead to negative consequences, creating a moderate threat to the project or organization.	3
	C: Critical	With significant negative consequences that will seriously impact the success of the organization or project (the need to close the project or a large number of negative events).	4
	CA: Catastrophic	With extremely negative consequences that can lead to the closure or long-term failure of the entire company. Requires the most attention and resources.	5
Risk level Output	IN: Insignificant	The risk is tolerable without any mitigation. Impact is minor and unlikely to occur. These types of threats are generally ignored.	1–4
	T: Tolerable	Partial mitigation may be required. The probability of occurrence does not allow them to be ignored, and the consequences may be tangible. If possible, measures should be taken to prevent the occurrence of medium risks, but it should be remembered that they are not a priority and cannot critically impact the success of an organization or project.	5–8
	SU: Substantial	Mitigation may be required. Such risks may have serious consequences and are likely to occur. They should be responded to in the near future.	9–12
	S: Significant	Mitigation measures must be taken to reduce the risk. Critical risks that have serious consequences and have a high probability of occurring. They have a high priority. Measures should be taken immediately to eliminate or reduce the possible consequences.	13–16
	INT: Intolerable	Risk mitigation measures must be implemented. These are catastrophic risks that have serious consequences and have a high probability of occurrence. They have the highest priority. Can threaten the existence of the organization or the success of most of the tasks. Measures should be taken immediately to eliminate or reduce the possible consequences.	17–25

For the functioning of the fuzzy logic system, referring to the standard risk matrix is required.

The risk matrix is a tool of the threat management process designed to increase the objectivity of its interpretation [17]. To place an item in the matrix, you must assign it a probability and damage rating.

The degree of risk is determined on the basis of the risk matrix [13] and, accordingly, this component of the developed fuzzy inference system for risk assessment is a knowledge base and fuzzy rules, including 25 fuzzy "if" rules, which are presented in Table 2.

Table 2. Mathematical model rule table.

No.	Description
Rule 1	If the likelihood is unlikely and the consequences are negligible, then the risk is negligible.
Rule 2	If the probability is unlikely and the consequences are catastrophic, then the risk is high.
. . .	. . .
Rule 25	If the probability is frequent and the consequences are critical, then the risk is unacceptable.

Tables 3 and 4 shows the indicators of the standard risk matrix.

Table 3. Risk Matrix.

Risk = P × I		Probability				
		IM	R	O	P	F
Impact	N	IN	IN	IN	IN	T
	M	IN	IN	T	T	SU
	MA	IN	T	SU	SU	S
	C	IN	T	SU	S	INT
	CA	T	SU	S	INT	INT

Table 4. Risk matrix with ranks.

Risk = P × I		Probability				
		IM	R	O	P	F
Impact	N	1	2	3	4	5
	M	2	4	6	8	10
	MA	3	6	9	12	15
	C	4	8	12	16	20
	CA	5	10	15	20	25

The next component of the developed fuzzy inference system for risk assessment is the fuzzy inference mechanism. The inference engine evaluates and makes logical inference to the rules using inference algorithms, and after the inference rules are aggregated by the defuzzifier block they are converted to an explicit or numeric value. The fuzzy inference mechanism is the Mamdani algorithm [17]. The optimum method is used to aggregate the output data, and the center of gravity method is used for defuzzification.

The fuzzy risk assessment index is considered as an output parameter, and varies from 0 to 5. In this article, the risk is divided into five equal parts, as shown in Figures 3–5. Risks are represented by fuzzy sets, the ranges of which coincide with the linguistic terms given in Table 1. Using the appropriate transformation scale, the linguistic terms are converted into fuzzy ratings. One of the key points in fuzzy modeling is the definition of fuzzy numbers, which are vague concepts and expressed in inaccurate terms in natural language [36].

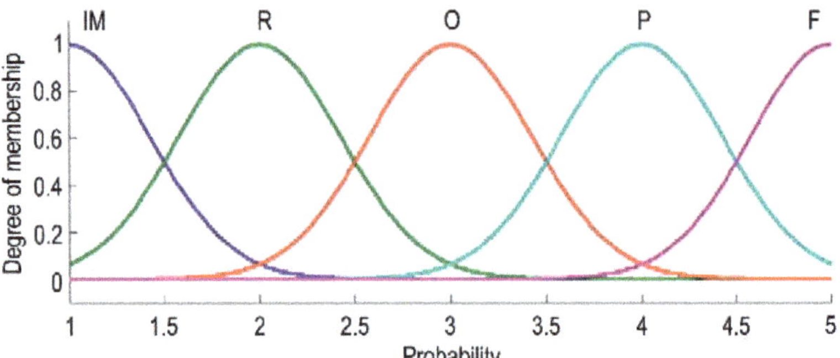

Figure 3. Membership function for the probability level.

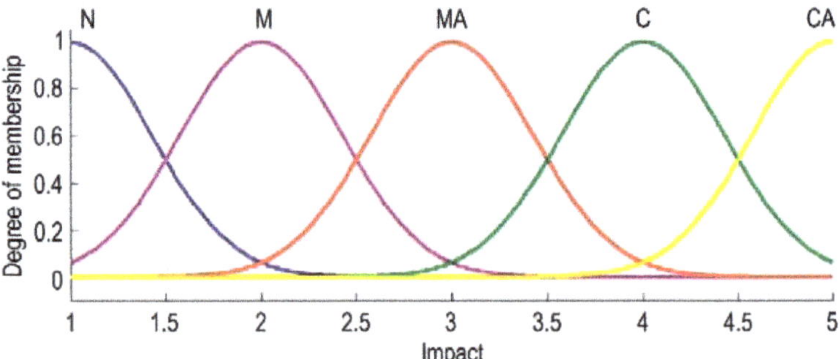

Figure 4. Membership function for the influence level.

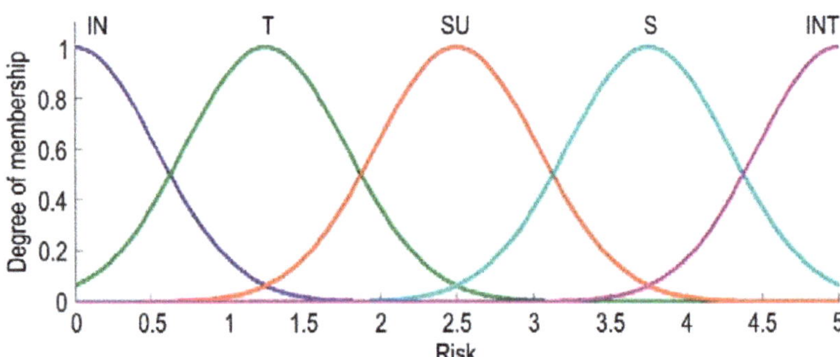

Figure 5. Membership function for the risk level.

In this work, fuzzification distributes system variables, including probability (P), impact level (I) and risk levels (R) with clear numbers. The structure of the fuzzy model is shown in Figure 6.

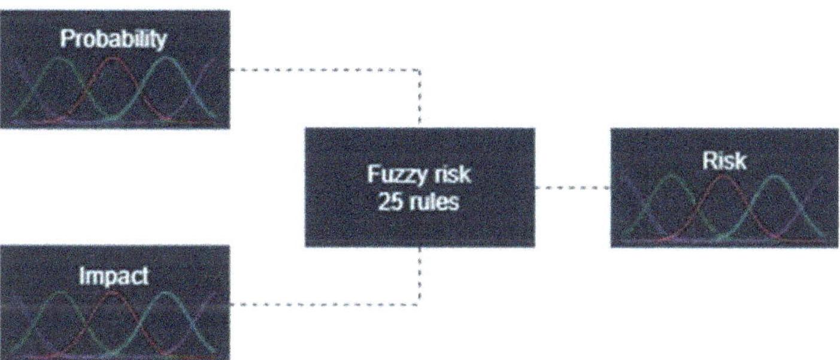

Figure 6. Fuzzy model inference structure.

Twenty-five rules were introduced to the mathematical model in Table 2, performing the defuzzification process [38,39]. Defuzzification in fuzzy inference systems is the process of transition from the membership function of an output linguistic variable to its clear (numerical) value. The purpose of defuzzification is to use the results of the accumulation of all output linguistic variables to obtain quantitative values for each output variable used by devices external to the fuzzy inference system [17,39].

The last step in the approximation is defuzzification. This step contains the process of replacing a fuzzy value with a clear inference, consisting of a procedure for weighing and averaging the outputs of all individual fuzzy rules. In total, there are six defuzzification methods [40]:

Centroid Average (CA)
Center of Gravity (COG)
Maximum Center Average (MCA)
Medium of the Maximum (MOM)
Smallest of the Maximum (SOM)
Largest of the Maximum (LOM)

Center of gravity (COG) is one of the most popular defuzzification methods, chosen because of its simple calculations and intuitive plausibility [41].

COG is defined by the following equation:

$$Z = \frac{\int \mu_i(x)x\,dx}{\int \mu_i(x)\,dx} \tag{1}$$

where:

Z—defuzzified result.
x—output variable.
$\mu_i(x)$—aggregated membership function.

The defuzzification process creates a clear value from fuzzy sets that reflect the risk of the project, as in Figure 7.

The data of mathematical calculation of expert assessments are presented in Table 5.

The DS method is a more general form of the Bayesian approach that retains all its advantages. For example, in the DS method, as in the Bayesian method, available a priori information can be included in the inaccurate output of uncertain indicators and inferred results. Nevertheless, the use of a priori information in the DS method is not mandatory. This is one of the advantages of the DS theory [42,43].

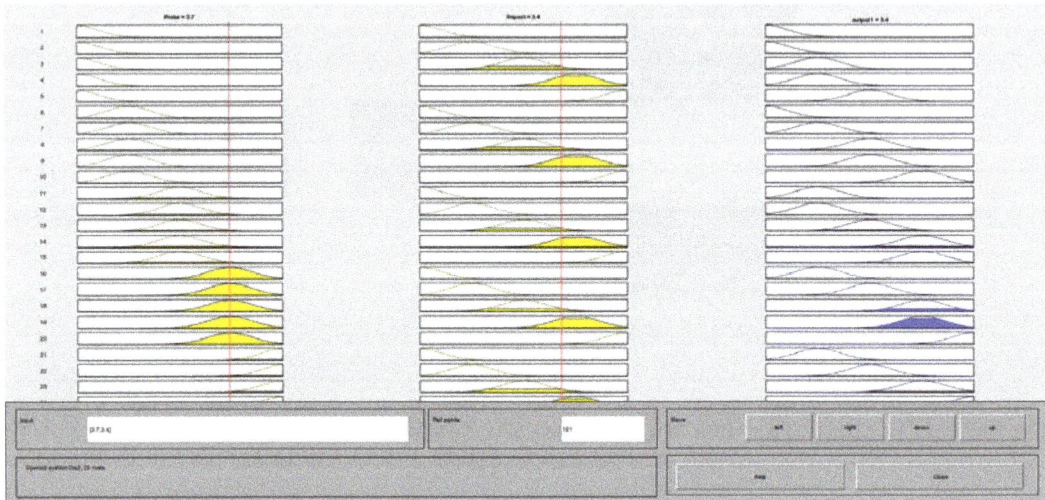

Figure 7. Output from Polyspace software package based on 25 rules.

$$DS = m(A) = \frac{n - \mathrm{minF}}{\mathrm{maxF} - \mathrm{minF}} \qquad (2)$$

where:

m(A)—degree of reliability.
maxF = max$\{f_j \mid j \in [1, n]\}$;
minF = min$\{f_j \mid j \in [1, n]\}$;
n—number of factors

Compared with other probabilistic methods, such as the Bayesian method, the DS method does not require the calculation of a priori probability; it has a flexible and understandable mass function, and the formation of a mass function is convenient and simple. The computational complexity of this method is much less than that of the Bayesian method [41,44].

For the processing of expert data, a risk matrix, the Dempster-Shaffer theory and a mathematical model of fuzzy logic were used. The processed results of the study are presented in Section 3.

Dempster-Shafer cell data were obtained by mathematical calculation according to formula 2. Fuzzy logic output data, defuzzification results, were obtained by mathematical modeling through the Polyspace software package. The inference algorithm was used. After aggregating the output rules by the defuzzifier block, we obtained an explicit numerical value as the result of fuzzy inference.

The data of mathematical calculations are presented in Section 3. The values of FLRC and FLRT are ranked in ascending order, according to the fuzzy inference group.

4. Result and Discussion

The following values of probability and impact on project parameters were obtained during the expert survey of Section 3.1. The results of the expert survey are presented in Table 5.

Table 5. Analysis of the expert survey results.

No.	Criteria	SUB—Criteria	Risk Factor	Probability P	Impact on the Cost of IC	Impact on Duration IT
F1			Increased seismicity at the construction site	2.8	3.1	2.8
F2			Precipitation	2.3	2.3	1.9
F3		Environment	Flooding	3	2.7	2
F4			Landscape (plain, hills, etc.)	2.2	2.8	2.2
F5			Climatic and natural conditions	2.7	3.4	3.4
F6		Substructure of the construction site	Area of archeological studies	2.4	3.5	4.3
F7	Construction site		Lack of construction site space	2.8	3.4	3
F8			High transport load	2.6	2.3	2.5
F9			Delays in obtaining permits	3.7	3.4	3.9
F10		Construction project	Evaluation of technical conditions results	2.5	2.6	2.8
F11			Infrastructure assessment results	2.5	2.3	2.6
F12			Security requirements and restrictions of nearby facilities	2.6	2.7	3
F13		Other	There are structures for demolition at the construction site	2.6	3.4	3
F14			A short construction period	3.9	4.4	3.1
F15			Labor qualification level of key employees	3.3	3.9	2.9
F16			Staff, qty (low number of employees)	3.3	3.1	3.1
F17			Projects with a positive expert opinion (experience of passing)	2.8	2.2	2.3
F18	The main party of the project	General Designer	Availability and number of subcontractors	3.7	3.2	3.2
F19			Current projects (company workload)	3.8	3	3.1
F20			Application of new technologies (lack of experience using technologies)	3.5	4.1	3.1
F21			Coordination of work with a subcontractor (no work model, no experience)	4.3	3.3	3.3
F22			Registration level of GOST documentation	2.1	1.5	1.6
F23		Initial permitting documentation	Quality of the conducted engineering-geological tests	3.3	2.1	2
F24			Completeness of required data for design	3.8	2.4	2.5
F25		Regulatory and technical support level for project preparation	The level of work with regulatory documentation at the international and federal level	3.1	2.4	2.1
F26		Results of engineering and geological surveys	Results of the assessment of geology, geodesy, ecology, hydrometeorology, geotechnical expertise of the IGI work program	2.7	2.1	2.1
F27	Formation of project documentation		Results of geotechnical research, assessment of the state of soil bases of buildings and structures	2.7	2.3	2
F28		Results of special types of engineering surveys	Results of local monitoring of environmental components, exploration of soil building materials, local surveys of contaminated soils and groundwater	2.4	1.9	1.8
F29			Results of the geotechnical examination of the project of subgardes and foundations,, scientific technical conclusion on the assessment of the karst-suffusion hazard of the construction site	2.6	1.9	2.2
F30		Assesment of engineering survey results	Results of engineering survey assessment	3.1	2.4	2.2

Table 5. *Cont.*

No.	Criteria	SUB—Criteria	Risk Factor	Probability P	Impact on the Cost of IC	Impact on Duration IT
F31			Labor qualification level	3	2.3	2.6
F32			Work experience	3.1	2.7	2.9
F33			Experience of passing the assessment	2.8	2	2.5
F34			Experience with residential facilities	2.8	2.3	2.2
F35			Uniqueness of the project (complexity of geometric forms of structures)	3.7	3.2	3.3
F36			Height of the project	3.6	3.5	3.4
F37			Registration level of GOST documentation	1.2	1.1	1
F38		Project documentation	Algorithm for transferring information between related sections of design and estimate documentation	1.9	1.3	1.6
F39			Results of taking into account natural and climatic conditions (seismicity of the region, zones with increased aggressive environment, precipitation, construction in the zone of negative and positive temperatures)	3.3	2.6	2.6
F40			Results of accounting for human-induced processes (industrial explosions, traffic, subway construction, operation of industrial equipment)	2.8	2.4	2.4
F41			Results of determining the scope of work	2.2	1.9	1.6
F42			Labor qualification level	3.1	2.3	2.4
F43		BIM Department	Work experience	3.2	2.1	2.5
F44			Staff, qty (low number of employees)	3.2	2.2	2.8
F45			Level of BIM model evaluation	3	2.4	2.7
F46	Formation of project documentation	Development of measures to ensure access for persons with disabilities	Labor qualification level	2.8	1.8	1.8
F47			Work experience	2.4	1.6	1.7
F48			Proficiency in BIM technologies	1.7	1.3	1.4
F49			Employee qualification	2.9	2.6	2
F50			Work experience	2.9	2.5	2.1
F51		Fire safety measures	Projects	2.5	2.3	2
F52			Proficiency in BIM technologies	1.9	2	1.7
F53			Results of Special Technical Regulations	3.1	2.6	2.3
F54		Assessment of documentation	Results of the project documentation assessment	2.7	2.4	2.2
F55			Labor qualification level	3	2.8	2.9
F56			Work experience	3	2.6	2.6
F57		Working documentation	Experience of passing the assessment	2.7	2.3	3
F58			Experience with residential facilities	2.5	2.2	2.1
F59			Registration level of GOST documentation	2.1	1.5	1.4
F60			Algorithm for transferring information between related sections of design and estimate documentation	2.1	1.5	1.7
F61		Other	Impact of related processes on the result of work (e.g., engineers made a mistake in the calculation of loads, shaft openings, entails adjustment of openings AR and CR)	3.5	3.1	2.7
F62			Availability of a common information platform for coordinating work between stakeholders	2.8	2.4	2.1
F63		Building an information model of a building	Model building experience	2.8	2.5	2.3
F64			Staff, qty (low number of employees)	2.4	2.1	2.7

To understand the operation of the mathematical model in the life cycle of a multistorey residential building, each factor at different stages of the project was considered and the rank of the factor was determined by fuzzy logic, as this was the main tool in our study, with 25 preprogrammed rules.

The data of the expert survey are the input data for the mathematical model presented in Section 3.2. The results of the mathematical model are presented in Table 6.

Table 6. Comparative analysis of the obtained data.

No.	P	IoC	IoT	Risk Matrix				Dempster–Shafer				Fuzzy Logic Output			
				RC	Rank	RT	Rank	DSRC	Rank	DSRT	Rank	FLRC	Rank	FLRT	Rank
F9	3.7	3.4	3.9	12.58	2	14.4	1	3.2	5	3.73	2	3.4	5	3.6	1
F5	2.7	3.4	3.4	9.18	4	9.18	4	3.2	3	3.2	3	3.4	3	3.4	2
F6	2.4	3.5	4.3	8.4	7	10.3	3	3.31	2	4.18	1	3.5	2	3.2	3
F14	3.9	4.4	3.1	17.16	1	12	2	4.29	1	2.9	4	4.4	1	3.15	4
F1	2.8	3.1	2.8	8.68	6	7.84	6	2.9	7	2.6	8	1.9	14	2.8	5
F8	2.6	2.3	2.5	5.98	12	6.5	10	2.13	13	2.32	11	2.3	9	2.4	6
F11	2.5	2.3	2.6	5.75	13	6.5	11	2.13	14	2.41	10	2.3	10	2.4	7
F4	2.2	2.8	2.2	6.16	11	4.84	13	2.6	8	2.04	12	2.2	12	2.2	8
F10	2.5	2.6	2.8	6.5	10	7	9	2.41	11	2.6	9	2.4	7	2.2	9
F3	3	2.7	2	8.1	8	6	12	2.51	9	1.86	13	2	13	2	10
F7	2.8	3.4	3	9.52	3	8.4	5	3.2	4	2.8	5	3.4	4	2	11
F12	2.6	2.7	3	7.02	9	7.8	7	2.51	10	2.8	6	2.3	11	2	12
F13	2.6	3.4	3	8.84	5	7.8	8	3.2	6	2.8	7	3.4	6	2	13
F2	2.3	2.3	1.9	5.29	14	4.37	14	2.13	12	1.77	14	2.3	8	1.9	14
F15	3.3	3.9	2.9	12.87	3	9.57	6	3.73	2	2.7	6	3.35	2	3.35	1
F18	3.7	3.2	3.2	11.84	4	11.8	2	3	4	3	2	3.2	4	3.2	2
F21	4.3	3.3	3.3	14.19	2	14.1	1	3.1	3	3.1	1	3.2	5	3.2	3
F16	3.3	3.1	3.1	10.23	6	10.2	5	2.9	5	2.9	3	3.15	6	3.15	4
F19	3.8	3	3.1	11.4	5	11.7	3	2.8	6	2.9	4	3.25	3	3.15	5
F20	3.5	4.1	3.1	14.35	1	10.8	4	3.95	1	2.9	5	3.45	1	3.15	6
F17	2.8	2.2	2.3	6.16	7	6.44	7	2.04	7	2.13	7	2.2	7	2.2	7
F24	3.8	2.4	2.5	9.12	4	9.5	3	2.22	12	2.32	13	3.65	1	3.65	1
F36	3.6	3.5	3.4	12.6	1	12.2	2	3.31	1	3.2	1	3.5	2	3.4	2
F39	3.3	2.6	2.6	8.58	5	8.58	8	2.41	6	2.41	11	3.35	4	3.35	3
F35	3.7	3.2	3.3	11.84	2	12.2	1	3	2	3.1	2	3.2	7	3.3	4
F23	3.3	2.1	2	6.93	18	6.6	19	1.95	27	1.86	29	3.35	3	3.25	5
F43	3.2	2.1	2.5	6.72	21	8	11	1.95	29	2.32	15	3.25	5	3.25	6
F44	3.2	2.2	2.8	7.04	16	8.96	6	2.04	25	2.6	6	3.25	6	3.25	7
F61	3.5	3.1	2.7	10.85	3	9.45	4	2.9	3	2.51	8	3.15	8	3.25	8
F64	2.4	2.1	2.7	5.04	32	6.48	21	1.95	30	2.51	9	2.1	21	2.3	9
F29	2.6	1.9	2.2	4.94	33	5.72	28	1.77	34	2.04	20	1.9	29	2.2	10
F33	2.8	2	2.5	5.6	29	7	16	1.86	31	2.32	14	2	23	2.2	11
F34	2.8	2.3	2.2	6.44	24	6.16	23	2.13	21	2.04	22	2.2	13	2.2	12
F40	2.8	2.4	2.4	6.72	20	6.72	18	2.22	15	2.22	16	2.2	14	2.2	13

Table 6. *Cont.*

No.	P	IoC	IoT	Risk Matrix				Dempster–Shafer				Fuzzy Logic Output			
				RC	Rank	RT	Rank	DSRC	Rank	DSRT	Rank	FLRC	Rank	FLRT	Rank
F54	2.7	2.4	2.2	6.48	23	5.94	25	2.22	17	2.04	23	2.3	11	2.2	14
F63	2.8	2.5	2.3	7	17	6.44	22	2.32	11	2.13	19	2.2	17	2.2	15
F26	2.7	2.1	2.1	5.67	28	5.67	29	1.95	28	1.95	25	2.1	18	2.1	16
F50	2.9	2.5	2.1	7.25	13	6.09	24	2.32	10	1.95	26	2.1	20	2.1	17
F58	2.5	2.2	2.1	5.5	30	5.25	31	2.04	26	1.95	27	2.2	15	2.1	18
F62	2.8	2.4	2.1	6.72	22	5.88	26	2.22	18	1.95	28	2.2	16	2.1	19
F27	2.7	2.3	2	6.21	25	5.4	30	2.13	19	1.86	30	2.3	9	2	20
F31	3	2.3	2.6	6.9	19	7.8	12	2.13	20	2.41	10	2	22	2	21
F45	3	2.4	2.7	7.2	14	8.1	9	2.22	16	2.51	7	2	24	2	22
F49	2.9	2.6	2	7.54	10	5.8	27	2.41	7	1.86	31	2.1	19	2	23
F51	2.5	2.3	2	5.75	27	5	33	2.13	23	1.86	32	2.3	10	2	24
F55	3	2.8	2.9	8.4	6	8.7	7	2.6	4	2.7	5	2	25	2	25
F56	3	2.6	2.6	7.8	9	7.8	13	2.41	9	2.41	12	2	26	2	26
F57	2.7	2.3	3	6.21	26	8.1	10	2.13	24	2.8	3	2.3	12	2	27
F25	3.1	2.4	2.1	7.44	11	6.51	20	2.22	13	1.95	24	1.9	27	1.9	28
F30	3.1	2.4	2.2	7.44	12	6.82	17	2.22	14	2.04	21	1.9	30	1.9	29
F32	3.1	2.7	2.9	8.37	7	8.99	5	2.51	5	2.7	4	1.9	31	1.9	30
F42	3.1	2.3	2.4	7.13	15	7.44	14	2.13	22	2.22	17	1.9	33	1.9	31
F53	3.1	2.6	2.3	8.06	8	7.13	15	2.41	8	2.13	18	1.9	35	1.9	32
F22	2.1	1.5	1.6	3.15	38	3.36	38	1.42	38	1.5	38	1	37	1	33
F28	2.4	1.9	1.8	4.56	34	4.32	34	1.77	33	1.68	33	1.9	28	1	34
F37	1.2	1.1	1	1.32	43	1.2	43	1.08	43	1	43	1.1	36	1	35
F41	2.2	1.9	1.6	4.18	35	3.52	37	1.77	35	1.5	40	1.9	32	1	36
F59	2.1	1.5	1.4	3.15	39	2.94	41	1.42	39	1.33	42	0.9	39	0.9	37
F47	2.4	1.6	1.7	3.84	36	4.08	35	1.5	37	1.59	35	0.9	38	0.8	38
F48	1.7	1.3	1.4	2.21	42	2.38	42	1.25	42	1.33	41	0.8	41	0.8	39
F60	2.1	1.5	1.7	3.15	40	3.57	36	1.42	40	1.59	37	0.9	40	0.8	40
F38	1.9	1.3	1.6	2.47	41	3.04	40	1.25	41	1.5	39	0.7	42	0.7	41
F46	2.8	1.8	1.8	5.04	31	5.04	32	1.68	36	1.68	34	0.7	43	0.7	42
F52	1.9	2	1.7	3.8	37	3.23	39	1.86	32	1.59	36	1.9	34	0.7	43

P—probability; IoC—impact on cost; IoT—impact on timeline; RC—risk cost; RT—risk of timeline; DCRS—Dempster Schafferis risk cost; DCRT—Dempster Schafferis risk of timeline; FLRC—fuzzy logic risk cost; FLRT—fuzzy logic risk of timeline.

After analyzing the results of mathematical calculations, a diagram with factors and their ranks can be constructed as shown in Figures 8 and 9. The data are presented without ranking by the magnitude of the influence.

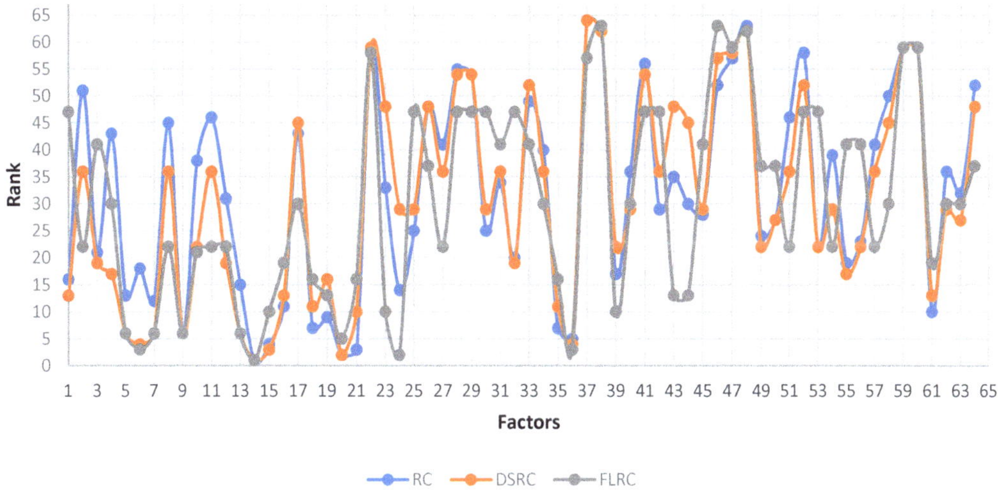

Figure 8. Distribution diagram of the impact of risk factor on the cost by ranks.

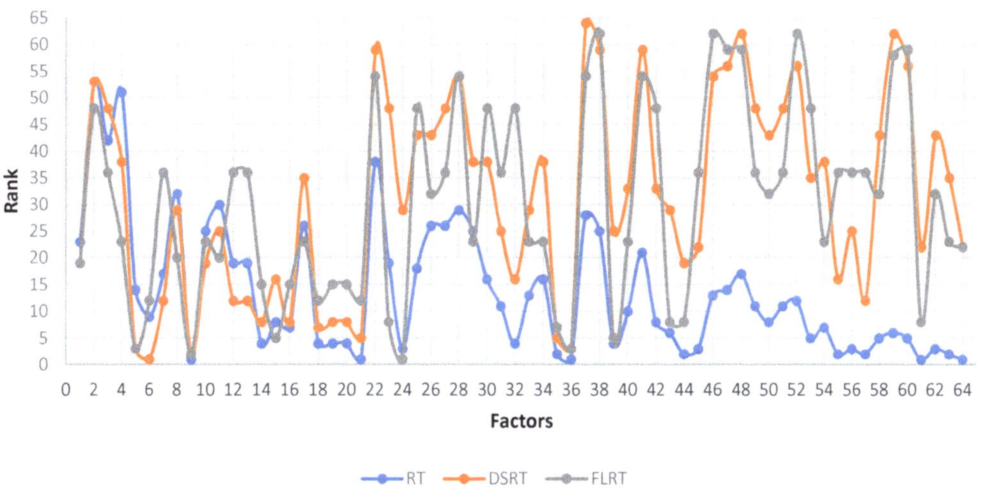

Figure 9. Diagram of the distribution of the impact of risk factor on the duration by ranks.

The diagram shows 64 factors, with each ranked in relation to another; due to this we see a clearer picture of the distribution of risk factors by measurement value, both in cost and in time.

The study identified the most dangerous risk factors that affect the key parameters of the life cycle of a multi-storey residential building.

Mathematical calculations showed that the most effective mathematical apparatus is fuzzy logic based on 25 given rules. Dempster Schaffer's theory has small deviations from fuzzy logic, but this spread is within the acceptable limit. The standard risk matrix has large deviations in data reliability, as it excludes the presence of the said risk factor definition rules.

The final step was to determine the magnitude of the impact of the main factors identified in Table 6, on the parameters of cost and duration of the construction project.

Table 7 shows the results of the analysis of the obtained data. The cost and duration values were determined by the experts in Supplementary, Section 3.1.

Table 7. Critical risk factor analysis.

No.	Stage	FLRC	FLRT	Increase in Cost, $ mln.	Increase in Duration, Months
F5	Planning	3.4	3.4	≈0.45	≈2
F6		3.5	3.2		
F7		3.4	2		
F9		3.4	3.6		
F13		3.4	2		
F14		4.4	3.15		
F15	Pre-project stage	3.35	3.35	≈0.45	≈2
F16		3.15	3.15		
F18		3.2	3.2		
F19		3.25	3.15		
F20		3.45	3.15		
F21		3.2	3.2		
F23	Project stage	3.35	3.25	≈0.6	≈3
F24		3.65	3.65		
F35		3.2	3.3		
F36		3.5	3.4		
F39		3.35	3.35		
F43		3.25	3.25		
F44		3.25	3.25		
F61		3.15	3.25		
Total:				≈1.5	≈7

The factors in Table 7 are in the Significant category of Table 1. Mitigation measures must be taken to reduce the risks. These are critical risks that have serious consequences and a high probability of occurrence. High priority means immediate action is required to eliminate or mitigate possible consequences.

Factors not included in the table are not excluded; they are part of the whole project system and are subject to the rules of Table 1.

The following data were obtained as a result of the analysis of key risk factors:

1. Twenty most hazardous risk factors categorized as "Significant" were identified.
2. A high level of increase in duration and costs was observed at the design stage.
3. The indicators of increase in the value at each stage of the project life cycle were determined; the amount of damage caused by the factors is ≈$1.5 mln.
4. The indicators of increase in duration at each stage of the project life cycle were determined; increase in duration is ≈7 months.

The difference in the rank values of the risk factors presented in Figures 8 and 9 shows that the choice of mathematical tool plays an important role in determining the rank of risk factors.

The results obtained during the study will help to predict project risks and allow taking the right steps in due time to manage them and to adjust the budget and resources.

5. Discussion

A key component of the experiment was focused on the analysis of the influence of various risk factors that affect the stages of important parameters of the building life cycle. The experiment showed the performance of the mathematical model and identified critical factors. This technique allows work on one structure, that is, the life cycle of an object with all its parameters and the mathematical apparatus for taking into account the influence of factors, which allows a response to their impact in a timely manner. In general, the study of the influence of factors on the life cycle of an object will allow creating a common interconnection environment focused on successful implementation and improvement of informed decisions that can bring maximum benefit to stakeholders.

The use of two mathematical models for assessing the risk factor is not comprehensive today, but it copes well with the tasks set; namely, it takes into account the requirements and rules laid down by the operator for each object. However, for future buildings, actual data on behavior is not available, there are no public registries, no record of the maximum influencing factors is kept, which is a hot topic these days, and often the data are confidential. Hence, co-modeling by integrating BIM with a robust risk analysis model is one of the most appropriate methods to solve this problem. It is convenient when each factor has its own individual number, tracked in real time at each stage of work in the BIM system.

This study has the following unsolved problems. The scope of the simulation experiment was limited both in terms of the simulation time period and the space coverage of the object data. Over time, more participation from experts from the construction industry is required. More designs, materials and design approaches need to be evaluated as the pace of construction continues to be high and every year we see new technologies emerging in the construction industry. Simulation results will be more coherent and informative if it is possible to expand the range of data collection on the objects under study; the functions of joint modeling can be improved as research progresses.

Since the proposed model is relatively new, it should be considered a starting point for a new assessment of the impact of risk factors on the project. The methodology is subject to improvement, and many aspects remain to be studied. Of course, this model will allow managers of organizations to significantly reduce costs, correctly form the tasks set, identify and eliminate risk factors in a timely manner, and identify weaknesses in the company that will lead to financial losses.

Future research in this area should focus on identifying risk factors and managing them during the project cycle. It is worth introducing an electronic database of risk factors, so the percentage of risks can be reduced and projects implemented more efficiently.

6. Conclusions

This article proposes a scientifically justified mathematical model of the life cycle of a multi-storey residential building. The model allows competent determination and ranking of the influence of risk factors at each stage of the project. The presented methodology was developed to assess the impact of risk factors on the main parameters of the project. The stages of the life cycle for a residential building were analyzed, the risk factors arising at each stage identified, and their impact assessed by an expert survey. The expert survey involved 60 experts who are professionals in the construction industry, with more than 10 years of experience. The experts were requested to assess the impact of factors on both the cost and the duration. As a result, the following conclusions can be made.

- The mathematical model based on the fuzzy set theory with 25 programmable rules identified critical project factors and shows a small deviation from the Dempster-Schafer theory.
- The most hazardous risk factors with the influence on the life cycle of the project, affecting the parameters of the duration and cost of the project, were identified and ranked. There are 31.25% of them in the life cycle. All factors should have an identi-

fication number to track them. This data will help to predict the consequences in a timely manner and take measures to eliminate them.

- Particular attention should be paid to the design phase, as the highest concentration of risk factors is observed in this category, i.e., 65.63%.
- Analysis of the data showed that under the influence of critical risk factors on the project, the cost of the project grows by 1.5 million dollars, and the duration increases by 7 months.

Supplementary Materials: The following are available online at https://www.mdpi.com/article/10.3390/buildings12040484/s1, Table S1: Expert Questionnaire.

Author Contributions: Conceptualization, A.L.; methodology, A.L. and D.T.; software, O.C.; data analysis, D.T.; investigation, O.C. and T.K.; data curation, D.T. and T.K.; writing—original draft preparation, O.C. and T.K.; writing—review and editing, O.C. and T.K.; final conclusions, O.C. and T.K. All authors have read and agreed to the published version of the manuscript.

Funding: This work was financially supported by the Ministry of Science and Higher Education of Russian Federation (grant NO. 075-15-2021-686).

Institutional Review Board Statement: Not applicable.

Informed Consent Statement: Not applicable.

Data Availability Statement: Not applicable.

Conflicts of Interest: The authors declare no conflict of interest.

References

1. Kevesh, A.L. Construction in Russia. *Stat. Sat./Rosstat M* **2018**, 56–60.
2. Asaul, A.N. *Risks in the Activity of a Construction Organization*; Economic Problems and Organizational Solutions to Improve Investment and Construction Activities: A Collection of Scientific Papers; State Architecture Builds University: Saint Petersburg, Russia, 2004; Volume 2, pp. 8–12.
3. Allahi, F.; Cassettari, L.; Mosca, M. Stochastic Risk Analysis and Cost Contingency Allocation Approac for Construction Projects Applying Monte Carlo Simulation. In Proceedings of the World Congress on Engineering, London, UK, 5–7 July 2017; Volume I, pp. 385–391.
4. Renuka, S.; Kamal, S.; Umarani, C. A model to estimate the time overrun risk in construction projects. *Empir. Res. Urban Manag.* **2017**, *12*, 64–76.
5. Islam, M.S.; Nepal, M.P.; Skitmore, M.; Drogemuller, R. Risk induced contingency cost modeling for power plant projects. *Autom. Constr.* **2021**, *123*, 103519. [CrossRef]
6. Assaf, S.A.; Al-Hejji, S. Causes of delay in large construction projects. *Int. J. Proj. Manag.* **2006**, *24*, 349–357. [CrossRef]
7. Andersson, R.; Kövecses, S.; Bargalló, E.; Nordt, A. Challenges in Technical Risk Management for High-Power Accelerators. *J. Phys. Conf. Ser.* **2018**, *1021*, 012003. [CrossRef]
8. Ishin, A.V. Obligatory certification of specialists of companies-a guarantee of safety and quality of their work. *Technol. Organ. Constr. Prod.* **2013**, *3*, 23–24.
9. Lapidus, A.A. The impact of modern and organizational measures on the achievement of the planned results of construction projects. *Technol. Organ. Constr. Prod.* **2013**, *2*, 1.
10. Sarkar, D.; Panchal, S. Integrated interpretive structural modeling and fuzzy approach for project risk management of ports. *Int. J. Constr. Proj. Manag.* **2015**, *1*, 17–31.
11. Lapidus, A.A.; Safaryan, G.B. Quantitative analysis of risk modeling of production and logistics processes in construction. *Technol. Organ. Constr. Prod.* **2017**, *3*, 4.
12. Lapidus, A.A.; Chapidze, O.D. Factors and sources of risk in housing consruction. *Constr. Prod.* **2020**, *3*, 2–9.
13. Kozień, E. Assessment of technical risk in maintenance and improvement of a manufacturing process. *Open Eng.* **2020**, *10*, 658–664. [CrossRef]
14. Darko, A.; Chan, A.P.; Yang, Y.; Tetteh, M.O. Building information modeling (BIM)-based modular integrated construction risk management—Critical survey and future needs. *Comput. Ind.* **2020**, *123*, 103327. [CrossRef]
15. Renuka, S.M.; Umarani, C.; Kamal, S. A Review on Critical Risk Factors in the Life Cycle of Construction Projects. *J. Civ. Eng. Res.* **2014**, *4*, 31–36. [CrossRef]
16. Zhao, Z.Y.; Lv, Q.L.; Zuo, J.; Zillante, G. Prediction System for Change Management in Construction Project. *J. Constr. Eng. Manag.* **2010**, *136*, 659–669. [CrossRef]
17. Yazdani-Chamzini, A. Proposing a new methodology based on fuzzy logic for tunnelling risk assessment. *J. Civ. Eng. Manag.* **2014**, *20*, 82–94. [CrossRef]

18. Elfahham, Y. Estimation and Prediction of Construction Cost Index Using Neural Net-works, Time Series, and Regression. *Alex. Eng. J.* **2019**, *58*, 499–506. [CrossRef]

19. Guan, L.; Liu, Q.; Abbasi, A.; Ryan, M.J. Developing a comprehensive risk assessment model based on fuzzy bayesian belief network (fbbn). *J. Civ. Eng. Manag.* **2020**, *26*, 614–634. [CrossRef]

20. Asadabadi, M.R.; Zwikael, O. Integrating risk into estimations of project activities' time and cost: A stratified approach. *Eur. J. Oper. Res.* **2019**, *291*, 482–490. [CrossRef]

21. Filippetto, A.S.; Lima, R.; Barbosa, J.L.V. A risk prediction model for software project management based on similarity analysis of context histories. *Inf. Softw. Technol.* **2020**, *131*, 106497. [CrossRef]

22. Nguyen, P.T.; Pham, C.P.; Phan, P.T.; Vu, N.B.; Duong, M.T.H.; Nguyen, Q.L.H.T.T. Exploring critical risk factors of office building projects. *J. Asian Financ. Econ. Bus.* **2021**, *8*, 309–315.

23. Osuszek, L.; Ledzianowski, J. Decision support and risk management in busi-ness context. *J. Decis. Syst.* **2020**, *29*, 413–424. [CrossRef]

24. An, J.; Mikhaylov, A. Russian energy projects in South Africa. *J. Energy S. Afr.* **2020**, *31*, 58–64. [CrossRef]

25. Lopatin, E. Methodological approaches to research resource saving industrial enterprises. *Int. J. Energy Econ. Policy* **2019**, *9*, 181–187. [CrossRef]

26. Schulte, J.; Villamil, C.; Hallstedt, S. Strategic Sustainability Risk Management in Product Development Companies: Key Aspects and Conceptual Approach. *Sustainability* **2020**, *12*, 10531. [CrossRef]

27. Ruposov, V.L.; Chernykh, A.A. Processing formalized swot-analysis expert evaluation data. *Proc. Irkutsk. State Tech. Univ.* **2017**, *21*, 81–89. [CrossRef]

28. Ruposov, V.L. *Methods of Determining the Number of Experts*; Herald of Irkutsk State Technical University: Eastern Siberia, Russia, 2015; Volume 3, pp. 286–292.

29. Fisher, R.A. *The Design of Experiments*; Oliver and Boyd Press: Edinburgh, UK, 1935.

30. Pietraszek, J. *Metody Planowania Badań Doświadczalnych Eksploatowanych Maszyn i Urządzeń*; Monografia nr 378; Wydawnictwo Politechniki Krakowskiej: Kraków, Poland, 2010.

31. Owen, A.B. *Empirical Likelihood*; CRC Press: Boca Raton, FL, USA, 2001. [CrossRef]

32. Pietraszek, J.; Dwornicka, R.; Krawczyk, M.; Kolomycki, M. The nonparametric approach to the quantification of the uncertainty in the design of experiments modelling. In *UNCECOMP 2017: Proceedings of the 2nd International Conference on Uncertainty Quantification in Computational Sciences and Engineering, Rhodes Island, Greece, 15–17 June 2017*; Papadrakakis, M., Papadopoulos, V., Stefanou, G., Eds.; Institute of Structural Analysis and Antiseismic Research, School of Civil Engineering, National Technical University of Athens: Athens, Greece, 2017; pp. 598–604.

33. Zadeh, L. Probability measures of Fuzzy events. *J. Math. Anal. Appl.* **1968**, *23*, 421–427. [CrossRef]

34. Kozien, E.; Kozien, M.S. Using the fuzzy logic description for the ex-ante risk assessment in the project. In *Economic and Social Development. Book of Proceedings: 35th International Scientific Conference on Economic and Social Development, Lisbon, Portugal, 15–16 November 2018*; Ribeiro, H., Naletina, D., da Silva, A.L., Eds.; Varazdin Development and Entrepreneurship Agency: Varazdin, Croatia, 2018; pp. 224–231.

35. Moreno-Cabezali, B.M.; Fernandez-Crehuet, J.M. Application of a fuzzy-logic based model for risk assessment in additive manufacturing R&D projects. *Comput. Ind. Eng.* **2020**, *145*, 106529. [CrossRef]

36. Singh, H.; Gupta, M.M.; Meitzler, T.; Hou, Z.G.; Garg, K.K.; Solo, A.M.; Zadeh, L.A. Real-life applications of fuzzy logic. *Adv. Fuzzy Syst.* **2013**. [CrossRef]

37. Urbina, A.G.; Aoyama, A. Measuring the benefit of investing in pipeline safety using fuzzy risk assessment. *J. Loss Prev. Process Ind.* **2017**, *45*, 116–132. [CrossRef]

38. Rubanov, V.G.; Filatov, A.G. Intelligent automatic control systems fuzzy control in technical systems. In *Tutorial*; BSTU im. V. G. Shukhova: Belgorod, Russia, 2010; p. 170.

39. Jaderi, F.; Ibrahim, Z.Z.; Zahiri, M.R. Criticality analysis of petrochemical assets using risk based maintenance and the fuzzy inference system. *Process Saf. Environ. Prot.* **2018**, *121*, 312–325. [CrossRef]

40. Nieto-Morote, A.; Ruz-Vila, F. A fuzzy approach to construction project risk assessment. *Int. J. Proj. Manag.* **2011**, *29*, 220–231. [CrossRef]

41. Wong, B.K.; Monaco, J.A. A bibliography of expert system applications for business (1984–1992). *Eur. J. Oper. Res.* **1995**, *85*, 416–432. [CrossRef]

42. Dempster, A.P. A generalization of Bayesian inference. *J. R. Stat. Soc. Ser. B Stat. Methodol.* **2001**, *64*, 205–232.

43. Shafer, G. *A Mathematical Theory of Evidence*; Princeton University Press: Princeton, NJ, USA, 1977; Volume 83, pp. 667–672.

44. Jamshidi, A.; Yazdani-Chamzini, A.; Yakhchali, S.H.; Khaleghi, S. Developing a new fuzzy inference system for pipeline risk assessment. *J. Loss Prev. Process Ind.* **2013**, *26*, 197–208. [CrossRef]

Article

Risk Factors That Lead to Time and Cost Overruns of Building Projects in Saudi Arabia

Saad Alshihri *, Khalid Al-Gahtani and Abdulmohsen Almohsen

Department of Civil Engineering, King Saud University, P.O. Box 800, Riyadh 11421, Saudi Arabia;
kgahtani@ksu.edu.sa (K.A.-G.); asmohsen@ksu.edu.sa (A.A.)
* Correspondence: toengsaad@gmail.com

Abstract: Rapid transformation across all sectors through Saudi Arabia's vision 2030 initiatives led to an increase in construction activities. However, the construction industry has been already facing huge cost and time overruns, affecting all stakeholders. The aim of this study is to identify and explore the influential risk factors that lead to completion delays and cost overruns of government-funded building construction projects in Saudi Arabia, all of which have been subjected to a traditional type of procurement method (Standard Public Works Contract). The literature examined in this study identified a total of 83 risk factors, which have been grouped into nine categories. A questionnaire-based survey was conducted to determine the participants' perspectives on the degree of probability of occurrence (P) of each risk and its potential impact on a project in terms of time (IT) and cost (IC). The questionnaire survey was distributed to 200 experts and professionals associated with Saudi building construction projects, which were grouped into four categories: clients, designers, consultants, and contractors. Fifty-five acceptable questionnaires were returned and analysed. The relative importance index (RII), and Risk Importance (RI) were used to identify the most influential risk factors, and an agreement test was conducted. The results of the survey revealed that the most significant risks factors contributing to the delay of building construction projects' completion are contractor's financial difficulties, owner's delay in making progress payments for completed works, contracts awarded to the lowest bidder, change orders during construction, ineffective project planning and scheduling by the contractor, shortage of manpower, and contractor's poor site management and supervision. In addition, change orders during construction and contracts awarded to the lowest bidder are the most significant risks factors of exceeding budgets. Based on the results, it is concluded that for achieving sustainable development, client, contractor, and labour-related risks must be effectively managed.

Keywords: Saudi Arabia; construction projects; time overrun; cost overrun; risks

Citation: Alshihri, S.; Al-Gahtani, K.; Almohsen, A. Risk Factors That Lead to Time and Cost Overruns of Building Projects in Saudi Arabia. *Buildings* **2022**, *12*, 902. https://doi.org/10.3390/buildings12070902

Academic Editors: Srinath Perera, Albert P. C. Chan, Dilanthi Amaratunga, Makarand Hastak, Patrizia Lombardi, Sepani Senaratne, Xiaohua Jin and Anil Sawhney

Received: 21 May 2022
Accepted: 19 June 2022
Published: 25 June 2022

Publisher's Note: MDPI stays neutral with regard to jurisdictional claims in published maps and institutional affiliations.

1. Introduction

The main evaluation dimension of the successful execution of construction projects is to examine the achievement of project objectives (time, cost, and quality) [1–3]. Previous research has elicited that construction projects experience underachievement in both developed and developing countries as a result of completion delays and cost-overruns, with resultant negative impacts experienced by all involved parties, including financial loss [4–6].

Project delay has been defined as 'the time overrun either beyond the completion date specified in the contract and the parties agreed upon for the delivery of the project, or a part of the project' [7,8]. The liability of the contract parties for construction projects delays can be classified into excusable with compensation delays, excusable without compensation delays, non-excusable delays or contractor responsible, and concurrent delays [9,10]. Construction cost overruns is the actual/final costs minus those estimated, presented as a percentage of the estimated costs [11].

Completion delays and cost-overruns typically stem from a multitude of severe risks and uncertainties [12]. Whilst an entire host of studies and research has sought to identify risk factors in the global construction industry, they have concurred that the risk factors are different from one project to another and also depend on the country, procurement route (i.e., PPP, design-bid-build, and design and build), and the type of construction project. In addition, the top causes of cost-overruns are subject to change over time (in each decade); therefore, knowledge of them needs to be kept up to date in order to manage complexity effectively so as to avoid or minimise risks [12]. There are four different categories of construction project, namely building construction, heavy/civil construction, industrial construction, and residential construction, with the foremost accounting for the highest segment at 35–40% of construction projects [13]. Therefore, it is important to limit identifying the risk factors to a certain category of construction project that experiences almost the same issues, challenges, and risks. However, there is a lack of research on identifying the risks and categorising them according to different types of projects [14]. Recent studies focused on specific types of projects such as oil and gas [15,16], manufacturing and buildings [17], and road projects [18,19], there is a need to increase the research in identifying the risk factors in different projects types and to assess the changes in risk factors importance and probability [20]. It is important to address this research gap, because these can have potentially serious consequences, such as cost and time overruns, and can add additional pressure to construction projects [21,22]. In this context, this study addresses the following research question.

RQ: What are the risk factors adversely affecting time and cost of execution of building construction projects?

Thus, this study addresses this research gap by identifying the influential risk factors that lead to completion delay and cost overrun specific to government-funded building construction projects (i.e., government buildings, hospitals, schools, and universities) in Saudi Arabia, all of which have been subjected to a traditional type of procurement method (Standard Public Works Contract). Accordingly, the following research objectives are outlined to address the RQ:

1. To explore and identify the influential risk factors leading to duration and cost overrun during the construction stage, with special consideration for building projects in Saudi Arabia (i.e., risk factors list and classification) through the completion of a comprehensive literature review.
2. To study, rank, and analyse the identified risk factors (i.e., 'risk impact × likelihood') by conducting questionnaires.

Addressing the RQ and the above-listed objectives could achieve interesting findings which can contribute to the literature in providing the risk factors by project type, i.e., government-funded building construction projects. It can also help decision makers in better understanding the risks in construction industry during COVID-19 in order to better formulate policies and decision making with respect to Vision 2030 objectives. Furthermore, the findings can aid the managers of building construction projects in designing effective risk management strategies.

2. Literature Review

2.1. Overview of Construction Industry

Construction, in simple terms, is the process of constructing an infrastructure that requires collaboration of multiple disciplines, including architectural design management, financial and legal management, engineering and technology, logistics and procurement, sustainability, risk management, project management, etc. Types of construction can be broadly classified into industry-specific, building, and residential constructions [23]. The construction industry is considered to be one of the sustainable and continuous businesses that has been recording steady growth in recent decades. However, there are various risk factors that influence this industry, such as geopolitics, economy, resources, technology, etc. The global construction output growth in 2019 reduced to 2.7%, which was less than 2018, and such deterioration was observed in many developing countries, especially the Middle

East, while developed countries, such as the USA and Australia, have struggled to maintain growth momentum [24].

Various findings have been identified in different studies [25–28], reflecting the complexity and different influencing factors in the construction industry. It has been estimated that there will be 85% growth (USD 15.5 trillion) in construction output by 2030 (3.9% growth per annuum), out of which 57% of growth was contributed by a developed country, the US, and developing countries including India and China [26]. Faster growth is predicted in the USA (5% per annuum) compared to China, followed by India and Japan. In a report by Robinson [27], the construction market is predicted to grow by USD 8 trillion by 2030. A KPMG [25] survey revealed that only 20% of the global constructive companies were innovative, 60% were followers, and 20% were behind the curve. In addition, disparities were observed in strategies, practices, and performance of the companies', reflecting gaps in the process. Deloitte identified seven factors that can have an impact on growth in construction industry, including the following: innovation, competitive dynamics and margin improvement; internationalism, compliance, regulation, and transparency; and sustainability [28]. The findings from these studies indicate the complexity in construction industries, with there being various influencing factors, including geopolitics, environmental, technology, strategies, innovation, etc. Furthermore, the COVID-19 impact has significantly affected the construction industry, with many companies facing liquidity problems. Reduced spending and consumption capacity, operating restrictions and fear of contagion, supply chain disruptions, and lack of labour have all contributed to the impact, which have affected the sustainability of many SMEs across the globe [29]. A recent report on the construction industry predicted that smaller businesses and sub-contractors may fail rapidly; contract management can be a major issue as customers may seek to terminate or renegotiate contracts; internationalisation may become less viable as companies may reconsider the regions in which they want to operate in [30].

2.2. Saudi Arabian Construction Industry

Saudi Arabia's construction industry was severely affected during 2015–2016 following the crash in oil prices, which reduced the capital flow; as a result, many projects were halted, postponed, or even cancelled. However, the construction industry in the country is expected to grow exponentially in the next few years, with it gearing up towards a post-oil era, when new major cities will be developed and constructed [31]. According to a report published by Mordor Intelligence [31], more than 5200 construction projects are currently ongoing in Saudi Arabia, valued at USD 819 billion, out of which 3727 are urban construction active projects, and these are valued at USD 386.4 billion. There 733 are utility sector projects valued at USD 95.6 billion and 500 relating to transportation, valued at USD 156.2 billion. The Saudi construction industry is highly competitive with major international players [32]. The market presents opportunities of growth, which is expected to increase the market competition further. However, with a few players holding a significant market share, the Saudi Arabian construction market has an observable level of consolidation [33]. Focusing on the type of construction, Saudi Arabia spent USD 575 billion on public construction projects between the years of 2008 and 2013 [34,35]. A recent report [36] has forecasted a growth of 2.9% in 2021 in the Saudi Arabian construction industry and CAGR of 4% during 2022–2025. Furthermore, a Public Investment Fund (PIF) of USD 800 billion was underlined by the Crown Prince for funding projects over the next decade. Moreover, year-on-year growth of construction contract awards in Saudi Arabia are forecasted to reach 96 percent in 2022, which is diversified over different types of projects [37]. For instance, the total value of planned building contract awards alone in the Saudi Arabia is predicted to be USD 10.95 billion in 2022 [38]. Given these forecasts, the construction industry will be growing rapidly in the next few years.

The Saudi Arabian construction market is expected to witness significant growth and offer lucrative potential, due to its Vision 2030, NTP (National Transformation Programme) 2020, and several ongoing reforms aimed at diversifying away from oil. The Vision 2030,

NTP 2020, and private sector investment boost as well as the ongoing reforms are likely to be the growth drivers for the Saudi construction market in 2018 and beyond. Vision 2030, along with a significant investment in housing and infrastructure development promoted across the country by local authorities, is revitalizing the construction industry and generating interest in a growing number of international players. Due to these programmes, the construction industry might have access to various opportunities; however, there are challenges associated with these programmes. Changes in regulations, policies, and the granting of planning approval may create complexity in the commencement of new projects and the completion of those already in progress, as they will have to be modified according to these new regulations. In this context, it is worth noting that the Saudi contractors' classification system functions within five grades according to the value they hold for a contract to be signed and 29 fields. The Example of the fields as following: buildings, roads, industrial works, marine works, dams, electrical works, and mechanical works [39]. In addition, according to the Government Tenders and Procurement Law in Saudi Arabia, all government bodies and agencies must use Saudi Arabia's Public Works Contract (SPWC) for all government-funded public construction projects. In addition, an increase in the projects will require growth in the work force, as a result of which companies may well have to depend on expatriates, which might result in acquiring an unskilled workforce lacking experience and facing issues in regard to cultural integration. In addition, without proper estimations of costs and risks, the contractors may end up suffering from financial losses.

2.3. Risk Factors Leading to Cost and Time Overruns

Studies have identified various critical success factors for construction projects. These included time, cost, quality, safety, client's satisfaction, employees' satisfaction, cash-flow management, profitability, environment performance, learning and development, etc. [40]. However, the majority of the past research have extensively focused on the three major factors for success in construction industry, which included cost, time, and quality [41]. It has been elicited that over 70% of public construction projects in Saudi Arabia have experience delays [42]. Various risk factors and challenges explaining the time and cost overruns for these projects have been uncovered. Baghdadi and Kishk [43] identified 54 risk factors in the context of external, internal, and force majeure in aviation construction projects, which were causing duration delays as well as cost overruns. Mahamid [44], focusing on the factors affecting performance in construction projects, identified various risks, including poor communication among project participants, poor labour productivity, poor planning and scheduling, payment delays, escalation of material prices, poor labour productivity, and poor site management. Regarding the causes of disputes, Mahamid [44] identified 29 direct and 32 indirect dispute causes, of which major direct dispute causes included delay in progress payment by the owner, unrealistic contract duration times, change orders, poor quality of completed work, and labour inefficiencies. Major indirect dispute causes included inadequate contractor experience, lack of communication between the construction parties, ineffective planning and scheduling of the project by the contractor, cash problems during construction, and poor estimation practices.

Focusing on the design risks, Sha'ar et al. [45] identified unstable client requirements, lack of proper coordination between the various disciplines of the design team, awarding the contract to the lowest price regardless of the quality of services, lack of skilled and experienced human resources in the design firms, lack of skilled human resources at the construction site, delaying of due payments, lack of a specialised quality-control team, lack of professional construction management, delaying the approval of completed tasks, and deficient drawings and specifications. Various other challenges, such as those related to subcontractors, labour, machinery, availability of materials, and quality; and client-related risks such as financial issues, issues related to design documents, change in codes and regulations, scope of work, accidents on site, lack of expertise, re-designing, unqualified workforce, organisational culture, and poor contract management were identified from various studies conducted on Saudi Arabian construction industry [46–48].

Furthermore, the causes of the cost and time overruns factors differ between various projects/buildings. For instance, when comparing the delay factors between road infrastructure and building projects, a recent study [49] found that the major critical delay factors for road infrastructure projects included inadequate contractor experience and payment delays to the contractor, while the shortage of materials and financial difficulties of contractor were most salient for building projects. For tall building projects, the major causes of delay and cost overruns identified in [50] included "client's cash flow problems/delays in contractor's payment", "contractor's financial difficulties", and "poor site organization and coordination between various parties". Another study focusing on regular manufacturing and building construction [17] identified delays in progress payments, difficulties in financing the project by contractor/manufacturer, slowness in decision making, late procurement of materials, and delay in approving design documents as the major causes of cost and time overruns. In specialised construction projects, such as railways, the causes were found to be related to "Client's decision-making process and changes in control procedures", "Design errors (including ambiguities and discrepancies of details/specifications)", "Labor skills level", "Design changes by client or consultant", and "Issues regarding permissions/approvals from other stakeholders" [51]. In addition, Allahaim [14] emphasised causes and classifications as differing by project type and stakeholder, with overall cost overrun depending on the type of project: power and health projects (60% cost overruns), transport and water projects (40% cost overruns), and education projects (30% cost overruns). Aljohani et al. [52] carried out a review of the literature and identified 173 causes of cost overrun in seventeen contexts, with the main ones being frequent design change, contractors' financing, payment delay for completed work, lack of contractor experience, poor cost estimation, poor tendering documentation, and poor materials management. The authors concluded that the main causes differed from country to country, and that it would be an inaccurate method to use only the global literature to identify the causes for a specific country [14]. In contrast, Ahady et al. [53] found that most of causes of cost overruns in construction industries of development countries are similar, and the causes are different for every project. The most significant causes of cost overruns were fluctuations and increases in material price. Appendix A shows that various risk factors associated with construction projects from 17 studies [2,4,14,17,44,51,52,52–65].

Hence, the factors causing cost and time overruns may change by the types of construction projects. Therefore, there is a need to focus the research on specific building projects in the context of Saudi Arabia. Furthermore, most studies in the literature probed the causes of either cost or time overruns for the construction industry, but very few considered both. Given these gaps, it is essential that risk factors and risk management techniques in Saudi Arabia have to be studied from time to time in order to prevent any damage/losses and avoid cost and duration overruns in construction projects. Accordingly, the purpose of this study is to identify the influential risk factors that lead to completion delays and cost overruns of government-funded building construction projects in Saudi Arabia, all of which have been subjected to a traditional type of procurement method Standard Public Works Contract (SPWC).

3. Research Methodology

For this study, the researchers adopted a cross-sectional questionnaire-based survey to identify risk factors related to government-funded building construction projects in Saudi Arabia. Figure 1 illustrates the adapted research methodology phases used to achieve the study objectives. This methodology includes four phases: the identification of initial risk factors from the literature, questionnaire design, data collection, and then data analysis.

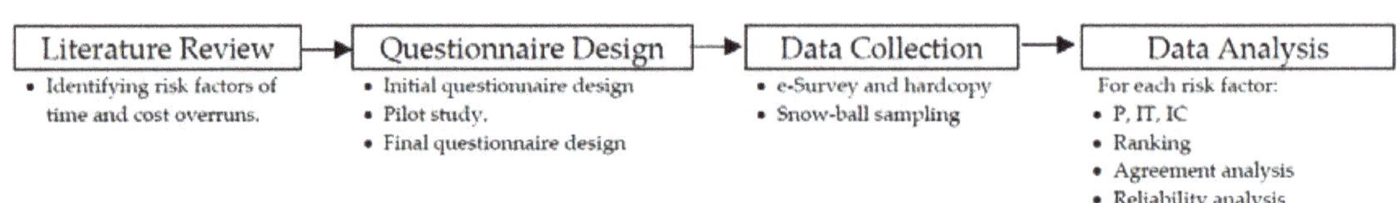

Figure 1. Research methodology process.

The first phase was the identification of initial risk factors from previous literature. A comprehensive literature review was carried out to uncover the various risk factors associated with construction projects. Then, the researchers identified the risk factors that were applicable in the context of Saudi Arabia. A final list of 83 risk factors, classified into nine different groups (client-related, designer-related, consultant-related, contractor-related, labour-related, material-related, equipment-related, external risks, and force majeure), was identified to be relevant for investigation in the context of this study, as shown in Appendix B.

The second phase was the questionnaire's design. The initial questionnaire was developed based on the findings in the previous phase. All these applicable risk factors were included in the questionnaire, which was divided into three sections. The first section of the questionnaire included the participants' demographic information, while the second focused on the level to which project delays and cost overruns affect construction projects. The third section pertained to identifying which risk factors caused project delays and cost overruns by asking three sub-questions for each risk. These included the probability of occurrence (P) in projects based on the respondents' perspective and experience, the negative impact of the risk on project's time (IT), and the negative impact of the risk on project's cost (IC). The questionnaire used a Likert scale of five ratings (1: very low; 2: low; 3: moderate; 4: high; 5: very high) and was designed in both English and Arabic to improve the participants' ease of accessibility and understandability. The researchers conducted a pilot study to validate the prepared questionnaire by distributing it to a set of experts in the construction field. The collected comments were reviewed to develop the final questionnaire.

The third phase of study was data collection. The questionnaire link was forwarded to the experts in construction industry who have been working in relevant building construction projects using various online networks. The researchers adopted snowball sampling [66], requesting the participants to forward the survey link to their colleagues and other relevant professionals. The survey was initially forwarded to 38 experts. Snowball sampling is a more conducive and practical technique for the research scope and to overcome the obstacle of the questionnaire's length, finding the target audience, and providing high-quality information. However, because of snowball sampling, 63 responses were received. After removing eight incomplete responses, the responses from 55 participants were included in the data analysis.

The fourth phase of study was the data analysis of the survey results using MS Excel. The relative importance index (RII) calculated the probability of occurrence (P) of each risk, the impact of the risk on project's time (IT), and the impact of the risk on project's cost (IC). Risk Importance (RI) was used to determine the level of importance of each identified risk associated with building construction projects by multiplying the probability and impact for each in terms of project time and cost. In addition, the reliability of factor analysis was used to measure the strength of the internal consistency of the identified risk factors, and an agreement analysis test (Cronbach's alpha) was conducted to measure the strength and direction of relationship between the parties involved in this study (client, contractor, and consultant).

3.1. Ranking of Risks

To carry out data analysis, the relative importance index (*RII*) for each risk was calculated by Equation (1) for the probability of occurrence (*P*) in projects based on the respondents' perspective and experience, and for negative impact (*I*) of the risk on project's time (*IT*) and for negative impact on project's cost (*IC*), using five point Likert scales:

$$RII = \sum_{i=0}^{n} \frac{W_i}{A \times N} = \sum_{i=0}^{n} \frac{5n_5 + 4n_4 + 3n_3 + 2n_2 + n_1}{5N} \tag{1}$$

where

RII—is the Relative Importance Index;
W_i—is the weight given to each factor by the respondents from 1, 2, 3, 4, and 5 for very low, low, moderate, high, and very high, respectively;
A—is the highest weight (i.e., 5 in five-point Likert scale);
N—is the total number of respondents for every variable.

To prioritise risks, the formula of Risk Importance (*RI*) was calculated by multiplying the probability and impact for each in terms of project time and project cost (see Equation (2)). Based on the calculations, risks were classified as "high", "moderate", or "low" importance. Risks that have an (*RI*) value equal to or greater than (0.6) were classified as "high" and were significantly important, and those between 0.6 and 0.4 were classified as "moderate" importance and less than 0.4 as "low" importance:

$$\text{Risk Importance;} \, RI = P \times I \tag{2}$$

where

RI—is the Risk Importance to determine the level of importance of each identified risk;
P—is the probability of risk occurrence;
I—is the impact of risk on time or cost.

3.2. Reliability of Factor Analysis

For this study, Cronbach's alpha (*Cα*) testing was used to measure the reliability and strength of the internal consistency of the identified risk factors. The *Cα* range is between 0 and 1, and the acceptable reliability number is typically 0.7 or higher as identified by [67]. The *Cα* formula for Likert scale is shown in Equation (3) below:

$$C\alpha = \frac{K}{K-1} \left[1 - \frac{\sum_{i=0}^{k} \sigma_b^2}{\sigma_t^2} \right] \tag{3}$$

where:

Cα—is Cronbach's alpha;
K—is many items;
σ_b^2—is the variance of test score;
σ_t^2—is the variance of item scores after weighing.

3.3. Agreement Analysis

Spearman's rank correlation coefficient (r_s) was used to measure the strength and direction of relationship between two ranked sets rather than the actual values. The coefficient was calculated by Equation (4) for ranked risk factors for pairs of the parties involved in this study (client, contractor, and consultant):

$$r_s = 1 - \frac{6\sum d^2}{n(n^2 - 1)} \tag{4}$$

where

r_s—is Spearman's rank correlation coefficient between two parties;
d—is the difference between ranks assigned to each risk;
n—is the number of pairs of rank.

4. Results and Discussion

4.1. Participants' Demographics

Out of 55 acceptable questionnaires, 30 respondents (54.55%) belonged to the public sector, whereas (34.55%) were from the private sector, and 5.45% belonged to semi-government sector; the remaining 5.45% belonged to academic and research institutions. Twenty-seven respondents (49.09%) designated themselves as the client (owner/government agency), eighteen respondents (32.73%) were designers and consultants, and eight respondents (14.55%) reported to be contractors. The majority indicated that they had a masters degree (MSc) (41.82%), and 23.64% responded that they held a PhD.

Furthermore, the majority of the participants (63%) in this study had an experience of more than 15 years on construction projects, and they were distributed across various areas in the construction sector, reflecting the quality inputs gathered from the diverse experts. The quality of the responses was considered reliable for the analysis due to personal level interaction, relevant experiences, and clear understanding of the questionnaire among the participates. Table 1 summarises the first part of the questionnaire responses, including the respondents' educational background and experience.

Table 1. Participants' demographic details.

Category	Respondent Number	Percentage	Category	Respondent Number	Percentage
Years of Experience			**Sector/Entity**		
Less than 5 years	1	1.82%	Public sector	30	54.55%
6–15 years	19	34.55%	Private sector	19	34.55%
16–25 years	26	47.27%	Semi-government sector	3	5.45%
More than 25 years	9	16.36%	Academic and research institutions	3	5.45%
		100.00%			100.00%
Educational Background			**Role**		
Civil Engineering	33	60.00%	Owner/government agency	27	49.09%
Architecture	7	12.73%	Designer	2	3.64%
Electrical Engineering	5	9.09%	Consultant	16	29.09%
Mechanical Engineering	10	18.18%	Contractor	8	14.55%
			Others	2	3.64%
		100.00%			100.00%

4.2. Delay and Cost Overrun in Construction Projects

Based on reported experience, more than 40% of projects had been subject to delays in the execution phase for thirty respondents (54.55%), and the percentage of project delays was more than 30%, as identified by 25 respondents. Fifty-four respondents had experienced project cost overruns in the execution phase and the average percentages of cost overruns were between 10% and 25% for 29 respondents, whereas 25 respondents (45.45%) have experienced projects cost overruns with less than 10% of average percentage of cost overruns. Table 2, below, summarises the results of second part of the questionnaire.

However, it has been documented that over 70% of the public projects in Saudi Arabia were delayed [68]. For instance, university construction projects were found to be experiencing delays from 50% to 150% [42]. The findings of this study indicate slightly fewer delays (45% of participants stating delays less than 40%) compared to previous studies [69,70], which have identified them as being from 70% to 75%.

Table 2. Performance of building construction projects.

Project Delays			Project Cost Overruns		
Category	Respondent Number	Percentage	Category	Respondent Number	Percentage
% Projects Exposed to Delays			**% of Projects Exposed to Cost Overruns**		
Never	0	0	Never	1	1.82%
Less than 10%	1	1.82%	Less than 10%	11	20.00%
11–20%	9	16.36%	11–20%	11	20.00%
21–30%	4	7.27%	21–30%	13	23.64%
31–40%	11	20.00%	31–40%	9	16.36%
More than 40%	30	54.55%	More than 40%	10	18.18
	55	100.00%		55	100.00%
Average delay %			**Average cost overruns %**		
Never	0	0%	Never	1	1.82%
Less than 10%	1	1.82%	Less than 5%	6	10.91%
11–20%	10	18.18%	6–10%	19	34.55%
21–30%	19	34.55%	11–15%	10	18.18%
31–40%	10	18.18%	16–20%	8	14.55%
More than 40%	15	27.27%	21–25%	8	14.55%
			More than 25%	3	5.45%
	55	100.00%		55	100.00%

4.3. Ranking of Risks

Risks that are associated with building construction projects in Saudi Arabia were assessed and ranked in terms of project delay and project cost overruns by calculating the Relative Importance Index (RII) of the probability of occurrence (P) for each risk, RII of impact of the risk on project time (IT), and (RII) of impact of the risk on project cost (IC). Then, Risk Importance (RI) was calculated for each risk in terms of project time (delay) and project cost (cost overruns), being subsequently ranked, as shown in Table A2 (Appendix B). The top ten risk factors that led to delay and cost overruns in building construction projects are shown in Table 3.

Table 3. Top 10 Risk factors that lead to delay and cost overruns in building construction projects.

No	Code	Risk Factors	Project Delay		Project Cost Overruns		Overall		Category
			RI	Rank	*RI*	Rank	*RI*	Rank	
38	G4R2	Contractor's financial difficulties (ineffective cash flow management)	0.692	1	0.597	3	0.692	1	Contractor-related
4	G1R4	Owner's delay in making progress payments for completed works (Payment delays)	0.672	2	0.525	12	0.672	2	Client-related
6	G1R6	Contract awarded to lowest bidder	0.631	3	0.601	2	0.631	3	Client-related
11	G1R11	Change orders during construction	0.627	4	0.622	1	0.627	4	
40	G4R4	Ineffective planning and scheduling of project by contractor	0.627	5	0.528	9	0.627	5	Contractor-related
55	G5R1	Shortage of manpower (skilled, semi-skilled, unskilled)	0.608	6	0.531	7	0.608	6	Labour-related
39	G4R3	Contractor's poor site management and supervision	0.601	7	0.526	11	0.601	7	Contractor-related
37	G4R1	Inadequate contractor experience (lack of experience, and managerial skills)	0.595	8	0.529	8	0.595	8	Contractor-related
42	G4R6	Delays in sub-contractors' work or suppliers	0.588	9	0.474	24	0.588	9	
56	G5R2	Unqualified/inexperienced workers.	0.588	10	0.548	6	0.588	10	Labour-related

As a result of Risk Importance classification (high, moderate, and low), seven risk factors that were the most significant risks factors contributing to completion delay of building construction projects were (1) contractor's financial difficulties (RI = 0.692), (2) owner's delay in making progress payments for completed works (RI = 0.672), (3) contract awarded to lowest bidder (RI = 0.631), (4) change orders during construction (RI = 0.627), (5) ineffective planning and scheduling of project by contractor (RI = 0.627), (6) shortage of manpower (RI = 0.608), and (7) contractor's poor site management and supervision (RI = 0.601). On the other hand, there were two significant risks factors contributing to cost overruns: (1) change orders during construction (RI = 0.622) and (2) contract awarded to lowest bidder (RI = 0.601). The top ten risk factors that led to delay and cost overruns in building construction projects are shown in Table 3.

The most significant risks factors identified in this study are related to contractors (financial difficulties, ineffective planning and scheduling of projects, and poor site management and supervision), clients (delay in making payments, awarding contracts to the lowest bidder, and changing orders during construction), and labour (shortage of manpower). Contractors' financial difficulties (ineffective cash flow management) was ranked as the first major risk factor in this study. Shash and Qarra [71] conducted a study that revealed that 40% of contractors in Saudi Arabia experience financial failure due to poor cash flow management. Saudi contractors' classification system classifies contractors to a five-grade scale. Although these grades determine the maximum project budget size that allow contractors to bid for (an upper limit), it does not consider the maximum number of projects (the total financial limit of all awarded projects) [39]. Consequently, some contractors use the cash flow of one project to finance different project deficits [71]. This result is in line with some of the investigated studies [2,71–73]. The Saudi contractors need to adopt effective cash flow management practices that require planning, monitoring, and controlling cash inflow and outflow at both the company and project levels to achieve financial success and avoid project deficits.

The second ranked risk factor is the owners' delay in making progress payments to the contractor for completed works (payment delays). Most Saudi construction contractors suffer from progress payment delays. Although Saudi contractors receive 5.0% of the contract price at the beginning as an advance payment from the project's owner, the progress payments are the key sources of cash inflow to resolve deficit cash flow and avoid or minimise outsource finance. Delayed progress payments and high expenses of construction project leads to delaying construction work progress and increasing the project costs unless the contractor is capable financially. Approval process (65%) and bureaucracy (25%) are the primary reasons for delays in owners' progress payments [71]. This result is supported by [52,60,71,74].

Contracts awarded to the lowest bidder was ranked as the third most significant risk to building construction projects in Saudi Arabia. This risk can be attributed to the government's tender and procurement system and the contractors' classification system in Saudi Arabia. This practice creates uncertainty due to a lack of experience, lack of financial capability, incompetent contractors, and suicide tendering. It is supported by studies in different contexts and was also identified by [7,60,74] in Saudi Arabia as the most important significant risk factor in Saudi Arabia.

Changing orders during construction were considered the fourth most important risk for project delay in this study. It was also identified by [7] Assaf in Saudi Arabia and by [75] in Kuwait as the most significant risk factor causing project delays. Change orders usually lead to change project schedules and contract prices, claims and disputes, and poor quality of work. Khalifa and Mahamid [20] identified the factors causing change orders in Saudi Arabia. The top causes of change orders are owners' additional work, design errors and omissions, lack of coordination, defective workmanship, owners' financial difficulties, and differing site conditions.

Ineffective project planning and scheduling by contractors was ranked as the fifth among the top risk factors in this study. It was also identified by [7,70,74] in Saudi Arabia and by [2] in Malaysia as the most important risk factor. The shortage of manpower (skilled, semi-skilled, and unskilled) was ranked the sixth major risk factor in this study, which is similar to the findings observed in [42]. However, studies [68,76] identified shortage of labour as being less influential compared to the other factors among the top ten terms of risk. Disruptions in supply chain and movement of labour due to the recent COVID-19 pandemic could be one of the reasons for the higher ranking for shortage of labour. Although COVID-19 may be considered as a force majeure risk, the impact it caused may affect all three stakeholders, including clients, consultants, and contractors. Furthermore, the number of risk factors identified in Saudi Arabia in previous studies [68,76,77] was from 45 to 60, and they were mostly related to owners (clients) and contractors. Finally, contractors' poor site management and supervision was ranked as the seventh most important risk factor in this study. It was identified by [72] in Vietnam and by [2] in Malaysia.

Changing orders during construction and contracts awarded to the lowest bidder were ranked as the first and second most significant risks to construction projects in Saudi Arabia that caused project cost overruns, which were client-related risks. This result is supported by previous research conducted by [52,54].

Furthermore, from the perspective of the three groups of respondents (clients, consultants, and contractors), they indicated the risks related to their areas with low *RI* compared to the other groups (as shown in Table A3, Appendix B). For instance, *RI* for almost all the client-related risks was less than 0.6, as rated by the participants who were in this category, whereas some of these risks were rated with an that was *RI* more than 0.6 by consultants and contractors. However, no major differences among the groups were identified in rating the risks pertaining to designer-related, labour-related, material-related, equipment-related, and external risk factors. Table A3 (Appendix B) presents the ranking according to the perspectives of the three groups of respondents.

Among the identified risk groups, contractor-related risks were identified to be the major risk factors causing both time and cost overruns. Considering the remaining categories, materials-related, labour-related, consultant-related, and external risks had greater impact on cost overruns; materials-related, force-majeure, and consultant-related risks had greater impact of time overruns. The findings clearly indicated the disruptions in supply chain, which may be attributed to the recent pandemic and issues in planning and implementation.

In addition, analysing the risks of each group in order to identify the most important group of risk in building projects in Saudi Arabia, as shown below in Table 4.

Table 4. The most important group of risk factors.

Group No.	Risk Factor Group	*RI*	Rank	Category
G1	Client-related Risks	0.55	1	Internal
G4	Contractor-related Risks	0.505	2	Internal
G5	Labour-related Risks	0.477	3	External
G2	Designer-related Risks	0.455	4	Internal
G6	Materials-related Risks	0.431	5	External
G3	Consultant-related Risks	0.421	6	Internal
G8	External Risks	0.397	7	External
G7	Equipment-related Risks	0.395	8	External
G9	Force Majeure Risks	0.342	9	External

The results revealed four groups as the most important groups with score more than 0.45, which include client-related risks, contractor-related risks, labour-related risks, and design-related risks, all of which were found to have a greater impact on both cost and time overruns. Client-related risks were ranked highest in government-funded projects. However, this finding contrasts with some studies on Saudi Arabian construction where contractor-related risks were given the highest importance [69,74], while in [22] client-related risks were identified as being in this place. Contractor-related risks have been elicited as being the second most important risk in this study, which contrasts with its rankings in other studies [69,70,74]; however, it was identified as being one of the most significant risks in [43]. Moreover, labour- and design-related risks were identified as being significant in studies [43,78] conducted in Saudi Arabia, while other studies [69,70,74] did not find this to be the case.

The risk factors identified in this study, although they reflected similar risks identified in other recent studies in different project types in Saudi Arabia, had few differences identified in terms of the nature of significant risks and their priority. For instance, in the study focusing on the oil and gas industry [15], client-related risks included changes in design and contractor-related risk, and poor planning and implementation were identified as the significant risks; On the other hand, in the study focusing on manufacturing and building projects [17], contractor-related risks including financial difficulties and delays in procurement of raw-materials were identified to be significant risks. In another project related to roads construction, poor planning and poor labour productivity and unskilled labour were identified to be the significant risks. Lean practices can be an effective approach in this context for improving the planning and implementation of construction projects in Saudi Arabia, as it can result in social, economic, and environmental benefits [79]. Although lean practices were identified to be effective in different countries [79,80], different barriers such as traditional practices, client related, technological, performance and knowledge, and cost-related barriers were identified, which limit the implementation of lean practices in Saudi Arabia [81]. Therefore, there is a need to address these barriers for effective implementation of lean practices for addressing the various types of risks in Saudi Arabian construction industries. It is evident from these studies that the nature of risks and its significance may change with the types of projects and countries; therefore, risk management strategies and approaches have to be adjusted accordingly.

These research findings provide a good lesson to not only Saudi Arabia but also the construction industries in other countries, especially the Middle East countries, where there is a lack of skilled resources, high dependency on expatriates, and rising demand for new construction projects. Furthermore, the findings in this study contrasted with studies conducted in other developing countries. For instance, in Malaysia [2,61], design and contract risks were identified to be of high priority, followed by labour risks. However, with increase in FDIs, the clients of the governments may require different changes or raise issues in agreements that may lead to an increase in such risks, as identified in this study in Saudi Arabia, which is focusing on acquiring huge FDIs. Similar results may be identified in China [59], where client risks and contractor-related risks were identified to be the significant risks. Therefore, for developing countries looking for FDIs in the construction industry, client-related risks may emerge as top risks in the near future. While other risks such as material and labour-related risks would be commonly identified in developing countries with limited technical and human resources [55].

4.3.1. Reliability of Factor Analysis

It was calculated for the nine groups and the overall factors, as shown in Table 3. The results of Cronbach's alpha were all more than 0.8, thus indicating an acceptable level of reliability was achieved, as shown in Table 5.

4.3.2. Agreement Analysis

As shown in Table 6, the results indicate positive agreement between the pairs of parties, with the highest level being between the client and consultant, at 82.8%, and 73.8% agreement between the consultant and contractor, and then 64.1% agreement between the client and contractor. The lowest degree of agreement appears to between client and contractor (43.3% with impact on project cost overruns and 34.1% with risk importance of cost overrun). The overall agreements between the parties in ranking the risk factors and other major findings in this study can be used for further research and analysis.

Table 5. Reliability analysis (Cronbach's alpha) for the risk factors.

Group No.	Risk Factor Group	No. of Risk Factors	Probability (*P*)		Impact on Time (*IT*)		Impact on Cost (*IC*)	
			Cα	Result	*Cα*	Result	*Cα*	Result
G1	Client-related Risks	16	0.825	Good	0.813	Good	0.847	Good
G2	Designer-related Risks	10	0.862	Good	0.824	Good	0.842	Good
G3	Consultant-related Risks	10	0.866	Good	0.865	Good	0.897	Good
G4	Contractor-related Risks	18	0.915	Excellent	0.906	Excellent	0.940	Excellent
G5	Labour-related Risks	8	0.860	Good	0.847	Good	0.901	Excellent
G6	Materials-related Risks	6	0.867	Good	0.884	Good	0.937	Excellent
G7	Equipment-related Risks	22	0.853	Good	0.837	Good	0.841	Good
G8	External Risks	9	0.873	Good	0.828	Good	0.877	Good
G9	Force Majeure Risks	4	0.839	Good	0.864	Good	0.862	Good
	Overall		**0.9858**	**Excellent**				

Table 6. Spearman's rank correlation coefficient between parties **.

Parties	Spearman Rank Correlation Coefficient
Client and Consultant	0.834
Client and Contractor	0.653
Consultant and Contractor	0.736

Parties	Probability (*P*)	Impact (*I*)		Risk Importance (*RI*)		
		Project Delay	Project Cost Overruns	Project Delay	Project Cost Overruns	Overall
Client and Consultant	0.814	0.778	0.788	0.817	0.830	**0.828**
Client and Contractor	0.650	0.633	0.433	0.655	0.341	**0.641**
Consultant and Contractor	0.756	0.683	0.610	0.765	0.548	**0.738**

** Correlation is significant at the 0.01 level (2-tailed).

Regarding the level of agreement amongst the different stakeholders, it is evident from Table 6 that client–consultant had the highest, while client–contractor had average levels of agreement, thus indicating the major issues relating to the clients–contractors' relationships and transactions. The low probability in client and contractor relationship can be understood in different perspectives and interests. The relationship between client and contractor can be influenced by various factors. For instance, commitments from the contractors and competence trust of the clients are very volatile, which can significantly affect the relationship between them [82]. While time, cost, and quality were considered as important client values, they were not considered as exclusive values for assessing contractors service, indicating the differences in the values, attitudes of both parties, and the relationships between them [83]. The major issues identified in this study and previous ones [68,76,77] have revealed that the majority of the risk factors of high priority pertain to client–contractor relationships. Hence, it can be concluded that the companies and consultants in Saudi Arabian construction industry should focus on improving the client/contractors' relationships, the tendering process, project planning and execution, and financing.

5. Conclusions

The construction industry in Saudi Arabia has suffered from completion delays and cost overruns, which have caused financial losses for all parties involved in such a competitive environment. The survey results revealed the seven risk factors that were the most significant risk factors contributing to the completion delays of building construction projects out of the eighty-three risk factors identified from literature review. These risk factors included contractors' financial difficulties, owners' delay in making progress payments for completed works, contracts awarded to the lowest bidder, change orders during construction, ineffective project planning and scheduling by contractor, shortage of manpower, and contractors' poor site management and supervision. Additionally, changing orders during construction and contracts awarded to the lowest bidder were the most significant risks that caused projects cost overrun, which were client-related risks. It revealed four risk groups as the most significant: client-related risks, contractor-related risks, labour-related risks, and design-related risks. Each group was found to have a notable impact on both cost and time overruns. The statistical analyses revealed an acceptable level of reliability of the identified risk factors and a positive agreement between the clients, consultants, and contractors.

The findings have revealed issues in the client/contractor relationship and tender allocation process, which may help industry experts and government agencies in future plans to mitigate the risks identified in this study. Furthermore, with uncertainty continuing due to the COVID-19 pandemic and the opening of the markets, future studies could focus on investigating the force majeure risks and the impact these have on the relationships between the stakeholders and supply chain systems in the Saudi Arabian construction industry. To achieve sustainable development, client-, contractor-, and labour-related risks must be effectively managed.

The novelty of contributions in this study can be reflected in the findings achieved in specific to government funded building construction projects in Saudi Arabia, which previous studies have not focused, although the difference in the risk factors with project types were highlighted in previous studies. Furthermore, the findings of this study are novel, due to the situations created by external factors such as COVID-19 pandemic, which has greatly affected resource management and continuity in construction. However, there are certain limitations that can be observed in this study. This study adapted snowball sampling methods and only considered government-funded building construction projects through SPWC processes, while there are also other project types. These limitations can be addressed in future research works. Future research can focus on another project types in the context of Saudi Arabia, such as roads, industrial projects, etc. However, various implications can be drawn from the findings in this study. Firstly, the results from this study aids decision makers to better understand the impact of the COVID-19 pandemic on the construction industry, based on which necessary policy-related decisions may be taken to strengthen the construction industry and better implement vision 2030 objectives. Secondly, the findings in this study contribute to the literature on the risk factors by project types, as this study focused only on government-aided building construction projects.

Author Contributions: S.A. was the primary researcher responsible for the designing the study and conducting the majority of the work. K.A.-G. was the principal supervisor who contributed to the research design and revised the paper on several occasions. A.A. contributed by revising the overall structure and content of the survey. All authors have read and agreed to the published version of the manuscript.

Funding: This research was funded by the Deanship of Scientific Research (DSR), King Saud University (KSU).

Institutional Review Board Statement: Not applicable.

Informed Consent Statement: Not applicable.

Data Availability Statement: The data presented in this study are available on request from the corresponding author.

Acknowledgments: The author would like to thank the Deanship of Scientific Research (DSR), King Saud University (KSU), for supporting and funding this research.

Conflicts of Interest: The authors declare no conflict of interest.

Table A1. List of major risk factors identified from the literature review.

Studies	Methodologies	Country	No. of Factors Identified	Major Risk Factors	Impact on Construction Projects
Mahamid et al. (2015) [60]	Survey and Literature Review	Saudi Arabia	31	Bid awarded for lowest price, changes in material types and specifications during construction, contract management, duration of contract period, fluctuation of prices of materials, frequent changes in design, improper planning, inflationary pressure, lack of adequate manpower, long period between design and time of implementation, payments delay, poor labour productivity and rework	Time overrun
Memon et al. (2012) [61]	Interviews and surveys	Malaysia	35	Design and documentation issues, financial resource management, project management and contract administration, contractors site management, information and communication technology, material and machinery resource, labour (human) resource, external factors	Time and cost overrun
Aljohani et al. (2017) [52]	Literature Review	Multiple countries	173	Increase in material cost, inaccurate material estimates, shortage of skilled labour, client's late contract award, project complexity, increase in labour costs, bidding differences, shortage manpower, design issues, poor planning and implementation, lack of talent resources	Cost overrun
Allahaim and Liu (2015) [14]	Literature Review and Surveys	Saudi Arabia	41	Market conditions, unrealistic estimations, decision-making errors, payment issues, project size, lack of talent resources, poor planning and implementation, waste on site, currency fluctuations, lack of technology and material resources, changes in design and scope, political obstacles, poor strategies	Cost overrun
Creedy et al. (2010) [57]	Case studies	Australia	37	Design issues, political issues, requirements changes, cultural issues, material costs increase, contract failures, administration issues, location changes, inflation	Time, cost, quality, performance
Jackson (2002) [58]	Interviews	UK	341	Design changes, project management, site conditions, commercial pressures, lack of talent resources, external factors, estimation methods, information availability	Time, cost, quality, performance, planning, project management
Baloyi and Bekker (2011) [55]	Literature Review and Surveys	South Africa	18	Increase in material cost, inaccurate material estimates, shortage of skilled labour, client's late contract award, project complexity, increase in labour costs, bidding differences, shortage manpower, design issues, poor planning and implementation, labour issues, poor information availability, delay in approvals	Time, cost, quality, performance, planning, project management
Subramani et al. (2014) [63]	Case studies and Literature Review	India	10	Slow decision making, poor schedule management, increase in material/machine prices, poor contract management, poor design/ delay in providing design, rework due to wrong work, problems in land acquisition, wrong estimation/ estimation method, and long period between design and time of bidding/tendering	Cost overrun
Memon et al. (2011) [2]	Case studies and Survey	Malaysia	30	Practice of assigning contract to lowest bidder, contractor's poor site management, cash flow and financial difficulties faced by contractors, ineffective planning and scheduling by contractors, problems with subcontractors, inadequate contractor experience, material procurement, poor estimate project duration, incompetent designers, shortage of site workers, lack of communication among parties, unforeseen ground condition, changes in scope of projects, low speed of decision making, frequent changes by owners, escalation of material prices, owner interference, change in management	Time overrun

Appendix A

Table A1. *Cont.*

Studies	Methodologies	Country	No. of Factors Identified	Major Risk Factors	Impact on Construction Projects
Mahamid et al. (2015) [60]	Survey and Literature Review	Saudi Arabia	31	Bid awarded for lowest price, changes in material types and specifications during construction, contract management, duration of contract period, fluctuation of prices of materials, frequent changes in design, improper planning, inflationary pressure, lack of adequate manpower, long period between design and time of implementation, payments delay, poor labour productivity and rework	Time overrun
Mahamid (2017) [44]	Case studies and survey	Saudi Arabia	34	Improper planning, poor labour productivity, additional works, rework, and lack of contractor experience, disputes, arbitration, litigation, and poor quality	Time overrun, Performance
Tebeje Zewdu and Teka Aregaw (2015) [64]	Survey and Literature Review	Ethiopia	41	Poor planning, fluctuation of price of materials, poor productivity, inflationary pressure and project financing, economic instability, lack of talented labour	Cost overrun, Performance
Sharma and Goyal (2019) [62]	Literature Review and Expert opinion	India	55	Fluctuation in price of materials, lowest bid procurement policy, inflation inappropriate govt. Policy, mistakes and discrepancies in the contract document, inaccurate time and cost estimate, additional work, frequent design change, unrealistic contract duration and financial difficulty faced by contractors	Cost overrun, Performance
Cirovic and Sudjic (2012) [56]	Literature Review and Case studies	Montenegro	16	Local market issues, Montenegrin legislation, local infrastructure, poor resources management, delay in approvals, lack of planning, lack of labour	Cost and duration overrun
Bahamid et al. (2019) [54]	Literature Review	Multiple countries	111	Inflation/price fluctuation, technology issues, incomplete design scope, changes, labour equipment, delays in approvals, financial issues, poor estimations, poor designs, political instability, criminal acts, poor communication, poor planning and control	Time, cost, quality, performance, planning, project management
Liu et al. (2016) [59]	Survey and Literature Review	China	20	Host government–related risk, contractor's lack of experience, and lack of managerial skills had significant effect on project cost, quality, and schedule objectives, resource price fluctuation	Time, cost, quality, performance, planning, project management
Iqbal et al. (2015) [4]	Survey	Pakistan	24	Financial issues for projects, accidents on site and defective design, subcontractors, labour, machinery, availability of materials and quality, issues related to design documents, changes in codes and regulations, political instability, and scope of work	Time, cost, quality, performance, planning
Zhao et al. (2016) [65]	Survey	Singapore	28	Inaccurate cost estimation, cost overrun, poor planning, external factors	Time, cost, quality

Table A2. *RII* for risk factors of building construction projects and ranking.

Appendix B

No	Code	Risk Factors	Probability		Impact on Time		Impact on Cost		Risk Importance (*RI*)					
									Project Delay		Cost Overrun		Overall *RI*	
			RII	Rank	*RII*	Rank	*RII*	Rank	*RI*	Rank	*RI*	Rank	*RI*	Rank
		Clients-Related Risks												
1	G1R1	Client's lack of experience about construction project.	0.665	33	0.724	40	0.713	25	0.482	35	0.474	25	0.482	39
2	G1R2	Excessive bureaucracy in owner's administration	0.735	11	0.742	33	0.665	44	0.545	18	0.489	19	0.545	18
3	G1R3	Client's financial difficulties	0.665	33	0.818	8	0.720	19	0.544	19	0.479	22	0.544	19
4	G1R4	Owner's delay in making progress payments for completed works (payment delays)	0.767	3	0.876	1	0.684	37	0.672	2	0.525	12	0.672	2
5	G1R5	Selecting consultant, or designer, based on the lowest price	0.760	4	0.756	29	0.742	11	0.575	14	0.564	4	0.575	14
6	G1R6	Contract awarded to the lowest bidder	0.818	1	0.771	24	0.735	13	0.631	3	0.601	2	0.631	3
7	G1R7	Unrealistic/Inadequate original contract duration (tight schedule)	0.702	19	0.822	5	0.720	19	0.577	13	0.505	16	0.577	13
8	G1R8	Political pressure to complete the project and speed up construction processes	0.549	71	0.644	70	0.753	9	0.353	68	0.413	49	0.413	58
9	G1R9	Delay in land acquisition/Handover to the contractor	0.571	65	0.684	55	0.524	77	0.390	61	0.299	76	0.390	64
10	G1R10	Difficulties in obtaining work permits from the authorities	0.640	44	0.760	27	0.589	69	0.486	31	0.377	55	0.486	35
11	G1R11	Change orders during construction	0.756	5	0.829	3	0.822	1	0.627	4	0.622	1	0.627	4
12	G1R12	Scope of work reduction by owner during execution phase	0.535	72	0.545	79	0.575	74	0.292	80	0.307	75	0.307	78
13	G1R13	Delay penalty clause in the Saudi Public Works Contract is inefficient	0.582	60	0.680	57	0.596	65	0.396	58	0.347	62	0.396	61
14	G1R14	Inaccurate estimation of construction cost by owner	0.684	24	0.669	63	0.771	3	0.457	43	0.527	10	0.527	27
15	G1R15	Project suspension by owner	0.556	68	0.789	17	0.684	37	0.439	46	0.380	53	0.439	47
16	G1R16	Delays due to dispute resolution	0.673	31	0.789	17	0.698	31	0.531	24	0.470	28	0.531	24
		Designer-Related Risks												
17	G2R1	Unclear and inadequate drawings, specifications, bills of quantities (BOQ)	0.724	13	0.804	12	0.767	4	0.582	11	0.555	5	0.582	11
18	G2R2	Discrepancies between project documents (contract, BOQ, specifications, and drawings)	0.647	41	0.749	31	0.738	12	0.485	32	0.478	23	0.485	36
19	G2R3	Mistakes and deficiencies in design documents	0.651	40	0.742	33	0.727	18	0.483	34	0.473	26	0.483	38
20	G2R4	Inaccurately estimated quantities (BOQ)	0.680	26	0.695	51	0.764	5	0.472	40	0.519	13	0.519	28

Table A2. *Cont.*

No	Code	Risk Factors	Probability		Impact on Time		Impact on Cost		Risk Importance (RI)					
									Project Delay		Cost Overrun		Overall RI	
			RII	Rank	RII	Rank	RII	Rank	RI	Rank	RI	Rank	RI	Rank
21	G2R5	All existing underground utilities information not available on the design documents (e.g., live cables and pipelines) for the contractor	0.687	23	0.782	21	0.705	28	0.537	20	0.485	20	0.537	20
22	G2R6	Inadequate geotechnical investigations report about ground conditions of the project site (if available)	0.647	41	0.731	39	0.778	2	0.473	39	0.504	17	0.504	32
23	G2R7	Complexity of design	0.495	79	0.665	64	0.665	44	0.329	74	0.329	68	0.329	74
24	G2R8	Absence of contractor's involvement in the design stage	0.615	49	0.509	82	0.524	77	0.313	76	0.322	70	0.322	76
25	G2R9	Speeding up of design phase's schedule	0.676	28	0.665	64	0.691	34	0.450	45	0.467	30	0.467	43
26	G2R10	Limited budget for design	0.647	41	0.662	66	0.676	42	0.428	50	0.438	38	0.438	48
		Consultant-Related Risks												
27	G3R1	Lack of experience and competence of consultant's staff.	0.691	22	0.764	26	0.680	39	0.528	26	0.470	27	0.528	26
28	G3R2	Slowness in decision making for approval (shop drawings, submittals, sample materials, change orders, etc.)	0.716	17	0.793	14	0.655	46	0.568	15	0.469	29	0.568	15
29	G3R3	Consultant's rejection of submittals (shop drawings, equipment, and material samples)	0.596	52	0.720	43	0.607	60	0.429	49	0.362	57	0.429	53
30	G3R4	Consultant's delay in performing inspection and testing, and giving instructions	0.567	66	0.695	51	0.571	75	0.394	59	0.324	69	0.394	62
31	G3R5	Poor coordination and communication between consultant and other parties	0.636	46	0.753	30	0.615	59	0.479	37	0.391	51	0.479	41
32	G3R6	Poor quality control and assurance	0.662	37	0.684	55	0.687	35	0.452	44	0.455	32	0.455	46
33	G3R7	Consultant's corruption	0.585	58	0.724	40	0.720	19	0.424	52	0.422	46	0.424	55
34	G3R8	Inflexibility (rigidity) of consultant	0.575	64	0.676	59	0.560	76	0.389	63	0.322	71	0.389	65
35	G3R9	Excessive safety consideration	0.447	81	0.505	83	0.491	83	0.226	83	0.220	83	0.226	83
36	G3R10	Internal company problems (at consultant company's head office)	0.535	72	0.604	77	0.509	80	0.323	75	0.272	80	0.323	75
		Contractor-Related Risks												
37	G4R1	Inadequate contractor experience (lack of experience, and managerial skills)	0.724	13	0.822	6	0.731	15	0.595	8	0.529	8	0.595	8
38	G4R2	Contractor's financial difficulties (ineffective cash flow management)	0.789	2	0.876	1	0.756	7	0.692	1	0.597	3	0.692	1
39	G4R3	Contractor's poor site management and supervision	0.731	12	0.822	6	0.720	19	0.601	7	0.526	11	0.601	7
40	G4R4	Ineffective planning and scheduling of project by contractor	0.756	5	0.829	3	0.698	31	0.627	4	0.528	9	0.627	4
41	G4R5	Ineffective control of the project progress	0.676	28	0.782	21	0.676	42	0.529	25	0.457	31	0.529	25

Table A2. *Cont.*

No	Code	Risk Factors	Probability		Impact on Time		Impact on Cost		Risk Importance (RI)					
									Project Delay		Cost Overrun		Overall *RI*	
			RII	Rank	*RII*	Rank	*RII*	Rank	*RI*	Rank	*RI*	Rank	*RI*	Rank
42	G4R6	Delays in sub-contractors' work or caused by suppliers	0.745	9	0.789	17	0.636	51	0.588	9	0.474	24	0.588	9
43	G4R7	Delay in preparation of shop drawings and material samples	0.709	18	0.782	21	0.604	62	0.554	16	0.428	45	0.554	16
44	G4R8	Delay in site mobilization.	0.662	37	0.698	49	0.524	77	0.462	41	0.347	63	0.462	44
45	G4R9	Rework and wastage on site, due to errors or quality of work (poor quality of workmanship)	0.665	33	0.749	31	0.680	39	0.498	30	0.453	34	0.498	34
46	G4R10	Variations in quantities	0.684	24	0.629	73	0.731	15	0.430	47	0.500	18	0.500	33
47	G4R11	Cost of penalties	0.589	53	0.582	78	0.622	54	0.343	71	0.366	56	0.366	69
48	G4R12	The contractor does not carry out a field visit to the site during the bidding process	0.655	39	0.724	40	0.695	33	0.474	38	0.455	33	0.474	42
49	G4R13	New existing underground utilities not mentioned on the design documents (e.g., live cables and pipelines)	0.676	28	0.793	14	0.756	7	0.536	21	0.512	15	0.536	21
50	G4R14	Health and Safety requirements (in light of COVID-19)	0.589	53	0.629	73	0.585	70	0.371	66	0.345	64	0.371	68
51	G4R15	Accidents on site	0.524	74	0.524	81	0.509	80	0.274	81	0.267	81	0.274	81
52	G4R16	Conflict between contractor and consultant	0.702	19	0.735	37	0.622	54	0.516	28	0.436	40	0.516	30
53	G4R17	Tender-winning prices are unrealistically low (suicide tendering)	0.640	44	0.804	12	0.749	10	0.514	29	0.479	21	0.514	31
54	G4R18	Unavailability of incentives for contractor for finishing ahead of schedule or to reduce the cost	0.720	15	0.673	61	0.625	52	0.484	33	0.450	35	0.484	37
	Labour-Related Risks													
55	G5R1	Shortage of manpower (skilled, semi-skilled, and unskilled)	0.753	7	0.807	10	0.705	28	0.608	6	0.531	7	0.608	6
56	G5R2	Unqualified/inexperienced workers	0.745	9	0.789	17	0.735	13	0.588	9	0.548	6	0.588	9
57	G5R3	Low productivity level of manpower/labourers	0.720	15	0.807	10	0.713	25	0.581	12	0.513	14	0.581	12
58	G5R4	Low payment for labour force	0.665	33	0.640	71	0.622	54	0.426	51	0.414	48	0.426	54
59	G5R5	Injuries to labourers on the construction site	0.556	68	0.545	79	0.505	82	0.303	77	0.281	79	0.303	79
60	G5R6	Delayed salary payments to staff by the contractor	0.753	7	0.735	37	0.596	65	0.553	17	0.449	37	0.553	17
61	G5R7	High turn-over of personnel	0.622	47	0.651	69	0.578	72	0.405	56	0.360	59	0.405	59
62	G5R8	Labour strikes	0.498	78	0.709	45	0.622	54	0.353	69	0.310	74	0.353	71
	Materials-Related Risks													
63	G6R1	Shortage of construction materials—special building materials not available in the local market	0.509	76	0.705	46	0.655	46	0.359	67	0.333	67	0.359	70
64	G6R2	Delay in delivery of materials	0.622	47	0.738	35	0.607	60	0.459	42	0.378	54	0.459	45

Table A2. *Cont.*

No	Code	Risk Factors	Probability		Impact on Time		Impact on Cost		Risk Importance (RI)					
									Project Delay		Cost Overrun		Overall RI	
			RII	Rank	RII	Rank	RII	Rank	RI	Rank	RI	Rank	RI	Rank
65	G6R3	Delay in the special manufacture of building materials/equipment	0.673	31	0.793	14	0.647	49	0.533	23	0.435	41	0.533	23
66	G6R4	Delay in procurement of materials	0.680	26	0.760	27	0.604	62	0.517	27	0.410	50	0.517	29
67	G6R5	Damage to material in storage/at site	0.487	80	0.618	76	0.644	50	0.301	78	0.314	73	0.314	77
68	G6R6	Rejecting materials' submittals	0.585	58	0.713	44	0.585	70	0.417	53	0.343	65	0.417	56
		Equipment-Related Risks												
69	G7R1	Inadequate or inefficient equipment, tools, and plants	0.582	60	0.687	53	0.604	62	0.400	57	0.351	61	0.400	60
70	G7R2	Equipment availability and failure	0.578	63	0.676	59	0.593	68	0.391	60	0.343	66	0.391	63
		External Risks												
71	G8R1	Economic instability	0.582	60	0.738	35	0.713	25	0.429	48	0.415	47	0.429	52
72	G8R2	High fluctuation in cost (e.g., money exchange rate; taxes and burdens; and interest rates charged by bankers on loan)	0.589	53	0.662	66	0.731	15	0.390	62	0.431	44	0.431	51
73	G8R3	Inflation (e.g., material, equipment, and labour prices)	0.607	50	0.633	72	0.720	19	0.384	64	0.437	39	0.437	49
74	G8R4	Changes in government regulations and laws (e.g., economy, tax, safety, environment, industrial, recruitment and workers' visas, and localization)	0.604	51	0.687	53	0.720	19	0.415	55	0.435	42	0.435	50
75	G8R5	Delay in connecting services with external parties (e.g., electricity, water, sewage, etc.)	0.695	21	0.771	24	0.625	52	0.535	22	0.434	43	0.535	22
76	G8R6	Delay in recruitment and workers' visa approval	0.589	53	0.705	46	0.596	65	0.416	54	0.351	60	0.416	57
77	G8R7	Corruption (fraudulent practices, kickbacks, and lack of respect for the law)	0.560	67	0.622	75	0.687	35	0.348	70	0.385	52	0.385	66
78	G8R8	Legal disputes between various parties	0.556	68	0.680	57	0.651	48	0.378	65	0.362	58	0.378	67
79	G8R9	Import/Export restrictions	0.516	75	0.662	66	0.618	58	0.342	73	0.319	72	0.342	73
		Force Majeure Risks												
80	G9R1	Earthquakes, fires, and floods	0.356	83	0.702	48	0.702	30	0.250	82	0.250	82	0.250	82
81	G9R2	Severe weather conditions	0.509	76	0.673	61	0.578	72	0.342	72	0.294	77	0.342	72
82	G9R3	Wars in region/Political instability	0.425	82	0.698	49	0.680	39	0.297	79	0.289	78	0.297	80
83	G9R4	Spreading of disease, epidemic or pandemic (e.g., COVID-19)	0.589	53	0.815	9	0.764	5	0.480	36	0.450	36	0.480	40

Table A3. Risk Importance (RI) for risk factors of building construction projects and ranking, according to the perspective of the three groups of respondents.

No	Code	Risk Factors	Client		Consultant		Contractor		RI Overall	
			RI	Rank	*RI*	Rank	*RI*	Rank	*RI*	Rank
		Client-Related Risks								
1	G1R1	Client's lack of experience about the construction project.	0.504	29	0.497	40	0.438	65	0.482	39
2	G1R2	Excessive bureaucracy in owner's administration.	0.538	16	0.529	32	0.581	29	0.545	18
3	G1R3	Client's financial difficulties.	0.489	34	0.566	23	0.713	6	0.544	19
4	G1R4	Owner's delay in making progress payments for completed works (payment delays).	0.632	1	0.680	2	0.809	1	0.672	2
5	G1R5	Selecting consultant, or designer, based on the lowest price.	0.527	17	0.640	9	0.700	7	0.575	14
6	G1R6	Contract awarded to the lowest bidder.	0.593	6	0.676	3	0.743	4	0.631	3
7	G1R7	Unrealistic/Inadequate original contract duration (tight schedule).	0.556	13	0.577	20	0.656	13	0.577	13
8	G1R8	Political pressure to complete the project and speed up construction processes.	0.397	58	0.403	62	0.474	55	0.413	58
9	G1R9	Delay in land acquisition/Handover to the contractor.	0.406	55	0.315	76	0.440	64	0.390	64
10	G1R10	Difficulties in obtaining work permits from the authorities.	0.450	43	0.495	42	0.595	25	0.486	35
11	G1R11	Change orders during construction.	0.591	9	0.657	6	0.675	11	0.627	4
12	G1R12	Scope of work reduction by owner during execution phase.	0.290	77	0.308	79	0.325	78	0.307	78
13	G1R13	Delay penalty clause in the Saudi Public Works Contract is inefficient.	0.508	27	0.308	78	0.281	81	0.396	61
14	G1R14	Inaccurate estimation of construction cost by owner.	0.523	20	0.530	31	0.578	31	0.527	27
15	G1R15	Project suspension by owner.	0.413	52	0.432	54	0.532	39	0.439	47
16	G1R16	Delays due to dispute resolution.	0.435	44	0.646	8	0.619	19	0.531	24
		Designer-Related Risks								
17	G2R1	Unclear and inadequate drawings, specifications, bills of quantities (BOQ)	0.592	7	0.586	16	0.638	15	0.582	11
18	G2R2	Discrepancies between project documents (contract, BOQ, specifications, and drawings).	0.472	38	0.511	37	0.574	33	0.485	36
19	G2R3	Mistakes and deficiencies in design documents.	0.486	35	0.505	38	0.520	42	0.483	38
20	G2R4	Inaccurately estimated quantities (BOQ).	0.524	19	0.505	38	0.638	15	0.519	28
21	G2R5	All existing underground utilities information not available on the design documents (e.g., live cables and pipelines) for the contractor.	0.508	27	0.596	13	0.495	50	0.537	20
22	G2R6	Inadequate geotechnical investigations report about ground conditions of the project site (if available).	0.495	32	0.566	23	0.371	72	0.504	32
23	G2R7	Complexity of design.	0.326	72	0.361	70	0.338	77	0.329	74
24	G2R8	Absence of contractor's involvement in the design stage.	0.362	66	0.260	82	0.316	79	0.322	76
25	G2R9	Speeding up of design phase's schedule.	0.516	23	0.459	48	0.374	71	0.467	43
26	G2R10	Limited budget for design.	0.429	46	0.459	47	0.471	57	0.438	48

Table A3. *Cont.*

No	Code	Risk Factors	Client		Consultant		Contractor		RI Overall	
			RI	Rank	RI	Rank	RI	Rank	RI	Rank
		Consultant-Related Risks								
27	G3R1	Lack of experience and competence of consultant's staff.	0.499	30	0.596	14	0.540	38	0.528	26
28	G3R2	Slowness decision making for approval (shop drawings, submittals, sample materials, change orders, etc.)	0.520	21	0.620	11	0.598	23	0.568	15
29	G3R3	Consultant's rejection of submittals (shop drawings, equipment, and material samples).	0.430	45	0.411	60	0.471	57	0.429	53
30	G3R4	Consultant's delay in performing inspection and testing and giving instructions.	0.378	64	0.404	61	0.453	59	0.394	62
31	G3R5	Poor coordination and communication between consultant and other parties.	0.427	48	0.586	16	0.531	40	0.479	41
32	G3R6	Poor quality control and assurance.	0.464	39	0.490	43	0.446	63	0.455	46
33	G3R7	Consultant's corruption.	0.394	60	0.517	36	0.413	68	0.424	55
34	G3R8	Inflexibility (rigidity) of consultant.	0.359	67	0.381	66	0.516	43	0.389	65
35	G3R9	Excessive safety consideration.	0.227	82	0.227	83	0.273	83	0.226	83
36	G3R10	Internal company problems (at consultant company's head office).	0.257	79	0.449	51	0.344	76	0.323	75
		Contractor-Related Risks								
37	G4R1	Inadequate contractor experience (lack of experience, and managerial skills).	0.569	11	0.666	4	0.574	33	0.595	8
38	G4R2	Contractor's financial difficulties (ineffective cash flow management).	0.627	2	0.798	1	0.763	2	0.692	1
39	G4R3	Contractor's poor site management and supervision.	0.592	7	0.664	5	0.553	36	0.601	7
40	G4R4	Ineffective planning and scheduling of project by contractor.	0.598	5	0.657	7	0.678	9	0.627	4
41	G4R5	Ineffective control of the project progress.	0.556	12	0.552	29	0.453	59	0.529	25
42	G4R6	Delays in sub-contractors work or due to suppliers.	0.570	10	0.562	27	0.675	11	0.588	9
43	G4R7	Delay in preparation of shop drawings and material samples.	0.526	18	0.595	15	0.616	20	0.554	16
44	G4R8	Delay in site mobilization	0.489	33	0.473	45	0.375	70	0.462	44
45	G4R9	Rework and wastage on site, due to errors or quality of work (poor quality of workmanship).	0.456	41	0.521	34	0.591	28	0.498	34
46	G4R10	Variations in quantities.	0.511	25	0.520	35	0.508	45	0.500	33
47	G4R11	Cost of penalties.	0.369	65	0.400	63	0.300	80	0.366	69
48	G4R12	The contractor does not carry out a field visit to the site during the bidding process.	0.499	30	0.465	46	0.488	53	0.474	42
49	G4R13	New existing underground utilities not mentioned on the design documents (e.g., live cables, pipelines).	0.518	22	0.566	23	0.580	30	0.536	21
50	G4R14	Health and Safety requirements (in light of COVID-19).	0.338	70	0.364	69	0.506	46	0.371	68
51	G4R15	Accidents on site.	0.250	81	0.284	80	0.413	67	0.274	81
52	G4R16	Conflict between contractor and consultant.	0.464	40	0.579	18	0.595	25	0.516	30

Table A3. *Cont.*

No	Code	Risk Factors	Client		Consultant		Contractor		RI Overall	
			RI	Rank	RI	Rank	RI	Rank	RI	Rank
53	G4R17	Tender-winning prices are unrealistically low (suicide tendering).	0.481	36	0.557	28	0.595	25	0.514	31
54	G4R18	Unavailability of incentives for contractor for finishing ahead of schedule or to reduce the cost.	0.410	54	0.497	40	0.744	3	0.484	37
		Labour-Related Risks								
55	G5R1	Shortage of manpower (skilled, semi-skilled, and unskilled).	0.616	3	0.612	12	0.678	9	0.608	6
56	G5R2	Unqualified/inexperienced workers.	0.610	4	0.579	19	0.630	17	0.588	9
57	G5R3	Low productivity level of manpower/labourers.	0.553	14	0.629	10	0.616	20	0.581	12
58	G5R4	Low payment for labour force.	0.429	47	0.422	57	0.523	41	0.426	54
59	G5R5	Injuries to labourers on the construction site.	0.292	76	0.309	77	0.359	74	0.303	79
60	G5R6	Delayed salary payments to staff by the contractor.	0.548	15	0.562	26	0.680	8	0.553	17
61	G5R7	High turnover of personnel.	0.419	49	0.387	65	0.489	52	0.405	59
62	G5R8	Labour strikes.	0.320	73	0.397	64	0.474	55	0.353	71
		Materials-Related Risks								
63	G6R1	Shortage of construction materials—special building materials not available in the local market.	0.327	71	0.373	68	0.495	50	0.359	70
64	G6R2	Delay in delivery of materials.	0.454	42	0.436	53	0.578	31	0.459	45
65	G6R3	Delay in the special manufacture of building materials/equipment.	0.512	24	0.568	22	0.598	23	0.533	23
66	G6R4	Delay in procurement of materials.	0.509	26	0.552	29	0.560	35	0.517	29
67	G6R5	Damage to material in storage/at site.	0.339	69	0.315	75	0.281	81	0.314	77
68	G6R6	Rejecting materials' submittals.	0.412	53	0.375	67	0.553	36	0.417	56
		Equipment-Related Risks								
69	G7R1	Inadequate or inefficient equipment, tools, and plants.	0.418	51	0.350	72	0.504	47	0.400	60
70	G7R2	Equipment availability and failure.	0.404	56	0.338	73	0.504	47	0.391	63
		External Risks								
71	G8R1	Economic instability.	0.386	62	0.442	52	0.613	22	0.429	52
72	G8R2	High fluctuation in cost (e.g., money exchange rate; taxes and burdens; and interest rates charged by bankers on loan).	0.378	63	0.422	58	0.630	17	0.431	51
73	G8R3	Inflation (e.g., material, equipment and labour prices).	0.403	57	0.459	49	0.503	49	0.437	49
74	G8R4	Changes in government regulations and laws (e.g., economy, tax, safety, environment, industrial, recruitment and workers' visas, and localisation).	0.418	50	0.450	50	0.453	59	0.435	50
75	G8R5	Delay in connecting services with external parties (e.g., electricity, water, sewage, etc.)	0.473	37	0.570	21	0.656	13	0.535	22

Table A3. *Cont.*

No	Code	Risk Factors	Client		Consultant		Contractor		RI Overall	
			RI	Rank	*RI*	Rank	*RI*	Rank	*RI*	Rank
76	G8R6	Delay in recruitment and workers' visa approval.	0.355	68	0.474	44	0.516	43	0.416	57
77	G8R7	Corruption (fraudulent practices, kickbacks, and lack of respect for law).	0.396	59	0.414	59	0.356	75	0.385	66
78	G8R8	Legal disputes between various parties.	0.313	75	0.432	54	0.488	53	0.378	67
79	G8R9	Import/export restrictions.	0.320	73	0.359	71	0.450	62	0.342	73
		Force Majeure Risks								
80	G9R1	Earthquakes, fires, and floods.	0.225	83	0.277	81	0.360	73	0.250	82
81	G9R2	Severe weather conditions.	0.267	78	0.424	56	0.435	66	0.342	72
82	G9R3	Wars in region/political instability.	0.252	80	0.334	74	0.407	69	0.297	80
83	G9R4	Spreading of disease, epidemic, or pandemic (e.g., COVID-19).	0.386	61	0.523	33	0.736	5	0.480	40

References

1. Luong, D.; Tran, D.; Nguyen, P. Optimizing multi-mode time-cost-quality trade-off of construction project using opposition multiple objective difference evolution. *Int. J. Constr. Manag.* **2018**, *21*, 271–283. [CrossRef]
2. Memon, A.H.; Rahman, I.A.; Abdullah, M.R.; Aziz, A.A.A. Time Overrun in Construction Projects from the Perspective of Project Management Consultant (PMC). *J. Surv. Constr. Prop.* **2011**, *2*, 1–13. [CrossRef]
3. Lotfi, R.; Yadegari, Z.; Hosseini, S.; Khameneh, A.; Tirkolaee, E.; Weber, G. A robust time-cost-quality-energy-environment trade-off with resource-constrained in project management: A case study for a bridge construction project. *J. Ind. Manag. Optim.* **2022**, *18*, 375. [CrossRef]
4. Iqbal, S.; Choudhry, R.; Holschemacher, K.; Ali, A.; Tamošaitienė, J. Risk management in construction projects. *Technol. Econ. Dev. Econ.* **2015**, *21*, 65–78. [CrossRef]
5. Maqsoom, A.; Choudhry, R.; Umer, M.; Mehmood, T. Influencing factors indicating time delay in construction projects: Impact of firm size and experience. *Int. J. Constr. Manag.* **2019**, *21*, 1251–1262. [CrossRef]
6. Yap, J.; Goay, P.; Woon, Y.; Skitmore, M. Revisiting critical delay factors for construction: Analysing projects in Malaysia. *Alex. Eng. J.* **2021**, *60*, 1717–1729. [CrossRef]
7. Assaf, S.A.; Al-Hejji, S. Causes of delay in large construction projects. *Int. J. Proj. Manag.* **2006**, *24*, 349–357. [CrossRef]
8. Kaliba, C.; Muya, M.; Mumba, K. Cost escalation and schedule delays in road construction projects in Zambia. *Int. J. Proj. Manag.* **2009**, *27*, 522–531. [CrossRef]
9. Arditi, D.; Akan, G.T.; Gurdamar, S. Reasons for delays in public projects in Turkey. *Constr. Manag. Econ.* **1985**, *3*, 171–181. [CrossRef]
10. Kraiem, Z.M.; Diekmann, J.E. Concurrent Delays in Construction Projects. *J. Constr. Eng. Manag.* **1987**, *113*, 591–602. [CrossRef]
11. Azhar, N.; Farouqui, R.U. Cost Overrun Factors in Construction Industry of Pakistan. In Proceedings of the First International Conference on Construction in Developing Countries (ICCIDC–I) "Advancing and Integrating Construction Education, Research & Practice", Karachi, Pakistan, 4–5 August 2008.
12. Wang, M.-T.; Chou, H.-Y. Risk Allocation and Risk Handling of Highway Projects in Taiwan. *J. Manag. Eng.* **2003**, *19*, 60–68. [CrossRef]
13. Barrie, D.S.; Paulson, B.C., Jr. *Professional Construction Management: Including Contracting CM, Design-Construct, and General Contracting*; McGraw-Hill: New York, NY, USA, 1992.
14. Allahaim, F.S.; Liu, L. Causes of cost overruns on infrastructure projects in Saudi Arabia. *Int. J. Collab. Enterp.* **2015**, *5*, 32. [CrossRef]
15. Seddeeq, A.; Assaf, S.; Abdallah, A.; Hassanain, M. Time and Cost Overrun in the Saudi Arabian Oil and Gas Construction Industry. *Buildings* **2019**, *9*, 41. [CrossRef]
16. Long, R.J. Typical Problems Leading to Delays, Cost Overruns, and Claims on Process Plant and O Shore Oil and Gas Projects. Available online: https://www.long-intl.com/wp-content/uploads/2014/02/Long-Intl-Typical-Problems-Leading-to-Delays-Cost-Overruns-Claims.pdf (accessed on 9 July 2021).
17. Abdellatif, H.; Alshibani, A. Major Factors Causing Delay in the Delivery of Manufacturing and Building Projects in Saudi Arabia. *Buildings* **2019**, *9*, 93. [CrossRef]
18. Mahamid, I.; Laissy, M. Time Overrun in Construction of Road Projects in Developing Countries: Saudi Arabia as a Case Study. *Int. J. Eng. Inf. Syst. (IJEAIS)* **2019**, *3*, 18–23.
19. Velumani, P.; Nampoothiri, N.; Urbański, M. A Comparative Study of Models for the Construction Duration Prediction in Highway Road Projects of India. *Sustainability* **2021**, *13*, 4552. [CrossRef]
20. Khalifa, W.M.A.; Mahamid, I. Causes of Change Orders in Construction Projects. *Eng. Technol. Appl. Sci. Res.* **2019**, *9*, 4956–4961. [CrossRef]
21. Dixit, S.; Sharma, K.; Singh, S. Identifying and Analysing Key Factors Associated with Risks in Construction Projects. In *Emerging Trends in Civil Engineering*; Springer: Singapore, 2020; pp. 25–32. [CrossRef]
22. Kartam, N.A.; Kartam, S.A. Risk and its management in the Kuwaiti construction industry: A contractors' perspective. *Int. J. Proj. Manag.* **2001**, *19*, 325–335. [CrossRef]
23. Ranns, R.H.B.; Ranns, E.J.M. *Practical Construction Management*, 1st ed.; Routledge: Oxfordshire, UK, 2016; p. 4.
24. Global Construction Outlook to 2023—Q3 2019 Update. Available online: https://www.reportlinker.com/p05582683/Global-Construction-Outlook-to-Q3-Update.html?utm_source=PRN (accessed on 16 December 2021).
25. Future-Ready Index. Global Construction Survey 2019. Available online: https://assets.kpmg/content/dam/kpmg/xx/pdf/2019/04/global-construction-survey-2019.pdf (accessed on 15 December 2021).
26. Global Construction 2030: A Global Forecast for the Construction Industry to 2030. Available online: https://www.pwc.com/vn/en/industries/engineering-and-construction/pwc-global-construction-2030.html (accessed on 12 December 2021).
27. Robinson, G. Global Construction Market to Grow $8 trillion by 2030: Driven by China, US and India. Global Construction Perspectives and Oxford Economics, UK. Available online: https://www.ice.org.uk/ICEDevelopmentWebPortal/media/Documents/News/ICE%20News/Global-Construction-press-release.pdf (accessed on 16 December 2021).
28. Meisels, M. 2018 Global Construction Industry Overview. Global Construction Industry: A Positive Outlook. Deloitte Report (GPOC). Available online: https://www2.deloitte.com/us/en/pages/energy-and-resources/articles/global-construction-industry-overview.html (accessed on 14 December 2021).

29. Impact of COVID-19 on the Construction Sector. 2021. Available online: https://www.ilo.org/wcmsp5/groups/public/---ed_dialogue/---sector/documents/briefingnote/wcms_767303.pdf (accessed on 28 August 2021).

30. COVID-19's Impact on the Engineering & Construction Sector | Deloitte Global. Deloitte. 2021. Available online: https://www2.deloitte.com/global/en/pages/about-deloitte/articles/covid-19/understanding-the-sector-impact-of-covid-19--engineering---const.html (accessed on 28 August 2021).

31. Mordor Intelligence. Saudi Arabia Construction Market—Growth, Trends, and Forecast (2019–2024). Available online: https://www.mordorintelligence.com/industry-reports/saudi-arabia-construction-market-growth-trends-and-forecast-2019-2024 (accessed on 19 December 2021).

32. Davids, G. Potential for Construction Sector in GCC—Linesight. Available online: https://meconstructionnews.com/36485/ksa-holds-greatest-potential-for-construction-sector-in-gcc-linesight (accessed on 19 December 2021).

33. Almutairi, Y.; Arif, M.; Khalfan, M. Moving towards managing offsite construction techniques in the Kingdom of Saudi Arabia: A review. *Middle East J. Manag.* **2016**, *3*, 164. [CrossRef]

34. Algahtany, A.; Alhammadi, Y.; Kashiwagi, D. Introducing a New Risk Management Model to the Saudi Arabian Construction Industry. *Procedia Eng.* **2016**, *145*, 940–947. [CrossRef]

35. Alrashed, I.; Alrashed, A.; Taj, S.A.; Kantamaneni, M.P.K. Risk Assessments for Construction projects in Saudi Arabia. *Res. J. Manag. Sci.* **2014**, *3*, 1–6.

36. Research & Markets. Construction in Saudi Arabia—Key Trends and Opportunities to 2025 (Q2 2021). 2021. Available online: https://www.researchandmarkets.com/reports/5367699/construction-in-saudi-arabia-key-trends-and?utm_source=BW&utm_medium=PressRelease&utm_code=rbmm66&utm_campaign=1566018+-+Saudi+Arabia+Construction+Market+Report+2021%3a+Key+Trends+and+Opportunities+to+2025%2c+Q2+2021+Update&utm_exec=chdo54prd (accessed on 7 September 2021).

37. Puri-Mirza, A. Saudi Arabia: Growth Value of Planned Construction Contract Awards 2022 | Statista. 2021. Available online: https://www.statista.com/statistics/1100182/saudi-arabia-growth-value-of-plannned-construction-contract-awards/ (accessed on 7 September 2021).

38. Puri-Mirza, A. Saudi Arabia: Value of Planned Buildings Contract Awards 2022 | Statista. 2021. Available online: https://www.statista.com/statistics/1100216/saudi-arabia-value-of-planned-buildings-contract-awards/ (accessed on 7 September 2021).

39. MOMRA. Contractors Classification Agency. Available online: https://contractors.momra.gov.sa (accessed on 12 December 2021).

40. Silva, G.A.; Warnakulasooriya, B.N.F.; Arachchige, B. Criteria for Construction Project Success: A Literature Review. In Proceedings of the University of Sri Jayewardenepura, Sri Lanka, 13th International Conference on Business Management (ICBM), Nugegoda, Sri Lanka, 8 December 2016; Available online: https://ssrn.com/abstract=2910305 (accessed on 17 December 2021).

41. Ramlee, N.; Tammy, N.J.; Raja Mohd Noor, R.N.; Ainun Musir, A.; Abdul Karim, N.; Chan, H.B.; Mohd Nasir, S.R. Critical success factors for construction project. *AIP Conf. Proc.* **2016**, *1774*, 030011.

42. Alzara, M.; Kashiwagi, J.; Kashiwagi, D.; Al-Tassan, A. Using PIPS to Minimize Causes of Delay in Saudi Arabian Construction Projects: University Case Study. *Procedia Eng.* **2016**, *145*, 932–939. [CrossRef]

43. Baghdadi, A.; Kishk, M. Saudi Arabian Aviation Construction Projects: Identification of Risks and Their Consequences. *Procedia Eng.* **2015**, *123*, 32–40. [CrossRef]

44. Mahamid, I. Schedule Delay in Saudi Arabia Road Construction Projects: Size, Estimate, Determinants and Effects. *Int. J. Arch. Eng. Constr.* **2017**, *6*, 51–58. [CrossRef]

45. Sha'Ar, K.; Assaf, S.; Bambang, T.; Babsail, M.; El Fattah, A.A. Design—Construction interface problems in large building construction projects. *Int. J. Constr. Manag.* **2016**, *17*, 238–250. [CrossRef]

46. Al-Emad, N.; Rahman, I.A.; Nagapan, S.; Gamil, Y. Ranking of Delay Factors for Makkah's Construction Industry. *MATEC Web Conf.* **2017**, *103*, 3001. [CrossRef]

47. Arditi, D.; Nayak, S.; Damci, A. Effect of organizational culture on delay in construction. *Int. J. Proj. Manag.* **2017**, *35*, 136–147. [CrossRef]

48. Elawi, G.S.A.; Algahtany, M.; Kashiwagi, D. Owners' Perspective of Factors Contributing to Project Delay: Case Studies of Road and Bridge Projects in Saudi Arabia. *Procedia Eng.* **2016**, *145*, 1402–1409. [CrossRef]

49. Sánchez, O.; Castañeda, K.; Mejía, G.; Pellicer, E. Delay Factors: A Comparative Analysis between Road Infrastructure and Building Projects. In Proceedings of the Construction Research Congress, Tempe, AZ, USA, 8–10 March 2020; pp. 223–231. [CrossRef]

50. Sanni-Anibire, M.O.; Zin, R.M.; Olatunji, S.O. Causes of Delay in Tall Building Projects in GCC Countries. In Proceedings of the 8th International Conference on Construction Engineering and Project Management, Hong Kong, China, 7–8 December 2020.

51. Mohammed Gopang, R.K.; Alias Imran, Q.B.; Nagapan, S. Assessment of Delay Factors in Saudi Arabia Railway/Metro Construction Projects. *Int. J. Sustain. Constr. Eng. Technol.* **2020**, *11*, 225–233. [CrossRef]

52. Aljohani, A. Construction Projects Cost Overrun: What Does the Literature Tell Us? *Int. J. Innov. Manag. Technol.* **2017**, *8*, 137–143. [CrossRef]

53. Ahady, S.; Gupta, S.; Malik, R.K. A critical review of the causes of cost overrun in construction industries of developing countries. *Int. Res. J. Eng. Technol.* **2017**, *4*, 2550–2558.

54. Bahamid, R.A.; Doh, S.I.; Al-Sharaf, M.A. Risk factors affecting the construction projects in the developing countries. *IOP Conf. Ser. Earth Environ. Sci.* **2019**, *244*, 012040. [CrossRef]

55. Baloyi, L.; Bekker, M. Causes of construction cost and time overruns: The 2010 FIFA World Cup stadia in South Africa. *Acta Structilia J. Phys. Dev. Sci.* **2011**, *18*, 51–67.

56. Cirovic, G.; Sudjic, S. Risk Assessment in Construction Industry. Available online: https://www.irbnet.de/daten/iconda/CIB158 54.pdf (accessed on 11 December 2019).

57. Creedy, G.D.; Skitmore, M.; Wong, J.K.W. Evaluation of risk factors leading to cost overrun in delivery of highway construction projects. *J. Constr. Eng. Manag.* **2010**, *136*, 528–537. [CrossRef]

58. Jackson, S. Project cost overruns and risk management. In Proceedings of the Association of Researchers in Construction Management 18th Annual ARCOM Conference, Newcastle, UK, 2–4 September 2002; Volume 1, pp. 99–108.

59. Liu, J.; Zhao, X.; Yan, P. Risk Paths in International Construction Projects: Case Study from Chinese Contractors. *J. Constr. Eng. Manag.* **2016**, *142*, 05016002. [CrossRef]

60. Mahamid, I.; Al-Ghonamy, A.; Aichouni, M. Risk Matrix for Delay Causes in Construction Projects in Saudi Arabia. *Res. J. Appl. Sci. Eng. Technol.* **2015**, *9*, 665–670. [CrossRef]

61. Memon, A.H.; Rahman, I.A.; Azis, A.A.A. Time and Cost Performance in Construction Projects in Southern and Central Regions of Peninsular Malaysia. *Int. J. Adv. Appl. Sci.* **2012**, *1*, 45–52. [CrossRef]

62. Sharma, S.; Goyal, P.K. Fuzzy assessment of the risk factors causing cost overrun in construction industry. *Evol. Intell.* **2019**, 1–13. [CrossRef]

63. Subramani, T.; Sruthi, P.; Kavitha, M. Causes of Cost Overrun in Construction. *IOSR J. Eng.* **2014**, *3*, 1–7. [CrossRef]

64. Zewdu, Z.T.; Aregaw, G.T. Causes of Contractor Cost Overrun in Construction Projects: The Case of Ethiopian Construction Sector. *Int. J. Bus. Econ. Res.* **2015**, *4*, 180. [CrossRef]

65. Zhao, X.; Hwang, B.G.; Gao, Y. A fuzzy synthetic evaluation approach for risk assessment: A case of Singapore's green projects. *J. Clean. Prod.* **2016**, *115*, 203–213. [CrossRef]

66. Saunders, M.; Lewis, P.; Thornhill, A. *Research Methods for Business Students*, 8th ed.; Pearson: New York, NY, USA, 2019.

67. Taber, K.S. The use of Cronbach's alpha when developing and reporting research instruments in science education. *Res. Sci. Educ.* **2018**, *48*, 1273–1296. [CrossRef]

68. Alsuliman, J.A. Causes of delay in Saudi public construction projects. *Alex. Eng. J.* **2019**, *58*, 801–808. [CrossRef]

69. Ikediashi, D.I.; Ogunlana, S.O.; Alotaibi, A. Analysis of project failure factors for infrastructure projects in Saudi Arabia: A multivariate approach. *J. Constr. Dev. Ctries.* **2014**, *19*, 35.

70. Al-Khalil, M.I.; Al-Ghafly, M.A. Important causes of delay in public utility projects in Saudi Arabia. *Constr. Manag. Econ.* **1999**, *17*, 647–655. [CrossRef]

71. Shash, A.A.; Qarra, A.A. Cash flow management of construction projects in Saudi Arabia. *Proj. Manag. J.* **2018**, *49*, 48–63. [CrossRef]

72. Le-Hoai, L.; Dai Lee, Y.; Lee, J.Y. Delay and cost overruns in Vietnam large construction projects: A comparison with other selected countries. *KSCE J. Civ. Eng.* **2008**, *12*, 367–377. [CrossRef]

73. Abd El-Razek, M.E.; Bassioni, H.A.; Mobarak, A.M. Causes of delay in building construction projects in Egypt. *J. Constr. Eng. Manag.* **2008**, *134*, 831–841. [CrossRef]

74. Albogamy, A.; Scott, D.; Dawood, N. Addressing construction delays in the Kingdom of Saudi Arabia. *Int. Proc. Econ. Dev. Res.* **2012**, *45*, 148–153.

75. Koushki, P.A.; Al-Rashid, K.; Kartam, N. Delays and cost increases in the construction of private residential projects in Kuwait. *Constr. Manag. Econ.* **2005**, *23*, 285–294. [CrossRef]

76. Assaf, S.A.; Al-Khalil, M.; Al-Hazmi, M. Causes of Delay in Large Building Construction Projects. *J. Manag. Eng.* **1995**, *11*, 45–50. [CrossRef]

77. Hammadi, S.; Nawab, M. Study of Delay Factors in Construction Projects. *Int. Adv. Res. J. Sci. Eng. Technol.* **2016**, *3*, 87–93. [CrossRef]

78. Arain, F.M.; Pheng, L.S.; Assaf, S.A. Contractors' Views of the Potential Causes of Inconsistencies between Design and Construction in Saudi Arabia. *J. Perform. Constr. Facil.* **2006**, *20*, 74–83. [CrossRef]

79. Babalola, O.; Ibem, E.; Ezema, I. Implementation of lean practices in the construction industry: A systematic review. *Build. Environ.* **2019**, *148*, 34–43. [CrossRef]

80. Bajjou, M.S.; Chafi, A. Lean construction implementation in the Moroccan construction industry: Awareness, benefits and barriers. *J. Eng. Des. Technol.* **2018**, *16*, 533–556. [CrossRef]

81. Sarhan, J.; Xia, B.; Fawzia, S.; Karim, A.; Olanipekun, A. Barriers to implementing lean construction practices in the Kingdom of Saudi Arabia (KSA) construction industry. *Constr. Innov.* **2018**, *18*, 246–272. [CrossRef]

82. Jagtap, M.; Kamble, S. The effect of the client–contractor relationship on project performance. *Int. J. Product. Perform. Manag.* **2019**, *69*, 541–558. [CrossRef]

83. Aliakbarlou, S.; Wilkinson, S.; Costello, S. Rethinking client value within construction contracting services. *Int. J. Manag. Proj. Bus.* **2018**, *11*, 1007–1025. [CrossRef]

Systematic Review

Predictive Analytics for Early-Stage Construction Costs Estimation

Sergio Lautaro Castro Miranda [1], Enrique Del Rey Castillo [1,*], Vicente Gonzalez [2] and Johnson Adafin [3]

[1] Department of Civil and Environmental Engineering, University of Auckland, Auckland 1010, New Zealand; scas669@aucklanduni.ac.nz

[2] Construction Engineering and Management, Faculty of Engineering—Civil and Environmental Engineering Department, University of Alberta, Edmonton, AB T6G 2R3, Canada; vagonzal@ualberta.ca

[3] Department of Quantity Surveying and Construction Management, Northland Polytechnic, Auckland 1010, New Zealand; jadafin@northtec.ac.nz

* Correspondence: e.delrey@auckland.ac.nz

Abstract: Low accuracy in the estimation of construction costs at early stages of projects has driven the research on alternative costing methods that take advantage of computing advances, however, direct implications in their use for practice is not clear. The purpose of this study was to investigate how predictive analytics could enhance cost estimation of buildings at early stages by performing a systematic literature review on predictive analytics implementations for the early-stage cost estimation of building projects. The outputs of the study are: (1) an extensive database; (2) a list of cost drivers; and (3) a comparison between the various techniques. The findings suggest that predictive analytic techniques are appropriate for practice due to their higher level of accuracy. The discussion has three main implications: (a) predictive analytics for cost estimation have not followed the best practices and standard methodologies; (b) predictive analytics techniques are ready for industry adoption; and (c) the study can be a reference for high-level decision-makers to implement predictive analytics in cost estimation. Knowledge of predictive analytics could assist stakeholders in playing a key role in improving the accuracy of cost forecast in the construction market, thus, enabling pro-active management of the project owner's budget.

Keywords: buildings; cost estimation; predictive analytics; systematic literature review

Citation: Castro Miranda, S.L.; Del Rey Castillo, E.; Gonzalez, V.; Adafin, J. Predictive Analytics for Early-Stage Construction Costs Estimation. *Buildings* **2022**, *12*, 1043. https://doi.org/10.3390/buildings12071043

Academic Editor: Osama Abudayyeh

Received: 8 May 2022
Accepted: 15 July 2022
Published: 19 July 2022

Publisher's Note: MDPI stays neutral with regard to jurisdictional claims in published maps and institutional affiliations.

1. Introduction

Cost management and knowing whether a final account is on budget or not is critical to measure a project's success [1]. As an example, the Project Management Institute [2] highlights the importance of monitoring and controlling costs using estimates as baselines to achieve budgeting goals. Cost estimation is the process of producing cost estimates by quantifying and valuing the necessary resources to develop a project [3]. The process is iterative in the sense that estimates are updated according to the level of information that becomes available during the inception and design stages, which is fundamental for the decision-making process. The estimation of costs enables the determining of the project's economic feasibility and the evaluation of alternatives, moreover, it can be a driver for the scope given the greater influence project owners have in the initial stages [2].

The most commonly used method to estimate costs in the early stages of building projects is the superficial area method [4]. This method, also called floor area method, consists of multiplying the total gross internal floor area (GIFA) by an appropriate cost/m^2, based on historical data [5]. This traditional method provides low accuracy ranging between -15% to $+25\%$ [6,7]. Increasing the accuracy and reliability of cost estimates is of utmost importance for the decision-maker's ability to optimally assess alternatives and improve investment decisions early on in projects.

Predictive analytics is a term that has been used since 2006 to find and exploit relationships in data [8]. Some methods, such as regression analysis, have been used in statistics for 200 years, starting with the early Legendre and Gauss Least Squares Method, used to determine orbits about the sun from astronomical observations [9]. Other more recent techniques, including Artificial Neural Networks (ANN), Decision Trees (DT), and Case-Based Reasoning (CBR), have evolved with the increase in computation capabilities and the growing volume of data stored [10]. Predictive analytics has been classified as a subset of data science [11], with the aim being to elaborate empirical predictions [12]. Predictive analytics started being applied in credit scoring in the decade beginning in 1950 and has increased its presence and benefits in the areas of fraud detection, healthcare, marketing, insurance, and retail [13,14].

In the process of creating predictive models, the initial stages consider the collection and preparation of observational data related to the desired phenomenon to forecast. The amount of data is critical to achieving higher accuracy in the results [12,15]. Given the data-intensive nature of predictive analytics, two characteristics of construction information can make predictive analytics suitable for cost estimation. First, construction projects consume a large amount of information in the form of drawings, schedules, contract documents, and specifications [10]. Secondly, project data, including cost, are becoming highly structured with the aim of 5D building information modelling, which provides quantities in real time from the information linked to virtual models [16]. The potential of predictive analytics in the construction industry has been widely supported by the research developed since 2000 [15,17].

A review of 27 studies on the use of artificial intelligence to construction-cost estimation has revealed three main drawbacks in the research area: (1) the need to consider more modeling parameters; (2) the need for standard validation methods to estimate the accuracy of models; and (3) ambiguity and opacity of the experimental results [17]. In a later review, the modeling process sorted by technique was identified by analysing more than 100 publications related to artificial intelligence and parametric estimation for construction cost [15]. Elfaki [17] and Elmousalami [15] focused on providing guidelines to improve the experimentation and the modelling process from a research perspective. Yet, explicit benefits and implications for practice, such as the accuracy levels, have not been addressed. Predictive analytics has tremendous potential to benefit construction projects, but the industry has not widely adopted this new technology [10].

In this paper, a systematic literature review based on the approach suggested by Kitchenham and Charters [18] was conducted to explore the applications of predictive analytic techniques on the early-stage cost estimation of building projects. This review aimed to investigate how predictive analytics can enhance the practice by: (1) exploring the model's input determination; (2) identifying the techniques used and accuracy of models; and (3) examining the direct benefits and challenges identified by the authors. The structure of the paper follows with a background of cost estimation and predictive analytics. Next, Section 3 reports the methodology, then the results and discussion are presented in Section 4. Finally, the conclusion is provided in Section 5.

2. Background

2.1. Cost Estimation

Industry organisations, such as the Royal Institute of Chartered Surveyors (RICS) in the UK and the Association for the Advancement of Cost Engineering (AACE) in the USA, have promoted the development of cost estimation, leading the engineering practice into the standardisation of cost-information management. The guides developed by the Royal Institution of Chartered Surveyors [5] have provided significant advances and contain sets of rules to estimate construction projects' costs. Researchers have also contributed to the knowledge domain by providing crucial educational training material on cost estimation, presenting it as a control measure for all the stages of construction projects [3,4,19,20].

Nevertheless, the need remains for improvements in the understanding of the key factors of construction costs and their estimates accuracy [4].

Researchers have encouraged paradigm shifts in the construction industry, especially in the area of cost estimation [21]. Brandon [22] stressed the importance of putting under scrutiny the philosophy of estimation, proposing that the advance in computer hardware and utilisation of large databases would provide means to reduce the limitations of human abilities and move into simulations to model the reality. In the same line, understanding of the construction activity through principles found in the Japanese industrial production has intensified the research within the construction industry [23,24]. The need for innovation towards lean construction has led to different proposals to manage costs in construction projects, such as Activity Based Costing (ABC) [25] or Target Costing [26]. Despite these promising advances, the traditional philosophy to estimate costs remains broadly utilised in practice.

The main objective of cost-estimation practice, since its establishment within the discipline of quantity survey in the decade beginning in 1950, has been to provide a basis to control project costs with the elaboration of cost estimates [4]. Framed within the knowledge area of cost management, different cost estimates provide the necessary information for the decision-making process in the development of projects [2]. With the same perspective, [19] argues that the Royal Institute of British Architects' (RIBA) Plan of Work (PoW) is conceived as an organised procedure for taking design decisions, with accompanying data to be included at various stages of the design evolution. And RICS New Rules of Measurement NRM 1 [5] identified the RIBA Plan of Work as a construction-industry-recognised model that organises the processes of designing and administering/managing building projects.

Given the nature of the link between cost estimations and the evolution of the projects' designs, the techniques used to estimate costs will depend on the objective of the stage at which the project is in and the level of information available. In the inception stage, when the information about the project is limited and the main goal is to determine feasibility and viability of projects, cost estimates provide the information for investment decisions and a cost reference for the initiation of the design stage. In this early stage, preliminary cost estimates, also called Order of Magnitude estimates or Rough Cost estimates, use the statistical square area (superficial) method, also called floor-area method [2,4,5]. The superficial method relies on statistical data from previous building projects that are adjusted according to the location and year of construction, and it is widely used due to its simplicity, quick calculation because most published cost data are expressed in this form (square area), and is easily understood by the architect/designers and client. Alternative methods, such as cube and storey enclosure methods, are available in the early stages, but they have not been widely adopted in the construction industry as they involve more rigorous calculations than any of the previous methods and historical rates for use are not usually published.

In the design stage, the objective is to create a building design within the scope defined by the owner's requirements and within the cost target defined in the earlier stages. This objective makes cost estimation a tool of control for the design in terms of cost. The estimate is called cost plan in the stage of design, and it evolves with the increasing level of detail in the design. This cost plan follows an analogous approach in which unitary costs from historical databases are assigned to the different project elements that are aggregated according to the total quantities and then adjusted using location and time indexes [4]. The subdivision of the buildings in elemental constituent parts, such as substructure, frame, upper floors, and roof, follow standard guidelines [5].

Contractors estimate costs in the tendering stage with the objective of elaborating budgets and controlling later expenses. Since the design is usually completed in the tender stage, it includes the details of the project, and, contrarily to the early stage Rough Cost estimate, the detailed cost-estimation process follows a bottom–up approach, in which the cost is estimated based on complete design documentation and by work packages associated with the work breakdown structure considering the necessary resources, e.g., labour, equipment, materials, and subcontractors [2].

Further, the RICS [5] illustrates the key components of a cost estimate. The base cost estimate is the total estimated cost of the building works, the main contractor's preliminaries, and the main contractor's margin (profit and overheads). Therefore, the base cost estimate contains no allowances for risk or inflation (that is, the risk-free estimate). Also, allowances for risk and inflation (i.e., fluctuations allowance in the basic prices of materials, labour, and plant during the period from the date of tender return to the mid-point of the construction period) are to be calculated separately and added to the base cost estimate to determine the client's cost limit for the building project. In comparison with the foregoing submission, Smith and Jaggar [27] categorised contingency factors, including the risks involved during design development stages, as:

- Planning contingency (e.g., planning restrictions, legal requirements, environmental concerns, and statutory constraints);
- Design contingency (e.g., inadequate brief, aesthetics and space concerns, changes in estimating data, incomplete drawings, and little or no information about M&E services).

In an attempt to address uncertainty in cost estimation, risk management recognises that factors may affect the design phase of the development process, and the traditional way of dealing with them is to make a percentage contingency allowance. For example, the RICS [5] identified contingency provision as a key element that could be incorporated into a cost estimate. These contingencies are to provide for risks associated with design development, construction, employer-driven changes, and other employer-restrictive concerns.

In the early stages of projects, accuracy remains a challenge [6]. The accuracy of final estimates falls within the range of $\pm5\%$ as the project approaches the tendering process [7]. Despite the critical importance of the early stages mentioned in the previous paragraphs and the low accuracy of traditional methods, alternatives supported by computational advances have not been widely adopted in the construction industry [4].

2.2. Predictive Analytics

The concept of predictive analytics can be understood as the systematic analysis of data to elaborate models for prediction using computational techniques. Predictive analytics has been used since the decade of the 1950s [28]. Shmueli [29] stated that predictive modelling aims to predict future observations as a process using data-mining algorithms or statistical models to data. Predictive analytics techniques have been applied successfully in different areas, such as marketing and finance [30], to prevent bank fraud, according to Boyacioglu [31], and in medical areas, for the prediction of diseases, such as diabetes [32]. The increasing capacity of data transmission, the increasing amount of data stored by organisations, and the higher processing capacities have boosted the use of predictive analytics in industry [33]. Despite these advances, the uptake in the construction industry is behind compared to other industries, such as financial services, transportation and logistics, and energy and resources [10,34].

A complete process of constructing predictive models consists of the steps shown in Figure 1, where the initial consideration in the modelling process is the appropriate identification of the main model's objective from a predictive perspective, followed by the data collection and study design. Large-size data and data of an observational nature within the same population are considered optimal for higher accuracies. The data-preparation step has two main issues. Missing information can be helpful if the data is informative enough of the output, but, if not, these data need to be handled by removing observations or parameters by utilising dummy variables or developing different models according to the missing data distribution [29]. The second issue relates to data partitioning for testing purposes. The data set should be randomly partitioned into two parts, one for training the model and the other one to evaluate the predictive performance of the final model.

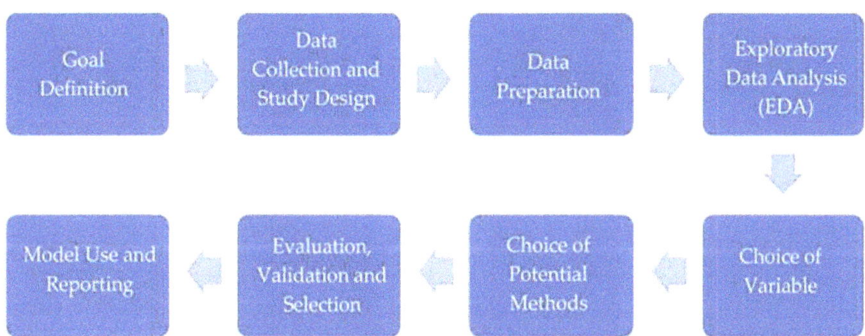

Figure 1. Empirical model-building steps schematic. Adapted from Shmueli and Koppius [12].

The Exploratory Data Analysis (EDA) follows the data-preparation step and is used informally in predictive analytics to synthesise the data graphically and numerically to capture unknown or not formulated relationships [12]. Additionally, EDA is used to reduce the dimensionality of the data by reducing the number of parameters and to reduce the sample variance. Some methods, such as Principal Component Analysis (PCA) and Factor Analysis, can be used to assess relations between parameters of potential models. Variable inputs or parameters are chosen considering the relation between input and output, the data quality, and the availability of the parameters at the moment of prediction. Although the accuracy of the models mainly influences the model's choice, techniques with higher accuracy sacrifice interpretability and objectivity of models. The many available techniques used in predictive analytics can be classified as linear and nonlinear models. Linear and logistic regressions are the most common techniques used for data modelling. Although, with higher chances of overfitting models, techniques such as Decision Trees, Artificial Neural Networks, Support Vector Machine (SVM), and Fuzzy Logic Systems (FLS) have the capacity of modelling nonlinear relationships [30]. Case-Based Reasoning (CBR) is also a common technique studied to elaborate predictive models.

The evaluation and validation are the main criteria for assessing the predictive power of a model [12]. The model selection aims at identifying the appropriate level of complexity leveraging bias and variance for higher accuracy. Model evaluation is conducted by assessing the accuracy of the models using out-of-sample data. The use of statistical significance variables such as R-squared are considered a minor role, while generic predictive measures on observational data such as Root Mean Square Error (RMSE) and Mean Absolute Percentage Error (MAPE) are more typical metrics of accuracy. The selection of out-of-sample data depends on the method of validation used for the model's evaluation. The two methods, hold-out cross-validation and k-fold cross-validation, are standard for validation of models [35]. The hold-out cross-validation method is the most straightforward approach and involves splitting the data into a training dataset and a testing dataset. In the second method, k-fold cross-validation, the same data is used to train and test several models. The data selected for testing and training purposes are different on each train session, but the average of the test results should provide better estimates than individual test results [35].The extreme case is when the number of subsets is the total number of data points, and it is called Leave One Out Cross Validation (LOOCV). Validation methods also help to overcome the challenge of model overfitting, which occurs when a model fits the data for training to the extreme of not being able to predict new data [12]. The model use and reporting stage relate closely to the predictions and the performance measures where results need to be translated into new knowledge following the initial objectives.

The following section describes the research method followed in this paper to investigate how predictive analytics can enhance the practice of cost estimation.

3. Methodology

Systematic literature reviews can support the development of a new knowledge base for practitioners and managers to provide collective insights [36]. According to Borrego [37], these rigorous reviews have become a significant source of evidence in medical research and are gaining importance in areas such as psychology and education. On the other hand, Denyer and Tranfield [38] highlighted the potential of systematic literature reviews as an evidence-based approach for management research. According to Pan [39], the two guidelines have become well-known guidelines for systematic reviews, the Preferred Reporting Items for Systematic Reviews and Meta-Analyses (PRISMA) and Kitchenham guide [18,40]. Although the PRISMA has been designed primarily for studies that evaluate the effects of health interventions, Page [40] argues that its check lists items are applicable to other areas and it has been adopted for global standards when conducting systematic literature reviews. However, Denyer and Tranfield [38] exposed that fit-for-purpose methodologies should be developed according to the unique characteristics of the study's design. The present review focused on implementing predictive analytics techniques, which have evolved in the area of informatics requiring intensive use of computation applications. Since the guidelines by Kitchenham and S. Charters [18] for systematic literature reviews have been adapted from the medical and psychology, and according to Ayodele [41], implemented in computer science, the study has followed such guidelines considering them appropriate to address the research objective. A step-by-step description of the methodology is illustrated in Figure 2. Overall, the review process consisted of three main stages—planning, conducting, and reporting the review.

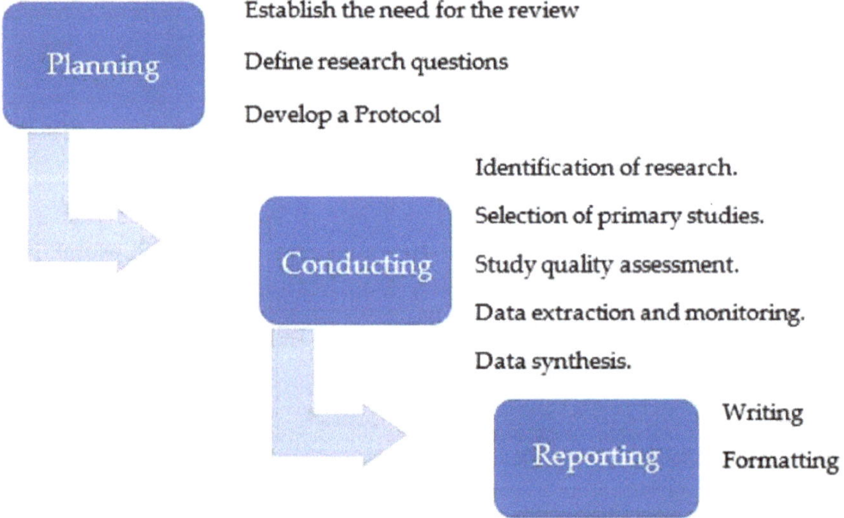

Figure 2. Methodology.

The planning stage was the most crucial part of the review because it provided a guide for the activities necessary to address the research objective. Accordingly, the first step in this stage was to identify the need for the review. For this purpose, a scoping review was conducted in the area of estimation, focusing on their challenges and future trends. A further review of cost modelling techniques allowed to establish the need to aggregate the individual results of the studies and transform them into recommendations for its uptake. In the second step, the consequent objective of investigating how predictive analytics can enhance cost estimation was divided into three questions:

Q1. How does predictive analytics determine the input parameters of models, and what are the parameters commonly used?

Q2. What is the predictive power of the predictive analytics techniques to forecast the construction cost in the early stages of building projects, and what are the most explored techniques?

Q3. What are the benefits and challenges in the use of predictive analytics techniques in cost estimation?

Following the suggestions on Kitchenham and Charters [18], the third step was to create a protocol for the inclusion of the fundamental procedures for the conduction of the review. This formal document is essential in systematic literature reviews because it is a plan helping to maintain objectivity in the research [36].

The second stage, conducting the review started with the identification of research. The database search engine selected was Scopus and the target material for the review was published applications of predictive analytics for estimating the costs of building construction projects in the early stages. The search syntax was TITLE ((cost OR costs) AND (estimation OR prediction OR modeling OR modelling OR model OR estimate) AND (buildings OR construction OR projects)) and it returned 1586 documents.

Aiming at finding resources to answer the research questions, the selection of primary studies was done based on the inclusion criteria which also considered as excluded from the review any study not fulfilling all the indicators. The following list contains the criteria used to include and exclude literature:

1. Only literature published between 1974 and May 2022;
2. Only studies from journals and conferences written in English;
3. Only studies focusing on early-stage cost estimation;
4. Only studies implementing predictive analytic models to estimate cost;
5. Only focusing on building projects;
6. Only studies using percentage error as accuracy measure of the final cost;
7. Only studies providing the accuracy results and parameters used; and
8. Only studies using real data of buildings.

The selection of primary studies was conducted in two phases, first, by analysing the titles and abstracts and, then, a second selection was made by fully reviewing the studies. In the first filter, candidates were excluded when their characteristics were clearly against the selection criteria. In the second filter, a study was selected only when it fulfilled all the selection criteria. The preselection narrowed the list of papers from 1586 down to 127, and then, the full review allowed to identify 30 papers. A backward and forward snowballing process was performed on the 30 articles following the previous approach and following the suggestions provided by Wohlin [42]. With this process 16 additional studies were identified, finalising with 46 papers in total.

Quality assessment of studies using a variety of empirical methods remains a major problem [43]. In order to control the quality of the studies in the review, the presence of their publication venues in the Scimago H index and Google h5 index, together with the number of citations on Google Scholar were part of a quality-monitoring process.

In data extraction and monitoring the necessary information from the articles was imported from the Scopus search list in an XML format extraction and stored in an Excel sheet. This information consisted of title, authors, year of publication, venue, and number of citations until May 2022. In addition to the bibliographical data, the following content data items were sought to answer the research questions.

- Venue type;
- Venue name;
- Country of study;
- Publication date;
- Number of citations;
- Scimago H index;
- Google h5 index;

- Type of buildings;
- Data source;
- Sample size of data set;
- Number of parameters used in the models;
- Mean absolute percentage error;
- Parameter identification method;
- Method to optimise parameters;
- Rankings of parameters;
- Type of technique;
- Sub technique compared;
- Component of the model improved;
- Techniques compared;
- Type of validation;
- Sample size;
- Benefits; and
- Challenges.

Systematic literature reviews typically use meta-analysis to combine and assess quantitative experimental results [44], but the present study used a statistical descriptive and content analysis approach. The bibliographic information was first analysed to have an overview of the publications and to understand the context of the research area. The compilation was synthesised into the items, date of publication, number of publications distributed in time, and origin country of the study.

The synthesis of the data to answer the research question one provided the number of techniques used in the process of selecting the initial parameters of the models and the parameters most used. To determine the parameters, the ranked lists of parameters provided in the studies were aggregated by the Borda–Kendall technique. This method was selected because its use has been widely implemented for rank aggregation and the derived techniques are intuitive and easy to understand [45–47].

The techniques implemented in the studies and the accuracy of the models were collected to answer the second research question. The numbers of techniques most explored were grouped as percentages. The accuracy of the models was summarised in averages and distributed in quartiles, while the second component of predictive power, validation methods, were grouped by type.

In answering research question three, benefits and drawbacks of the utilisation of predictive analytics techniques in cost modelling were compiled using reciprocal translation, which allowed integrating different terms describing the same meaning [18]. The ideas were extracted only from the discussion and conclusion sections to ensure they were derived from the experimentation. These were tabulated and ranked according to the number of authors mentioning them. The last stage of systematic literature reviews is the report. For this purpose, the report followed the protocol structure since it contains the fundamental elements of the review.

4. Results and Discussion

This section presents a synthesis and discussion of the data extracted from the 46 studies selected in the systematic literature review. The first subsection provides an overview of the bibliographical features of the publications, followed by a discussion of the input parameters, the predictive power, the techniques used, and the benefits and challenges of predictive analytics techniques implemented in the studies.

4.1. Studies Description

From the 46 selected studies five were from conference papers and 41 from journals. The largest number of publications corresponded by far to the *Journal of Construction Engineering and Management* with 11 studies (24% of the total). The studies dated from 1974 to 2022, but only two of them were published before 2000, Elhag and Boussabaine [48] and Karshenas [49]. These papers have seminal material in the area of cost modelling of building projects. As can be seen in Figure 3, the number of publications in the research area increased from 2000 and until 2014–2015, presenting a spike in 2004–2005. From 2014–2015 until 2018–2019 the research activity decreased, and in the last period of 2020–2022 the publications increased. The reduction of publications suggested that the research area may have reached a maturity level, where a next stage in the research area may be appropriate to be explored. The graph of the same figure presents Korea as the most prolific country after the United States with 17 and five studies, respectively. The Korean presence in the research area can be explained by the dedication of researchers, such as Gwang-Hee Kim and Sae-Hyun Ji, who together are authors of 13 of the 17 studies.

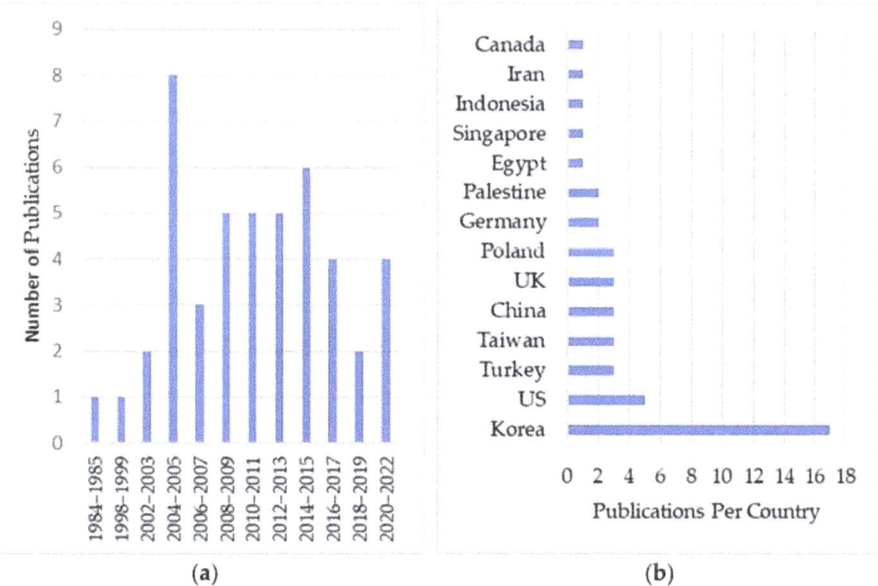

Figure 3. Statistical properties of the publications: (**a**) biannual distribution of publications of the review; and (**b**) distribution of publications per country.

The top 10 most cited documents in Google Scholar are shown below in Table 1. Kim et al. [50] present the highest number of citations, 617, and was the first publication comparing the most promising techniques for cost estimation, Multiple Regression Analysis (MRA), Artificial Neural Networks (ANN), and Case-Based Reasoning (CBR). In this study the high accuracy achieved by the three techniques, and, particularly, the transparency of CBR in explaining the results, suggest predictive analytics techniques can be a feasible alternative to traditional cost estimation in the early stages of projects. Kim et al. [50] and the rest of the top 10 publications, having over 100 citations each, have become a reference in the research area of cost modelling not only for building projects but for general construction projects.

4.2. Models Input Parameters

Even though the performance of cost models heavily relies on the appropriate identification of the cost drivers, the data available is the fundamental input to elaborate the models. This section starts presenting the relevant features of the data used in the studies, such as data source, type of buildings, and quantity of data. Next, two approaches used to identify and select the parameters from the data are presented. Then, the most predominant parameters used in the studies are shown in the form of an aggregated ranking.

Table 1. Most cited papers.

No.	Authors/Year	Title	Country	Citations
1	Kim et al. [50]	Comparison of construction-cost-estimating models based on regression analysis, neural networks, and case-based reasoning.	Korea	617
2	Günaydin and ΞDoğan [51]	A neural network approach for early cost estimation of structural systems of buildings.	Turkey	314
3	Lowe et al. [52]	Predicting construction cost using multiple regression techniques.	UK	320
4	An et al. [53]	A case-based reasoning cost-estimating model using experience by analytic hierarchy process.	Korea	248
5	Emsley et al. [54]	Data modelling and the application of a neural network approach to the prediction of total construction costs.	UK	192
6	Sonmez [55]	Conceptual cost estimation of building projects with regression analysis and neural networks.	US	176
7	Cheng et al. [56]	Conceptual cost estimates using evolutionary fuzzy hybrid neural network for projects in the construction industry.	Taiwan	176
8	Kim et al. [57]	Neural network model incorporating a genetic algorithm in estimating construction costs.	Korea	173
9	Chan and Park [58]	Project cost estimation using principal component regression.	Singapore	147
10	Doğan et al. [59]	Determining attribute weights in a case-based reasoning model for early cost prediction of structural systems.	Turkey	139

4.2.1. Data Utilised in the Studies

In predictive analytics, the data used for modelling should, ideally, be extracted from a population of similar characteristics to achieve more accurate predictions (Shmueli and Koppius [12]. In this sense, prediction accuracy is strongly linked to the data characteristics. The general type of buildings identified in the systematic literature review was multistorey, and subclassifications were identified according to their use, e.g., residential, schools, office use, or mixed. Also, seven studies specified the structure type of the building used. The source of data was also not uniform. Twenty-three studies expressed that its data origin were general contractors, public databases, theses, and other public and private organsations. General contractors and databases were the most commonly used data sources, and 22 did not provide details about the source of data. Transparency in this regard is an issue to improve in the research domain due to the fact that reliability of the input data is crucial to achieve reliable results [10].

4.2.2. Qualitative Identification/Selection Approach

Selecting the initial parameters is a fundamental step in the modelling process. Shmueli and Koppius [12] and Elmousalami [15] have identified the first of two phases as a qualitative process in which combining domain knowledge, theory, and exploratory analysis is fundamental to give grounds for the inclusion of inputs. The method to identify the potential parameters and the number of related studies is shown in Table 2, where 23 studies identified potential parameters from literature reviews or/and expert knowledge, and six used the researchers' criteria. Two studies selected the parameters from the data available, and the rest did not specify the process to select them. Notably, publications from journals provided initial parameters for the studies [53,54,60–64]. The compilation of expert knowledge was realised by interviews and questionnaire surveys. Elaborated techniques to acquire information, such as a Likert Scale, Delphi method, and Analytic Hierarchy Process, are standard according to Elmousalami (2020), but only five studies implemented them.

Table 2. Number of methods to identify the parameters.

Parameter Identification Method	Number of Studies
Not mentioned	14
Literature review	10
Literature review and expert survey	9
Author criteria	6
Expert survey	4
From data available	2
Expert survey and MCA	1
Grand Total	46

The process followed in the studies to identify potential parameters can be improved by the use of both expert knowledge and previous literature, in order to increase the credibility of the outcomes and to improve the model's performance. Predictive analytics is a relatively new area of research that has evolved with the developments in informatics. Therefore, its guidelines are still being tested, but robustness in research needs to be a priority regardless of the innovations in technology. Secondly, experts in the area of cost estimation and architects were surveyed, but developers' knowledge was considered only in Stoy et al. [65], where the developers are the individuals making crucial decisions regarding investment options in the early stages of projects.

4.2.3. Quantitative Identification/Selection Approach

Dimension reduction is a method within exploratory data analysis used to reduce the number of parameters and to increase predictive accuracy [12,15]. In this regard, of the 46 studies, 27 utilised exploratory methods, used also to weight the parameters in the CBR models [59,66–69]. Table 3 shows the optimization parameters methods reviewed and the number of related studies. Nine of the studies implemented stepwise regression analysis. Methods such as PCA, Correlation Analysis, and Factor Analysis are commonly used to analyse cause–effect relationships, but these also provide a reduction in the number of parameters to achieve more accurate models. Although the main objective of predictive analytics is to produce models that forecast costs, the techniques used in the studies can determine the strength of the relationship between parameters and also the relative strength of its effect on the output. This information can serve decision-makers as guides in the subsequent stages to optimise the building features in the design stage.

Table 3. Methods used to optimise the parameters.

Parameter Identification Method	Studies	Number of Studies
Stepwise Regression Analysis	[52,55,70–76]	9
Principal Component Analysis	[58,77,78]	3
Correlation Analysis	[67,79,80]	3
ANOVA	[50,65]	2
Genetic Algorithm	[59,81]	2
Attribute Impact	[66]	1
Shapley Additive Explanations	[82]	1
MRA Standard Coefficients	[69]	1
Analytic Hierarchy Process	[53]	1
Boosting Regression Trees	[83]	1
Rough Set	[84]	1
Multifactor Evaluation	[85]	1
Factor Analysis	[54]	1
Decision Tree	[68]	1

4.2.4. Parameters Used

The size of the data has significant effects on the accuracy of the model. The more extensive databases are, the less sample variance and model bias are obtained. In addition, testing the modelling process requires the use of additional data. Shmueli and Koppius [12] stated that guidelines to set the minimum data size are difficult to define, although a commonly used rule of thumb of using 10 times the number of parameters is considered reasonable in computer experiments [86]. Following this criterion, 19 of the 46 studies had less than 10 data points per parameter, 24 had 10 or more data points per parameter, and three did not mention the total number of datapoints. Meta-analysis was not performed in this review, but the average MAPE of studies using 10 or more data points by parameter was 7.6%. On the other hand, the studies using less than 10 data points per parameter achieved 10.7% of average MAPE. This situation suggests that more extensive data relative to the number of parameters may produce better results.

The studies considered different parameters for their models, classifying them as quantitative and qualitative. Twenty-seven of the 46 studies (59%) provided the parameters used in the models in the form of ranks. The different authors developed these lists with the different methods from the quantitative approach and mean sensitivity ANN analyses from the results of the modelling processes. The Borda–Kendall technique, was used to synthesise the lists of the individual rankings into one aggregated ranking list. This method was used to acquire a generic view of the relative importance of the parameters within the studies.

For the calculation of the ranking of parameters the Borda rule represented as the vector of weights:

$$w = (n, n - 1, \dots, 2, 1), \tag{1}$$

which applies to a set of complete or partial ranked lists of n alternatives where w_i is the weight attached to an alternative located at the ith rank in any given list. Then, the cumulative score Cs for the ith alternative is given by:

$$Cs_i = \sum w_{ij}, \tag{2}$$

which is the weighted sum over all the lists, j, corresponding to the rank in each list for the ith alternative [87].

In the study, 78 were the total alternative parameters n from 27 lists, so the parameters in the first place of the lists had a score of 78, the ones in the second, a score of 77 and so forth. Then, the sum of scores by parameter allowed to elaborate the rank.

Note that the ranking corresponds to data from different locations, and it would require further examination to consider it a representative ranking of general buildings in different locations.

The rank aggregation provided a rank of 78 parameters. The 10 parameters with the highest scores are shown in Table 4. The Gross Floor Area (GFA) and the number of floors are the two most important parameters, having scores significantly higher. The rest of the parameters may not be the principal source of costs, but their consideration in the cost models elaboration may increase their predictive power. Notably, the parameters of foundation type, type of roof, structure type, and location are measured in categorical scales. Therefore, the ability of predictive analytics to deal with categorical scales enhances its usability for cost estimation.

Table 4. Ranked parameters.

Parameter	Rank	Score
GFA	1	1301
Number of floors	2	1137
Foundation type	3	803
Number of units	4	647
Number of elevators	5	589
Type of roof	6	506
Structure type	7	434
Duration	8	373
Number of unit floor households	9	304
Location	10	299

4.3. Predictive Power

Predictive accuracy, also known as predictive power, is the model's ability to elaborate accurate predictions of new observations [12]. Two criteria need to be met for an adequate test of predictive performance: assessment of the model's accuracy using adequate predictive measures, and determination of the appropriate validation method [12]. Root Mean Square Error (RMSE), Mean Square Error (MSE), and MAPE were commonly used generic predictive measures, but the first two are scale-dependent and should not be used when comparing across datasets that have different scales [88]. MAPE, being scale-independent, was an appropriate measurement to analyse the studies' models under a standard accuracy measurement. For the second criterion, the review synthesised the method of validation, which defines how the data is partitioned and tested for accuracy. The following subsection introduces accuracy measurements in the studies, followed by the validation methods.

4.3.1. Accuracy

The most critical feature of models for predicting events is its accuracy. It is fundamental, especially for decision-makers, when assessing investment opportunities with rather limited information. The average accuracy error of all the models included was under 10%, with a standard deviation of 5%, as shown in Figure 4. The use of ANN resulted in a slightly more dispersed distribution of the second and third quartile compared to MRA and CBR, but its overall dispersion is smaller than MRA. On the other hand, CBR presented the narrowest overall and second-third quartile distribution of MAPE, additionally, the range position of the two quartiles and its mean are lower than those of ANN and CBR. Although additional studies would deliver more substantial grounds to advocate for a particular technique, the collected data suggest that the CBR technique tends to provide higher accuracies than others. The MAPE of the overall models ranged between 2 and 21%, with the second and third quartile between 5 and 13%, respectively. Considering that the accuracy error in traditional cost estimation ranges from -15% to $+25\%$, which, in absolute terms, is 35%, the three techniques can perform significantly better, presenting errors under 21%, indicating that the absolute limit of 21% can serve as a baseline for an acceptance range of error for building projects' cost estimations in the early stages.

4.3.2. Validation

The method of validation in the studies was collected to assess the satisfaction of the second criterion stated by [12]. As part of the modelling process exposed earlier, models need an appropriate assessment of their accuracy using an independent data set. Forty-five of the studies considered out-of-sample data for testing, and only Chan and Park [58] did not specify whether a subset was set aside or not. Hold-out cross-validation, k-fold cross-validation, and Leave One Out Cross Validation (LOOCV) were used on 33, eight, and four studies, respectively. Two considerations were pondered to assess suitability of the method used. First, for small samples, k-fold cross validation would be pertinent because it should provide better estimates of accuracy according to [35]. A second consideration was extracted from Shmueli and Koppius [12], where a sample size of 213 data points was considered small in the modeling process, and cross-validation was preferred to a simple hold-out. Therefore, in this research the method of hold-out is considered appropriate for samples of more than 213 data points. Accordingly, only 20 of the studies in this review conducted appropriate validation methods utilizing cross-validation or hold-out for data samples bigger than 213 data points, 22 studies did not implement the best validation method, and four studies did not indicate the type of validation nor the sample size. These results agree with Elfaki et al. [17] by evidencing a urgent need for standard validation methods to determine the level of accuracy of models and ease the implementation of predictive analytics.

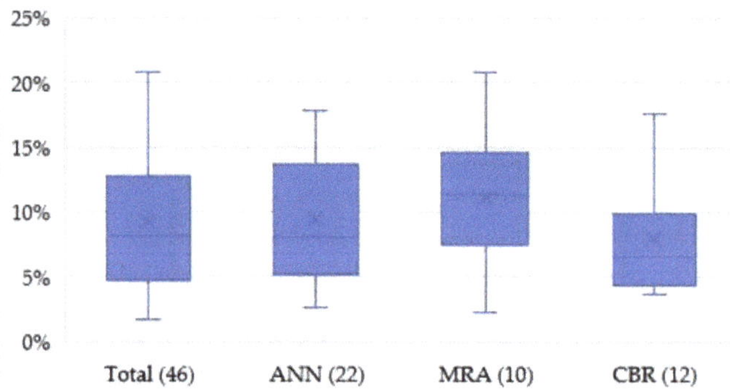

Figure 4. Box and whiskers chart of the average MAPE by technique.

4.4. Modelling Techniques

The five main techniques applied in the studies for the estimation of building construction costs at the early stages were:

- Artificial Neural Networks (ANN);
- Case-Based Reasoning (CBR);
- Multiple Regression Analysis (MRA);
- Boosting Regression Trees (BRT); and
- Support Vector Machine (SVM).

ANN, CBR, and MRA were the predominant techniques used to elaborate the cost-prediction models. ANNs were used in 48% of the studies, while MRA and CBR were used in 22% and 26%, respectively. The other two techniques, BRT and SVM, represented only 4% each. Three approaches were followed by the reviewed papers to evaluate the techniques. The first approach used a single technique to develop a model, such as Chan and Park [58], who proposed a technique based on Principal Component Analysis to identify the most significant parameters to develop a linear function to model the costs of buildings. In the second approach, the studies compared different alternatives to improve a single technique. For example, Kim et al. [57] incorporated genetic algorithms to optimise the architecture of the artificial neural network model, and Doğan et al. [59] used genetic algorithms in a case-based model to determine the optimal weights of the case attributes. The third approach considered the comparison of different techniques, e.g., Kim et al. [50] based its research methodology comparing ANN, CBR, and MRA in cost modelling of buildings. Overall, 24% of the studies developed models without performing comparisons, 50% evaluated alternatives enhancing a single technique, and 26% compared different techniques. The studies comparing variations of one technique provided valuable outcomes regarding the component on which technique has the potential to increase the accuracy of the models. The areas to improve and the methods successfully used are shown in the following subsections.

4.4.1. Artificial Neural Networks

In 22 studies, ANNs were considered the primary technique. Seven of the 22, compared the ANN models with other techniques, such as MRA, CBR, and SVM. In six studies there were no comparisons, and the main objective was only to introduce ANN as an accurate technique for cost estimation. The comparisons between different ANNs were considered in nine of the publications listed in Table 5, which shows that the improvements of the models were achieved predominately by optimising the ANN architecture by different techniques or methods. Generally, Genetic Algorithms (GA) were utilised to improve the ANN architecture components. Kim et al. [52] optimised the number of neurons in the hidden layer and the learning rate of the neural network. On the other hand, Elhag and Boussabaine [48] compared two ANNs, using 13 parameters and using only four.

Table 5. Improvements in ANN models from studies.

Author	Year	Model Component Improvement	Technique or Method Used
Elhag and Boussabaine [48]	1998	Input parameters	Inclusion of additional parameters
Kim et al. [52]	2004	ANN Architecture	GA
Kim et al. [89]	2005	ANN Architecture	GA
Cheng et al. [90]	2009	ANN Architecture	FL/GA
Cheng et al. [56]	2010	ANN Architecture	High Order NN/FL/GA
Sonmez [91]	2011	Input parameters/ANN Architecture	Bayesian regularisation/Bootstraps prediction intervals
Rafiei and Adeli [92]	2018	Model architecture	DBM combination
Jumas et al. [93]	2018	Input parameters	MRA
Badawy [94]	2020	Model architecture	MRA combination

4.4.2. Case-Based Reasoning

From the 12 studies implementing CBR to model the costs of building projects, only Kim et al. [72] conducted a comparison with a different technique—ANN. The 11 other papers shown in Table 6 presented attribute weight and case similarity measures as the primary concern at the time of developing improvements in CBR, utilising GA and MRA to assign the optimum weight of the attributes.

Table 6. Improvements on CBR models from studies.

Author	Year	Model Component Improvement	Technique or Method Used
An, et al. [53]	2006	Attribute weighting	Analytic Hierarchy Process (AHP)
Doğan et al. [59]	2006	Attribute weighting	GA
Doğan et al. [68]	2008	Attribute weighting	Decision Trees
Ji et al. [81]	2011	Case Similarity Measurement Attribute weighting	Euclidean distance-based similarity function GA
Jin et al. [69]	2012	Result error	MRA-based revision method
Ji et al. [77]	2012	Case adaptation	MRA
Jin et al. [75]	2014	Input parameters	Inclusion of categorical attributes
Ahn et al. [66]	2014	Attribute weighting	Attribute impact method
Ahn et al. [67]	2017	Case Similarity Measurement	Euclidean distance Mahalanobis distance Arithmetic summation Fractional function
Ahn et al. [79]	2020	Input parameters	GA Euclidean distance
Jung et al. [95]	2020	Attribute weighting	GA Local search technique

4.4.3. Multiple-Regression Analysis

The use of multiple-regression analysis as a primary technique was utilised in 10 of the 46 articles. Five of them did not create additional models to compare results. Sonmez [55] and Dursun and Stoy [73] compared their accuracy with models developed with ANN, and Li et al. [74] compared an MRA model with the Unit Area Cost method. Lowe et al. [52] and Ji et al. [71] utilised techniques of Stepwise Regression and Principal Component Analysis to select the optimal parameters, respectively. Although MRA was not the most explored technique by the studies, it can support other techniques and enhance their effectiveness, e.g., it was used in CBR modelling to improve the adaptation capability [77]. Additionally, MRA is a technique more accessible for cost-estimation practitioners because it has broadly studied and implemented in statistics.

4.5. Benefits and Challenges

The commonly reported benefit in virtually all studies was the higher accuracy of the models in comparison to the traditional cost estimation techniques. This benefit has not been included in the benefits and challenges analysis because it was included in the Predictive Power section, where it was quantitatively analysed. The next two most mentioned benefits were (1) the suitability of the techniques for real practice, and (2) the possibility of improvement by combining them with other techniques. Cheng et al. [56] concluded that the techniques implemented were suitable for practice, where the authors highlighted that the model can enhance the ability of designers, owners, and contractors in the decision-making process leading to higher possibilities to achieve project success. Regarding the improvement in the techniques, Sonmez [55] concluded that the simultaneous use of ANN and MRA could provide satisfactory conceptual models.

Some authors of the publications have found limitations that make predictive analytics in cost estimation an area still in development with drawbacks to address. The main challenges expressed were (1) the need for more data, (2) to generalise models towards location and different project types, and (3) the improvement of attribute weighting. Predictive analytics bases its performance on data. Therefore, it becomes essential for cost modelling to have access to building-projects data. Models use input data to learn and larger data sets would increase their performance [51]. Since construction is an economic activity, the nature of competition does not incentivise sharing information because it is an element of competitive advantage, but individual companies may be able to implement predictive analytics by themselves. Ngo et al. [10] found that construction companies in Singapore do have pertinent data to implement predictive analytics. In this sense, the availability of data is a drawback in research, but, from the perspective of companies, it can be considered as a benefit due to a large amount of data they store from previous projects in the form of contract documents, schedules, drawings, specifications, and images. The second area to overcome, according to researchers, is the need for generalisation about location and typologies. Generalisation means an increase in the number of input parameters, and, therefore, more parameters require more data [86]. So, the increase in generalisation is strongly related to the first challenge—data availability. The third challenge perceived in the studies is the need to improve the techniques. The studies exposed that ANNs need improvement in the methods to optimise the network architecture and CBR needs to address attribute weighting, but other techniques not yet explored in the cost estimating of buildings may provide alternatives that suit the particular circumstances of the estimation case.

5. Conclusions

Several emergent techniques from predictive analytics have become a major area for researchers seeking to improve the practice of construction-cost estimation in the early stages of projects. Advances in methodology and techniques have become available in the last 20 years, but the explicit benefits and implications for cost-estimation practice have not been sufficiently highlighted to ignite the uptake by the industry. As an initial stimulus for the adoption, a systematic literature review was conducted in this study to investigate how predictive analytics can enhance early-stage cost estimation of buildings, resulting in three main contributions to the body of research:

1. An extensive database of 46 relevant publications on the use of predictive analytics for construction-costs estimations at the early stages of the development process was compiled and analysed;
2. A large number of cost-drivers were identified and ranked;
3. The various predictive analytics tools were compared to understand their applicability and ability to predict construction costs at the early stages of the development process.

We found that previously published research identified structured processes to apply predictive analytics on cost estimation, and that the accuracy of the models developed has surpassed that of the traditional practices of building construction-cost estimation. Additionally, the practices for modelling costs with predictive analytics have been structured and well documented. Three main implications can be drawn from this discussion:

1. Predictive analytics for cost-estimation research has not widely followed the best practices and standard methodologies. By following more strict parameters identification methods, using better data and predictive power considerations, models would produce more reliable predictions. Methodologies to apply predictive analytics for cost estimation have been recently standardised by Elmousalami [15] and Elfaki et al. [17];
2. The already accurate predictive analytics techniques investigated in previous studies and the tested modelling methodologies represent the necessary evidence to lead research into the next stage of progress, focusing on adoption and implementation of predictive analytics by the industry;
3. The study serves as a reference for high-level decision-makers in organisations developing building projects, providing them with the incremental developments in predictive analytics applications to promote a change of paradigm in the practice of cost estimation.

Future research perspectives relate to implementation issues of predictive analytics in cost estimation, focusing on investigating the current state of uptake in the industry, and the necessary ground conditions in organisations to deploy them, such as necessary skills of practitioners and decision-makers' awareness regarding the implications of predictive analytics for construction project success. The main limitation possibly influencing the results of the review was identified. There was a possibility of not having found all the relevant papers due to the different words used to describe a concept within predictive analytics in cost estimation. The implementation of backward and forward snowballing contributed to addressing the first limitation identifying papers out of the search performed using the search engines.

Author Contributions: Conceptualization, S.L.C.M. and E.D.R.C.; methodology, S.L.C.M., E.D.R.C. and V.G.; validation, S.L.C.M., E.D.R.C., V.G. and J.A.; formal analysis, S.L.C.M. and E.D.R.C.; investigation, S.L.C.M. and E.D.R.C.; resources, S.L.C.M., E.D.R.C., V.G. and J.A.; data curation, S.L.C.M., E.D.R.C., V.G. and J.A.; writing—original draft preparation, S.L.C.M. and E.D.R.C.; writing—review and editing, S.L.C.M., E.D.R.C., V.G. and J.A.; visualization, S.L.C.M. and E.D.R.C.; supervision, E.D.R.C. and V.G.; project administration, E.D.R.C.; funding acquisition, E.D.R.C. All authors have read and agreed to the published version of the manuscript.

Funding: This research was funded by the New Zealand Earthquake Commission (grant number 18/U777).

Institutional Review Board Statement: Not applicable.

Data Availability Statement: No new data were created or analyzed in this study. Data sharing is not applicable to this article.

Conflicts of Interest: The authors declare no conflict of interest. The funders had no role in the design of the study; in the collection, analyses, or interpretation of data; in the writing of the manuscript; or in the decision to publish the results.

References

1. Sanvido, V.; Grobler, F.; Parfitt, K.; Guvenis, M.; Coyle, M. Critical Success Factors for Construction Projects. *J. Constr. Eng. Manag.* **1992**, *118*, 94–111. [CrossRef]
2. Project Management Institute (PMI). *Construction Extension to the PMBOK®Guide*, 2nd ed.; Project Management Institute: Newtown Square, PA, USA, 2016.
3. Amos, S. *Skills & Knowledge of Cost Engineering: A Project of the Education Board of AACE International*, 5th ed.; AACE International: Morgantown, WV, USA, 2004.
4. Ashworth, A.; Perera, S. *Cost Studies of Buildings*, 6th ed.; Routledge: Abingdon Oxon, UK; New York, NY, USA, 2015.
5. Royal Institution of Chartered Surveyors. RICS New Rules of Measurement. In *NRM 1, Order of Cost Estimating and Cost Planning for Capital Building Works*; RICS: London, UK, 2012.
6. Abourizk, S.M.; Babey, G.M.; Karumanasseri, G. Estimating the cost of capital projects: An empirical study of accuracy levels for municipal government projects. *Can. J. Civ. Eng.* **2002**, *29*, 653–661. [CrossRef]
7. Ashworth, A. *Pre-Contract Studies: Development Economics, Tendering, and Estimating*; Blackwell: Oxford, UK; Melden, MA, USA, 2008.
8. Nisbet, R.; Miner, G.; Yale, K. The Data Mining and Predictive Analytic Process. In *Handbook of Statistical Analysis and Data Mining Applications*; Nisbet, R., Miner, G., Yale, K., Eds.; Academic Press: Cambridge, MA, USA, 2018; pp. 39–54. [CrossRef]
9. Yan, X.; Su, X. *Linear Regression Analysis: Theory and Computing*; World Scientific: Singapore, 2009.
10. Ngo, J.; Hwang, B.-G.; Zhang, C. Big Data and Predictive Analytics in the Construction Industry: Applications, Status Quo, and Potential in Singapore's Construction Industry. In Proceedings of the Construction Research Congress 2020: Computer Applications, Tempe, AZ, USA, 8–10 March 2020; American Society of Civil Engineers (ASCE): Reston, VA, USA, 2020; pp. 715–724.
11. Waller, M.A.; Fawcett, S.E. Data Science, Predictive Analytics, and Big Data: A Revolution That Will Transform Supply Chain Design and Management. *J. Bus. Logist.* **2013**, *34*, 77–84. [CrossRef]
12. Shmueli, G.; Koppius, O.R. Predictive Analytics in Information Systems Research. *MIS Q. Manag. Inf. Syst.* **2011**, *35*, 553–572. [CrossRef]
13. Mishra, N.; Silakari, S. Predictive Analytics: A Survey, Trends, Applications. *Int. J. Comput. Sci. Inf. Technol.* **2012**, *3*, 4434–4438.
14. Shah, N.D.; Steyerberg, E.W.; Kent, D.M. Big Data and Predictive Analytics: Recalibrating Expectations. *JAMA—J. Am. Med. Assoc.* **2018**, *320*, 27–28. [CrossRef]
15. Elmousalami, H.H. Artificial Intelligence and Parametric Construction Cost Estimate Modeling: State-of-The-Art Review. *J. Constr. Eng. Manag.* **2020**, *146*, 03119008. [CrossRef]
16. Forgues, D.; Iordanova, I.; Valdivesio, F.; Staub-French, S. Rethinking the Cost Estimating Process through 5D BIM: A Case Study. *Constr. Res. Congr.* **2012**, 778–786. [CrossRef]
17. Elfaki, A.O.; Alatawi, S.; Abushandi, E. Using Intelligent Techniques in Construction Project Cost Estimation: 10-Year Survey. *Adv. Civ. Eng.* **2014**, *2014*, 107926. [CrossRef]
18. Kitchenham, B.; Charters, S.M. *Guidelines for Performing Systematic Literature Reviews in Software Engineering*; Keele University: Keele, UK; University of Durham: Durham, UK, 2007.
19. Kirkham, R.; Brandon, P.; Ferry, D. *Ferry and Brandon's Cost Planning of Buildings*; Blackwell: Oxford, UK; Malden, MA, USA, 2007.
20. Potts, K. *Construction Cost Management: Learning from Case Studies*; Taylor & Francis: London, UK; New York, NY, USA, 2008.
21. Fellows, R. New research paradigms in the built environment. *Constr. Innov.* **2010**, *10*, 5–13. [CrossRef]
22. Brandon, P. Building Cost Research: Need for a Paradigm Shift? In *Building Cost Techniques: New Directions*; E&FN Spon: London, UK, 1982.
23. Contasfor; Egan, S.J.; Williams, D. [Summary of] "Rethinking construction"—The report of the construction task force. Ice briefing sheet. *Proc. Inst. Civ. Eng. Munic. Eng.* **1998**, *127*, 199–203. [CrossRef]
24. Koskela, L. Theory of Lean Construction. In *Lean Construction*; Tzortzopoulos, P., Kagioglou, M., Koskela, L., Eds.; Routledge: Oxfordshire, UK, 2020; pp. 2–13. [CrossRef]
25. Kim, Y.; Ballard, G. Activity-Based Costing and Its Application to Lean Construction. In Proceedings of the 9th Annual Conference of the International Group for Lean Construction, Singapore, 6–8 August 2001; pp. 6–8.
26. Ballard, G. The Lean Project Delivery System: An Update. *Lean Constr. J.* **2008**, *1*, 1–19.
27. Smith, J.; Jaggar, D. *Building Cost Planning for the Design Team*, 2nd ed.; Elsevier: Oxford, UK, 2007.
28. Finlay, S. *Predictive Analytics, Data Mining and Big Data*; Springer: Amsterdam, The Netherlands, 2014. [CrossRef]
29. Shmueli, G. To Explain or to Predict? *Stat. Sci.* **2010**, *25*, 289–310. [CrossRef]
30. Wu, J.; Coggeshall, S. *Foundations of Predictive Analytics*; CRC Press: Boca Raton, FL, USA, 2012. [CrossRef]
31. Boyacioglu, M.A.; Kara, Y.; Baykan, K. Predicting bank financial failures using neural networks, support vector machines and multivariate statistical methods: A comparative analysis in the sample of savings deposit insurance fund (SDIF) transferred banks in Turkey. *Expert Syst. Appl.* **2009**, *36 Pt 2*, 3355–3366. [CrossRef]
32. Mamuda, M.; Sathasivam, S. Predicting the Survival of Diabetes Using Neural Network. *AIP Conf. Proc.* **2017**, *1870*, 040046. [CrossRef]
33. Saggi, M.K.; Jain, S. A survey towards an integration of big data analytics to big insights for value-creation. *Inf. Process. Manag.* **2018**, *54*, 758–790. [CrossRef]
34. Blanco, J.L.; Fuchs, S.; Parsons, M.; Ribeirinho, M.J. Artificial intelligence: Construction technology's next frontier. *Build. Econ.* **2018**, 7–13. Available online: https://search.informit.org/doi/abs/10.3316/informit.048712291685521 (accessed on 7 May 2022).

35. Russell, S.; Norvig, P. *Artificial Intelligence a Modern Approach*, 3rd ed.; Pearson: London, UK, 2010.
36. Tranfield, D.; Denyer, D.; Smart, P. Towards a Methodology for Developing Evidence-Informed Management Knowledge by Means of Systematic Review. *Br. J. Manag.* **2003**, *14*, 207–222. [CrossRef]
37. Borrego, M.; Foster, M.J.; Froyd, J.E. Systematic Literature Reviews in Engineering Education and Other Developing Interdisciplinary Fields. *J. Eng. Educ.* **2014**, *103*, 45–76. [CrossRef]
38. Denyer, D.; Tranfield, D. Producing a Systematic Review. In *The SAGE Handbook of Organizational Research Methods*; SAGE: Thousand Oaks, CA, USA, 2009; pp. 671–689. [CrossRef]
39. Pan, M.; Yang, Y.; Zheng, Z.; Pan, W. Artificial Intelligence and Robotics for Prefabricated and Modular Construction: A Systematic Literature Review. *J. Constr. Eng. Manag.* **2022**, *148*, 03122004. [CrossRef]
40. Page, M.J.; McKenzie, J.E.; Bossuyt, P.M.; Boutron, I.; Hoffmann, T.C.; Mulrow, C.D.; Shamseer, L.; Tetzlaff, J.M.; Akl, E.A.; Brennan, S.E.; et al. The PRISMA 2020 Statement: An Updated Guideline for Reporting Systematic Reviews. *Syst. Rev.* **2021**, *10*, 89. [CrossRef] [PubMed]
41. Ayodele, O.A.; Chang-Richards, A.; González, V. Factors Affecting Workforce Turnover in the Construction Sector: A Systematic Review. *J. Constr. Eng. Manag.* **2020**, *146*, 03119010. [CrossRef]
42. Wohlin, C. Guidelines for Snowballing in Systematic Literature Studies and a Replication in Software Engineering. In Proceedings of the 18th International Conference on Evaluation and Assessment in Software Engineering, London, UK, 13–14 May 2014; pp. 1–10. [CrossRef]
43. Kitchenham, B.; Brereton, P.A. Systematic Review of Systematic Review Process Research in Software Engineering. *Inf. Softw. Technol.* **2013**, *55*, 2049–2075. [CrossRef]
44. Rosenthal, R.; DiMatteo, M.R. Meta-Analysis: Recent Developments in Quantitative Methods for Literature Reviews. *Annu. Rev. Psychol.* **2001**, *52*, 59–82. [CrossRef]
45. Lin, S. Rank Aggregation Methods. *Wiley Interdiscip. Rev. Comput. Stat.* **2010**, *2*, 555–570. [CrossRef]
46. Fields, E.B.; Okudan, G.E.; Ashour, O.M. Rank aggregation methods comparison: A case for triage prioritization. *Expert Syst. Appl.* **2013**, *40*, 1305–1311. [CrossRef]
47. Wang, Y.-M.; Yang, J.-B.; Xu, D.-L. A preference aggregation method through the estimation of utility intervals. *Comput. Oper. Res.* **2005**, *32*, 2027–2049. [CrossRef]
48. Elhag, T.M.S.; Boussabaine, A.H. An Artificial Neural System for Cost Estimation of Construction Projects. In Proceedings of the 14th Annual ARCOM Conference, Reading, UK, 1 September 1998; Volume 1, pp. 219–226.
49. Karshenas, S. Predesign Cost Estimating Method for Multistory Buildings. *J. Constr. Eng. Manag.* **1984**, *110*, 79–86. [CrossRef]
50. Kim, G.-H.; An, S.-H.; Kang, K.-I. Comparison of construction cost estimating models based on regression analysis, neural networks, and case-based reasoning. *Build. Environ.* **2004**, *39*, 1235–1242. [CrossRef]
51. Günaydın, H.M.; Doğan, S.Z. A neural network approach for early cost estimation of structural systems of buildings. *Int. J. Proj. Manag.* **2004**, *22*, 595–602. [CrossRef]
52. Lowe, D.J.; Emsley, M.W.; Harding, A. Predicting Construction Cost Using Multiple Regression Techniques. *J. Constr. Eng. Manag.* **2006**, *132*, 750–758. [CrossRef]
53. An, S.-H.; Kim, G.-H.; Kang, K.-I. A case-based reasoning cost estimating model using experience by analytic hierarchy process. *Build. Environ.* **2007**, *42*, 2573–2579. [CrossRef]
54. Emsley, M.W.; Lowe, D.J.; Duff, A.R.; Harding, A.; Hickson, A. Data modelling and the application of a neural network approach to the prediction of total construction costs. *Constr. Manag. Econ.* **2002**, *20*, 465–472. [CrossRef]
55. Sonmez, R. Conceptual cost estimation of building projects with regression analysis and neural networks. *Can. J. Civ. Eng.* **2004**, *31*, 677–683. [CrossRef]
56. Cheng, M.-Y.; Tsai, H.-C.; Sudjono, E. Conceptual cost estimates using evolutionary fuzzy hybrid neural network for projects in construction industry. *Expert Syst. Appl.* **2010**, *37*, 4224–4231. [CrossRef]
57. Kim, G.-H.; Yoon, J.-E.; An, S.-H.; Cho, H.-H.; Kang, K.-I. Neural network model incorporating a genetic algorithm in estimating construction costs. *Build. Environ.* **2004**, *39*, 1333–1340. [CrossRef]
58. Chan, S.L.; Park, M. Project cost estimation using principal component regression. *Constr. Manag. Econ.* **2005**, *23*, 295–304. [CrossRef]
59. Doğan, S.Z.; Arditi, D.; Günaydın, H.M. Determining Attribute Weights in a CBR Model for Early Cost Prediction of Structural Systems. *J. Constr. Eng. Manag.* **2006**, *132*, 1092–1098. [CrossRef]
60. Park, U.Y.; Kim, G.H. A Study on Predicting Construction Cost of Apartment Housing Projects Based on Support Vector Regression at the Early Project Stage. *J. Archit. Inst. Korea* **2007**, *23*, 165–172.
61. Skitmore, M. The Effect of Project Information on the Accuracy of Building Price Forecasts. In *Building Cost Modelling and Computers*; E& FN Spon: London, UK, 1987; pp. 327–336.
62. Picken, D.H.; Ilozor, B.D. Height and construction costs of buildings in Hong Kong. *Constr. Manag. Econ.* **2003**, *21*, 107–111. [CrossRef]
63. Elhag, T.M.S.; Boussabaine, A.H.; Ballal, T.M.A. Critical determinants of construction tendering costs: Quantity surveyors' standpoint. *Int. J. Proj. Manag.* **2005**, *23*, 538–545. [CrossRef]
64. Wheaton, W.C.; Simonton, W.E. The Secular and Cyclic Behavior of "True" Construction Costs. *J. Real Estate Res.* **2007**, *29*, 1–25. [CrossRef]

65. Stoy, C.; Pollalis, S.; Schalcher, H.-R. Drivers for Cost Estimating in Early Design: Case Study of Residential Construction. *J. Constr. Eng. Manag.* **2008**, *134*, 32–39. [CrossRef]
66. Ahn, J.; Ji, S.-H.; Park, M.; Lee, H.-S.; Kim, S.; Suh, S.-W. The attribute impact concept: Applications in case-based reasoning and parametric cost estimation. *Autom. Constr.* **2014**, *43*, 195–203. [CrossRef]
67. Ahn, J.; Park, M.; Lee, H.-S.; Ahn, S.J.; Ji, S.-H.; Song, K.; Son, B.-S. Covariance effect analysis of similarity measurement methods for early construction cost estimation using case-based reasoning. *Autom. Constr.* **2017**, *81*, 254–266. [CrossRef]
68. Doğan, S.Z.; Arditi, D.; Günaydin, H.M. Using Decision Trees for Determining Attribute Weights in a Case-Based Model of Early Cost Prediction. *J. Constr. Eng. Manag.* **2008**, *134*, 146–152. [CrossRef]
69. Jin, R.; Cho, K.; Hyun, C.; Son, M. MRA-based revised CBR model for cost prediction in the early stage of construction projects. *Expert Syst. Appl.* **2012**, *39*, 5214–5222. [CrossRef]
70. Kim, G.-H.; Shin, J.-M.; Kim, S.; Shin, Y. Comparison of School Building Construction Costs Estimation Methods Using Regression Analysis, Neural Network, and Support Vector Machine. *J. Build. Constr. Plan. Res.* **2013**, *1*, 29576. [CrossRef]
71. Ji, S.-H.; Park, M.; Lee, H.-S. Data Preprocessing–Based Parametric Cost Model for Building Projects: Case Studies of Korean Construction Projects. *J. Constr. Eng. Manag.* **2010**, *136*, 844–853. [CrossRef]
72. Kim, S.-Y.; Choi, J.-W.; Kim, G.-H.; Kang, K.-I. Comparing Cost Prediction Methods for Apartment Housing Projects: CBR versus ANN. *J. Asian Arch. Build. Eng.* **2005**, *4*, 113–120. [CrossRef]
73. Dursun, O.; Stoy, C. Conceptual Estimation of Construction Costs Using the Multistep Ahead Approach. *J. Constr. Eng. Manag.* **2016**, *142*, 04016038. [CrossRef]
74. Li, H.; Shen, Q.; Love, P.E. Cost modelling of office buildings in Hong Kong: An exploratory study. *Facilities* **2005**, *23*, 438–452. [CrossRef]
75. Jin, R.; Han, S.; Hyun, C.; Kim, J. Improving Accuracy of Early Stage Cost Estimation by Revising Categorical Variables in a Case-Based Reasoning Model. *J. Constr. Eng. Manag.* **2014**, *140*, 04014025. [CrossRef]
76. Sonmez, R. Parametric Range Estimating of Building Costs Using Regression Models and Bootstrap. *J. Constr. Eng. Manag.* **2008**, *134*, 1011–1016. [CrossRef]
77. Ji, S.-H.; Park, M.; Lee, H.-S. Case Adaptation Method of Case-Based Reasoning for Construction Cost Estimation in Korea. *J. Constr. Eng. Manag.* **2012**, *138*, 43–52. [CrossRef]
78. Juszczyk, M. Application of PCA-Based Data Compression in the ANN-Supported Conceptual Cost Estimation of Residential Buildings. In Proceedings of the AIP Conference, Rhodes, Greece, 22 September 2015; American Institute of Physics: College Park, MD, USA, 2016; Volume 1738, p. 200007. [CrossRef]
79. Ahn, J.; Ji, S.-H.; Ahn, S.J.; Park, M.; Lee, H.-S.; Kwon, N.; Lee, E.-B.; Kim, Y. Performance evaluation of normalization-based CBR models for improving construction cost estimation. *Autom. Constr.* **2020**, *119*, 103329. [CrossRef]
80. Hyung, W.-G.; Kim, S.; Jo, J.-K. Improved similarity measure in case-based reasoning: A case study of construction cost estimation. *Eng. Constr. Arch. Manag.* **2019**, *27*, 561–578. [CrossRef]
81. Ji, S.-H.; Park, M.; Lee, H.-S. Cost estimation model for building projects using case-based reasoning. *Can. J. Civ. Eng.* **2011**, *38*, 570–581. [CrossRef]
82. Wang, R.; Asghari, V.; Cheung, C.M.; Hsu, S.-C.; Lee, C.-J. Assessing effects of economic factors on construction cost estimation using deep neural networks. *Autom. Constr.* **2021**, *134*, 104080. [CrossRef]
83. Shin, Y. Application of Boosting Regression Trees to Preliminary Cost Estimation in Building Construction Projects. *Comput. Intell. Neurosci.* **2015**, *2015*, 149702. [CrossRef]
84. Feng, K.; Xiaojuan, W.; Liya, C. Application of RS-SVM in Construction Project Cost Forecasting. In Proceedings of the 2008 4th International Conference on Wireless Communications, Networking and Mobile Computing, Dalian, China, 12–17 October 2008; pp. 3–6. [CrossRef]
85. Wang, W.-C.; Bilozerov, T.; Dzeng, R.-J.; Hsiao, F.-Y.; Wang, K.-C. Conceptual Cost Estimations Using Neu-ro-Fuzzy and Multi-Factor Evaluation Methods for Building Projects. *J. Civ. Eng. Manag.* **2017**, *23*, 1–14. [CrossRef]
86. Loeppky, J.L.; Sacks, J.; Welch, W.J. Choosing the Sample Size of a Computer Experiment: A Practical Guide. *Technometrics* **2009**, *51*, 366–376. [CrossRef]
87. Fraenkel, J.; Grofman, B. The Borda Count and its real-world alternatives: Comparing scoring rules in Nauru and Slovenia. *Aust. J. Polit. Sci.* **2014**, *49*, 186–205. [CrossRef]
88. Hyndman, R.; Koehler, A.B. Another look at measures of forecast accuracy. *Int. J. Forecast.* **2006**, *22*, 679–688. [CrossRef]
89. Kim, G.H.; Seo, D.S.; Kang, K.I. Hybrid Models of Neural Networks and Genetic Algorithms for Predicting Preliminary Cost Estimates. *J. Comput. Civ. Eng.* **2005**, *19*, 208–211. [CrossRef]
90. Cheng, M.-Y.; Tsai, H.-C.; Hsieh, W.-S. Web-based conceptual cost estimates for construction projects using Evolutionary Fuzzy Neural Inference Model. *Autom. Constr.* **2009**, *18*, 164–172. [CrossRef]
91. Sonmez, R. Range estimation of construction costs using neural networks with bootstrap prediction intervals. *Expert Syst. Appl.* **2011**, *38*, 9913–9917. [CrossRef]
92. Rafiei, M.H.; Adeli, H. Novel Machine-Learning Model for Estimating Construction Costs Considering Economic Variables and Indexes. *J. Constr. Eng. Manag.* **2018**, *144*, 04018106. [CrossRef]
93. Jumas, D.; Mohd-Rahim, F.A.; Zainon, N.; Utama, W.P. Improving accuracy of conceptual cost estimation using MRA and ANFIS in Indonesian building projects. *Built Environ. Proj. Asset Manag.* **2018**, *8*, 348–357. [CrossRef]

94. Badawy, M. A hybrid approach for a cost estimate of residential buildings in Egypt at the early stage. *Asian J. Civ. Eng.* **2020**, *21*, 763–774. [CrossRef]
95. Jung, S.; Pyeon, J.-H.; Lee, H.-S.; Park, M.; Yoon, I.; Rho, J. Construction Cost Estimation Using a Case-Based Reasoning Hybrid Genetic Algorithm Based on Local Search Method. *Sustainability* **2020**, *12*, 7920. [CrossRef]

 *sustainability*

Article

Predicting the Impact of Construction Rework Cost Using an Ensemble Classifier

Fatemeh Mostofi [1], Vedat Toğan [1,*], Yunus Emre Ayözen [2] and Onur Behzat Tokdemir [3]

[1] Civil Engineering Department, Karadeniz Technical University, 61080 Trabzon, Türkiye
[2] Strategy Development Department, Ministry of Transport and Infrastructure, 06338 Ankara, Türkiye
[3] Civil Engineering Department, Istanbul Technical University, 34469 Istanbul, Türkiye
* Correspondence: togan@ktu.edu.tr

Abstract: Predicting construction cost of rework (COR) allows for the advanced planning and prompt implementation of appropriate countermeasures. Studies have addressed the causation and different impacts of COR but have not yet developed the robust cost predictors required to detect rare construction rework items with a high-cost impact. In this study, two ensemble learning methods (soft and hard voting classifiers) are utilized for nonconformance construction reports (NCRs) and compared with the literature on nine machine learning (ML) approaches. The ensemble voting classifiers leverage the advantage of the ML approaches, creating a robust estimator that is responsive to underrepresented high-cost impact classes. The results demonstrate the improved performance of the adopted ensemble voting classifiers in terms of accuracy for different cost impact classes. The developed COR impact predictor increases the reliability and accuracy of the cost estimation, enabling dynamic cost variation analysis and thus improving cost-based decision making.

Keywords: construction rework; cost estimation; nonconformance report; voting classifier; ensemble learning; machine learning

Citation: Mostofi, F.; Toğan, V.; Ayözen, Y.E.; Behzat Tokdemir, O. Predicting the Impact of Construction Rework Cost Using an Ensemble Classifier. *Sustainability* **2022**, *14*, 14800. https://doi.org/10.3390/su142214800

Academic Editors: Albert P. C. Chan, Srinath Perera, Xiaohua Jin, Patrizia Lombardi, Dilanthi Amaratunga, Anil Sawhney, Makarand Hastak and Sepani Senaratne

Received: 29 September 2022
Accepted: 7 November 2022
Published: 9 November 2022

Publisher's Note: MDPI stays neutral with regard to jurisdictional claims in published maps and institutional affiliations.

1. Introduction

A successful construction project is delivered on time and within budget, conforming to the specified quality. To achieve this, potential construction errors and violations are managed by applying an adequate construction quality management (CQM) system. An indispensable procedure within CQM is quality control (QC), which involves ensuring construction activity delivery at a specified standard, appraising its conformance, and maintaining continuous quality improvement. In construction projects, the arrays of errors, omissions, negligence, changes, failures, and violations resulting from poor management, communication, and coordination, or the materialization of potential risks are solved through rework. Thus, it is necessary to put in place a construction QC mechanism that not only prevents the need for rework but also prepares for accepting, acting on, and coping with required rework. Hence, the cost of rework (COR) is an inseparable component of overall construction costs, and its reduction directly improves construction cost and quality performance.

Although construction rework has been addressed in the literature, it remains a widespread [1] and prevalent problem [2,3], and poses a real challenge [4–6]. Despite all the advances in philosophies such as lean and total quality management (TQM) in preventing construction errors, COR still accounts for a considerable portion of the total project cost [2,7–9] and affects the construction schedule and quality [10]. Construction rework directly impacts the contract value by 5% to 20% [2], which can lead to complete project failure. Measuring COR enables the CQM system to control the construction budget and improve cost performance while allowing construction professionals to better understand the magnitude of the rework, its causes, and decisions on rework prevention measures [9]. Identifying the impact of COR and its sources enables reductions in the amount of rework and improvements in construction cost performance [11]. It is noteworthy that anticipating COR facilitates the utilization of QC techniques, such as Pareto analysis and pie charts. These QC techniques are dynamically used throughout the construction lifecycle to predict the construction rework items with a high-cost impact, which, in turn, allows for the timely adjustment of the associated construction schedule, budget, quality, human resources, and communication plans for the appropriate countermeasures. It is also noteworthy that obtaining COR is a key to understanding the cost of quality (COQ), i.e., the conformance costs, and the nonconformance costs, also referred to as cost of poor quality (COPQ) [12]. The ability of construction firms to measure COQ is essential for their survival in today's competitive environment [13].

While the construction management literature agrees on the important contribution of COR to total construction cost, it is not consistent with respect of the magnitude of COR's impact on overall construction cost, estimates of which vary between as much as 0.5% and 20% of the overall contract value [14]. This range reduces the practicality of COR in implementing effective countermeasures during the stages of a project. Additionally, in practice, countermeasures should focus on the construction work items with the higher impact on the total construction cost. It is not always feasible to implement preventive countermeasures or rework management strategies for all rework items associated with different construction activities. Evaluating the cost impact of construction rework supports early decision making on high-impact nonconformance items. For example, ranking the most impactful building defects offers construction companies insight for selecting the most appropriate strategy to continuously improve their construction activities and in turn to support sustainable decision making for the design and operation of buildings [15]. Furthermore, unless the COR for each construction activity is measured, it cannot be compared with the cost of the associated prevention or control plan. COR influences the construction budget, risk, and quality plans, which, in turn, affects the decision making associated with other project management knowledge areas. In order to improve the CQM, budget, and schedule plan, therefore, it is necessary to estimate the COR for each construction activity. This increases the error preparedness of construction organizations, which enhances decision-making resilience and plan accuracy while facilitating the prompt implementation of appropriate countermeasures, and thus also allows for appropriate contingency plans to be developed while helping the manager prevent later issues in other construction project phases [16].

In this study, COR is predicted for the total construction cost, which is a critical decision-making parameter. Experiments with advanced machine learning (ML) models, such as the ensemble method for predicting construction cost and COR, are lacking in the literature. Therefore, our work uses ensemble learning applied to the widely used construction nonconformity reports (NCRs) to ensure the robustness of the created COR predictor. As outlined in Figure 1, the main objective of this study is to assist construction quality managers and cost managers in including COR in their evaluations of different construction activities.

The remainder of this study is organized as follows. The next section discusses construction rework and its impact on overall construction cost. In addition, ensemble learning as a subdiscipline of ML is introduced, and, since the literature is limited to ML-based COR estimators, the ML applications for CQM and cost estimation are reviewed. The following section describes the NCRs obtained from different construction projects in Turkey. Then, the adopted methodology and ensemble COR predictor details are presented, and the benchmark ML predictor configuration is outlined. The next section gives the results that were obtained and discusses the practical implementations of the COR predictor that was developed and its contribution to the existing research on construction management. Finally, a brief conclusion reviews the study findings.

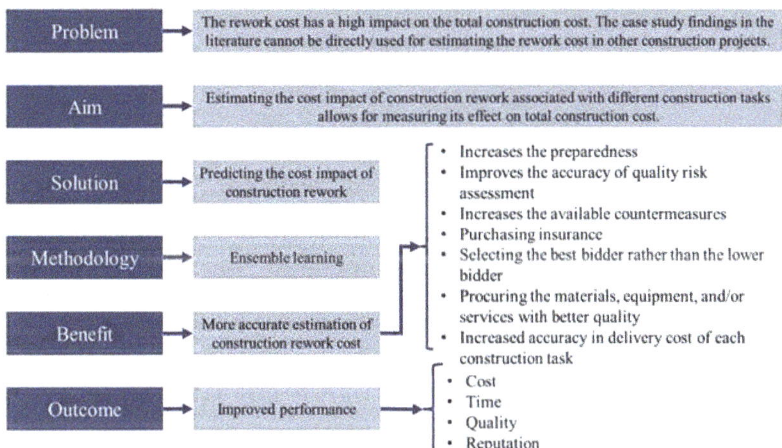

Figure 1. Study outline.

2. Research Background

2.1. Construction Rework

The conventional construction rework procedure based on nonconformities raised within the NCRs is outlined in Figure 2.

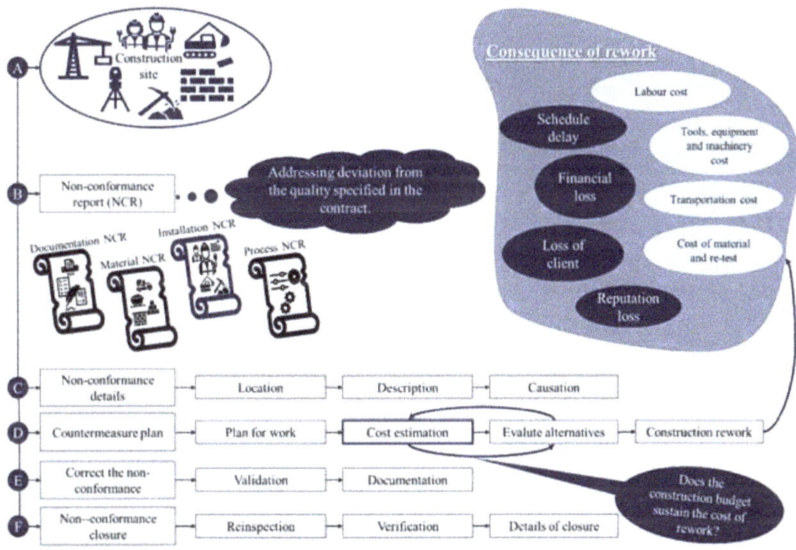

Figure 2. Construction rework procedure outline.

Site accidents, errors, failures, and violations are all causes of delay in the construction schedule and increase costs. Subsequent issues are often raised during the quality inspection of the construction activities performed, whereby the results are recorded in NCRs. The raised nonconformity should be addressed by the contractor, mainly in the form of rework. The topic of construction rework, including its causes, consequences, and prevention measures, has been widely addressed [17–20] using different terms and interpretations [9], such as quality deviation [21], construction nonconformance [22], defects [23], quality failure [24], and rework [25]. These have all emphasized the importance of factoring in construction rework during the early stages of construction planning in order to mitigate its consequences, which are mainly cost overruns.

2.2. The Cost Impact of Construction Rework

Despite the attention given to construction cost estimation in the previous research, the prediction of construction cost overrun has received relatively little consideration [16]. Similarly, the estimation of cost overruns resulting from the cost of construction rework has not been adequately addressed. Accordingly, the literature on estimating the cost impact of COR is reviewed here, along with construction cost estimation methods within a broader framework. The associated literature [13,26–28] has covered the broader topics of COQ and COPQ. In the literature on CQM, Love and different co-authors have explored construction quality from different perspectives, including construction error [29–33] and rework management [32,34], its impact on construction safety [35–37], and cost [9,38]. Hall and Tomkins [39] included the prevention and appraisal costs required to achieve a 'complete' COQ for buildings in the UK, while Love and Li [40] extended their earlier work on rework causation to quantify the magnitude of COR for Australian construction projects [41].

The research agrees on the negative impact of COR on overall construction cost while attaining varying impact percentages for this according to the demographics and types of evaluated construction projects. Davis et al. [42] found the nonconformance cost to be responsible for over 12% of the total contract value. Love [8,9], through a questionnaire survey on different project types and procurement routes, identified the direct and indirect impact of rework on total construction cost as being 26% and 52%, respectively. Rework costs drag down construction productivity by damaging the associated plans related, for example, to time, cost, and human resources, and this causes financial and reputation loss for the project participants. Hwang et al. [11] evaluated the contribution of COR to the total construction cost of 359 projects, along with its impact on both client and contractor. They found that construction owners are absorbing twice as much impact from COR than contractors. To reduce the magnitude of this problem, contractors often apply an internal quality control and assurance system, and they also often implement proactive measures to anticipate possible rework and associated costs.

In addition to the negative effects of construction rework, there is a possible positive impact on the project cost and quality. Ye et al. [2] investigated 277 construction projects in China to identify the main areas of rework and showed that active rework can improve construction cost, time, and quality. This study further suggests that by implementing a reward strategy and value management tools, required rework can be identified early, enabling timely decision making about the rework, time, cost, and quality benefits for the construction project. A statistical evaluation of 78 data points obtained from construction professionals by Simpeh et al. [7] revealed a mean 5.12% contribution of COR to total contract value and a 76% probability of exceeding its average value. This study also found that rework prediction facilitates quantitative risk assessment and, subsequently, the identification of alternative countermeasures for rework prevention. A more recent study by Love and Smith [4] evaluated the literature and put the impact of COR at between less than 1% and more than 20% of the total contract value. The literature is not consistent in specifying the conditions according to which the impact of COR should be measured, which hinders its practical implementation. The most recent study stated that the COR can vary from 0.5% to 20% of the total contract value [14]. Thus, these studies have provided in-depth investigations on the cost impact of rework, but they are not consistent when it comes to the magnitude of that impact.

The literature on construction management has recorded different contribution percentages for the impact of COR on overall construction cost. Since studies are conducted on projects of different size and type, and within different demographics, the cost impact figures obtained cannot be directly extended to other projects. Although the literature shows the importance of the early identification of COR for improving construction cost performance, the uncertainty about the magnitude of the impact hinders decision making when selecting the most advantageous countermeasures. Moreover, it is necessary to reach a different COR impact figure for each construction activity in order to prioritize activities with a higher cost impact, since it is not always feasible to implement preventive countermeasures or rework management strategies for all rework items. Furthermore, unless the COR for each work item is measured, it cannot be compared with the rework prevention or control cost.

Thus, to enhance the quality of decision making and quality planning, as well as to increase the chance of construction project success, it is important to estimate the COR for each work item. To translate the literature results on the impact of COR into the context of different construction projects, ML offers a data-oriented solution that can be utilized in different construction project contexts. ML approaches can predict COR by learning the complex patterns within the quality dataset.

2.3. ML for Construction Cost Prediction

ML uses historical evidence to offer a reliable solution that facilitates informed decision making. The literature on ML applications utilizing different types of datasets is growing in various fields [43–56]. Different ML approaches, such as artificial neural network (ANN), deep neural network (DNN), and support vector machine (SVM) are employed due to their ability to understand the complicated, non-linear patterns of real-world datasets. In this regard, the two ML approaches used for the cost estimation of construction projects were ANN [57,58] and SVM [55,59]. Even though other ML approaches, such as k-nearest neighbors (KNN) and decision trees (DT) share similarities with the ANN and SVM algorithms, they have yet to be investigated in the construction management literature [48]. Overall, construction cost estimation studies of more advanced ML approaches are scarce.

The literature on construction quality has mostly focused on quality assurance and quality control, using visual defect detection methodologies for a variety of tasks, including crack identification [60,61], damage localization on wooden building elements [62], and evaluation of pavement conditions [63]. ML approaches have also been used for the identification of rework or defect construction items. To this end, Fan [64] recently constructed a hybrid ML model using association rule mining (ARM) and a Bayesian network (BN) approach identify quality determinants and gain more effective evaluations of defect risk and its occurrence. In a related study, Kim et al. [65] utilized SVM, random forest (RF), and logistic regression (LR) along with three natural language processing (NLP) methods on 310,000 defect cases from South Korea to assign defect items to the appropriate repair task. Shoar et al. [16] used RF to estimate the COR of engineering services in construction to be used for devising appropriate contingency plans. Their study found using RF as a cost estimator to be an efficient approach for screening and prioritizing from the standpoint of cost overrun within construction projects, and that it can be used to devise related contingency plans.

Regarding the present study, the most relevant study is that conducted by Doğan [66] to predict the cost impact of construction nonconformities using case-based reasoning (CBR). His results indicated that the ability of CBR to predict the cost impact of quality problems is higher in construction NCRs. Reviewing the construction management literature, one may say that the development of ML-based cost estimators is still at an early stage. There is a lack of advanced ML approaches, such as ensemble learning methods. Although studies have established the usefulness of these ML methods, they have not elaborated on the robustness of the developed estimators, that is, on the ability to use the systems developed for other datasets. Thus, there is a research gap in the implementation of advanced ML-based techniques for predicting the COR associated with different construction activities.

2.4. Ensemble Learning

Single ML classifiers, such as SVM, KNN, NB, and DT, are trained with labeled datasets through various approaches to predict an output label class. Ensemble classifiers, however, combine the best predictions of these single ML approaches to improve the final prediction accuracy with improved stability and robustness [67]. The ensemble methods vary according to how they combine the results of single ML classifiers, while their performance depends on the number of individual members along with their prediction accuracy [67]. There are three popular ensemble techniques: (i) stacked, (ii) voting classifiers, and (iii) tree-based. Kansara et al. [68] applied the stacked ensemble (XGBoost regression) and tree-based ensemble (RF) approaches to improve the price prediction accuracy for real estate datasets. However, stacked ensemble approaches have the disadvantages of additional complexity and high computational time. Thus, they are feasible only when other ensemble approaches are not applicable.

Overall, due to their improved accuracy [51,68,69], studies have adopted ensemble learning methods for different prediction activities within different fields. Therefore, ensemble predictors are expected to provide more accurate cost predictions. In addition, the mechanism of the ensemble classifier benefits from both strong and weak predictors, where the latter is used to improve the prediction of the underrepresented classes. The literature on construction quality has still not matured with respect to cost estimation using both single ML predictors and ensemble learning predictors. Furthermore, the COR for different construction activities is not addressed in the construction quality literature. Therefore, because of the superior performance of ensemble learning over single ML models [51,68,69], this study adopts two such techniques, referred to as *soft* and *hard voting classifiers*, and compares them with three conventional tree-based ensemble classifiers (RF, gradient boosting (GB), and AdaBoosting (AB)) along with four single ML classifiers (DT, naïve Bayes (NB9), Logistic Regression (LR), and SVM). Accordingly, Figure 3 presents a simplified form of the procedure for the hard and soft voting classifiers adopted in this study.

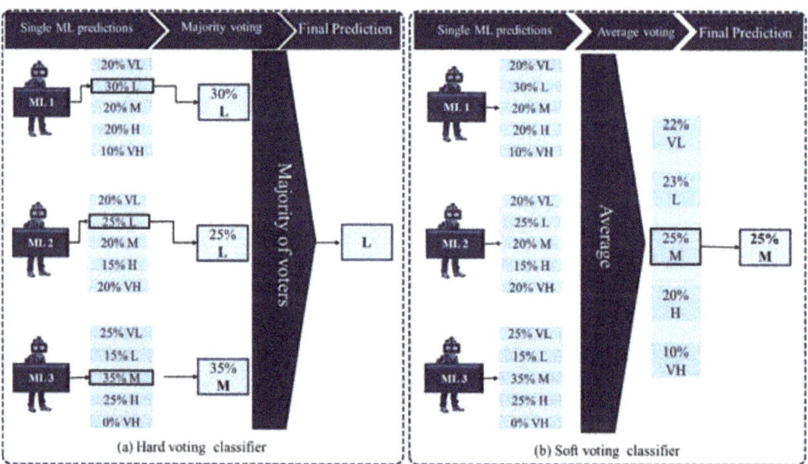

Figure 3. Simplified hard and soft voting COR classifiers.

As shown in Figure 3, each single classifier (ML 1–3) is referred to as a member that predicts an output class label, referred to as a vote. The hard voting classifier selects the label voted for by the majority of the members. The hard voting ensemble classifier uses the average of the predicted probabilities of all the members. For example, in Figure 3, ML 1 and ML 2 both classify the impact of COR as two, while ML 3 classifies it as three, so the hard voting model predicts the COR impact as two. Soft voting is less straightforward, since it uses the probability of each of the five classes and finds the average probability of all the classifiers within each class to select the final label. Voting classifiers can benefit from the voting of both single and ensemble classifiers. Tree-based ensemble approaches, such as

GB and AB, have been utilized to predict the rental price of apartments, showing better prediction accuracy than single ML approaches [69]. In addition to voting classifiers, bagging and boosting tree-based ensemble approaches are experimented with in this study. Figure 4 outlines the bagging and boosting mechanism within the tree-based ensemble approaches.

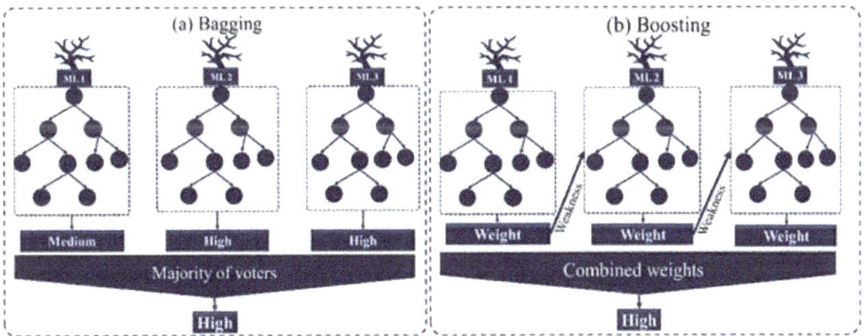

Figure 4. Simplified RF and boosting (AB and GB) ensemble mechanisms.

The bagging (i.e., RF) and boosting (i.e., AB and GB) mechanisms are the main ensemble approaches used within tree-based ensemble models, taking advantage of the best predictions of single DTs.

3. Data Description from Construction Nonconformance Report

This study uses the nonconformance items from diverse construction projects undertaken by international construction companies, collected in a study by Doğan [66] in 2021. The dataset comprises 2527 nonconformance items recorded by inspecting the different activities throughout the construction phase. A histogram associated with the construction activities and the frequency of recorded nonconformity is given in Figure 5, with activities having less than 20 occurrences aggregated under the 'other project activities' group. The collected nonconformance items were assigned to the different causation attributes through interviews.

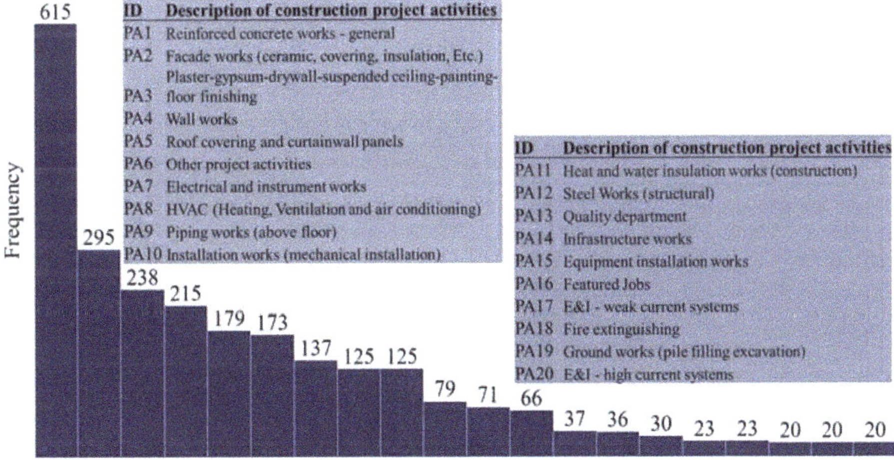

Figure 5. Details of construction activities registered in the NCRs.

Since the dataset was collected during the construction phases, attributes related to the pre-construction, design, and tendering phases, such as those related to clients and subcontractors, are omitted. The obtained NCR dataset is described using the stacked histogram, which details different construction project types. The NCRs include details of the causation of each recorded item, divided into material, design, operation, and construction causation. In addition, the cost impact of COR (y) is assigned as an output feature column. This assigns each input feature a cost impact of between one and five, corresponding to very low (VL), low (L), medium (M), high (H), and very high (VH).

In addition, the output cost impact is recategorized into three, four and five cost impact classes to evaluate the class prediction accuracy of the adopted ensemble approaches (Figure 6).

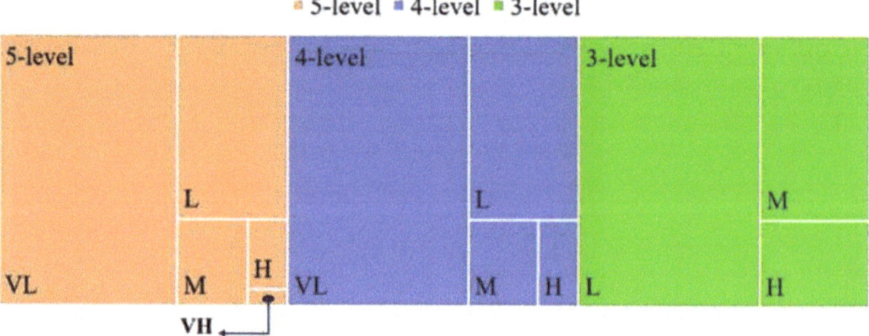

Figure 6. Cost impact classes.

As Figure 6 shows, critical nonconformance items with high-cost impact classes are underrepresented, there being few records of these compared with lower impact cost classes. Figure 7 shows the frequency of material-related nonconformance attributes used in this study. The observations from Figure 6 highlight the class imbalance among the different cost impact groups. The ability of ML to represent the under-represented cost impact classes is reduced.

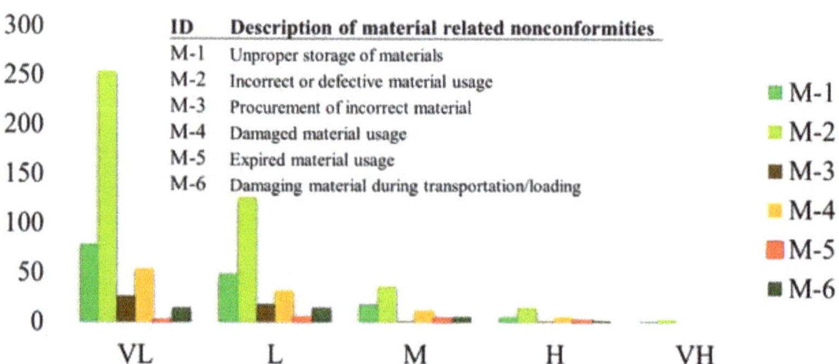

Figure 7. Material-related nonconformance attributes.

Furthermore, as shown in Figure 7, the collected NCRs also include the stage at which the nonconformance issue was initiated. Thus, each nonconformance attribute is linked to installation, documentation, material, or process damage. For example, 10 nonconformance issues with a cost impact of three are recorded as caused by damaged material usage (M-4), where the damage is initiated at the installation stage. Likewise, Figure 8 shows the nonconformity attributes related to design and operation. The design-related attributes recorded during the construction phase are limited because the collected NCRs were only gathered from the construction site.

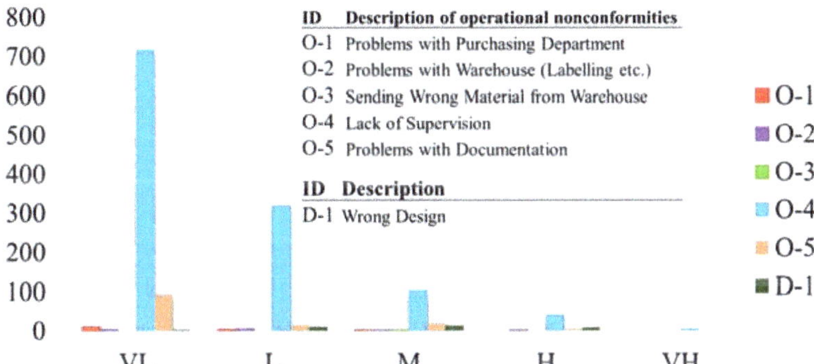

Figure 8. Design- and operation-related nonconformance attributes.

As Figure 8 shows, the design-related nonconformance attributes mostly occurred during the processing stage, while the operation-related issues were mostly associated with a lack of supervision (O-4) during the installation phase. As this study uses NCRs from construction sites, the frequency of nonconformances within the installation stage significantly increases in construction-related attributes (Figure 9).

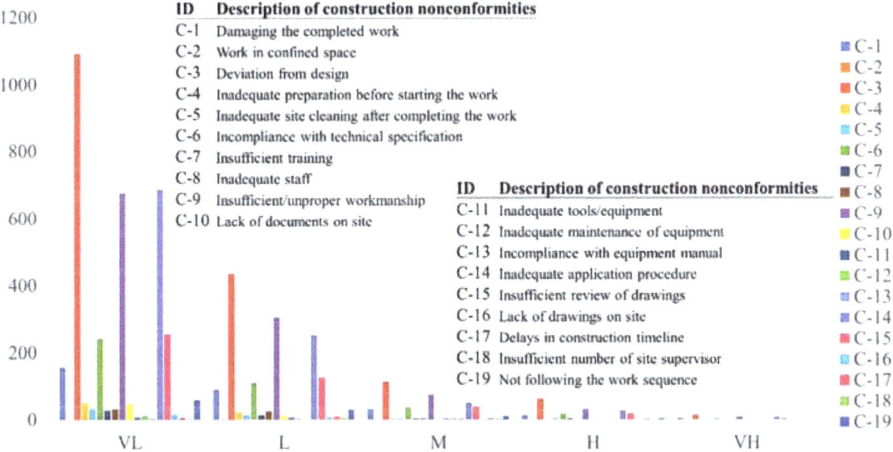

Figure 9. Construction-related nonconformance attributes.

4. Methodology

This study adopts ensemble ML classifiers to determine the impact of COR on overall project cost. The methodology is outlined in Figure 10.

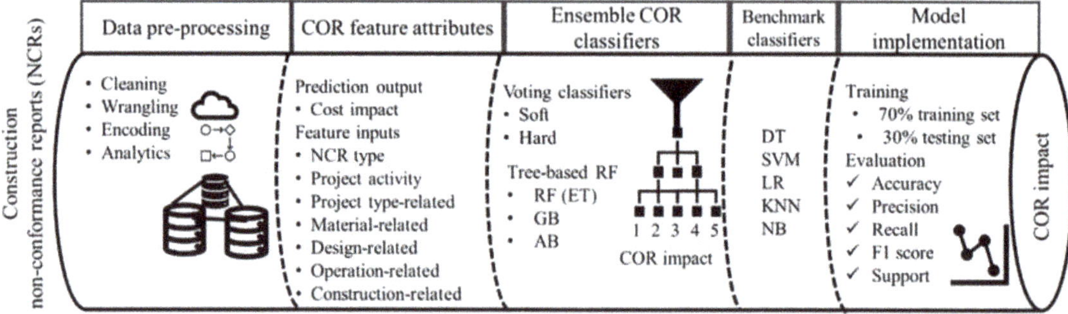

Figure 10. COR classifier methodology.

The methodology described was implemented within the Python programming environment. Most of the data preprocessing, analysis, and associated ML configuration was performed with widely used libraries, such as the Pandas [70], NumPy [70], and Scikit-Learn [71,72] packages. The nonconformance dataset was utilized with an ensemble classifier to predict the impact of COR on overall construction cost, while the results were compared with the single ML predictors.

4.1. Data Preprocessing

The dataset (D) was obtained from the construction NCRs with 39 feature columns and 2527 rows (Equation (1)):

$$D = \{x1,\ x2,\ \dots,\ x31,\ x32,\ \dots,\ x36, x37, x38, y\} \tag{1}$$

The material-, design-, operation-, and construction-related nonconformity items were used as 31 binary input feature columns. The project types were presented by five binary columns ($x32 - x36$), associated with industrial, hospital, high-rise, housing, and other building construction types. The NCR type column that shows the initiation area of the recorded nonconformity item was used as another input feature ($x37$) with four categories: installation, documentation, material inspection, and processes. To translate each category into the ML language, the dummy encoding method was used, which converts the NCR type into four columns, each showing a single category with either zero or one. For example, $\{1, 0, 0, 0\}$ shows the NCR type as installation, while $\{0, 0, 1, 0\}$ stands for material NCR type. In addition, the construction activity associated with each nonconformance item was used as a categorical input column ($x38$) with 20 categories, as depicted in Figure 5. Again, dummy encoding was used to translate the construction activity into binary format, this time within 20 feature columns. This resulted in an encoded dataset with 61 columns and 2527 nonconformance rows. Finally, 70% of the dataset was used for training and the rest (30%) was kept for performance evaluation.

4.2. Configuring Voting Classifiers

There is a wide range of single and ensemble ML algorithms that voting classifiers can use for a given prediction. This study explores different single and ensemble ML methods used in the literature to reach the best combination for the given study (Figure 11).

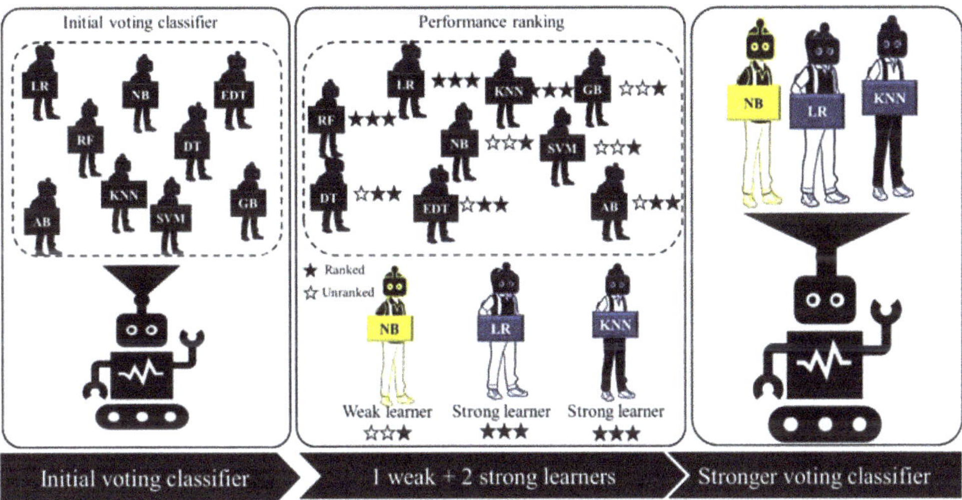

Figure 11. Configuration of soft and hard voting classifiers.

This study aimed at developing a COR impact predictor that achieves good prediction accuracy with simple implementation. Therefore, dimension reduction, class imbalances, and optimization techniques were not used to ascertain the best hyperparameters. Instead, the layers of different ML approaches were placed within a voting classifier and, based on the observed performance, a final voting classifier was configured with fewer ML members to accelerate the training procedure. The compulsory parameters, including the number of estimators (for RF, AB, and GB) and the number of neighbors (for KNN), were set roughly close to the benchmark models, while the other options were left as default. Although feature engineering and optimization techniques would have improved the performance of the ensemble predictor, their implementation was beyond the scope of the present study.

The radial basis function (RBF) was used as the SVM kernel while enabling a balanced class weight option. KNN was trained with 23 neighbors. The LR was adjusted with a LIBLINEAR optimizer and an l1 regularization. To ensure the creation of a weak learner, NB was used in its default form. The number of estimators for AB and RF was adjusted to 300 trees. The GB used 100 estimators while its learning rate was customized at 0.1. Afterwards, in order to reduce computational cost, the number of voting classifier members was reduced to three. In this respect, two strong learners were combined with a weak learner to simultaneously ensure accuracy and the elimination of bias. Combining strong and weak ML learners enhances the prediction accuracy of models for different rework cost impact classes while boosting the model's overall performance in terms of generalization ability and computational cost. Therefore, LR and KNN were used as strong learners, while NB was used as a weak learner for both the soft and hard voting classifiers.

4.3. Configuring Benchmark Classifiers

Unlike the voting classifiers, which were configured without any particular attention to the fine-tuning of their hyperparameters, each of the benchmark ML approaches was specifically fine-tuned to ensure a fair comparison between the ensemble voting classifiers and single ML predictions (Table 1).

Table 1. Hyperparameters of benchmark COR classifiers.

ML	Hyperparameters	Iterated Values	Selected Value
KNN	Number of neighbors	(2, 25)	23
LR	Optimizer	LIBLINEAR	LIBLINEAR
	Regularization	L1 L2	L1
SVM	Kernel	RBF	RBF
	C	1, 5, 10, 15	15
	Gamma	0.0001, 0.0005, 0.001, and 0.005	0.005
	Class weight	Imbalanced Balanced	Balanced
DT	Number of features	1589	15
	Tree depth	(1–37)	3
RF	Number of trees	50, 100, 150, 200, 300, 400	300
RF (ET)	Number of trees	115, 125, 135, 145, 155, 165, 175, 180, 185, 190, 195, 205	145
GB	Number of estimators (trees)	10, 20, 30, 40, 50, 60, 70, 80, 90, 100	10
	Learning rate	0.1, 0.01, 0.001	0.1
AB	Number of estimators (trees)	100, 150, 200	100
	Learning rate	0.01, 0.001	0.1

4.4. Evaluation Metrics

The prediction performance of the OCR impact predictors was evaluated using conventional accuracy, precision, accuracy, and F1 scores. For this, the number of correct predictions of each cost class (true positive (TP)) and correct assignments of the sample to the rest of the subclasses (true negative (TN)) was obtained. Likewise, the incorrect predictions within each subclass (false positive (FP), and false negative (FN)) were also recorded. Accuracy, F1 scores, precision, and recall [73] were obtained using Equations (2)–(5).

$$Accuracy = \frac{TP + TN}{TP + TN + FP + FN} \tag{2}$$

$$F1\ score = \frac{2 \times (Precision \times Recall)}{Precision + Recall} \tag{3}$$

$$Percsion_{Multiclass} = \frac{\sum_{i=2}^{classes} TP_i}{\sum_{i=2}^{classes} TP_i + FP_i} \tag{4}$$

$$Recall_{Multiclass} = \frac{\sum_{i=2}^{classes} TP_i}{\sum_{i=2}^{classes} TP_i + FN_i} \tag{5}$$

Accuracy did not provide a satisfactory evaluation for an imbalanced dataset as it does not consider FP and FN. Thus, the F1 score was preferred for the imbalanced dataset as it combines precision and recall. However, for a reliable COR predictor, the accuracy of the subclasses also needed to be evaluated.

5. Results and Discussion

The prediction performance of the soft and hard voting classifiers along with the benchmark ML approaches are shown in Table 2.

Table 2. COR prediction performance results.

ML Type	OCR Classifier		Accuracy	Precision	Recall	F1 Score
Ensemble ML (voting)	Hard voting	5-level	0.59	0.49	0.51	0.59
		4-level	0.59	0.52	0.5	0.59
		3-level	0.61	0.54	0.53	0.61
	Soft voting	5-level	0.54	0.51	0.48	0.54
		4-level	0.53	0.71	0.48	0.53
		3-level	0.5	0.57	0.51	0.5
Single ML	KNN	5-level	0.6	0.48	0.5	0.6
		4-level	0.61	0.55	0.55	0.61
		3-level	0.59	0.54	0.55	0.59
	LR (lr)	5-level	0.61	0.47	0.5	0.61
		4-level	0.62	0.53	0.52	0.62
		3-level	0.62	0.54	0.53	0.62
	LR (l1)	5-level	0.55	0.43	0.48	0.55
		4-level	0.56	0.44	0.48	0.56
		3-level	0.6	0.56	0.55	0.6
	LR (l2)	**5-level**	0.63	0.52	0.56	**0.63**
		4-level	0.61	0.52	0.54	0.61
		3-level	0.62	0.49	0.49	0.62
	SVM	5-level	0.38	0.55	0.44	0.38
		4-level	0.42	0.55	0.47	0.42
		3-level	0.43	0.55	0.47	0.43
	DT	5-level	0.38	0.55	0.44	0.38
		4-level	0.62	0.49	0.48	0.62
		3-level	0.63	0.71	0.49	**0.63**
	NB	5-level	0.09	0.49	0.12	0.09
		4-level	0.09	0.6	0.09	0.09
		3-level	0.35	0.58	0.33	0.35
Ensemble ML (tree-based)	RF	5-level	0.62	0.61	0.61	0.62
		4-level	0.57	0.51	0.52	0.57
		3-level	0.57	0.48	0.51	0.57
	RF (ET)	5-level	0.56	0.47	0.5	0.56
		4-level	0.57	0.48	0.51	0.57
		3-level	0.57	0.51	0.53	0.57
	GB	5-level	0.63	0.62	0.55	0.63
		4-level	0.63	0.46	0.49	**0.63**
		3-level	0.62	0.39	0.48	0.62
	AB	5-level	0.62	0.38	0.47	0.62
		4-level	0.62	0.45	0.48	0.62
		3-level	0.62	0.45	0.49	0.62

As Table 2 shows, LR outperformed all predictors in terms of F1 score for the five-level cost impact prediction. Additionally, DT and GB displayed better F1 scores for the three- and four-level COR impact predictions. However, the accuracy and F1 scores can be misleading when working with an imbalanced dataset, so to investigate the practicality of the predictors, their ability in predicting each cost impact class needs to be investigated. This is best achieved by measuring the F1 score of each of the cost impact classes. Figure 12 presents the prediction performances of the best-performing classifiers within the benchmark model with the hard and soft voting classifiers.

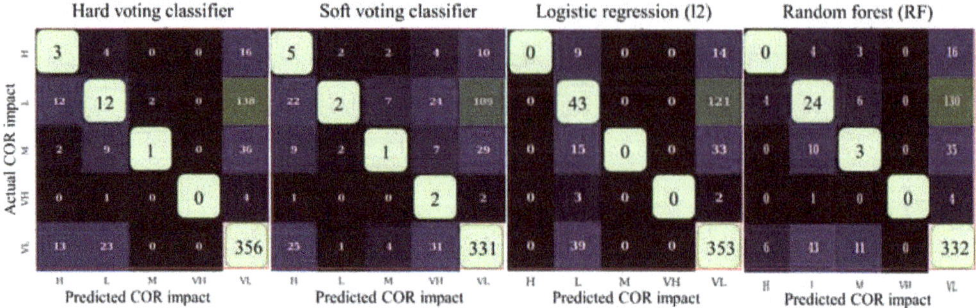

Figure 12. Prediction performance of 5-level classifiers.

As Figure 12 shows, both the soft and the hard voting classifiers were able to detect the high-cost impact items with only a 4% occurrence (support = 23). Among the COR classifiers, however, only one soft voting classifier detected the rework with a cost impact of five (VH) with a low occurrence of 1% (support = five).

A high accuracy with DT was only associated with nonconformance items with a very low cost impact, and it performed poorly for other underrepresented but more important cost impact classes (Figure 13). Likewise, voting classifiers exhibited better performance for medium- and high-impact cost estimation compared with GB, which completely failed to predict COR with a medium impact (Figure 14). Despite its poor class performance in the four-level classification, RF resulted in the best prediction for high-impact COR items without sacrificing overall accuracy. On the other hand, the soft voting classifier proved to be consistent in its precision accuracy for different classification levels.

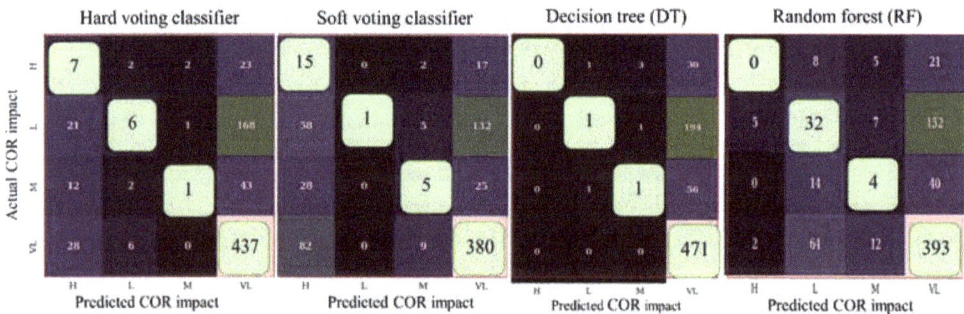

Figure 13. Prediction performance in 4-level classifiers.

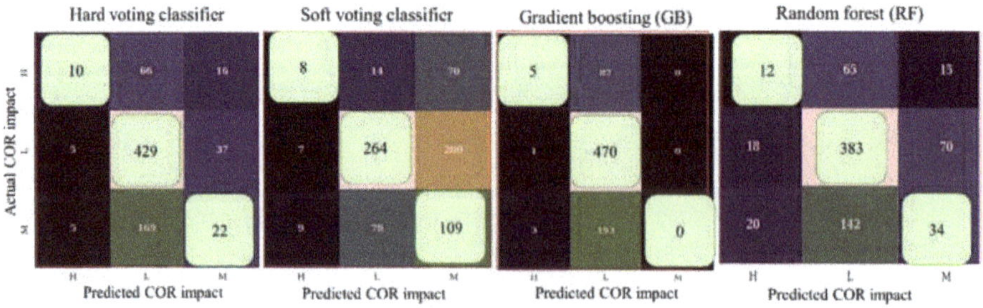

Figure 14. Prediction performance in 3-level classifiers.

To achieve a robust COR predictor, it is important to evaluate the ability of ML to solve the problem. Most single and tree-based ensemble ML predictors failed to estimate COR with high (four) or very high (five) impacts. The superior accuracy of benchmark ML approaches is due to their predicting low-cost impact rework items. However, they are incapable of predicting high-impact cost items. The prediction of low-cost impact rework items cannot reduce the deviation between the as-planned and as-built costs. On the other hand, voting classifiers are more successful in predicting high-impact COR items. The soft voting classifiers were the most robust for COR, displaying a significantly good performance in detecting underrepresented cost impact classes.

To better illustrate the practical implementation of the model, a trained soft voting classifier was used to predict an unseen user input from the test dataset. For example, the user may want to evaluate the material-related nonconformities of a high-rise building project using the available construction team status report and site conditions. The user defines a scenario in which negligence in the initial material inspection due to a lack of site supervision (O-4) results in the receipt of defective material (M-2) from the supplier. The defective material, accompanied by an insufficient review of the design documents (C-15), causes a deviation from the design (C-3). Thus, with respect to the scenario specified by the user, the system can predict the cost impacts of different construction activities under a user-defined scenario. Once the user specifies different activities, such as the facade works (ceramic, coating, insulation, etc.), the system uses the trained soft voting classifier to evaluate its impact on the overall construction cost. In this example, the soft voting classifier can accurately predict a cost impact of three (medium) on the overall construction cost.

6. Contribution to the Body of Knowledge

Despite its importance, construction organizations are rarely aware of the rework impact on their budgets and on safe and environmental performances [34]. Small-sized construction companies still do not appreciate the magnitude of profit loss due to poor quality as they do not usually allocate CQM within a budget [13]. CQM and QC are not limited to error and violation prevention measures, but can also contribute to coping strategies, such as planning alternative countermeasures. The ability to predict COR allows for timely decision making for the required countermeasures, which can also improve the construction time, cost, and quality performance [2]. If a construction budget cannot sustain COR, it is hard to correct a construction error.

In this context, an ensemble COR predictor allows for dynamic and fast cost impact estimations throughout a project and offers more reliable cost impact estimations than the existing, single ML approaches. This is especially useful for cost variation and content analysis using Pareto and pie chart techniques. Throughout the construction lifecycle, the accumulation of project experiences adds to the knowledge of the ensemble models, which can further enhance their cost impact prediction accuracy and thus facilitate enhanced strategic planning by prioritizing the quality control items with the greatest cost impact. Therefore, the proposed COR impact estimator can enhance decision making and the associated planning for construction professionals. Specifically, relationship-style construction contracting models, such as alliance contracts, incorporate an element of error though procuring the construction project under a 'no blame, no fault' culture.

Generally, estimating COR improves cost, schedule, and resource-allocation planning, enhances the creation of the associated contingency plans [16], and also increases the visibility of the expected failure scenarios when purchasing the rework insurance, which is usually added to the general liability policy. This study has focused on estimating the impact of COR on overall construction cost, so the ability of ensemble construction cost predictors to improve the cost impact estimation in these areas remains to be addressed in future research. From a technical perspective, the ensemble method adopted needs to be further improved using different engineering features and optimization techniques to enhance the model prediction accuracy, especially for the underrepresented cost classes.

7. Conclusions

The early estimation of COR offers several benefits for construction professionals in terms of increasing the preparedness of the construction budget for dealing with risk (i.e., COR). On the one hand, the construction quality management literature is still limited with regard to the ML-based COR predictors, while on the other hand, the developed single ML predictors within the other cost estimation fields are not responsive for underrepresented classes with limited data records. However, a COR predictor does not solve this problem unless it can predict all the cost impact classes. Therefore, this study has proposed a robust ensemble ML predictor for estimating the cost of construction rework.

The adopted ensemble voting classifiers proved to be more effective in predicting the underrepresented high-cost impact construction rework activities than the benchmark models. Both single ML and tree-based predictors failed to estimate COR with (very) high-cost impacts on overall construction budgets. Additionally, the soft voting classifier proved to be consistent in the accuracy of its prediction outcomes and was able to classify all the different COR impacts for three-, four-, and five-level classification tasks. The developed COR impact predictor increases the reliability and accuracy of the cost impact estimation, which, in turn, enables dynamic cost variation analysis and thus improves cost-based decision making.

COR has many undesirable effects, from cost fluctuations to the waste of material and labor and equipment hours. Ultimately, it is one of the crucial aspects of sustainability in construction. Thus, the early identification of high-cost impact rework items allows for a focus on countermeasures to prevent critical rework items. This in turn reduces the waste in construction flow, time, and material consumption while enhancing the different aspects of project performance, such as budget and quality performance. Finally, the discussed aspects of the project improve its overall sustainability level in terms of quality, economy, and waste criteria. Therefore, we recommend the further exploration of the use of different ML methods to predict and reduce COR.

Author Contributions: Conceptualization, V.T. and F.M.; formal analysis, V.T. and F.M.; writing—original draft preparation, V.T. and F.M.; writing—review and editing, V.T., F.M., Y.E.A. and O.B.T.; data curation, Y.E.A. and O.B.T.; resources, Y.E.A. and O.B.T. All authors have read and agreed to the published version of the manuscript.

Funding: This research received no external funding.

Institutional Review Board Statement: Not applicable.

Informed Consent Statement: Not applicable.

Data Availability Statement: Some or all data, models, or codes that support the findings of this study are available from the corresponding author upon reasonable request.

Conflicts of Interest: The authors declare no conflict of interest.

References

1. Mohamed, H.H.; Ibrahim, A.H.; Soliman, A.A. Toward Reducing Construction Project Delivery Time under Limited Resources. *Sustainability* **2021**, *13*, 11035. [CrossRef]
2. Ye, G.; Jin, Z.; Xia, B.; Skitmore, M. Analyzing Causes for Reworks in Construction Projects in China. *J. Manag. Eng.* **2015**, *31*, 04014097. [CrossRef]
3. Hwang, B.-G.; Zhao, X.; Goh, K.J. Investigating the Client-Related Rework in Building Projects: The Case of Singapore. *Int. J. Proj. Manag.* **2014**, *32*, 698–708. [CrossRef]
4. Love, P.; Smith, J. Unpacking the Ambiguity of Rework in Construction: Making Sense of the Literature. *Civ. Eng. Environ. Syst.* **2018**, *35*, 180–203. [CrossRef]
5. Asadi, R.; Wilkinson, S.; Rotimi, J.O.B. Towards Contracting Strategy Usage for Rework in Construction Projects: A Comprehensive Review. *Constr. Manag. Econ.* **2021**, *39*, 953–971. [CrossRef]
6. Al-Janabi, A.M.; Abdel-Monem, M.S.; El-Dash, K.M. Factors Causing Rework and Their Impact on Projects' Performance in Egypt. *J. Civ. Eng. Manag.* **2020**, *26*, 666–689. [CrossRef]
7. Simpeh, E.K.; Ndihokubwayo, R.; Love, P.E.D.; Thwala, W.D. A Rework Probability Model: A Quantitative Assessment of Rework Occurrence in Construction Projects. *Int. J. Constr. Manag.* **2015**, *15*, 109–116. [CrossRef]

8.   Love, P.E.D.; Sing, C.-P. Determining the Probability Distribution of Rework Costs in Construction and Engineering Projects. *Struct. Infrastruct. Eng.* **2013**, *9*, 1136–1148. [CrossRef]

9.   Love, P.E.D. Influence of Project Type and Procurement Method on Rework Costs in Building Construction Projects. *J. Constr. Eng. Manag.* **2002**, *128*, 18–29. [CrossRef]

10.   Khalesi, H.; Balali, A.; Valipour, A.; Antucheviciene, J.; Migilinskas, D.; Zigmund, V. Application of Hybrid SWARA–BIM in Reducing Reworks of Building Construction Projects from the Perspective of Time. *Sustainability* **2020**, *12*, 8927. [CrossRef]

11.   Hwang, B.-G.; Thomas, S.R.; Haas, C.T.; Caldas, C.H. Measuring the Impact of Rework on Construction Cost Performance. *J. Constr. Eng. Manag.* **2009**, *135*, 187–198. [CrossRef]

12.   Schiffauerova, A.; Thomson, V. A Review of Research on Cost of Quality Models and Best Practices. *Int. J. Qual. Reliab. Manag.* **2006**, *23*, 647–669. [CrossRef]

13.   Abdelsalam, H.M.E.; Gad, M.M. Cost of Quality in Dubai: An Analytical Case Study of Residential Construction Projects. *Int. J. Proj. Manag.* **2009**, *27*, 501–511. [CrossRef]

14.   Love, P.E.D.; Matthews, J.; Sing, M.C.P.; Porter, S.R.; Fang, W. State of Science: Why Does Rework Occur in Construction? What Are Its Consequences? And What Can Be Done to Mitigate Its Occurrence? *Engineering*, 2022; *in press*. [CrossRef]

15.   Milion, R.N.; Alves, T.; da C. L. Alves, T.; Paliari, J.C.; Liboni, L.H.B. CBA-Based Evaluation Method of the Impact of Defects in Residential Buildings: Assessing Risks towards Making Sustainable Decisions on Continuous Improvement Activities. *Sustainability* **2021**, *13*, 6597. [CrossRef]

16.   Shoar, S.; Chileshe, N.; Edwards, J.D. Machine Learning-Aided Engineering Services' Cost Overruns Prediction in High-Rise Residential Building Projects: Application of Random Forest Regression. *J. Build. Eng.* **2022**, *50*, 104102. [CrossRef]

17.   Knyziak, P. The Impact of Construction Quality on the Safety of Prefabricated Multi-Family Dwellings. *Eng. Fail. Anal.* **2019**, *100*, 37–48. [CrossRef]

18.   Love, P.E.D.; Ika, L.; Luo, H.; Zhou, Y.; Zhong, B.; Fang, W. Rework, Failures, and Unsafe Behavior: Moving Toward an Error Management Mindset in Construction. *IEEE Trans. Eng. Manag.* **2022**, *69*, 1489–1501. [CrossRef]

19.   Fayek, A.R.; Dissanayake, M.; Campero, O. Developing a Standard Methodology for Measuring and Classifying Construction Field Rework. *Can. J. Civ. Eng.* **2004**, *31*, 1077–1089. [CrossRef]

20.   Sugawara, E.; Nikaido, H. Properties of AdeABC and AdeIJK Efflux Systems of Acinetobacter Baumannii Compared with Those of the AcrAB-TolC System of Escherichia Coli. *Antimicrob. Agents Chemother.* **2014**, *58*, 7250–7257. [CrossRef]

21.   Burati, J.L.; Farrington, J.J.; Ledbetter, W.B. Causes of Quality Deviations in Design and Construction. *J. Constr. Eng. Manag.* **1992**, *118*, 34–49. [CrossRef]

22.   Abdul-Rahman, H. The Cost of Non-Conformance during a Highway Project: A Case Study. *Constr. Manag. Econ.* **1995**, *13*, 23–32. [CrossRef]

23.   Josephson, P.E.; Hammarlund, Y. Causes and Costs of Defects in Construction a Study of Seven Building Projects. *Autom. Constr.* **1999**, *8*, 681–687. [CrossRef]

24.   Barber, P.; Graves, A.; Hall, M.; Sheath, D.; Tomkins, C. Quality Failure Costs in Civil Engineering Projects. *Int. J. Qual. Reliab. Manag.* **2000**, *17*, 479–492. [CrossRef]

25.   Ashford, J.L. *The Management of Quality in Construction*; Routledge: London, UK, 2002; ISBN 9781135833862.

26.   Campanella, J. Principles of Quality Costs: Principles, Implementation, and Use. In Proceedings of the Annual Quality Congress Proceedings, Milwaukee, WI, USA, 4–8 May 1998; p. 507.

27.   Bajpai, A.K.; Willey, P.C.T. Questions about Quality Costs. *Int. J. Qual. Reliab. Manag.* **1989**, *6*, 6. [CrossRef]

28.   Ziegel, E.R. Juran's Quality Control Handbook. *Technometrics* **1990**, *32*, 97–98. [CrossRef]

29.   Love, P.E.D.; Smith, J. Error Management: Implications for Construction. *Constr. Innov.* **2016**, *16*, 418–424. [CrossRef]

30.   Love, P.E.D.; Lopez, R.; Edwards, D.J. Reviewing the Past to Learn in the Future: Making Sense of Design Errors and Failures in Construction. *Struct. Infrastruct. Eng.* **2013**, *9*, 675–688. [CrossRef]

31.   Love, P.E.D. Creating a Mindfulness to Learn from Errors: Enablers of Rework Containment and Reduction in Construction. *Dev. Built Environ.* **2020**, *1*, 100001. [CrossRef]

32.   Love, P.E.D.; Matthews, J.; Fang, W. Rework in Construction: A Focus on Error and Violation. *J. Constr. Eng. Manag.* **2020**, *146*, 1901. [CrossRef]

33.   Love, P.E.D.; Smith, J.; Teo, P. Putting into Practice Error Management Theory: Unlearning and Learning to Manage Action Errors in Construction. *Appl. Ergon.* **2018**, *69*, 104–111. [CrossRef]

34.   Matthews, J.; Love, P.E.D.; Porter, S.R.; Fang, W. Smart Data and Business Analytics: A Theoretical Framework for Managing Rework Risks in Mega-Projects. *Int. J. Inf. Manag.* **2022**, *65*, 102495. [CrossRef]

35.   Love, P.; Ika, L.; Matthews, J.; Fang, W.; Carey, B. The Duality and Paradoxical Tensions of Quality and Safety: Managing Error in Construction Projects. *IEEE Trans. Eng. Manag.* **2022**, 1–8. [CrossRef]

36.   Love, P.E.D.; Smith, J.; Ackermann, F.; Irani, Z.; Fang, W.; Luo, H.; Ding, L. Houston, We Have a Problem! Understanding the Tensions between Quality and Safety in Construction. *Prod. Plan. Control* **2019**, *30*, 1354–1365. [CrossRef]

37.   Love, P.E.D.; Teo, P.; Morrison, J. Unearthing the Nature and Interplay of Quality and Safety in Construction Projects: An Empirical Study. *Saf. Sci.* **2018**, *103*, 270–279. [CrossRef]

38.   Love, P.E.D.; Smith, J.; Ackermann, F.; Irani, Z.; Teo, P. The Costs of Rework: Insights from Construction and Opportunities for Learning. *Prod. Plan. Control* **2018**, *29*, 1082–1095. [CrossRef]

39. Hall, M.; Tomkins, C. A Cost of Quality Analysis of Building Project: Towards a Complete Methodology for Design and Build. *Constr. Manag. Econ.* **2001**, *19*, 727–740. [CrossRef]
40. Love, P.E.D.; Manual, P.; Li, H. Determining the Causal Structure of Rework Influences in Construction. *Constr. Manag. Econ.* **1999**, *17*, 505–517. [CrossRef]
41. Peter, E.; Love, D.; Heng, L.I. Quantifying the Causes and Costs of Rework in Construction. *Constr. Manag. Econ.* **2000**, *18*, 479–490. [CrossRef]
42. Davis, K.; Ledbetter, W.B.; Burati, J.L. Measuring Design and Construction Quality Costs. *J. Constr. Eng. Manag.* **1989**, *115*, 385–400. [CrossRef]
43. Zhao, Y.; Kok Foong, L. Predicting Electrical Power Output of Combined Cycle Power Plants Using a Novel Artificial Neural Network Optimized by Electrostatic Discharge Algorithm. *Measurement* **2022**, *198*, 111405. [CrossRef]
44. Nejati, F.; Tahoori, N.; Sharifian, M.A.; Ghafari, A.; Nehdi, M.L. Estimating Heating Load in Residential Buildings Using Multi-Verse Optimizer, Self-Organizing Self-Adaptive, and Vortex Search Neural-Evolutionary Techniques. *Buildings* **2022**, *12*, 1328. [CrossRef]
45. Chong, M.; Abraham, A.; Paprzycki, M. Traffic Accident Analysis Using Machine Learning Paradigms. *Informatica* **2005**, *29*, 89–98. [CrossRef]
46. Liang, Y.; Reyes, M.L.; Lee, J.D. Real-Time Detection of Driver Cognitive Distraction Using Support Vector Machines. *IEEE Trans. Intell. Transp. Syst.* **2007**, *8*, 340–350. [CrossRef]
47. Hacıefendioğlu, K.; Mostofi, F.; Toğan, V.; Başağa, H.B. CAM-K: A Novel Framework for Automated Estimating Pixel Area Using K-Means Algorithm Integrated with Deep Learning Based-CAM Visualization Techniques. *Neural Comput. Appl.* **2022**, *34*, 17741–17759. [CrossRef]
48. Bodendorf, F.; Merkl, P.; Franke, J. Intelligent Cost Estimation by Machine Learning in Supply Management: A Structured Literature Review. *Comput. Ind. Eng.* **2021**, *160*, 107601. [CrossRef]
49. Moayedi, H.; Mosavi, A. An Innovative Metaheuristic Strategy for Solar Energy Management through a Neural Networks Framework. *Energies* **2021**, *14*, 1196. [CrossRef]
50. Zhu, M.; Li, Y.; Wang, Y. Design and Experiment Verification of a Novel Analysis Framework for Recognition of Driver Injury Patterns: From a Multi-Class Classification Perspective. *Accid. Anal. Prev.* **2018**, *120*, 152–164. [CrossRef]
51. Meharie, M.G.; Mengesha, W.J.; Gariy, Z.A.; Mutuku, R.N.N. Application of Stacking Ensemble Machine Learning Algorithm in Predicting the Cost of Highway Construction Projects. *Eng. Constr. Archit. Manag.* **2021**; *ahead-of-print*. [CrossRef]
52. Matías, J.M.; Rivas, T.; Martín, J.E.; Taboada, J. A Machine Learning Methodology for the Analysis of Workplace Accidents. *Int. J. Comput. Math.* **2008**, *85*, 559–578. [CrossRef]
53. Mostofi, F.; Toğan, V.; Başağa, H.B. House Price Prediction: A Data-Centric Aspect Approach on Performance of Combined Principal Component Analysis with Deep Neural Network Model. *J. Constr. Eng. Manag. Innov.* **2021**, *4*, 106–116. [CrossRef]
54. Ding, C.; Wu, X.; Yu, G.; Wang, Y. A Gradient Boosting Logit Model to Investigate Driver's Stop-or-Run Behavior at Signalized Intersections Using High-Resolution Traffic Data. *Transp. Res. Part C Emerg. Technol.* **2016**, *72*, 225–238. [CrossRef]
55. Wang, Y.-R.; Yu, C.-Y.; Chan, H.-H. Predicting Construction Cost and Schedule Success Using Artificial Neural Networks Ensemble and Support Vector Machines Classification Models. *Int. J. Proj. Manag.* **2012**, *30*, 470–478. [CrossRef]
56. Farid, A.; Abdel-Aty, M.; Lee, J. A New Approach for Calibrating Safety Performance Functions. *Accid. Anal. Prev.* **2018**, *119*, 188–194. [CrossRef] [PubMed]
57. Jiang, Q. Estimation of Construction Project Building Cost by Back-Propagation Neural Network. *J. Eng. Des. Technol.* **2019**, *18*, 601–609. [CrossRef]
58. Bala, K.; Bustani, S.A.; Waziri, B.S. A Computer-Based Cost Prediction Model for Institutional Building Projects in Nigeria. *J. Eng. Des. Technol.* **2014**, *12*, 519–530. [CrossRef]
59. Peško, I.; Mučenski, V.; Šešlija, M.; Radović, N.; Vujkov, A.; Bibić, D.; Krklješ, M. Estimation of Costs and Durations of Construction of Urban Roads Using ANN and SVM. *Complexity* **2017**, *2017*, 2450370. [CrossRef]
60. Wu, P.; Liu, A.; Fu, J.; Ye, X.; Zhao, Y. Autonomous Surface Crack Identification of Concrete Structures Based on an Improved One-Stage Object Detection Algorithm. *Eng. Struct.* **2022**, *272*, 114962. [CrossRef]
61. Zheng, Y.; Gao, Y.; Lu, S.; Mosalam, K.M. Multistage Semisupervised Active Learning Framework for Crack Identification, Segmentation, and Measurement of Bridges. *Comput.-Aided Civ. Infrastruct. Eng.* **2022**, *37*, 1089–1108. [CrossRef]
62. Hacıefendioğlu, K.; Ayas, S.; Başağa, H.B.; Toğan, V.; Mostofi, F.; Can, A. Wood Construction Damage Detection and Localization Using Deep Convolutional Neural Network with Transfer Learning. *Eur. J. Wood Wood Prod.* **2022**, *80*, 791–804. [CrossRef]
63. Sholevar, N.; Golroo, A.; Esfahani, S.R. Machine Learning Techniques for Pavement Condition Evaluation. *Autom. Constr.* **2022**, *136*, 104190. [CrossRef]
64. Fan, C.-L. Defect Risk Assessment Using a Hybrid Machine Learning Method. *J. Constr. Eng. Manag.* **2020**, *146*, 04020102. [CrossRef]
65. Kim, E.; Ji, H.; Kim, J.; Park, E. Classifying Apartment Defect Repair Tasks in South Korea: A Machine Learning Approach. *J. Asian Archit. Build. Eng.* **2021**, *21*, 2503–2510. [CrossRef]
66. Doğan, N.B. Predicting the Cost Impacts of Construction Nonconformities Using CBR-AHP And CBR-GA Models. Master's Thesis, Middle East Technical University, Ankara, Turkey, 2021.

67. Saqlain, M.; Jargalsaikhan, B.; Lee, J.Y. A Voting Ensemble Classifier for Wafer Map Defect Patterns Identification in Semiconductor Manufacturing. *IEEE Trans. Semicond. Manuf.* **2019**, *32*, 171–182. [CrossRef]
68. Kansara, D.; Singh, R.; Sanghvi, D.; Kanani, P. Improving Accuracy of Real Estate Valuation Using Stacked Regression. *Int. J. Eng. Dev. Res.* **2018**, *6*, 571–577.
69. Neloy, A.A.; Haque, H.M.S.; Islam, M.M.U. Ensemble Learning Based Rental Apartment Price Prediction Model by Categorical Features Factoring. In Proceedings of the 2019 11th International Conference on Machine Learning and Computing—ICMLC '19, Zhuhai, China, 22–24 February 2019; ACM Press: New York, NY, USA, 2019; Volume Part F1481, pp. 350–356.
70. Harris, C.R.; Millman, K.J.; van der Walt, S.J.; Gommers, R.; Virtanen, P.; Cournapeau, D.; Wieser, E.; Taylor, J.; Berg, S.; Smith, N.J.; et al. Array Programming with NumPy. *Nature* **2020**, *585*, 357–362. [CrossRef]
71. Pedregosa, F.; Varoquaux, G.; Gramfort, A.; Dubourg, V.; Thirion, B.; Grisel, O.; Brucher, M.; Prettenhofer, P.; Weiss, R.; Dubourg, V.; et al. Scikit-Learn: Machine Learning in Python. *J. Mach. Learn. Res.* **2011**, *12*, 2825–2830.
72. Buitinck, L.; Louppe, G.; Blondel, M.; Pedregosa, F.; Müller, A.C.; Grisel, O.; Niculae, V.; Prettenhofer, P.; Gramfort, A.; Grobler, J.; et al. API Design for Machine Learning Software: Experiences from the Scikit-Learn Project. *arXiv* **2013**, arXiv:1309.0238. [CrossRef]
73. Zhou, P.; El-Gohary, N. Domain-Specific Hierarchical Text Classification for Supporting Automated Environmental Compliance Checking. *J. Comput. Civ. Eng.* **2016**, *30*, 513. [CrossRef]

 *buildings*

Article

Exploring the Project-Based Collaborative Networks between Owners and Contractors in the Construction Industry: Empirical Study in China

Fangliang Wang, Min Cheng * and Xiaotong Cheng

Department of Management Science and Engineering, School of Management, Shanghai University, Shanghai 200444, China
* Correspondence: chengmin@shu.edu.cn

Abstract: In the project-based construction industry, organizations build collaborative relationships through specific projects. The owners and contractors who are the key project stakeholders have gradually formed a complex project-based industry-level collaborative network in many different projects, closely related to knowledge exchange and industry development. Based on the data set of the National Quality Engineering Award (NQEA) projects in China from 2013 to 2021, we empirically analyze the characteristics and evolution of project-based collaborative networks between owners and contractors in the construction industry by using social network analysis (SNA) and network motif analysis (NMA) method. The results show that (1) the owner–contractor collaborative network exhibits small-world network characteristics. The island effect caused by small groups in the network makes the overall connectivity of the network low. During the study period, the collaborative network became more compact. (2) State-owned construction companies, such as China Construction Third Engineering Bureau Corporation Limited, China Construction Eighth Engineering Bureau Corporation Limited, and China Construction Second Engineering Bureau Corporation Limited, with high degree centrality and betweenness centrality, are the core companies in the collaborative network. In China, state-owned construction enterprises are favored by owners and have established collaborative relationships with many owners and contractors. (3) There are two local collaborative patterns in the collaborative network: motif and anti-motif. Motifs include some triangle-based tight collaborative patterns, while anti-motifs involve some loose binary collaborative patterns. The results help understand the structure and evolution of the industry-level collaborative relationship network between owners and contractors and can provide references for owners and contractors to develop relationship cultivation strategies more effectively.

Keywords: social network analysis (SNA); network motif analysis (NMA); collaborative relationship; owner; contractor

Citation: Wang, F.; Cheng, M.; Cheng, X. Exploring the Project-Based Collaborative Networks between Owners and Contractors in the Construction Industry: Empirical Study in China. *Buildings* **2023**, *13*, 732. https://doi.org/10.3390/buildings13030732

Academic Editors: Rafiq Muhammad Choudhry and Annie Guerriero

Received: 18 December 2022
Revised: 16 February 2023
Accepted: 5 March 2023
Published: 10 March 2023

1. Introduction

The construction industry is a project-based industry. Construction projects are complex and usually involve multiple stakeholders. Among them, the project owner is the initiator of the construction project, and the contractor undertakes the construction tasks of the project [1,2]. The owner and the contractor collaborate on a specific project, and the collaborative relationship between them is essential to the success of the construction activity [3,4]. Due to the temporary nature of construction projects, an owner or a contractor constantly develops new collaborative relationships with new owners or contractors in new projects. The number of construction projects in China has grown substantially over the past decade. The contract value of new projects in 2021 was USD 5.12 trillion, three times that of 2011 [5]. The massive increase in the number of projects has involved more and more owners and contractors. Lee et al. [6] believed that owners and contractors had gradually formed a complex collaborative relationship network based on their intertwined collaboration in different projects. An organization's position in this collaborative network reflects its competitiveness in the construction market and its influence on the industry and determines its ability to access external resources and information [3,7].

As new projects are implemented, new participants and relationships are continuously embedded into the network. Therefore, the collaborative relationship network is dynamic, and this change will affect the exchange of information between owners and contractors and their future collaborative relationships [8,9]. According to the Industrial Marketing and Procurement (IMP) group, organizations should build long-term collaborative relationships to achieve mutual benefits and enhance competitiveness [10]. Studying the characteristics and evolution of the relationship network formed by owners and contractors in a certain period from a dynamic perspective helps understand their collaboration mechanism and the change of an organization's position in the construction market to provide a basis for formulating future collaboration strategies.

However, previous studies mainly focused on the one-time and short-term collaborative relationship between owners and contractors in particular projects [11,12]. There is a lack of industry-level exploration of the structural characteristics and evolution of the collaborative networks formed by numerous owners and contractors when they are involved in different projects. Although some studies on collaborative networks formed by different types of stakeholders in various projects involved owners and contractors, they focused on specific types of projects, such as skyscraper projects, BIM projects, and green building projects [13–15]. In fact, owners and contractors have formed intricate collaborative relationships based on their involvement in different types of construction projects. Studies based on broader boundaries can provide a more comprehensive insight into their collaboration.

Collaborative relationship network analysis is the foundation of organizational network governance, which is a long-term, selective, structured, and autonomous collection of organizations [16,17]. Unlike an organization concerned with maximizing its interests, organizational network governance focuses on the interactions between organizations and their performance in the network. Social network analysis (SNA) is a commonly used method to explore the macro-structural features of complex collaborative networks [18]. Meanwhile, the network motif analysis (NMA) method can be applied to analyze the local topology and micro-structural features of collaborative networks [19]. To bridge the knowledge gap in industry-level owner–contractor collaborative network research, we combined SNA and NMA to study the structural characteristics and dynamic evolution of the collaborative networks, which were established based on thousands of collaborative relationships among owners and contractors in the construction projects that won China's National Quality Engineering Award (NQEA) from 2013 to 2021.

This study aims to characterize the macro-structure and micro-structure of the collaborative networks formed by numerous owners and contractors involved in different projects and how they evolved over time by using the data of NQEA projects in China and combining SNA and NMA methods in order to help owners and contractors clarify their position in the partnership network and provide a reference for them to formulate future relationship cultivation strategies. The remainder of this study is organized as follows. In Section 2, the literature on the owner–contractor relationship and the collaborative network analysis in the construction field are reviewed. Section 3 presents the research methodology, and Section 4 explains the analytical procedures and data collection. In Section 5, the characteristics and evolution of the collaborative network formed by owners and contractors are analyzed based on SNA and NMA, followed by a discussion and some managerial implications.

2. Literature Review

2.1. Owner–Contractor Relationships

The collaborative relationship between owners and contractors affects the implementation of construction projects. Previous studies have analyzed the relationship between the owner and the contractor from three aspects.

First, some scholars explored the owner–contractor relationship in different delivery systems adopted for construction projects. For example, Li and Feng [20] explored the strategies for enhancing the trust relationship between owners and contractors in project management contracting (PMC) projects. Sun et al. [21] argued that effective collaboration between owners and general contractors improved the level of BIM adoption in engineering, procurement, and construction (EPC) projects. Collecting questionnaires from 243 Chinese project professionals, Zhang et al. [22] demonstrated that the level of design provided by the owner had an impact on the quality of the contractor's design in the design–build (DB) projects.

Second, some scholars explored the factors that influenced collaborative relationships between owners and contractors. For example, Suprapto et al. [23] revealed that relational attitudes, collaborative practices, and teams' joint capability influenced the collaborative relationship between the owner and the contractor. Jiang et al. [24] found that reputation, competence, honesty, communication, reciprocity, and contracts effectively influenced the establishment of trust relationship between owners and contractors. Zhang and Qian [25] analyzed how the mediated power influenced opportunism in owner–contractor relationships. Tai et al. [26] analyzed the factors influencing owners' trust in contractors in construction projects. Suprapto et al. [27] found that shared team responsibility, execution-focused teams, common capability and structures, and senior leadership pair can be effective in improving the relationship between owners and contractors.

Finally, the interaction mechanism between the owner and the contractor is also one of the research focuses. For example, Zhang et al. [28] discussed the combined influence of the owner's power and contractual mechanism on the behavior of contractors in China. Qian et al. [29] emphasized that there is both cooperation and competition in the relationship between the owner and the contractor and that when the two are balanced against each other, greater value can be created in the project for maximum benefit. Based on a contract management perspective, Nasir and Hadikusumo [30] developed an integrated model to manage the relationship between the owner and the contractor. Zhao et al. [31] believed that there was a strong reciprocity relationship between the owner and the contractor, and the parties accepted and maintained specific cooperation.

Although previous studies support understanding the collaborative mechanism between the owner and the contractor, their relationships were typically regarded as a binary structure or explored in the context of a specific project. A construction project is implemented by a temporary organizational alliance, which forms a temporary collaborative network [32]. The owners and contractors are constantly expanding into new project-based partnerships as they engage in new projects. From an industry perspective, they gradually form a long-term organizational network with a specific structure in project-based collaboration [33]. Project-based industry-level collaborative networks are more complex and dynamic than project-level networks, which characterize collaborative relationships within a single project. Analysis of project-based industry-level collaborative networks not only helps to understand an organization's position in the industry and its competitiveness in the construction market, but also reveal the organization's collaboration preferences, which can provide a basis for the organization to choose partners [34,35]. However, there is still a lack of research on the characteristics of owner–contractor industry-level collaborative relationship networks and the evolution of the network structures.

2.2. Collaborative Network Analysis in the Construction Field

In today's business environment, collaboration is seen as a way for organizations to acquire new business opportunities and facilitate the formation of a networked society [36].

Organizations can improve their market competitiveness by strengthening their position in the network [9]. As a result, the strategic focus of organizations has shifted from focusing solely on their operational performance to network-based collaboration and competition [37,38]. It is increasingly important to understand the relationship structure of the collaboration network and the position of the organization in the network.

Some scholars have adopted the social network analysis (SNA) method to study the collaborative network in the construction field from the industry perspective. The collaborative network formed by multiple different types of stakeholders is one of the research focuses. For example, Han et al. [13] studied the structural characteristics of the collaboration network formed by different owners, general contractors, design firms, and project managers involved in 422 skyscraper projects worldwide from 1990 to 2010. Tang et al. [14] explored the collaborative relationship network formed by owners, design consultants, and major contractors in Hong Kong's BIM projects from 2002 to 2017. Qiang et al. [15] explored how the collaborative networks formed by owners, contractors, and designers in the implementation of multiple green building projects evolve over time based on the SNA method. In addition, some scholars have studied the collaborative network formed by one or two kinds of stakeholders (such as the contractor–subcontractor collaboration network and the contractor–contractor collaboration network). For example, Tang et al. [35] studied the collaborative relationship between contractors and subcontractors in China's construction industry based on the data set of projects that won the China Construction Engineering Luban Prize, and the results provided a reference for contractors to choose subcontractors. Akgul et al. [39] used the SNA method to investigate the collaborative relationship of contractors in Turkey while participating in overseas projects based on the data from 449 projects in 46 countries. Liu et al. [40] used some indicators of the SNA method to characterize the collaborative network among contractors in China's construction industry. Liu et al. [41] analyzed the characteristics of the collaborative network among China's construction firms using the SNA method based on 251 international construction projects constructed by China's 156 construction firms in cooperation. Park et al. [42] investigated the collaborative networks of Korean construction firms formed in 389 overseas projects using the SNA method.

The above studies used the SNA method to describe the complex relationship and macro-structure characteristics of the collaborative network in the construction field. The research results can clarify the organization's influence in the owner–contractor collaborative network and provide a reference for organizations to develop cooperation strategies. The existing research mainly focuses on contractor–subcontractor collaborative networks, contractor–contractor collaborative networks, and collaborative networks among multiple types of stakeholders. The owner and the contractor are the main stakeholders in the construction project. They gradually form a certain relationship network by participating in multiple projects. Understanding the characteristics of the relationship network between the two is helpful for the owner to select contractors and the contractor's bidding decision. Although previous studies on collaborative networks of multiple types of stakeholders involved owners and contractors, they focused on a specific project type. At present, there is still a lack of research on the relationship network of owners and contractors at the industry level based on extensive project data. Moreover, the collaborative networks studies based on SNA focus on exploring the macro-structural features of the network but cannot reveal the local patterns of collaboration.

To explore the micro-structure of complex networks more deeply, researchers shift their attention from focusing on the global properties of the network to local properties. Milo et al. [43] proposed network motifs to reflect a particular local pattern of network interactions, providing new insights for understanding the network structure and relational characteristics. Network motif analyses (NMA) have been applied to explore biochemical networks, ecological networks, neurobiological networks, traffic networks, and energy networks [44–47]. There are also a few scholars who use it to explore the local structural characteristics in organizational networks [48]. It would be meaningful and interesting to explore the local relationship patterns of owner–contractor collaboration networks based on the method of NMA to discover the evolution of their collaborative patterns from a local network perspective. The introduction of NMA can overcome the limitation of SNA, which focuses on exploring the macro-structure characteristics of the network rather than the local properties. Therefore, we will combine SNA and NMA to analyze the owner–contractor collaboration network and to reveal the characteristics and evolution of the macro-structure of the overall network and the micro-structure of the local network.

3. Methods

The collaborative relationship between owners and contractors based on different projects involves a complex network connecting different organizations. To fully understand this relationship's structural characteristics and evolution laws, SNA and NMA are applied in the study. Specifically, SNA is used to capture the overall structural characteristics and node locations of the owner–contractor collaborative network from a macro-level. MNA is applied to discover the structure of subgroups in networks and reveal local collaborative patterns from a micro-level.

3.1. Social Network Analysis (SNA)

The measurement for the macro-structure of the owner–contractor collaborative network based on SNA includes the network-level and the node-level indicators.

3.1.1. Network-Level Measurement

In the SNA method, four network indicators, including density, average degree, average distance, and clustering coefficient, are generally used to analyze the characteristics and evolution of the network.

(1) Density

Density refers to the ratio of the actual number of connections to the maximum possible number of connections in the network and can measure the degree of interconnection between nodes in the network [49,50]. The value of density is between 0 and 1. When all nodes in the network are connected to each other, the value of density is 1. When the nodes in the network are all isolated, the density value is 0 [51]. The calculation formula of density D is as follows.

$$D = \frac{2E}{N(N-1)} \tag{1}$$

where E refers to the number of connections between these nodes, and N refers to the number of all nodes in the network.

(2) Average degree

The average degree refers to the average number of connections to a node in the network, reflecting the network's tightness [34]. The larger the value of the average degree in the collaborative network, the tighter the network is [52]. The average degree AD is formulated as follows.

$$AD = \frac{E}{N} \tag{2}$$

(3) Average distance

In the undirected network, the number of connections in a path between two nodes is defined as the length of the path, and the length of the shortest path is defined as the distance between the two nodes [53]. The average distance of a network is the average of the shortest path length between pairs of nodes in the network, and it is used to measure the ease of communication between nodes [54]. The average distance L is calculated as follows.

$$L = \frac{\sum_{i \geq j} d_{ij}}{N(N+1)/2} \tag{3}$$

where d_{ij} represents the shortest path length from node i to node j.

(4) Clustering coefficient

The clustering coefficient of a node is the ratio of the number of actual connections between the node and its neighbors to the number of the maximum possible connections between those nodes. The clustering coefficient of the whole network is the average of the clustering coefficients of all the nodes [55]. The clustering coefficient is used to describe the extent to which a node is embedded in the network's local group and reflects the aggregation extent of networks [56]. The value of the clustering coefficient ranges from 0 to 1. The clustering coefficient (CC) is 1 when all nodes are interconnected in the network, while CC is 0 when all nodes are not connected. CC is expressed as follows.

$$CC = \sum_{i=1}^{N} \frac{2e_i}{k_i(k_i - 1)} \tag{4}$$

where k_i is the number of neighbors of the ith node, and e_i refers to the number of connections between these neighbors.

3.1.2. Node-Level Measurement

In the network, the transmission of information is affected by the location of nodes. Centrality is a commonly used indicator to measure the location and status of nodes in the network, which helps figure out the core nodes, i.e., nodes that are relatively more connected with other nodes [57]. Betweenness centrality and degree centrality are two indicators commonly used for centrality analysis [58,59].

(1) Degree centrality

Degree centrality refers to the number of direct connections a node has to other nodes [60]. Additionally, the normalized degree centrality is defined as the ratio of the number of direct connections of a node and the total number of connections in the network [61]. Generally, the higher the degree centrality a node has, the greater its influence is on the network [62]. The normalized degree centrality $C_D(i)$ is calculated as follows.

$$C_D(i) = \frac{\sum_{j=1}^{N} e_{i,j}}{N - 1} \tag{5}$$

where $e_{i,j}$ is the number of connections between node i and node j, and N is the total number of nodes in the network.

(2) Betweenness centrality

Betweenness centrality reflects the extent to which a node is located on the shortest paths between pairs of other nodes [63]. The greater the betweenness centrality of a node, the more it can influence the connections between other nodes [64]. The normalized betweenness centrality $C_B(i)$ can be calculated as follows.

$$C_B(i) = \sum_{j<k} \frac{g_{jk}(i)}{g_{jk}} \bigg/ \frac{(N-1)(N-2)}{2} \tag{6}$$

where $g_{jk}(i)$ refers to the number of shortest paths traversing node i, and g_{jk} is the total amount of the shortest paths between node j and node k.

3.2. Network Motif Analysis (NMA)

Network motifs are small connected subgraphs of 3–7 nodes that occur in real networks, the number of which is significantly higher than that in random networks [65,66]. Conversely, subgraphs that appear less frequently than in random networks are defined as anti-motifs [43]. The Z-Score is a statistical significance indicator, which is often used to determine the network motif and assess the importance of the motif structure in the network [67]. The Z-Score for each subgroup is represented as follows.

$$Z_i = \frac{N_{real_i} - N_{rand_i}}{\sigma_{rand_i}} \tag{7}$$

where N_{real_i} represents the number of occurrences of subgraph i in the real network; N_{real_i} represents the mean of the number of occurrences of subgraph i in the iterated random network; σ_{rand_i} represents the standard deviation of the number of occurrences of subgraph i in the random network.

Typically, $Z_i > 0$ represents that the number of occurrences of subgraph i in the owner–contractor collaborative network is greater than that in the corresponding random network. In this case, subgraph i is defined as a motif; otherwise, i is an anti-motif [43].

3.3. Analytical Procedures

We collected a longitudinal data set of the projects that won the National Quality Engineering Award (NQEA) in China and used the SNA and NMA methods to study the structural characteristics and evolutionary laws of owner–contractor collaborative relationship networks. Figure 1 depicts the analytical procedures, which consist of four steps: (i) Obtain the information on the owners and contractors of NQEA award-winning projects, (ii) Construct the owner–contractor relationship matrix based on the processed data and develop the owner–contractor snapshot network, (iii) Analyze the macro-structural characteristics of owner–contractor collaborative networks by using SNA, (iv) Discover the local collaborative patterns by using NMA.

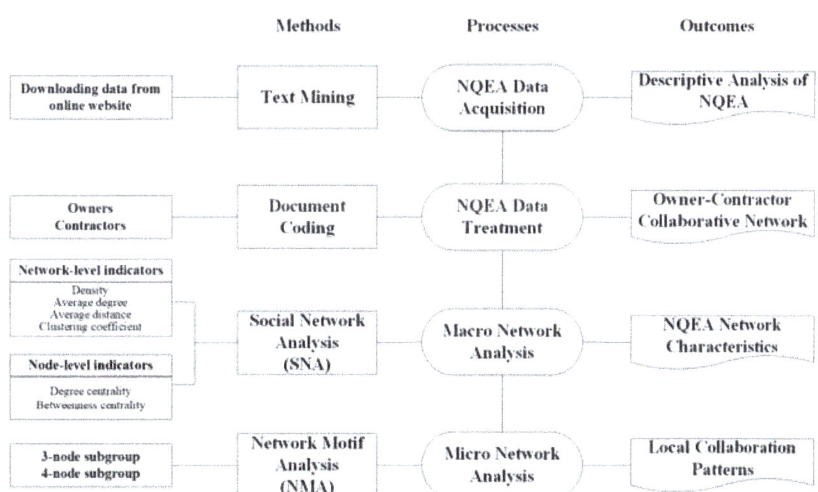

Figure 1. Analytical procedures.

3.4. Data Collection and Processing

China's NQEA is an award established to encourage construction companies to improve project quality. It was established in 1981 as China's construction industry's earliest and highest level national quality award. Generally, NQEA is awarded annually. The applicants include owners, contractors, designers, and some other enterprises participating in projects. The award projects must meet requirements such as excellent design, high construction quality, effective management, advanced technology, energy saving, and environmental protection. Projects in seven fields, including construction engineering, industrial engineering, traffic engineering, water conservancy engineering, and municipal engineering, are involved in NQEA. Among them, construction engineering projects account for 70% of the total number of award projects. The information on award projects, including project name, project type, year of winning the award, and project participants, is available on the official website (http://www.cacem.com.cn/ (accessed on 15 July 2022)) of the China Association of Construction Enterprise Management.

The NQEA project information provides a valuable data set for exploring the owner–contractor collaborative network. In this study, the data of 1371 construction projects that won NQEA from 2013 to 2021 were used to analyze the characteristics and evolution of the owner–contractor collaborative network in China's construction industry. These projects involved 1283 owners and 1560 contractors. The number of owners is smaller than that of projects because some NQEA projects have the same owners. In total, 1560 contractors are all the contractors included in the NQEA projects data set. Since some projects involve multiple contractors, the number of contractors is greater than that of projects. The descriptive statistics of the projects are shown in Table 1.

Table 1. Basic information on awarded projects.

Year	Number of Awarded Projects	Number of Awarded Owners	Number of Awarded Contractors
2013	104	105	185
2014	108	109	194
2015	133	135	256
2016	130	127	245
2017	164	161	301
2018	179	178	378
2019	173	173	342
2020	182	188	402
2021	198	201	452

We processed the project information collected in accordance with the following principles. First, for some large contractors with multiple tiers of subsidiaries, only the first-level subsidiaries were regarded as network nodes in this study. For example, China Construction Second Engineering Bureau Ltd., China Construction Third Engineering Bureau Ltd., China Construction Seventh Engineering Bureau Ltd., and China Construction Eighth Engineering Bureau Ltd. are all first-tier subsidiaries of China State Construction Engineering Corporation, one of China's largest construction companies. Therefore, they were displayed as different nodes in the collaborative relationship network. Second, we regarded the collaborative network between owners and contractors as an undirected and unweighted network. In other words, we only considered whether there was a collaborative relationship between an owner and a contractor, regardless of how many times they had collaborated. Third, we coded the enterprises in the network with an "O" for owners and a "C" for contractors and used different numbers to represent different enterprises. For example, C1446 represented China Construction Third Engineering Bureau Co., Ltd., and O68 represented Beijing Wangjing Souhou Real Estate Co., Ltd.

To understand the evolution of the network, a dynamic analysis of the network is required. For analyzing longitudinal networks, it is crucial to determine the optimal window size, which refers to the time interval between two snapshots. The NQEA is awarded annually, so we set each year as a time window to generate nine network snapshots over the study period from 2013 to 2021. Each network snapshot contains the awarded projects and the owners and contractors involved in that year. We constructed a two-mode network at each snapshot point. The network nodes were divided into two different sets in a two-mode network: the project set and the organization set. Figure 2a shows a schematic diagram of a two-mode network, where the square nodes represent the awarded projects, and the round nodes represent the awarded organizations. If a circular node is interconnected with a square node, the award-winning organization is involved in the construction project. Since project implementation depends on the organizations' collaboration, there are interconnections between the organizations involved in the same project. In addition, an organization may be involved in multiple projects and form a complex network of relationships with other organizations through different projects. For example, the black node C3 in Figure 2a is involved in both projects P1 and P2. Since this study aims to explore the collaborative relationship network between organizations, we converted the two-mode network consisting of the project set and organization set into the one-mode network containing only the organization set (see Figure 2b). Then, we established nine owner–contractor collaborative relationship matrices with the row and column 290 × 290, 303 × 303, 391 × 391, 372 × 372, 462 × 462, 556 × 556, 515 × 515, 590 × 590, 653 × 653, respectively. If there was a collaborative relationship between company i and company j at the snapshot time point, $r_{ij} = 1$; otherwise $r_{ij} = 0$.

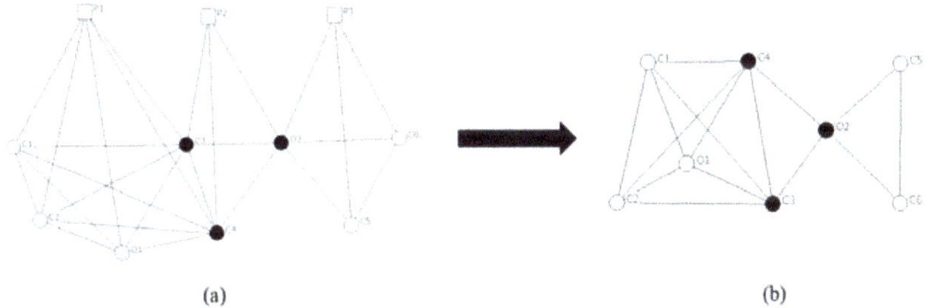

(a) (b)

Figure 2. Schematic diagram of project organization network: (**a**) two-mode network; (**b**) one-mode network.

4. Results and Discussion

4.1. Whole Network Topology

According to the established owner–contractor adjacency matrix, the topological diagrams of the collaborative relationship in nine snapshots are produced by Gephi software to show the evolution of the collaborative network (as shown in Figure 3). In Figure 3, the color of the nodes represents the type of companies, with green nodes representing the owners and pink nodes representing the contractors. The size of the node reflects the number of connections to this node. Specifically, the larger the node, the higher the number of connections to this nod, and vice versa. We can see from Figure 3 that the network structure at different snapshots is quite different. The number of nodes and connections in the collaborative network increases over time, which results in a larger network size. Furthermore, several significant components with many connected nodes can be found in each network.

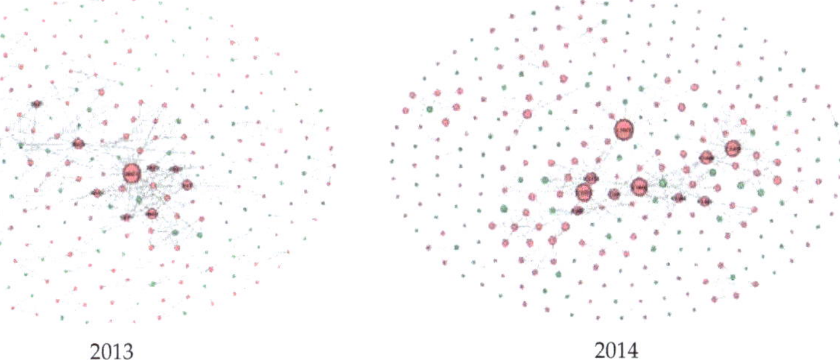

2013 2014

Figure 3. *Cont.*

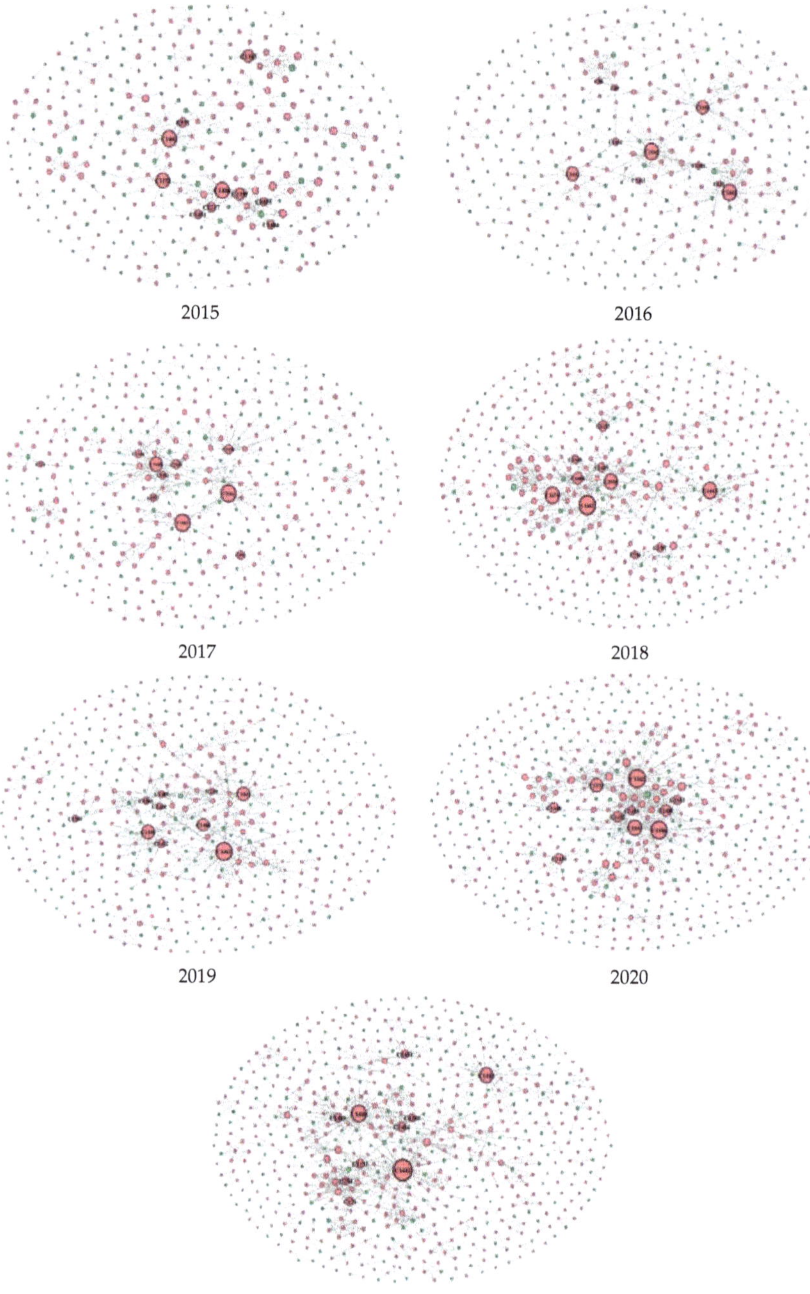

Figure 3. The topology diagram of different network snapshots.

4.2. Network-Level Analysis

4.2.1. Density

As an important indicator of the SNA method, network density reflects the connectivity of the network [51,52]. Figure 4 displays the evolution of the density of the owner–contractor collaborative relationship network over time. It can be seen that the density values for 2013, 2014, 2015, 2016, 2017, 2018, 2019, 2020, and 2021 are 0.013, 0.013, 0.010, 0.011, 0.008, 0.008, 0.008, 0.008, and 0.007, respectively, indicating that the density was low and decreased during the study period. This is different from some research conclusions on the collaborative network [8,34,35]. For example, Tang et al. [34] studied the collaborative relationships between contractors and subcontractors, and the results showed that during the study period, the contractor–subcontractor collaborative network became denser and more connected between nodes. This may be because it is more flexible for contractors and subcontractors to establish collaborative relationships, and it is easy to collaborate multiple times. However, the owners of different projects are often different, and the contractors are usually selected by means of bidding, which makes establishing a collaborative relationship between the owners and the contractors more restricted. Therefore, the value of density of the owner–contractor collaborative network did not increase over time. A low network density indicates that some owners and contractors have little communication, which is not conducive to exchanging information and sharing knowledge in the collaborative network [58]. The gradual decrease in network density reflects the worsening of network connectivity. This is because some groups have appeared in the evolution of the network. The organizations within the group are closely connected, but they are less connected with the organizations outside the group, causing an island effect, which results in low connectivity of the network. Most of these groups with island characteristics are composed of medium-sized contractors that won few NQEA. Organizations in the island group should fully understand their dilemmas and strengthen cooperation with other organizations to improve the overall connectivity of the collaborative network.

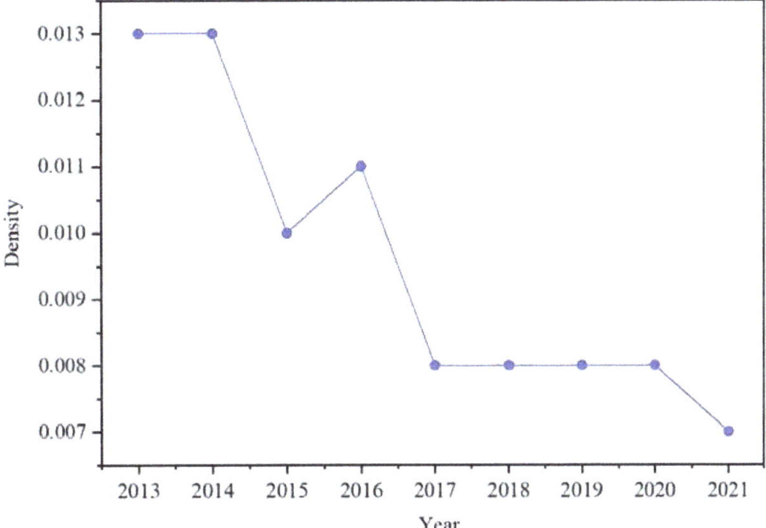

Figure 4. The density of the collaborative networks in 2013–2021.

4.2.2. Average Degree

The average degree is an indicator, which describes the compactness of the network and is the average number of connections (collaborative relationships) of a node (owner or contractor) in the collaborative network [68]. The higher the average degree, the more compact the network [69]. Figure 5 depicts the number of owners, contractors, and connections in the collaborative network from 2013 to 2020. During the study period, the number of nodes (contractors and owners) and connections in the network had increased, and the number of connections had increased more than that of nodes. This may be because the connection between nodes involves not only the collaboration between the newly joined contractors and owners but also the collaboration between the existing contractors in the collaborative network and the newly joined owners.

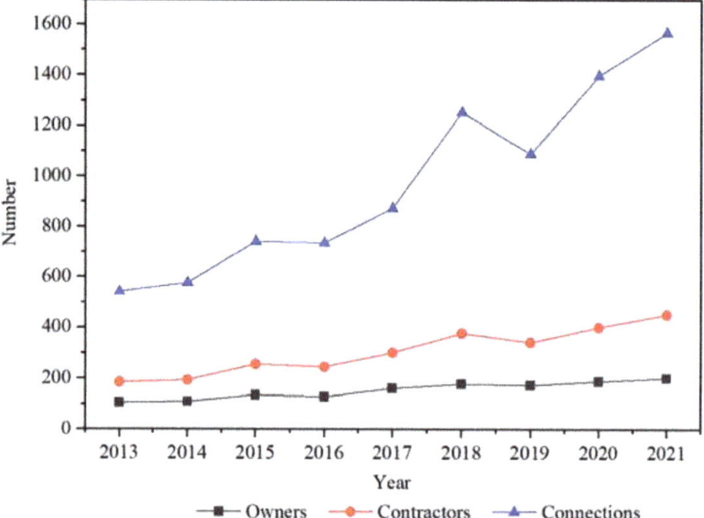

Figure 5. The number of owners, contractors, and connections in 2013–2021.

Figure 6 shows the change in average degree during the study period. It can be seen that the average degree of the collaborative network had increased over time, indicating that the collaborative network was becoming more and more compact. In 2021, the average degree of the owner–contractor collaborative network was 4.802, indicating that each node collaborated with at least four nodes, on average. Liu et al. [40]'s study on the collaborative relationships between contractors showed that the average degree of contractors' collaborative network in 2011 was 11.20, which is higher than the average degree of the owner–contractor collaborative network obtained in this study. Generally, a contractor can undertake several projects simultaneously or participate in a project together with other contractors. With the increase in the number of projects, the number of connections between different contractors also increased. Therefore, the average degree of the collaborative network of contractors is relatively high. However, for the collaborative network of owners and contractors, although the contractors of different projects may be the same, the owners are often different. This results in a relatively small number of relationships embedded in the collaborative network between owners and contractors, with a low average degree of the network.

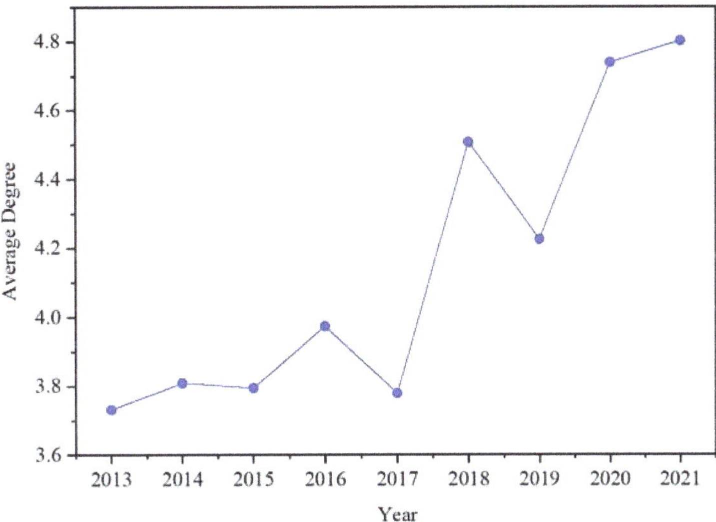

Figure 6. The average degree of the collaborative networks in 2013–2021.

4.2.3. Average Distance

In the organization network, the average distance refers to the average of the shortest path length between two organizations, which reflects the difficulty in communication between the two organizations and the possibility of information exchange [70]. Figure 7 depicts the variation of the average distance in the collaborative relationship network, as shown by the black line. In 2021, the average distance of the network was 3.719, which meant that it took about four steps from one node to another node. We can see from Figure 7 that the average distance of the collaborative network between owners and contractors in 2013, 2014, 2015, 2016, 2017, 2018, 2019, 2020, and 2021 was 3.298, 3.269, 3.864, 3.230, 3.615, 3.557, 3.580, 3.389, 3.719, respectively. The value of the average distance in 2021 is greater than that in 2013, which means that compared to 2013, more intermediate nodes are needed to establish connections between two companies in the collaborative network in 2021. This is because some newly joined companies in the collaborative network happened to be located on the shortest communication path between the other two companies, resulting in the need for the two companies to communicate through more intermediaries.

4.2.4. Clustering Coefficient

The clustering coefficient can be used to reflect the degree to which nodes in the network are clustered. In general, nodes clustered in a group can communicate and collaborate more effectively [71]. Figure 8 depicts the change in the clustering coefficient of the collaborative relationship network during the study period. The clustering coefficient gradually increased from 2013 to 2021. In 2021, the clustering coefficient value was 0.935, which meant that the owner–contractor network was highly clustered. This is because most of the NQEA projects are large in scale, and the owners usually contract out the construction tasks to several contractors to complete together, which makes them form a closely collaborative group, and many highly aggregated groups improve the aggregation degree of the whole network.

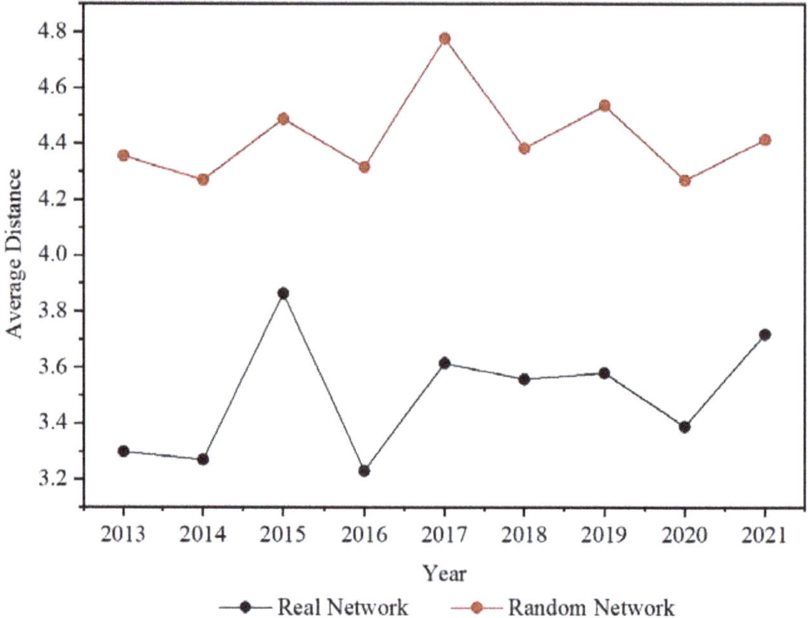

Figure 7. Evolution of the average distance of the collaborative networks in 2013–2021.

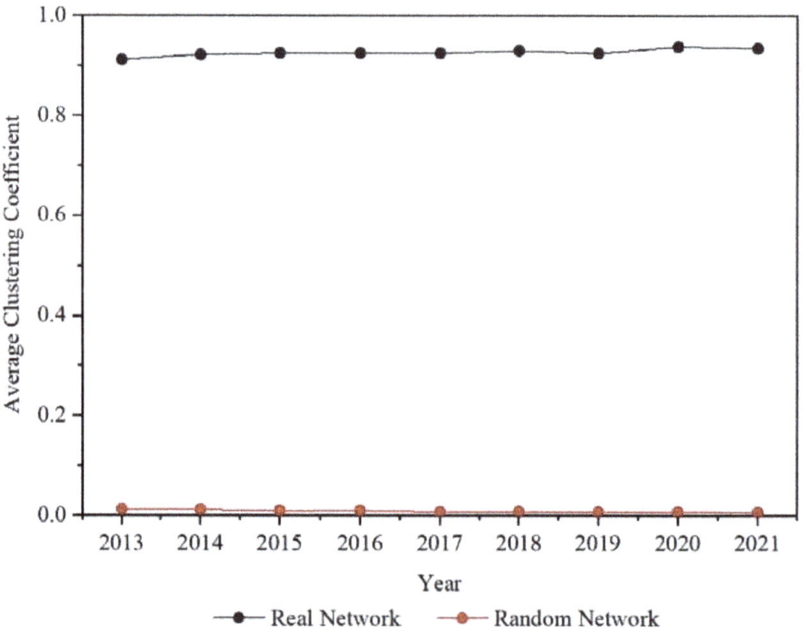

Figure 8. Change of the clustering coefficient of the collaborative networks in 2013–2021.

We further analyzed whether there was a small-world network in the owner–contractor collaborative network. Watts and Strogatz [72] and Neal [73] pointed out that if a network formed based on a specific rule had a larger clustering coefficient and a lower average distance than those of a random network with the same number of connected nodes and density, this indicated that the network had small-world characteristics. We randomly generate 100 networks with the same nodes and density as the owner–contractor collaborative network and calculate their average distances and clustering coefficients. Figures 7 and 8 show the mean distance and clustering coefficients of these 100 random networks, respectively, as shown by the red line. It can be seen that compared to random networks, the owner–contractor collaborative networks have lower average distances and higher clustering coefficients, that is, the collaborative networks have the characteristics of a small-world network. In a small-world network, the connection between two organizations requires only a few intermediary organizations, facilitating technology dissemination, capital accumulation, and personnel collaboration between the owners and contractors.

4.3. Node-Level Analysis

4.3.1. Degree Centrality

Degree centrality is the number of adjacent connections a node has in the network, reflecting the direct connection between nodes and other nodes [74]. In a collaborative network, the node with a high degree centrality has robust interactivity, significant influence, high participation degree, and it is at the core of the network [75].

Table 2 shows the top 15 companies ranked by the degree centrality of the collaborative network in nine snapshots. It can be seen from Table 2 that the degree centrality of C1442, C1446, and C1443 has consistently been ranked in the top three during the study period. This indicates that they are at the network's core and can be called core nodes. Compared with the other contractors, they have more experience in collaborating with owners and contractors. These companies exhibit the preference attachment effect, i.e., contractors who have won the NQEAs are more likely to acquire new projects and form partnerships with new owners.

Table 2. Top 15 companies ranked by degree centrality (DC).

No.	2013	2014	2015	2016	2017	2018	2019	2020	2021
1	C1446	C1442	C1442	C1446	C1443	C1442	C1442	C1442	C1442
2	C1451	C1446	C1446	C1442	C1442	C1474	C1446	C1446	C1446
3	C1443	C1153	C1153	C1443	C1446	C1443	C1443	C1443	C1443
4	C1484	C1448	C1443	C1153	C1451	C1446	C1484	C1153	C1484
5	C1105	C1444	C1105	C1051	C1035	C1484	C1452	C1485	C1153
6	C1051	C1474	C1237	C1451	C992	C1153	C1485	C1452	C1451
7	C1442	C485	C571	C641	C1073	C1485	C1501	C1448	C1485
8	C1452	C1443	C1451	C96	C1448	C1501	C1500	C1451	C171
9	C1363	C1484	C1484	C85	C1452	C1448	C1448	C1043	C1389
10	C1073	C462	C1035	C1452	C514	C754	C218	C1484	C754
11	C1410	C1069	C1410	C1035	C1484	C628	C1405	C588	C1474
12	C238	C782	C517	C1039	C1501	C1451	C1451	C1389	C1452
13	C1159	O68	C83	C1105	C13	C1452	C1073	C41	C588
14	O353	C573	C1363	O403	C750	C1449	C664	C1474	C750
15	C57	C992	C803	C508	C1438	C1511	C998	C750	C714

C1446, C1442, and C1443 refer to China Construction Third Engineering Bureau Corporation Limited, China Construction Eighth Engineering Bureau Corporation Limited, and China Construction Second Engineering Bureau Corporation Limited, respectively, all of which are the subsidiaries of China State Construction Engineering Corporation Ltd. (CSCE). CSCE is one of the largest construction contractors in China, ranking seventh in ENR's 2021 Top 250 International Contractors list. C1446, C1442, and C1443 are the three subsidiaries with the most potent comprehensive competitiveness of CSCE. In 2021, the newly signed contract values of C1446, C1442, and C1443 were around USD 88 billion, USD 94 billion, and USD 59 billion, respectively, and the operating income was around USD 44 billion, USD 53 billion, and USD 30 billion, respectively. The three companies have branches in many cities in China, which provide conditions for extensive participation in project bidding and establishing collaborative relationships with owners. They all have advanced technology and excellent R&D talents and have won many high-quality engineering awards. From 2013 to 2021, C1446, C1442, and C1443 have won 106, 116, and 112 NQEAs, respectively.

4.3.2. Betweenness Centrality

Betweenness centrality is an indicator, which measures the degree to which a node acts as an intermediary, that is, the degree to which a node influences the flow of information between other nodes [76]. The higher the betweenness centrality of a node, the greater its influence on the information flow between other nodes [77].

Table 3 lists the top 15 organizations in terms of betweenness centrality at each snapshot point. It can be seen from Table 3 that C1442, C1443, and C1446 always had a high value of betweenness centrality over time. This means that these companies act as bridges in the owner–contractor collaborative network. It is worth noting that C1153 (Suzhou Golden Mantis Building Decoration Co., Ltd.), as a private company, also had a high betweenness centrality in the owner–contractor collaborative network. In 2015, C1153 had the most significant betweenness centrality. C1153 (Suzhou Golden Mantis Building Decoration Co., Ltd.) is a large listed company with building decoration and renovation as its primary business. It has been ranked as one of the top 100 building decoration companies in China for 19 consecutive years, and its business has spread to various cities in China and many overseas markets. From 2013 to 2021, the number of NQEA won by C1153 has increased yearly, totaling 41 awards. This is due to its continuous development in building decoration with a large team of interior designers and excellent decoration and renovation construction teams. These advantages increase C1153's bidding competitiveness and make it easy to be favored by owners.

Table 3. Top 15 companies ranked by betweenness centrality (BC).

No.	2013	2014	2015	2016	2017	2018	2019	2020	2021
1	C1446	C1442	C1153	C1446	C1442	C1442	C1442	C1442	C1442
2	C1451	C1446	C571	C1442	C1443	C1443	C1484	C1443	C1446
3	C1443	C1448	C1442	C1153	C1446	C1446	C1446	C1446	C1443
4	C1410	C1153	C714	C1443	C1451	C1153	C1443	C1452	C1451
5	C1442	C1069	C1446	C1451	C1448	C1474	C1501	C1153	C1389
6	C1363	C485	C1105	C1452	C992	C1485	C1452	C1451	C1452
7	C1051	C1474	C750	C85	C1452	C1452	C1063	C1448	C1474
8	C1073	C1443	C1410	C1501	C1501	C754	C1500	C588	C1485
9	O250	C1445	C83	C1051	C64	C1448	C664	C750	C1484
10	C1452	C573	C1237	C1039	C1035	C1451	C750	C1474	C74
11	C1105	C819	C1484	C951	C1073	C1501	C430	C1405	C1512
12	C1237	C1484	C803	C540	C238	C992	C1512	C1043	C171
13	C1064	C1451	C517	C1105	C1153	C1484	C1485	C1485	C1153
14	C962	C1534	C1363	C641	C754	C1389	C1100	C737	C1448
15	C533	C171	C785	C96	C996	C540	C1405	C1389	C1258

We further analyze the attributes of the top 15 contractors ranked by degree centrality and betweenness centrality, as shown in Table 4. As can be seen from Table 4, among the top 15 contractors, there are more state-owned enterprises (SOEs) than private enterprises (PEs). This indicates that SOEs play a dominant and critical role in the owner–contractor collaborative network. Han et al. [78] also found that SOEs have high centrality and are the primary carrier of technological innovation in China's construction industry in the study on the collaborative innovation network of China's construction industry. State-owned construction enterprises often have substantial financial resources, government support, and extensive experience in contracting large-scale engineering projects. These advantages make it easier to develop collaborative relationships with owners and often collaborate with other subcontractors as a general contractor.

Table 4. Number of SOEs and PEs in the top 15 contractors in terms of degree centrality and betweenness centrality in 2013–2021.

No.	Degree Centrality		Betweenness Centrality	
	SOEs	PEs	SOEs	PEs
2013	8	7	8	7
2014	10	5	8	7
2015	5	10	4	11
2016	8	7	9	6
2017	11	4	10	5
2018	11	4	11	4
2019	12	3	10	5
2020	11	4	10	5
2021	10	5	10	5

4.4. Subgroup-Level Analysis

A network consists of several subgroups. Exploring the subgroup structure based on the NMA method helps gain a deeper understanding of the characteristics and evolution of the owner–contractor collaborative network. Unlike SNA, which focuses on the overall network structure and the role of nodes, NMA focuses on investigating the subgroup structure of the network. According to the number of award-winning projects and the number of organizations involved in the network, an average of three to four organizations (owners and contractors) are involved in each project. Therefore, we focus on the three-node subgroups and four-node subgroups of the owner–contractor collaborative network.

For the undirected unweighted network, there are two structural forms for the three-node subgroup and six structural forms for the four-node subgroup. Their topologies are shown in Figure 9.

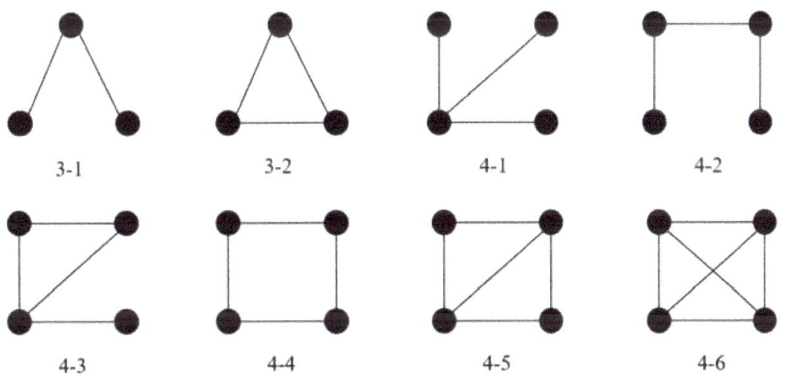

Figure 9. Topology of three-node and four-node subgroups.

We imported the data for each snapshot point into the Mfinder 1.2 software and performed 100 iterations, producing motif results for different subgraphs. The number of occurrences of a certain type of subgroup in the real network and the random network is shown in Figure 10. The Z-Scores of the three-node subgroups and the four-node subgroups in different snapshots are shown in Table 5.

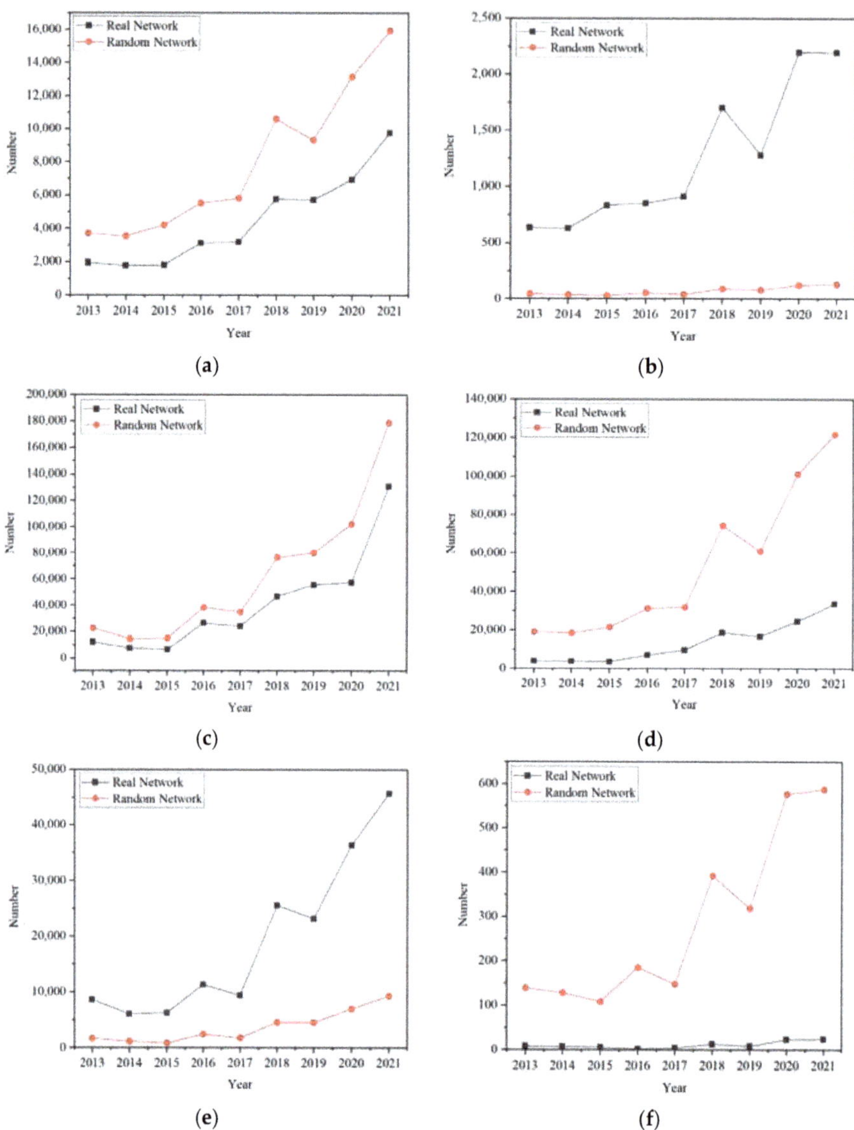

Figure 10. *Cont.*

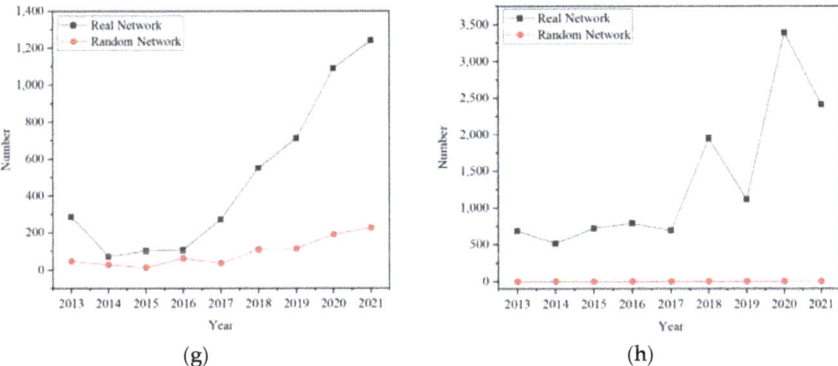

Figure 10. The number of three-node subgroups and four-node subgroups in real networks and random networks: (**a**) subgroup 3-1; (**b**) subgroup 3-2; (**c**) subgroup 4-1; (**d**) subgroup 4-2; (**e**) subgroup 4-3; (**f**) subgroup 4-4; (**g**) subgroup 4-5; (**h**) subgroup 4-6.

Table 5. Z-Score of all three-node subgroups and four-node subgroups in different snapshots.

ID	2013	2014	2015	2016	2017	2018	2019	2020	2021
3-1	−82.31	−80.88	−131.20	−86.88	−114.53	−154.19	−109.40	−131.68	−148.79
3-2	82.31	80.88	131.20	86.88	114.53	154.19	109.40	131.68	148.79
4-1	−31.39	−29.44	−46.49	−24.22	−25.65	−40.14	−27.96	−38.11	−30.86
4-2	−26.09	−29.19	−35.16	−25.84	−19.73	−31.30	−29.99	−34.85	−28.43
4-3	24.26	22.93	31.94	19.79	19.62	31.42	24.01	28.21	26.05
4-4	−7.45	−7.38	−6.75	−8.05	−6.96	−9.78	−9.78	−11.17	−10.56
4-5	11.75	3.10	11.13	1.54	12.60	12.18	12.97	14.08	12.83
4-6	921.86	850.89	2458.93	673.29	841.17	1098.79	614.04	971.35	674.41

It can be seen in Figure 10 that the number of three-node subgroups and four-node subgroups both increased in the real network and the random network during the study period. This means that the structure of owner–contractor collaborative networks in China's construction industry is becoming increasingly complex, and the collaboration between organizations is becoming more and more diverse. For the three-node subgroup, subgroup 3-1 always appeared more frequently in the random network than in the real network, while subgroup 3-2 is on the contrary. Table 4 shows that the Z-Score of subgroup 3-2 is positive, while that of subgroup 3-1 is negative. Thus, subgroup 3-2 is the network motif, while subgroup 3-1 is the network anti-motif. For the four-node subgroups, subgroup 4-1, subgroup 4-2, and subgroup 4-4 always appeared more frequently in the random network than in the real network, while subgroup 4-3, subgroup 4-5, and subgroup 4-6 appeared much more frequently in the real network than in the random network. It can also be seen in Table 5 that the Z-Scores of subgroup 4-3, subgroup 4-5, and subgroup 4-6 are all greater than 0, while those of subgroup 4-1, subgroup 4-2, and subgroup 4-4 are all less than 0. Therefore, for the four-node subgroup, subgroups 4-3, 4-5, and 4-6 are network motifs, and subgroups 4-1, 4-2, and 4-4 are network anti-motifs.

Motifs are fundamental patterns that recur in networks, and their frequency in real networks is significantly higher than in random networks with the same number of nodes and connections. Anti-motifs are just the opposite. The above results show that subgroups 3-2, 4-3, 4-5, and 4-6 are the motifs in the owner–contractor collaborative relationship network, that is, there are many local collaborative relationships of these forms in the network. Among them, subgroups 3-2 have the largest Z-Score in the three-node subgroup, and subgroups 4-6 have the largest Z-Score in the four-node subgroup. That is, subgroups 3-2 and 4-6 are the two most dominant subgroup structures in the owner–contractor collaborative network. As seen in Figure 9, these four forms are all generated based on the complete collaboration of the three organizations (containing at least one triangle). This subgroup structure facilitates the organization's efficient collaboration and the network's development. Subgroups 3-1, 4-1, 4-2, and 4-4 are anti-motifs in the owner–contractor collaborative relationship network. As can be seen in Figure 9, these four forms are mainly binary cooperation between organizations (containing no triangle) and do not have the basis for multiple collaborations. They are undesirable because they reduce the connectivity and cohesiveness of the network.

4.5. Managerial Implications

Based on the above results, the structural characteristics of the collaborative networks formed by owners and contractors involved in projects that won NQEA and the evolution of each organization's position in the network from 2013 to 2021 can be identified. Accordingly, the following management insights can be proposed for organizations to improve their collaborative relationships and thus contribute to the development of the organizations and the industry.

(1) The results regarding centrality indicated that some contractors, such as C1442, C1443, and C1446, always had a high degree centrality and betweenness centrality during the study period. These contractors are all subsidiaries of CSCE, one of China's largest contractors. The high value of degree centrality indicates that these large contractors have a lot of experience in collaborating with owners or other contractors and are at the core of the collaborative relationship network, while high betweenness centrality means that they have a significant impact on the owner–contractor collaborative relationship network. It can be seen that these organizations have played an essential role in the construction of high-quality projects in China. In the future, they should further play their leading role in developing the construction industry. Specifically, they can make efforts from the following three aspects. First, since new technology, such as blockchain and artificial intelligence, can effectively promote the high-quality development of construction industry, these companies can positively explore the application of these new technologies in construction projects. Second, completing high-quality projects requires constant materials and construction techniques innovation. Therefore, these companies need to increase the R&D efforts of new technologies and new materials and promote their application in construction. Third, these leading enterprises can actively participate in formulating relevant industry norms and technical standards to promote the industry's overall development.

(2) The results of density, average degree, average distance, and clustering coefficient showed that the collaborative relationship network became more and more compact during the study period. A compact collaborative network is beneficial for sharing new policies, technologies, and ideas, promoting industrial upgrading and high-quality development. Thus, more frequent interaction is required for the owners and contractors in China's construction industry to develop strong collaborative relationships. To this end, construction industry associations can often hold some technology-sharing activities to provide a good platform for promoting exchanges, giving small and medium contractors more opportunities to collaborate with owners and large contractors.

(3) As previously mentioned, some contractors who have won NQEAs in the past, such as C1142 and C1143, showed a preference attachment effect, which means that contractors who have won NQEAs were more likely to obtain new projects and establish partnerships with new owners. NQEA-winning collaboration projects must satisfy some requirements, including reliable quality, leading design ideas, and significant technological innovation achievement. Therefore, contractors who have won this award generally have good management capability, technical innovation ability, and construction levels. Awarding the NQEA further enhances the credibility of companies, which in turn helps the companies obtain new construction projects. This virtuous cycle can promote the development of enterprises. Consequently, construction

companies should pay special attention to improving their management capabilities and technical levels and follow the principle of excellence when undertaking projects to establish their industry reputation and lay the foundation for market expansion and collaborative relationship establishment.

(4) The results of NMA indicate that there are two local collaborative patterns in the network, i.e., motif and anti-motif. Motifs are mainly triangle-based collaborative patterns, and anti-motifs are binary collaborative patterns. Generally, project organizations can reduce the uncertainty in the search for collaboration and increase the likelihood of successful collaboration based on previous collaboration experience (relational embeddedness) and common third collaborator (structural embeddedness) [79,80]. The triangle-based collaborative pattern is beneficial for the organization to obtain information about indirect partners and can help the organization to establish contacts with indirect partners through direct partners, thus effectively expanding the scope of collaboration. Triangle-based motifs reflect collaborative patterns with structurally embedded features. The owners and contractors can deepen their partnership with more indirect partners to create opportunities to participate in more large-scale projects.

5. Conclusions

Collaboration between the owners and contractors is key to the success of a construction project. Recently, with the increase in construction projects in China, a complex collaborative relationship has formed between the owners and contractors. It is necessary to systematically and deeply understand the complexity and dynamics of this collaborative relationship. Based on the data of NQEA projects from 2013 to 2021, we adopted the SNA and NMA methods to establish a collaborative network between the owners and contractors in China's construction industry and analyzed the structural characteristics and dynamic evolution of the collaborative network.

The main findings of the study are as follows. (1) The collaborative networks formed by owners and contractors that have won NQEA in China became larger and more complex in structure during the study period. This indicates that the large number of construction projects in China has led to the involvement of an increasing number of owners and contractors who have formed an intricate network of relationships. In the evolution of the network, there have been island-like groups, where the organizations within the group are closely connected, but they are less connected to organizations outside the group. This result can help organizations in the isolated groups to understand their dilemma and suggest that they need to expand their cooperation with other organizations. (2) The results of the centrality analysis indicate that most of the organizations at the core of the network are large state-owned contractors who have rich resources and strong power. Their central position in the network indicates that they have had cooperative relationships with many owners and other contractors and have a large industry influence. There is a need to strengthen the driving role of these state-owned contractors in the development of China's construction industry. In addition, the results of centrality also show that most of the organizations at the core of the network have repeatedly won NQEA, such as China Construction Third Engineering Bureau, China Construction Eighth Engineering Bureau, and China Construction Second Engineering Bureau. This suggests that there is a preference attachment effect in the construction market, i.e., winning NQEAs for quality construction work gives them more opportunities to undertake new projects and form new partnerships with new owners and other contractors. (3) The results of NMA show that the collaboration patterns between owners and contractors have become complex and diverse over time, consisting mainly of sparse binary collaboration patterns and tight multiparty collaboration patterns. Multiparty collaboration patterns are increasingly present in the network, making the network more locally clustered. This indicates that more and more organizations form multiparty collaboration networks with other organizations when participating in new projects, laying the foundation for future participation in the fierce competition in the construction market. Organizations with only a binary cooperation model can also draw inspiration from this finding that they need to establish broader cooperative relationships with other organizations by contracting more projects.

This study brings the following knowledge contributions. First, this study enriches the existing body of knowledge on owner–contractor collaborative relationships by shifting the focus from one-off and short-term cooperation in a specific project to the collaborative relationship network at the industry level. Second, the evolution mechanism of the owner–contractor industry-level network is studied from a dynamic network perspective, which expands the study of the organizational network in the construction field and fills the gap in the research on the owner–contractor collaborative network. Third, the findings provide valuable insights into understanding the evolution of China's owner–contractor collaborative relationship and provide a basis for collaborative network governance and organizational collaboration strategy formulations. Fourth, this study proposes a method, which combines SNA and NMA to reveal the characteristics and evolution of the network's overall macro-structure and local micro-structure, providing a research idea for mining cooperation information from industry-level project-based social networks. The interdisciplinary network motif concept is introduced to characterize the local relationship patterns of the network and collaboration mechanisms of subgroups, and SNA is used to explore the network's overall characteristics and nodes' position. SNA and NMA complement each other and advance the understanding of the network properties and structural embeddedness. The method and ideas in this study can also be used to explore the collaborative relationships between stakeholders in the construction industry of other countries.

Although the study brings the above contributions, there are still some limitations. First, we only studied the collaborative network of owners and contractors based on the data of the NQEA project in China. In the future, data sources can be expanded for more in-depth research. Second, we only focused on the collaborative relationship between owners and contractors. However, a construction project involves many participants, such as subcontractors and designers. Future studies can include these stakeholders to better understand the collaborative relationships between different stakeholders. Third, we only discovered the three-node subgroups and four-node subgroups; however, some large projects may involve more owners and contractors. The structure of more node subgroups can be further explored in the future.

Author Contributions: Conceptualization, F.W. and M.C.; methodology, F.W.; software, F.W.; validation, M.C. and X.C.; data curation, F.W. and X.C.; writing—original draft preparation, F.W.; writing—review and editing, M.C. and X.C.; visualization, F.W. and X.C.; supervision, M.C.; funding acquisition, M.C. All authors have read and agreed to the published version of the manuscript.

Funding: This research was funded by The National Social Science Fund of China, grant No. 22BJY232.

Data Availability Statement: Data sharing not applicable.

Acknowledgments: We are thankful to the editor and anonymous referees for their valuable comments and guidance.

Conflicts of Interest: On behalf of all co-authors, the corresponding author states that there are no conflict of interest.

References

1. Krane, H.P.; Olsson, N.O.E.; Rolstadas, A. How project manager-project owner interaction can work within and influence project risk management. *Proj. Manag. J.* **2012**, *43*, 54–67. [CrossRef]
2. Olanipekun, A.O.; Xia, B.; Hon, C.; Hu, Y. Project owners' motivation for delivering green building projects. *J. Constr. Eng. Manag.* **2017**, *143*, 04017068. [CrossRef]
3. Cao, D.; Li, H.; Wang, G.; Luo, X.; Tan, D. Relationship network structure and organizational competitiveness: Evidence from BIM implementation practices in the construction industry. *J. Manag. Eng.* **2018**, *34*, 04018005. [CrossRef]
4. Li, J.; Jiang, W.; Zuo, J. The effects of trust network among project participants on project performance based on SNA approach: A case study in China. *Int. J. Constr. Manag.* **2020**, *20*, 837–847. [CrossRef]

5. Statistical Analysis of Construction Industry Development in 2021. Available online: https://mp.weixin.qq.com/s?__biz= MzUyNjM4NzkzOQ==&mid=2247490151&idx=1&sn=ddc2f4415f2dc28907d9341129bb89d3&chksm=fa0ec5ebcd794cfd28160 3a9e628d1d8438eaa338a267df4e713cd157fee259e93fc0096b750#rd (accessed on 25 August 2022).

6. Lee, Y.-S.; Kim, J.-J.; Lee, T.S. Topological competiveness based on social relationships in the Korean construction management industry. *J. Constr. Eng. Manag.* **2016**, *142*, 05016014. [CrossRef]

7. Lu, Y.; Liu, B.; Li, Y. Collaboration networks and bidding competitiveness in megaprojects. *J. Manag. Eng.* **2021**, *37*, 04021064. [CrossRef]

8. Cao, D.; Li, H.; Wang, G.; Luo, X.; Yang, X.; Tan, D. Dynamics of project-based collaborative networks for BIM implementation: Analysis based on stochastic actor-oriented models. *J. Manag. Eng.* **2017**, *33*, 04016055. [CrossRef]

9. Sedita, S.R.; Apa, R. The impact of inter-organizational relationships on contractors' success in winning public procurement projects: The case of the construction industry in the Veneto region. *Int. J. Proj. Manag.* **2015**, *33*, 1548–1562. [CrossRef]

10. Hakansson, H.; Snehota, I. No business is an Island: The network concept of business strategy. *Scand. J. Manag.* **1989**, *5*, 187–200. [CrossRef]

11. Pinto, J.K.; Slevin, D.P.; English, B. Trust in projects: An empirical assessment of owner/contractor relationships. *Int. J. Proj. Manag.* **2009**, *27*, 638–648. [CrossRef]

12. Lu, S.; Hao, G. The influence of owner power in fostering contractor cooperation: Evidence from China. *Int. J. Proj. Manag.* **2013**, *31*, 522–531. [CrossRef]

13. Han, Y.; Li, Y.; Rao, P.; Taylor, J.E. Global collaborative network of skyscraper projects from 1990 to 2010: Temporal evolution and spatial characteristics. *J. Constr. Eng. Manag.* **2021**, *147*, 04021095. [CrossRef]

14. Tang, Y.; Wang, G.; Li, H.; Cao, D.; Li, X. Comparing project-based collaborative networks for BIM implementation in public and private sectors: A longitudinal study in Hong Kong. *Adv. Civ. Eng.* **2019**, *2019*, 6213694. [CrossRef]

15. Qiang, G.; Cao, D.; Wu, G.; Zhao, X.; Zuo, J. Dynamics of collaborative networks for green building projects: Case study of Shanghai. *J. Manag. Eng.* **2021**, *37*, 05021001. [CrossRef]

16. Jones, C.; Hesterly, W.S.; Borgatti, S.P. A general theory of network governance: Exchange conditions and social mechanisms. *Acad. Manag. Rev.* **1997**, *22*, 911–945. [CrossRef]

17. Jayaraj, S.; Doerfel, M.; Williams, T. Clique to win: Impact of cliques, competition, and resources on team performance. *J. Constr. Eng. Manag.* **2022**, *148*, 04022047. [CrossRef]

18. Oraee, M.; Hosseini, M.R.; Edwards, D.; Papadonikolaki, E. Collaboration in BIM-based construction networks: A qualitative model of influential factors. *Eng. Constr. Archit. Manag.* **2022**, *29*, 1194–1217. [CrossRef]

19. Stone, L.; Simberloff, D.; Artzy-Randrup, Y. Network motifs and their origins. *PLoS Comput. Biol.* **2019**, *15*, e1006749. [CrossRef]

20. Li, H.Y.; Feng, J.C. Study on the improvement strategy of trust level between owner and PMC contractor based on system dynamics model. *Buildings* **2022**, *12*, 1163. [CrossRef]

21. Sun, C.; Wang, M.; Zhai, F. Research on the collaborative application of BIM in EPC projects: The perspective of cooperation between owners and general contractors. *Adv. Civ. Eng.* **2021**, *2021*, 4720900. [CrossRef]

22. Zhang, S.B.; Liu, X.Y.; Gao, Y.; Ma, P. Effect of level of owner-provided design on contractor's design quality in DB/EPC projects. *J. Constr. Eng. Manag.* **2019**, *145*, 04018121. [CrossRef]

23. Suprapto, M.; Bakker, H.L.M.; Mooi, H.G. Relational factors in owner-contractor collaboration: The mediating role of teamworking. *Int. J. Proj. Manag.* **2015**, *33*, 1347–1363. [CrossRef]

24. Jiang, W.P.; Lu, Y.J.; Le, Y. Trust and project success: A twofold perspective between owners and contractors. *J. Manag. Eng.* **2016**, *32*, 04016022. [CrossRef]

25. Zhang, L.; Qian, Q. How mediated power affects opportunism in owner-contractor relationships: The role of risk perceptions. *Int. J. Proj. Manag.* **2017**, *35*, 516–529. [CrossRef]

26. Tai, S.; Sun, C.; Zhang, S. Exploring factors affecting owners' trust of contractors in construction projects: A case of China. *SpringerPlus* **2016**, *5*, 1783. [CrossRef]

27. Suprapto, M.; Bakker, H.L.M.; Mooi, H.G.; Moree, W. Sorting out the essence of owner-contractor collaboration in capital project delivery. *Int. J. Proj. Manag.* **2015**, *33*, 664–683. [CrossRef]

28. Zhang, S.; Fu, Y.; Kang, F. How to foster contractors' cooperative behavior in the Chinese construction industry: Direct and interaction effects of power and contract. *Int. J. Proj. Manag.* **2018**, *36*, 940–953. [CrossRef]

29. Qian, Q.Z.; Zhang, J.Y.; Cao, T.T. Effect of behavior tension on value creation in owner-contractor relationships: Moderating role of dependence asymmetry. *Eng. Manag. J.* **2021**, *33*, 220–236. [CrossRef]

30. Nasir, M.K.; Hadikusumo, B.H.W. System dynamics model of contractual relationships between owner and contractor in construction projects. *J. Manag. Eng.* **2019**, *35*, 04018052. [CrossRef]

31. Zhao, L.; Wang, Y.; Sun, D.; Yang, Y. Cooperative behavior analysis of owners and contractors based on brain neurobehavioral mechanism. *NeuroQuantology* **2018**, *16*, 205–214. [CrossRef]

32. Zheng, X.; Le, Y.; Chan, A.P.C.; Hu, Y.; Li, Y. Review of the application of social network analysis (SNA) in construction project management research. *Int. J. Proj. Manag.* **2016**, *34*, 1214–1225. [CrossRef]

33. Lu, Y.J.; Wei, W.; Li, Y.K.; Wu, Z.L.; Jin, H. The formation and evolution of interorganisational business networks in megaprojects: A case study of Chinese skyscrapers. *Complexity* **2020**, *2020*, 2727419. [CrossRef]

34. Li, X.; Li, H.; Cao, D.; Tang, Y.; Luo, X.; Wang, G. Modeling dynamics of project-based collaborative networks for BIM implementation in the construction industry: Empirical study in Hong Kong. *J. Constr. Eng. Manag.* **2019**, *145*, 05019013. [CrossRef]

35. Tang, Y.; Wang, G.; Li, H.; Cao, D. Dynamics of collaborative networks between contractors and subcontractors in the construction industry: Evidence from National Quality Award Projects in China. *J. Constr. Eng. Manag.* **2018**, *144*, 05018009. [CrossRef]

36. Achrol, R.S. Changes in the theory of interorganizational relations in marketing: Toward a network paradigm. *J. Acad. Mark. Sci.* **1996**, *25*, 56–71. [CrossRef]

37. Ritter, T.; Gemünden, H.G. Interorganizational relationships and networks: An overview. *J. Bus. Res.* **2003**, *56*, 691–697. [CrossRef]

38. Akintoye, A.; Main, J. Collaborative relationships in construction: The UK contractors' perception. *Eng. Constr. Archit. Manag.* **2007**, *14*, 597–617. [CrossRef]

39. Akgul, B.K.; Ozorhon, B.; Dikmen, I.; Birgonul, M.T. Social network analysis of construction companies operation in international markets: Case of Turkish contractors. *J. Civ. Eng. Manag.* **2017**, *23*, 327–337. [CrossRef]

40. Liu, L.; Han, C.; Xu, W. Evolutionary analysis of the collaboration networks within National Quality Award Projects of China. *Int. J. Proj. Manag.* **2015**, *33*, 599–609. [CrossRef]

41. Liu, C.; Cao, J.; Wu, G.; Zhao, X.; Zuo, J. Interenterprise collaboration network in international construction projects: Evidence from Chinese construction enterprises. *J. Manag. Eng.* **2022**, *38*, 05021018. [CrossRef]

42. Park, H.; Han, S.H.; Rojas, E.M.; Son, J.; Jung, W. Social network analysis of collaborative ventures for overseas construction projects. *J. Constr. Eng. Manag.* **2011**, *137*, 344–355. [CrossRef]

43. Milo, R.; Shen-Orr, S.; Itzkovitz, S.; Kashtan, N.; Chklovskii, D.; Alon, U. Network motifs: Simple building blocks of complex networks. *Science* **2002**, *298*, 824–827. [CrossRef] [PubMed]

44. Sahoo, A.; Pechmann, S. Functional network motifs defined through integration of protein-protein and genetic interactions. *PeerJ* **2022**, *10*, e13016. [CrossRef] [PubMed]

45. Simmons, B.I.; Cirtwill, A.R.; Baker, N.J.; Wauchope, H.S.; Dicks, L.V.; Stouffer, D.B.; Sutherland, W.J. Motifs in bipartite ecological networks: Uncovering indirect interactions. *Oikos* **2019**, *128*, 154–170. [CrossRef]

46. Patra, S.; Mohapatra, A. Disjoint motif discovery in biological network using pattern join method. *IET Syst. Biol.* **2019**, *13*, 213–224. [CrossRef]

47. Shen, G.; Zhu, D.; Chen, J.; Kong, X. Motif discovery based traffic pattern mining in attributed road networks. *Knowl.-Based Syst.* **2022**, *250*, 109035. [CrossRef]

48. Liu, L.; Zhao, M.; Fu, L.; Cao, J. Unraveling local relationship patterns in project networks: A network motif approach. *Int. J. Proj. Manag.* **2021**, *39*, 437–448. [CrossRef]

49. Yu, T.; Shen, G.Q.; Shi, Q.; Lai, X.; Li, C.Z.; Xu, K. Managing social risks at the housing demolition stage of urban redevelopment projects: A stakeholder-oriented study using social network analysis. *Int. J. Proj. Manag.* **2017**, *35*, 925–941. [CrossRef]

50. Yang, M.; Chen, H.; Xu, Y. Stakeholder-associated risks and their interactions in PPP projects: Social network analysis of a water purification and sewage treatment project in China. *Adv. Civ. Eng.* **2020**, *2020*, 8897196. [CrossRef]

51. Abbasianjahromi, H.; Etemadi, A. Applying social network analysis to identify the most effective persons according to their potential in causing accidents in construction projects. *Int. J. Constr. Manag.* **2019**, *22*, 1065–1078. [CrossRef]

52. Kong, X.; Shi, Y.; Wang, W.; Ma, K.; Wan, L.; Xia, F. The evolution of turing award collaboration network: Bibliometric-level and network-level metrics. *IEEE Trans. Comput. Soc. Syst.* **2019**, *6*, 1318–1328. [CrossRef]

53. Kereri, J.O.; Harper, C.M. Social networks and construction teams: Literature review. *J. Constr. Eng. Manag.* **2019**, *145*, 03119001. [CrossRef]

54. Hattab, A.M.; Hamzeh, F. Using social network theory and simulation to compare traditional versus BIM-lean practice for design error management. *Autom. Constr.* **2015**, *52*, 59–69. [CrossRef]

55. Castillo, T.; Herrera, R.F.; Alarcon, L.F. The quality of small social networks and their performance in architecture design offices. *J. Constr. Eng. Manag.* **2023**, *149*, 04022162. [CrossRef]

56. Wang, Y.; Thangasamy, V.K.; Hou, Z.; Tiong, R.L.K.; Zhang, L. Collaborative relationship discovery in BIM project delivery: A social network analysis approach. *Autom. Constr.* **2020**, *114*, 103147. [CrossRef]

57. Abbsaian-Hosseini, S.A.; Liu, M.; Hsiang, S.M. Social network analysis for construction crews. *Int. J. Constr. Manag.* **2019**, *19*, 113–127. [CrossRef]

58. Dadpour, M.; Shakeri, E.; Nazari, A. Analysis of stakeholder concerns at different times of construction projects using social network analysis (SNA). *Int. J. Civ. Eng.* **2019**, *17*, 1715–1727. [CrossRef]

59. Abbasi, A.; Altmann, J.; Hossain, L. Identifying the effects of co-authorship networks on the performance of scholars: A correlation and regression analysis of performance measures and social network analysis measures. *J. Informetr.* **2011**, *5*, 594–607. [CrossRef]

60. Lyu, L.; Wu, W.; Hu, H.; Huang, R. An evolving regional innovation network: Collaboration among industry, university, and research institution in China's first technology hub. *J. Technol. Transf.* **2019**, *44*, 659–680. [CrossRef]

61. Freeman, L.C. Centrality in social networks conceptual clarification. *Soc. Netw.* **1978**, *1*, 215–239. [CrossRef]

62. Hu, X.; Chong, H.-Y. Integrated frameworks of construction procurement systems for off-site manufacturing projects: Social network analysis. *Int. J. Constr. Manag.* **2020**, *22*, 2089–2097. [CrossRef]

63. Xue, X.; Zhang, R.; Wang, L.; Fan, H.; Yang, R.J.; Dai, J. Collaborative innovation in construction project: A social network perspective. *KSCE J. Civ. Eng.* **2018**, *22*, 417–427. [CrossRef]

64. Yang, R.J.; Zou, P.X.W.; Wang, J. Modelling stakeholder-associated risk networks in green building projects. *Int. J. Proj. Manag.* **2016**, *34*, 66–81. [CrossRef]
65. Shen-Orr, S.S.; Milo, R.; Mangan, S.; Alon, U. Network motifs in the transcriptional regulation network of escherichia coli. *Nat. Genet.* **2002**, *31*, 64–68. [CrossRef]
66. Tavella, J.; Windsor, F.M.; Rother, D.C.; Evans, D.M.; Guimaraes, P.R., Jr.; Palacios, T.P.; Lois, M.; Devoto, M. Using motifs in ecological networks to identify the role of plants in crop margins for multiple agriculture functions. *Agric. Ecosyst. Environ.* **2022**, *331*, 107912. [CrossRef]
67. Milo, R.; Itzkovitz, S.; Kashtan, N.; Levitt, R.; Shen-Orr, S.; Ayzenshtat, I.; Sheffer, M.; Alon, U. Superfamilies of evolved and designed networks. *Science* **2004**, *303*, 1538–1542. [CrossRef]
68. Takahashi, M.; Indulska, M.; Steen, J. Collaborative research project networks: Knowledge transfer at the fuzzy front end of innovation. *Proj. Manag. J.* **2018**, *49*, 36–52. [CrossRef]
69. Newman, M.E.J. Scientific collaboration networks. I. network construction and fundamental results. *Phys. Rev. E* **2001**, *64*, 016131. [CrossRef]
70. Sasidharan, S.; Santhanam, R.; Brass, D.J.; Sambamurthy, V. The effects of social network structure on enterprise systems success: A longitudinal multilevel analysis. *Inf. Syst. Res.* **2012**, *23*, 658–678. [CrossRef]
71. Loosemore, M.; Braham, R.; Yuan, Y.; Bronkhorst, C. Relational determinants of construction project outcomes: A social network perspective. *Constr. Manag. Econ.* **2020**, *38*, 1061–1076. [CrossRef]
72. Watts, D.J.; Strogatz, S.H. Collective dynamics of 'small-world' networks. *Nature* **1998**, *393*, 440–442. [CrossRef] [PubMed]
73. Neal, Z. Is the urban world small? The evidence for small world structure in urban networks. *Netw. Spat. Econ.* **2018**, *18*, 615–631. [CrossRef]
74. Jafari, P.; Mohamed, E.; Lee, S.; Abourizk, S. Social network analysis of change management processes for communication assessment. *Autom. Constr.* **2020**, *118*, 103292. [CrossRef]
75. Wambeke, B.W.; Liu, M.; Hsiang, S.M. Using pajek and centrality analysis to identify a social network of construction trades. *J. Constr. Eng. Manag.* **2012**, *138*, 1192–1201. [CrossRef]
76. El-adaway, I.H.; Abotaleb, I.S.; Vechan, E. Social network analysis approach for improved transportation planning. *J. Infrastruct. Syst.* **2017**, *23*, 05016004. [CrossRef]
77. Kim, Y.; Choi, T.Y.; Yan, T.; Dooley, K. Structural investigation of supply networks: A social network analysis approach. *J. Oper. Manag.* **2011**, *29*, 194–211. [CrossRef]
78. Han, Y.; Li, Y.; Taylor, J.E.; Zhong, J. Characteristics and evolution of innovative collaboration networks in architecture, engineering, and construction: Study of National Prize-Winning Projects in China. *J. Constr. Eng. Manag.* **2018**, *144*, 04018038. [CrossRef]
79. Ahuja, G. Collaboration networks, structural holes, and innovation: A longitudinal study. *Adm. Sci. Q.* **2000**, *45*, 425–455. [CrossRef]
80. Cowan, R.; Jonard, N.; Zimmermann, J.B. Bilateral collaboration and the emergence of innovation networks. *Manag. Sci.* **2007**, *53*, 1051–1067. [CrossRef]

sustainability

MDPI

Article

Configuration Analysis of Integrated Project Delivery Principles' Obstacle to Construction Project Level of Collaboration

Tingting Mei [1,*], Shuda Zhong [2], Huabin Lan [1], Zeng Guo [3,*] and Yi Qin [1]

[1] School of Civil Engineering and Architecture, Wuhan Institute of Technology, Wuhan 430070, China
[2] School of Resources & Safety Engineering, Wuhan Institute of Technology, Wuhan 430070, China
[3] School of Entrepreneurship, Wuhan University of Technology, Wuhan 430070, China
* Correspondence: 20050101@wit.edu.cn (T.M.); 12981@whut.edu.cn (Z.G.)

Abstract: Integrated Project Delivery (IPD) with collaborative work as its core is supported by increasing numbers of scholars and practitioners, due to the performance improvement of project construction and projects' success promotion. However, some factors such as the contract, the technology, and the personnel behaviors hinder the application of IPD, which has negative impacts on the collaboration level of construction projects. On the basis of the configuration analysis, the purpose of this paper is to increase the effectiveness of collaborative management of construction projects by encouraging the application of IPD principles. This is achieved by introducing the proof of contradiction and thoroughly examining the impact of the application of IPD principles' barrier with the level of collaboration. Added to that, the research necessity of configuration analysis on IPD principles' obstacle to construction project collaboration is demonstrated through bibliometric analysis; thus, a questionnaire survey is applied to collect opinions related to IPD principles from 235 industry practitioners. Fuzzy set qualitative comparative analysis (fsQCA) is deployed to gather IPD principles' obstacles for construction project collaboration. The results show that (1) the absence of contractual and behavioral principles obstructs significantly the level of collaboration of construction projects in several cases, (2) catalysts for IPD have no significant impact in most cases, and (3) the unfamiliarity with IPD has negative impacts on the application of its principles. The theoretical contribution consists of filling the gap in IPD's collaborative management research and improving the research method in related fields. As for the practical contribution, it aims to prioritize the importance of IPD principles and provide valuable suggestions.

Keywords: Integrated Project Delivery (IPD); construction project; level of collaboration; path of obstruction; fuzzy set qualitative comparative analysis

Citation: Mei, T.; Zhong, S.; Lan, H.; Guo, Z.; Qin, Y. Configuration Analysis of Integrated Project Delivery Principles' Obstacle to Construction Project Level of Collaboration. *Sustainability* **2023**, *15*, 3509. https://doi.org/10.3390/su15043509

Academic Editors: Srinath Perera, Albert P. C. Chan, Dilanthi Amaratunga, Makarand Hastak, Patrizia Lombardi, Sepani Senaratne, Xiaohua Jin and Anil Sawhney

Received: 21 November 2022
Revised: 3 February 2023
Accepted: 9 February 2023
Published: 14 February 2023

1. Introduction

The construction of a project is carried out by several participants at various stages [1], its performance depends largely on the participants' collaboration, which is critical to improving efficiency and delivering successfully the construction project [2]. With the increase in technical complexity and the diversification of specifications of these construction projects, the delivery is becoming increasingly fragmented [3,4]. To overcome this problem, a new delivery method, called Integrated Project Delivery (IPD), has emerged.

Moreover, the delivery of an integrated project is based on collaboration [5], consisting of integrating the personnel, the system, the business structure, and the practice in one whole process [6]. In this process, all participants will harness sufficiently their talents and insights to optimize project performance, increase value to the owner, reduce waste, and maximize efficiency through all the project phases of design, manufacturing, and construction [7]. Building Information Modeling (BIM) is growing up as the basis for the rapid development of IPD, which is advocated as a technological tool to promote work with IPD, to provide opportunities for broader collaboration, promote the integration of architectural professionals to the greatest extent, and realize information sharing and efficient team collaboration [8,9].

However, due to several factors such as contracts, technology, and personnel behavior, the application of the IPD in traditional construction projects has slowed down [10,11]. In other words, the pure IPD cannot be applied directly in some countries, such as China [1]. However, some scholars and practitioners try to introduce the concept of IPD in traditional construction projects to improve the level of collaboration and management efficiency [12–14]. Added to that, the degree of application of the IPD principle will generate different effects on the level of collaboration in the project [15]. In view of the rapid development of the Chinese construction industry, the requirements for integrating project collaboration among the different groups are increasing. Considering the application of BIM technology in China and the barriers denying the direct introduction of pure IPD [16–18], this paper considers China as an example to study, analyze, and discuss the impact of IPD principles on construction project collaboration.

This paper studies systematically the effects of applying IPD principles at the level of construction project collaboration in detail on the basis of a questionnaire survey. Firstly, this paper uses bibliometric evaluation to analyze the literature related to IPD and construction collaboration. According to the research hotspots and deficiencies, it is found that the research needs to create a path for IPD principles for construction project collaboration depending on certain configuration analyses. Secondly, the technical route of detailed analysis is introduced. Based on the 15 IPD principles and 3 levels of collaboration adapted from NASFA et al. [8], the variables are selected and the questionnaire is designed and sent to practitioners experienced in BIM technology, and fuzzy set qualitative comparative analysis (fsQCA) is introduced as the research method. Thirdly, referring to the collaboration level of construction projects as the result variable and the three kinds of IPD principles (including contractual principles, behavioral principles, and catalysts) as the condition variable, the configuration analysis was carried out through fsQCA. Fourthly, based on the obtained results of configuration analysis and the previous research results, the reasons for the configuration formation were discussed and suggestions were put forward to reduce the obstruction along the configuration path. Finally, some contributions and future ideas are summarized. This study not only fills the gap in the field of IPD collaborative management obstacles, but also improves the research methodology based on fuzzy set qualitative comparative analysis and model asymmetry analysis. At the same time, the importance of IPD principles in the process of collaborative management is prioritized to provide a reference for improving the efficiency of collaborative management in practical engineering.

2. Literature Review

Bibliometric analysis refers to the cross science of quantitative analysis of all knowledge carriers by means of mathematics and statistics. The bibliometric analysis of literature keywords reflects the research hotspot and trend of the research field.

2.1. Research Hotspot Analysis Based on Bibliometric Analysis

Using China National Knowledge Infrastructure (CNKI) as the data source, the relevant literature search (including academic conference papers, journals, books, and dissertations) was carried out using the keywords "IPD" and "collaboration" as the search subject. A total of 150 non-repetitive literatures were retrieved, and 117 complete information literatures including 56 academic journals and 61 dissertations were retained to analyze the research status. VOS viewer was deployed to draw the keyword clustering diagram. As shown in Figure 1, the emerging clusters are mainly *IPD*, *BIM*, *collaborative management*, and *collaborative work*, indicating that these research directions in China's IPD collaboration field have attracted the attention of researchers and practitioners in the last 5 years.

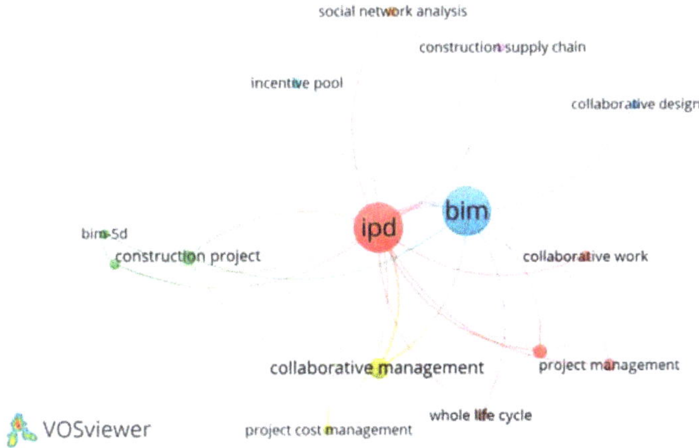

Figure 1. Cluster diagram of IPD collaborative keywords in China (keyword threshold is set to 3).

In opposite, using the Web of Science core collection as the data source, the relevant literature search was carried out with the two already listed keywords (e.g., IPD and collaboration). A total of 107 non-repetitive and complete information articles were retrieved, but 67 articles unrelated to the topic were excluded and 40 articles related to the topic were retained to analyze the research status. The keyword clustering diagram was drawn using VOS viewer. As shown in Figure 2, the emerging clusters are mainly *IPD*, *collaboration*, *BIM*, *performance*, and *management*, indicating that these research directions in IPD collaboration fields have attracted extensive attention in the world.

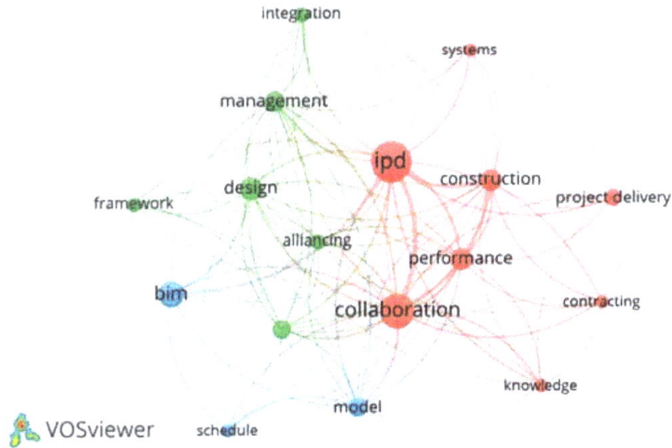

Figure 2. Cluster diagram of IPD collaboration keywords in the world (keyword threshold is set to 3).

Thus, to sum up, almost the same results have been obtained through the bibliometric analysis of literatures on related topics from two well-known databases, concerning the number of articles and the associated keywords found in the clustering of the retained papers.

2.2. Summary of Missing Parts in the Current Field of Research

What was mentioned above shows that BIM has become one of the research hotpots in the world as it serves as a technical basis for the rapid development of IPD [8,9]. Furthermore, IPD is used in construction projects to increase the project performance through a highly collaborative process [19,20], so collaborative management is one of the main research hotspots. As the keywords threshold was set at three, rare literature references about IPD collaboration's obstacles were not shown in the cluster diagrams of Figures 1 and 2. Based on the Analytic Hierarchy Process (AHP) model, the BIM software function, the mode reorganization cost, and the number of BIM practitioners were found as the main factors that obstruct the development of the IPD collaborative model [21]. In addition, referring to Structural Equation Modeling (SEM), the factors that obstruct the application of IPD were studied [11]. Based on the related investigation, good team relationships and willingness to cooperate between teams were found to reduce significantly the risk of IPD [22].

As IPD principles involve many conditions and multi-party teams, the risk-sharing principle, the partnership of the participants, and the legal contract framework create some obstacles to the IPD's application [7]. This application is not limited to partial behavioral principles and catalysts, so the study of the interaction between various IPD principles and the degree of project cooperation in IPD mode has certain significance. However, the application of the IPD in construction projects and related cases is limited in China [16–18]. Thus, the main objective of this paper is to get research data based mainly on a questionnaire survey.

3. Research Design and Method

3.1. Technical Route

FsQCA is the main method that will be used throughout this study, and the technical route of this paper includes mainly four parts, as shown in Figure 3. These parts will be detailed here below: Part 1: Questionnaire design, involving mainly the selection of questionnaire variables, questionnaire design, distribution, and recovery. The selection of questionnaire variables and the questionnaire design are based on previous research and industrial standards, whereas its distribution and recovery are realized through the questionnaire star platform. Part 2: Research results, involving mainly the test of data and the configuration of the analysis results. The first one consists of testing raw data through SPSS, validating data rational distribution through descriptive statistics, and dividing the sample groups according to the "degree of familiarity with IPD". As for the configuration analysis, it includes the variable calibration, the necessity analysis for single antecedent variables, the conditional configuration analysis, and the robustness test. Part 3: Discussion, comparing the main results of this study with those of previous studies, the similarities and the differences were summarized. Part 4: Research conclusion, summarizing the main theoretical and practical contributions, as well as the research limitations.

3.2. Questionnaire Survey

A questionnaire survey is a method to collect data by designing detailed questionnaires and asking respondents to answer accordingly. This paper designs and recovers questionnaires to collect the data for analysis.

3.2.1. Selection of Variables

From the owner's point of view, IPD collaboration is divided into three levels, including typical collaboration, enhanced collaboration, and required collaboration [8]. The first two levels refer to projects that adopt IPD as a concept, whereas the third level refers to projects that adopt IPD as a delivery method [8]. IPD principles include a total of 15 principles, including contractual principles, behavioral principles, and catalysts for IPD. Among them, the contractual principles can be included in the agreement. As for behavioral principles, they are necessary for project optimization, but they are based ultimately on choice. Finally, catalysts for IPD are very useful for optimizing project results. To sum up, this paper chooses contractual principles, behavioral principles, and catalysts for IPD as antecedent variables [8] and the level of collaboration will be considered as the outcome variable [6–8]. Based on configuration analysis and the proof of contradiction, the paper reveals the multiple concurrent paths that IPD principles obstruct the level of collaboration. Added to that, the paper summarizes the management enlightenment of the level of collaboration of construction projects in order to provide references for enhancing the construction project collaborative management efficiency under the application of IPD principles.

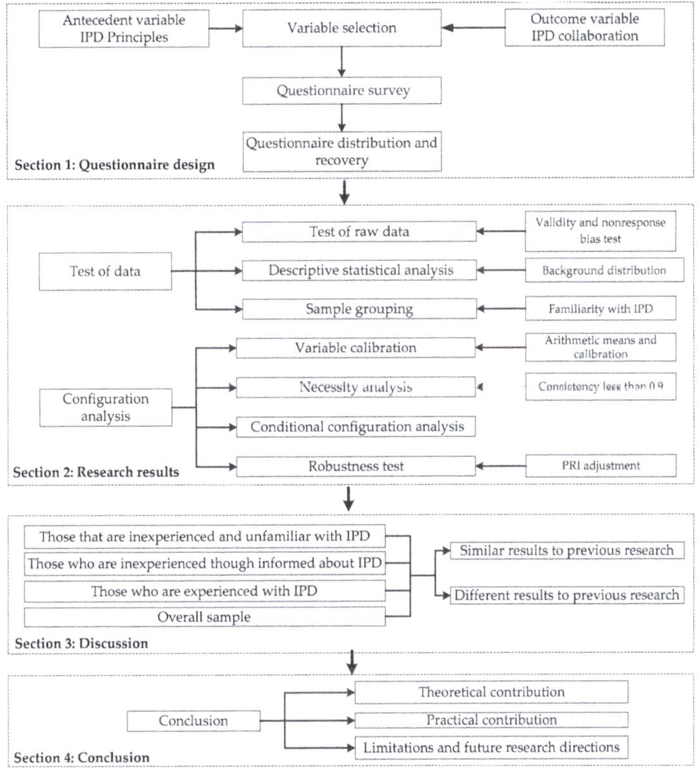

Figure 3. Technical route.

3.2.2. Questionnaire Survey

Through the questionnaire survey, this paper used the questionnaire star platform (https://www.wjx.cn/vm/waUUwNB.aspx (accessed on 31 October 2022) (Appendix A) to obtain the raw data. As the core of BIM technology and IPD mode consists of collaborative work, and the application of BIM technology belongs to the category of IPD principles, the research objective needs to have certain BIM experience. Thus, the contents of the questionnaire include mainly the following sections.(1) Background survey of the respondents: it includes mainly the type of institution the respondents work in, the number of years they have worked in the construction industry, the number of BIM projects they have participated in, their willingness to use BIM technology, and their familiarity with IPD mode. (2) Investigation on the influence of IPD principles on the level of collaboration of construction projects: it includes mainly a matrix scale designed based on Likert five points scale, with 15 secondary indexes of IPD principles presented in Table 1 as the vertical axis and five options as the horizontal axis (the options at the horizontal axis include the following: large negative influence, small negative influence, no influence, small positive influence, and large positive influence). The respondents were asked to select the most consistent parameters with the IPD principles' influence on the collaboration of construction projects based on their feelings and the implementation of the projects. (3) Investigation on the level of collaboration of IPD in China: taking three kinds of IPD collaboration levels in Table 2 as options, a single choice question was set up to require respondents to choose the most consistent construction project collaboration level with the current situation of China's construction industry according to their true feelings.

Table 1. Selection of antecedent variables.

Antecedent Variables	Secondary Indexes	Observed Variables	Assignment
Contractual Principles (*X1~X8*)	Key Participants Bound Together as Equals	*X1*	1–5
	Liability Waivers between Key Participants	*X2*	
	Early Involvement of Key Participants	*X3*	
	Fiscal Transparency between Key Participants	*X4*	
	Jointly Developed Project Target Criteria	*X5*	
	Shared Financial Risk and Reward Based on Project Outcome	*X6*	
	Intensified Design	*X7*	
	Collaborative Decision-Making	*X8*	
Behavioral Principles (*X9~X11*)	Mutual Respect and Trust	*X9*	1–5
	Willingness to Collaborate	*X10*	
	Open Communication	*X11*	
Catalysts for IPD (*X12~X15*)	Multi-Party Agreement	*X12*	1–5
	Building Information Modeling (BIM)	*X13*	
	Lean Design and Construction	*X14*	
	Co-location of Team	*X15*	

Table 2. Selection of outcome variable.

Outcome Variable		Variable Description	Assignment	Reference
Collaboration Levels of IPD	Typical	Collaboration not contractually required	1	[6–8]
	Enhanced	Some contractual collaboration requirements	2	
	Required	Collaboration required by a multi-party Contract	3	

3.2.3. Questionnaire Distribution and Recovery

According to the literature induction, this paper adopts initially 15 IPD principles that reflect the level of collaboration of construction projects. The preliminary designed questionnaire was distributed to two construction units for trial filling, and the questionnaire was revised according to the feedback opinions of the filling personnel. Finally, a formal questionnaire was developed and distributed. Three main ways to distribute the questionnaire were adopted in Table 3:

Table 3. Three main ways to distribute the questionnaire.

Questionnaire Survey Objects	Ways of Questionnaire Invitation
Corresponding authors in the literature related to the subject from CNKI, WANFANG, CQVIP, and other core journals	Email
Practitioners and researchers participating in relevant conferences and forums in the construction industry	Combination of online and offline distribution
The staff of the professional practice base or the previous graduates engaged in the industry	Questionnaire link sharing

The main objective was to collect 400 questionnaires, and data collection started on 1 December 2021 and lasted till 31 October 2022, where a total of 372 questionnaires were collected. The questionnaire data was screened by the missing value test and abnormal value test, and finally 352 complete and valid responses were retained. Therefore, the effective recovery rate of the questionnaire was 88%. As the core of BIM technology and IPD is collaborative work, and in order to be more consistent with the actual situation of IPD application, only 235 valid questionnaires from respondents with BIM experience were analyzed in this paper. Moreover, the reliability of 235 questionnaires represents a value of 0.946; thus, the high sample size and reliability make it possible for further analysis.

3.3. Fuzzy Set Qualitative Comparative Analysis

FsQCA takes a holistic view and conducts a case-oriented comparative analysis, where each case is viewed as a configuration of conditional variables [23]. The purpose of fsQCA is to find a causal relationship between the conditional configuration and the outcome by comparing different cases and the corresponding configuration that causes the outcome to appear or not. The concept that the social phenomenon is linked to the circumstances is considered when taking into account the condition configuration as a whole. For the cases whose antecedent conditions is n, any antecedent condition includes two states (present and absent), and the possible configuration number of the logical combination of antecedent conditions is 2^n. Through fsQCA, the qualified configuration can be found from 2^n configurations. Consistency and coverage are the two main indictors to reflect the reliability of the results. Consistency refers to the degree of consistency between the conditional variable or path and the result. Coverage refers to the extent to which a condition or path subset physically covers the conditions or paths sets.

FsQCA is considered the most appropriate approach for this study [24], as it: (1) Allows for the exploration of conditional (pathways) combinations that combine the obtained result of specific outcomes. (2) Allows for equivalence, yielding in different paths that lead to the same result [23]. (3) Distinguishes between sufficient conditions (a single condition sufficient to predict the outcome), the necessary conditions (which must be included in each potential pathway to the given outcome), and the INUS (an insufficient but necessary part of a condition which is itself unnecessary but sufficient for the result) conditions (which are part of one of the possible pathways to the outcome). (4) Allows for asymmetry, which means that conditions can lead to results and the reverse of condition needs do not lead to the opposite results [24].

4. Research Findings

After screening the collected questionnaires, this paper carries out the following tests and configuration analysis on the questionnaire data.

4.1. Test of Data

The test of data aims to verify the reliability of the questionnaire data, demonstrating that the data can satisfy the requirements for further configuration analysis.

4.1.1. Test of Raw Data

As participants can't submit the questionnaire unless it is fully filled out, the questionnaire data has no missing values and outliers. At the same time, the validity of the 15 principles was tested, the overall Kaiser–Meyer–Olkin (KMO) value was 0.944, and the KMO value of the secondary indicators of three principles was greater than 0.7 [25]. Thus, the questionnaire is valid, as the data is suited for factor analysis. The survey was also separated into two stages based on various response years. The years of working in the construction industry and the number of BIM projects conducted by the participant involved in the two indicators of the chi-square test significant values are larger than 0.05, removing the potential of non-response bias [26–28]. As a result, the data from the questionnaire have a high level of validity and may be used for subsequent analysis.

4.1.2. Descriptive Statistical Analysis

The background distribution of respondents was analyzed, as shown in Table 4. Respondents were mainly chosen from construction units, design units, and research institutions, and the proportion of every unit is relatively balanced and covers a wide range. Moreover, more than two-thirds of the respondents were engaged in the construction industry for more than three years and nearly half of them have participated in more than three BIM projects; respondents with long-term industry experience and sufficient BIM project experience make their responses have reference significance for related research. In addition, nearly 90% of the respondents have a strong desire to use BIM technology. As

BIM is the basis for the rapid development of the IPD, the participants working on BIM projects and their strong willingness to use BIM have a strong reference for the analysis of the IPD project collaboration.

Table 4. Descriptive statistics.

Variables		Frequency	Percentage
Employment units	Real estate units	21	8.94%
	Construction units	73	31.06%
	Design units	52	22.13%
	Consulting units	28	11.91%
	Supervision units	0	0.00%
	Suppliers	1	0.43%
	Research Institutions	52	22.13%
	Others	8	3.40%
Years of working in the construction industry	≤ 3	71	30.21%
	3~5	34	14.47%
	5~8	47	20.00%
	>8	83	35.32%
Number of experienced BIM projects	1~2	126	53.62%
	3~5	42	17.87%
	6~10	20	8.51%
	>10	47	20.00%
Willingness to use BIM technology	0	1	0.43%
	1	7	2.98%
	2	18	7.66%
	3	45	19.15%
	4	45	19.15%
	5	119	50.64%

4.1.3. Sample Grouping

As can be seen from Table 5, respondents that are inexperienced and unfamiliar with IPD accounts for 38.3% of the total, respondents who are inexperienced though informed about IPD accounts for 44.7% of the total, and respondents who are experienced with IPD accounts for 17% of total. The ratio of respondents' "familiarity with IPD from low to high" is about 4:4:2, and most of them either have no direct experience with IPD or are not familiar with its concept, which is consistent with previous research conclusions [7]. To study the configuration path of construction project collaboration more systematically and comprehensively, in addition to the overall sample analysis of 235 valid questionnaires, this paper will evaluate the three groups of samples. Through the comparative investigation of multiple concurrent paths under multiple samples and tracing the antecedent conditions from the reductionism perspective, this paper will explain the reasons why IPD mode affects the level of collaboration of the construction projects in a more scientific and reasonable way.

Table 5. Sample distribution.

Variable		Frequency	Percentage	Sample Grouping
Respondents' familiarity with IPD	Those that are inexperienced and unfamiliar with IPD	90	38.3%	1
	Those who are inexperienced though informed about IPD	105	44.7%	2
	Those who are experienced with IPD	40	17.0%	3

4.2. Configuration Analysis

Based on configuration analysis, the necessary conditions and configuration paths of low degree of cooperation can be found. The robustness of configuration analysis can also be tested through the robustness test.

4.2.1. Variable Calibration

In this study, fsQCA was used for analysis purposes. The arithmetic means of the secondary indexes in Table 1 are taken as three principles' scores [29], and the higher the score is, the higher the degree of influence will be. The objective of fsQCA is to calibrate and normalize the variables involved in the calculation so that scores can be converted into fuzzy scores between 0 and 1 to improve the interpretability of the results [30]. In this paper, complete subordination crossing points and complete non-subordination are located at 5, 3, and 1 for the three principles, and 3, 2, and 1 for level of collaboration, respectively.

4.2.2. Necessity Analysis of Single Antecedent Variable

The necessary conditions of a single factor are obtained, as shown in Table 6. Consistency is similar to the coefficient significance degree (p-value) in regression analysis, which refers to what extent a certain result requires the existence of a certain variable. Coverage refers to the extent to which a subset physically covers the target set, which is a direct indicator of the empirical importance of antecedent conditions. In fact, when the consistency is below 0.9, neither sample has a bottleneck, yet there is little collaboration [31,32]. As can be seen from Table 6, the necessary conditions were absent in both the overall sample and sample 1. Although in samples 2 and 3, the necessary conditions (contractual principles, behavioral principles, and catalysts for IPD in sample 2; catalysts for IPD in sample 3) of low collaboration levels exist, the low coverage of necessary conditions, less than 0.8, means that these subsets don't account for a large proportion of the total. Therefore, the multiple antecedent conditions need to be combined for configuration analysis in this study.

Table 6. Analysis of antecedents' necessary condition under multiple samples.

Outcome Variable	Condition Variable	Overall Sample		Sample 1		Sample 2		Sample 3	
		CY	CE	CY	CE	CY	CE	CY	CE
Low level of collaboration	Contractual Principles	0.88	0.75	0.83	0.79	0.93	0.76	0.89	0.63
	~Contractual Principles	0.25	0.89	0.28	0.85	0.22	0.93	0.26	0.92
	Behavioral Principles	0.89	0.75	0.85	0.8	0.93	0.75	0.89	0.63
	~Behavioral Principles	0.24	0.89	0.27	0.84	0.21	0.95	0.24	0.88
	Catalysts for IPD	0.88	0.75	0.83	0.79	0.91	0.75	0.9	0.63
	~ Catalysts for IPD	0.26	0.91	0.29	0.86	0.24	0.95	0.24	0.93

Note: A sideways tilde ~ indicates the absence or negation of the causal condition. CY indicates the consistency between the condition variable and low level of collaboration, and CE indicates the coverage between the condition variable and low level of collaboration.

4.2.3. Conditional Configuration Analysis

In this study, due to the values of samples 2 and 3, multiple antecedent variables need to be combined for analysis to explore the influence of the combination path on the outcome variables. The cutoff value for analysis at each sample size was set as follows: the acceptable number of cases was set at 1, the consistency threshold was set at 0.8 [33], and the Proportional Reduction in Inconsistency (PRI) was set at 0.7 [34]; thus, the complex solution, parsimonious solution, and intermediate solution can be obtained. This paper chooses the intermediate solution and the parsimonious solution to explain the configuration path model: low levels of collaboration (equal to f) (contractual principles, behavioral principles, and catalysts for IPD). The overall coverage of the overall sample, sample 1, sample 2, and sample 3 are 0.278718, 0.245517, 0.247738, and 0.279661 respectively, and the overall consistency values are 0.883432, 0.849868, 0.944798, and 0.884718 respectively. The combined path interpretability of the four groups of samples is relatively high.

According to the results of fsQCA, among all the antecedent variable combinations, the results of four sampling size studies show that there are seven obstacles relative to the low collaboration level in construction projects. In this study, the antecedent variable configurations are shown in Table 7.

In the overall sample, which includes all respondents, there exist two configuration paths. Path S1A shows that contractual principles are absent as the core condition, whereas behavioral principles and catalysts for IPD have no impact on low collaboration levels. Concerning the path S1B, it indicates that behavioral principles are absent as the core condition, catalysts for IPD exist as the edge condition, and contractual principles have no impact on low collaboration levels.

Table 7. Antecedent variables with low collaboration levels under multiple samples.

Configurations	Overall Sample		Sample 1	Sample 2		Sample 3	
	S1A	S1B	S2	S3A	S3B	S4A	S4B
Contractual Principles	☆		☆		☆	☆	
Behavioral Principles		☆	☆	☆	■	■	☆
Catalysts for IPD		■	☆	■	☆		■
Raw Coverage	0.253329	0.227442	0.245517	0.214753	0.192206	0.238136	0.237288
Unique Coverage	0.051276	0.025389	0.245517	0.0555323	0.0329853	0.0423729	0.0415255
Consistency	0.890225	0.885845	0.849868	0.953646	0.954388	0.913821	0.893142
Solution Coverage	0.278718		0.245517	0.247738		0.279661	
Solution Consistency	0.883432		0.849868	0.9447982		0.884718	

Note: Among them, ☆ indicates the absence of the core condition, which indicates the existence of a strong causal relationship between the condition and the concerned result, and ■ indicates the presence of the edge condition, which indicates the weak causal relationship between the condition and the result. A blank indicates that the presence or absence of this condition has no effect on the level of collaboration [35].

In sample 1, which includes the respondents that are inexperienced and unfamiliar with IPD, there exists 1 configuration path. Path 2 shows that contractual principles, behavioral principles, and catalysts for IPD are all absent as the core conditions for low collaboration levels.

In sample 2, which includes the respondents who are inexperienced though informed about IPD, there exist two configuration paths. Path S3A indicates that behavioral principles are absent as the core condition, catalysts for IPD exist as the edge condition, and contractual principles have no impact on low collaboration levels. Path S3B shows that contractual principles and catalysts for IPD are absent as the core condition and behavioral principles exist as the edge condition.

In sample 3, which includes the respondents who are experienced with IPD, there exist two configuration paths. Path S4A shows that contractual principles are absent as the core condition, behavioral principles exist as the edge condition, and catalysts for IPD have no impact on low collaboration level. Path S4B indicates that behavioral principles are absent as the core condition, catalysts for IPD exist as the edge condition, and contractual principles have no impact on low collaboration levels.

4.2.4. Robustness Test

In this paper, the PRI threshold was set to 0.75 [36] in the analysis of the truth table of the four groups of samples. In addition, the robustness test results showed that the configuration path of the new model for the three samples (overall sample, sample 1, and sample 2) is completely consistent with the configuration path of the original model. However, through one of the two configurations of the new model of sample 3, it is completely consistent with the original model's S4A, where the other path only replaces the core missing condition with the misleading contractual principles under the comparison with S4B, and the overall change of the path is not much, indicating that the research conclusion is relatively robust [29,31].

5. Discussion

Most of the previous studies were to explore the IPD in well-determined construction application projects and the impact of the construction factors, as well as to demonstrate the significance of a single factor. However, there are few studies considering the obstruction of IPD principles to the collaborative management of construction projects, and the influencing factors involved are relatively limited. There are also some deficiencies in the applied method, and the combined influence of multiple influencing factors has not been considered so far from a configuration perspective. The necessity analysis results do not have the necessary conditions with relatively high coverage, and corresponding to the configuration paths in one sample does not have the same conditional level of necessary conditions, so necessity analysis has no practical significance for the interpretation of the final configuration path. In four groups of samples in this paper, under seven concurrent obstruction paths were obtained by fsQCA. In fact, there are only five different paths obstructing the level of collaboration. The first path only consists of the absence of contractual principles, which are the core condition. The second path consists of the absence of behavioral principles, which are the core condition, and the presence of catalysts for IPD, which are the edge condition. The third path consists of the absence of three kinds of principles, which are all core conditions. The fourth path consists of the absence of contractual principles and catalysts for IPD, which are both core conditions, and the presence of behavioral principles, which are the edge condition. The fifth path only consists of the absence of contractual principles, which are the core condition, and the presence of behavioral principles, which are the edge condition. Compared to previous research findings, this paper presents its outcomes under two main aspects that will be detailed in the next two paragraphs.

5.1. Similar Results to Previous Research

The absence of behavioral principles can hinder significantly construction projects collaboration with the presence of catalysts for IPD. The paths S1B, S3A, and S4B in Table 7 are identical. In these paths, the absence of behavioral principles and the presence of catalysts for IPD represent the core and the edge conditions that obstruct the construction project collaboration, respectively. However, the presence or absence of contractual principles will not obstruct the level of collaboration of construction projects. The paths in samples 2 and 3 indicate that practitioners with IPD experience (and knowledge of IPD) believe that the catalysts for IPD, without matching behavioral principles, will obstruct construction projects collaboration regardless of the existence of contractual principles. In addition, the responses in sample 1 (from participants that do not know IPD) had a certain negative impact on the path in the overall sample; therefore, the S1B path consistency in the entire sample was lower than S3A and S4B of the same path.

In addition, the behavioral principles provide rules for the integration, the communication between parties, and the ability of team members to trust and support each other through cooperation, which is critical to eliminate the segregated roles of traditional contracting processes to reduce risks [23], increase value to the client, and reduce the amount of construction waste [37,38]. Moreover, appropriate catalysts, such as BIM and lean construction, which mainly provides a virtual design before the actual construction begins and enables the project stakeholders to see the building clearly [37,38], are the least important factors affecting the level of collaboration of IPD application [7] compared to contractual principles and behavioral principles. Behavioral principles can remove separations, consequently improving the collaboration environment that BIM implementation necessitates [39]. Therefore, even if there are catalysts for IPD, such as BIM technology, and if the behavioral principles cannot be guaranteed, it is difficult for construction projects under IPD mode to achieve good collaboration performance.

5.2. Different Results from Previous Research

(1) The absence of contractual principles will obstruct significantly the collaboration of construction projects. Thus, this absence in path S1A is the core condition to obstruct the level of collaboration of construction projects. However, the existence of behavioral principles and catalysts for IPD will not eventually obstruct the construction projects collaboration. One of the paths of the overall sample, S1A, reflects that the absence of the contractual principles at the overall sample level will seriously obstruct the level of collaboration of the construction projects. Hence, IPD acting as a collaborative contract approach changed the basic business and organizational and legal structure of the project to reduce dysfunction and improve performance [40]. Added to that, contractual principles are often associated with legal provisions and cannot be undermined to improve integration

and create a trust-based work environment by decentralizing project risks and outputs to all construction participants [41]. In addition, referring to the contractual relationships, early determination of project objectives and early formation of teams are essential keys to the IPD's success [6,42] where the principle of shared risk in the contract makes the construction industry more suspicious of the application of IPD [7]. Therefore, the absence of contractual principles can seriously obstruct the construction projects collaboration by hindering the mediating effect of IPD application;

(2) The absence of contractual principles can obstruct significantly the collaboration of construction projects under the assumption of the presence of behavioral principles. Thus, in path S3b, the absence of contractual principles and catalysts for IPD is the core condition that deteriorated the level of collaboration of the construction project; however, the existence of the behavioral principle is the edge condition. As an example, in path S4A, the absence of the contractual principle and the presence of the behavior principle are respectively the core and edge conditions of obstructing construction project collaboration, as the presence or absence of catalysts for IPD will not hinder the level of construction project collaboration. The contractual principles consist the premise that the relationship between team members becomes reliable, so that they respect each other and cooperate [19,20], whereas the absence of contractual principles will make the realization of the behavioral principles' lack of protection. At the same time, some principles are responsible for some characteristic improvement [43]. For example, in terms of the team aspect, the shared risk and reward principle can generate mutual goal achievement [44]. In fact, people who do not have IPD project experience in China believe that technological tools, such as BIM, have been applied in actual projects. Thus, the absence of catalysts for IPD will hinder seriously project collaboration, even though practitioners with IPD experience believe that the presence or absence of catalysts for IPD will not obstruct the level of collaboration of construction projects. In previous research, the practitioners with IPD experience had a higher ability to use BIM, so they believed that BIM should be applied to IPD projects to promote better application of the IPD principles [7], fitting with path S3 but contradicting S4A. As data from the previous research were derived mainly from various US research associations, and BIM is still not used widely in China [45,46], the background of BIM application varies so that the appearance of contradictions between this study and previous studies in the catalysts for IPD is logical. Combined with the current situation of BIM application in China, the S4A path is more in line with reality;

(3) The absence of contractual principles, behavioral principles, and catalysts for IPD can complicate significantly the collaboration of construction projects under the premise of behavioral principles. Thus, path S2 shows that the three principles of IPD are of equal importance (as considered by the participants who do not understand IPD), and the absence of these three kinds of principles are all core conditions that may obstruct the collaboration of construction projects. By comparison, it is more necessary to popularize the application of the IPD concept and its principles in construction projects [7].

6. Conclusions

This paper aims to promote the application of IPD principles to improve the level of collaboration and efficiency of collaborative management in construction projects. Based on the bibliometric analysis, this paper summarized the current hot spots of the IPD collaboration research field, such as IPD and BIM technology integration application, and the integration of IPD and collaborative management ideas. Through the proof by contradiction, the fsQCA method was used to analyze the obstacle configuration path of IPD principles at the projects' construction collaboration levels under multiple samples. Therefore, this paper globes both theoretical and practical contributions.

6.1. Theoretical Contribution

This paper expanded the scope of the IPD collaborative management research field. Based on the proof of contradiction, this paper studied the constraints of applying the IPD principles to the level of collaboration of construction projects and expanded the research to show the problems of IPD in the field of collaborative management. In addition, this paper considered the influence of the combination of contractual principles, the behavioral principles, and the catalysts for IPD on the level of collaboration in project construction.

Additionally, the research methods in the field of IPD collaborative management were enriched. This paper is result-oriented, as it analyzes the influencing factors that obstruct the level of collaboration of construction projects from the perspective of "induction-tracing". In addition, configuration analysis is an asymmetric analysis, and it shows that the combination of one or more conditions constitutes the antecedent of a specific result. Added to that, the nonlinear superposition of causality and the multiple concurrent mechanisms are considered. Referring to complex system theory perspectives, this paper analyzed the complex system of construction project with multi-stages, multi-participants, and multi-management elements, which makes up for the deficiency of traditional deductive methods in studying causality and improving the research methods.

6.2. Practical Contribution

Based on the results of the configuration analysis, considering the actual situation of the project, to better implement IPD principles in construction projects and improve the level of collaboration in construction projects, this paper puts forward the following suggestions.

Some construction industry practitioners lack understanding of IPD mode, so the concept and related principles of IPD should be popularized in the industry. Appropriately increasing the familiarity of practitioners with IPD will promote for better application of IPD in construction projects, and their experience and suggestions can also promote the pertinence and accuracy of related research.

The contractual principles take priority over the behavioral principles. The absence of the first principle will obstruct seriously the collaboration performance of IPD mode projects; thus, ensuring that the contractual principles can be met first when the IPD mode is applied to construction projects is of major importance. The contractual principles guarantee that the participants in a construction project are connected on an equal footing, establish common standards for the project objectives, and share risks when they arise. The satisfaction of the contractual principles is the basis of the realization of the behavioral principle. The willingness to cooperate and mutual respect and trust are based on the contractual principles.

The behavioral principles take precedence over the principle of catalysts for IPD. The absence of the behavioral principles will restrict the collaboration performance of IPD mode projects. Thus, ensuring that the behavioral principles can be satisfied when the IPD mode is applied to construction projects is very important. The realization of the behavioral principles means that the project participants can reach the goal of cooperation based on respect and trust and create the premise for the implementation of the catalysts for IPD, such as the co-location of team and the multi-party agreement.

6.3. Limitations and Future Research Directions

This study still has the following limitations: (1) The original data came from the questionnaire, which may have subjective bias; thus, future works can try to use practical engineering cases to study the obstructing influence of IPD principles on the level of collaboration in construction projects. (2) The sample data of multi-contrast is limited; thus, future works can enlarge the investigation scope and the sample size so that the research results can adapt to a wider range of research objects and improve the pertinence of the research results. (3) In this study, the proof by contradiction is used to study the obstructed path of contractual principles, behavioral principles, and catalysts for IPD to define the level of collaboration in construction projects under IPD mode. However, future works can explore the configuration path at high levels of collaboration in construction projects under IPD mode.

Author Contributions: Conceptualization, T.M. and S.Z.; methodology, S.Z.; software, T.M.; funding acquisition, T.M.; validation, T.M., Z.G. and H.L.; formal analysis, H.L.; investigation, Y.Q.; data curation, Z.G.; writing—original draft preparation, T.M. and S.Z.; writing—review and editing, T.M. and S.Z. All authors have read and agreed to the published version of the manuscript.

Funding: This research was funded by "2021 Hubei Province Construction Science and Technology Plan Project" and "2021 Internal Scientific Research Fund Project of Wuhan Institute of Technology (K2021033)".

Institutional Review Board Statement: Not applicable.

Informed Consent Statement: Not applicable.

Data Availability Statement: Some or all data, models, or code that support the findings of this study are available from the corresponding author upon reasonable request (list items).

Conflicts of Interest: The authors declare no conflict of interest.

Appendix A. Questionnaire on the Influences of IPD Principles on Collaboration Level (I) [English Version]

Dear experts/professionals,

Thank you very much for taking the time to participate in this questionnaire. Thank you for your support and help. This questionnaire is only for academic research, and does not have any commercial use, nor will it disclose any of your privacy. We will obey the requirements of information confidence, and the questionnaires are anonymous. Please feel free to fill in the anonymous questionnaire! Your selection has a decisive impact on this study. Therefore, please spare your precious time in your busy schedule to answer the relevant questions of this questionnaire and correct the shortcomings. Thank you for your support and wish you a happy work!

Part 1 Basic Information

1. What is your age? ( )

○ 20~25 years old   ○ 26~30 years old   ○ 31~40 years old   ○ 41 years old and more

2. What is your educational background? ( )

○ junior college and below   ○ undergraduate   ○ master's degree   ○ doctor's degree or above

3. What's your professional title? ( )

○ junior   ○ intermediate   ○ senior   ○ others

4. What is your employment unit? ( )

○ real estate units   ○ construction units   ○ design units   ○ consulting units
○ supervision units   ○ suppliers   ○ research institutions   ○ others

5. How many years have you worked in the construction industry? ( )

○ less than 3 years   ○ 3~5 years   ○ 5~8 years   ○ 8 years and above

6. How many BIM projects have your experienced? ( )

○ never   ○ 1~2   ○ 3~5   ○ 6~10   ○ more than 10

7. What is your willingness to use BIM technology? ( )

○ 0   ○ 1   ○ 2   ○ 3   ○ 4   ○ 5

Part 2 IPD Information Survey

Integrated Project Delivery (IPD) is a project delivery approach that integrates people, systems, business structures, and practices into a process that collaboratively harnesses the talents and insights of all participants to optimize project results, increase value to the owner, reduce waste, and maximize efficiency through all phases of design, fabrication and construction.

1. How familiar are you with IPD? ( )

○ those who are experienced with IPD   ○ those who are inexperienced, though informed about IPD   ○ those that are inexperienced and unfamiliar with IPD

2. Based on your experience in the BIM project and the implementation of the actual project, what do you think is the impact of the following IPD principles on information collaboration? ( )

	Very negative	More negative	No effect	More positive	Very positive
Key participants bound together as equals					
Liability waivers between key participants waivers					
Early involvement of key participants					
Fiscal transparency between key participants					
Jointly developed project target criteria					
Shared financial risk and reward based on project outcome					
Intensified design					
Collaborative decision-making					
Mutual respect and trust					
Willingness to collaborate					
Open communication					
Multiparty agreement					
BIM					
Lean design and construction					
Co-location of team					

3. What stage of collaboration do you think our industry is in now ( )

○ Typical (Collaboration not contractually required)   ○ Enhanced (Some contractual collaboration requirements)   ○ Required (Collaboration required by a multi-party Contract)

References

1. Mei, T.; Guo, Z.; Li, P.; Fang, K.; Zhong, S. Influence of integrated project delivery principles on project performance in china: An sem-based approach. *Sustainability* **2022**, *14*, 4381. [CrossRef]
2. Sabet, P.G.P.; Chong, H.-Y. Interactions between building information modelling and off-site manufacturing for productivity improvement. *Int. J. Manag. Proj. Bus.* **2019**, *13*, 233–255. [CrossRef]
3. Fischer, M.; Ashcraft, H.; Reed, D.; Khanzode, A. *Integrating Project Delivery*; John Wiley & Sons: Hoboken, NJ, USA, 2017. [CrossRef]
4. Gallaher, M.P.; O'Connor, A.C.; Dettbarn, J.L.; Gilday, L.T. *Cost Analysis of Inadequate Interoperability in the U.S. Capital Facilities Industry*; National Institute of Standards and Technology (NIST): Gaithersburg, MD, USA, 2004; Volume 11, pp. 223–252.
5. Piroozfar, P.; Farr, E.R.; Zadeh, A.H.; Inacio, S.T.; Kilgallon, S.; Jin, R. Facilitating building information modelling (bim) using integrated project delivery (ipd): A uk perspective. *J. Build. Eng.* **2019**, *26*, 100907. [CrossRef]

6.   AIA. *Integrated Project Delivery: A Guide*; The American Institute of Architects: Washington, DC, USA, 2007; Available online: https://www.docin.com/p-659123461.html (accessed on 1 November 2021).

7.   Kent, D.C.; Becerik-Gerber, B. Understanding construction industry experience and attitudes toward integrated project delivery. *J. Constr. Eng. Manag.* **2010**, *136*, 815–825. [CrossRef]

8.   NASFA, COAA, APPA, AGC, AIA. *Integrated project delivery for public and private owners*. Available online: https://doc.mbalib.com/view/c61c17f6b9e3184bcb4b905322e66c67.html (accessed on 1 November 2021).

9.   Chang, C.Y.; Pan, W.; Howard, R. Impact of Building Information Modeling Implementation on the Acceptance of Integrated Delivery Systems: Structural Equation Modeling Analysis. *J. Constr. Eng. Manag.* **2017**, *143*, 04017044. [CrossRef]

10. Ghassemi, R.; Becerik-Gerber, B. Transitioning to integrated project delivery: Potential barriers and lessons learned. *Lean Constr. J.* **2011**, *2011*, 32–52.

11. Durdyev, S.; Hosseini, M.R.; Martek, I.; Ismail, S.; Arashpour, M. Barriers to the use of integrated project delivery (ipd): A quantified model for malaysia. *Eng. Constr. Archit.* **2019**, *27*, 186–204. [CrossRef]

12. Kelly, D. Investigating the Relationships of Project Performance Measures with the Use of Building Information modeling (BIM) and Integrated Project Delivery (IPD). Ph.D. Thesis, Eastern Michigan University, Ypsilanti, MI, USA, 2015. Available online: https://commons.emich.edu/theses/599/ (accessed on 1 November 2021).

13. Elghaish, F.; Hosseini, M.R.; Talebi, S.; Abrishami, S.; Martek, I.; Kagioglou, M. Factors driving success of cost management practices in integrated project delivery (ipd). *Sustainability* **2020**, *12*, 9539. [CrossRef]

14. Choi, J.; Yun, S.; Leite, F.; Mulva, S.P. Team integration and owner satisfaction: Comparing integrated project delivery with construction management at risk in health care projects. *J. Manag. Eng.* **2019**, *35*, 05018014. [CrossRef]

15. Hanna, A.S. Benchmark performance metrics for integrated project delivery. *J. Constr. Eng. Manag. -ASCE* **2016**, *142*, 04016040. [CrossRef]

16. Mei, T.T.; Wang, Q.K.; Xiao, Y.P.; Yang, M. Rent-seeking behavior of bim & ipd-based construction project in china. *Eng. Constr. Archit.* **2017**, *24*, 514–536.

17. Xu, Y.Q.; Kong, Y.Y. The research status and forecast of ipd in china. *J. Eng. Manag.* **2016**, *30*, 12–17.

18. He, Q.; Wang, G.; Luo, L.; Shi, Q.; Xie, J.; Meng, X. Mapping the managerial areas of building information modeling (bim) using scientometric analysis. *Int. J. Proj. Manag.* **2017**, *35*, 670–685. [CrossRef]

19. El Asmar, M.; Hanna, A.S.; Loh, W.-Y. Quantifying performance for the integrated project delivery system as compared to established delivery systems. *J. Constr. Eng. Manag.* **2013**, *139*, 04013012. [CrossRef]

20. Whang, S.W.; Park, K.S.; Kim, S. Critical success factors for implementing integrated construction project delivery. *Eng. Constr. Archit. Manag.* **2019**, *26*, 2432–2446. [CrossRef]

21. Deng, W.L. Analysis of impediments to the development of IPD collaborative mode based on BIM---Based on AHP model. *Mod. Bus. Trade Ind.* **2018**, *39*, 204–206.

22. Ma, Q.; Li, S.; Cheung, S.O. Unveiling embedded risks in integrated project delivery. *J. Constr. Eng. Manag.* **2022**, *148*, 04021180. [CrossRef]

23. Rihoux, B.; Ragin, C.C. *Configurational Comparative Methods: Qualitative Comparative Analysis (qca) and Related Techniques*; Sage Publications: London, UK, 2008.

24. De Rooij, M.M.; Janowicz-Panjaitan, M.; Mannak, R.S. A configurational explanation for performance management systems' design in project-based organizations. *Int. J. Proj. Manag.* **2019**, *37*, 616–630. [CrossRef]

25. Bartlett, M.S. The effect of standardization on a $\chi 2$ approximation in factor analysis. *Biometrika* **1951**, *38*, 337–344. [CrossRef]

26. Armstrong, J.S.; Overton, T.S. Estimating nonresponse bias in mail surveys. *J. Mark. Res.* **1977**, *14*, 396–402. [CrossRef]

27. Shiau, W.-L.; Luo, M.M. Factors affecting online group buying intention and satisfaction: A social exchange theory perspective. *Comput. Hum. Behav.* **2012**, *28*, 2431–2444. [CrossRef]

28. Shiau, W.-L.; Yuan, Y.; Pu, X.; Ray, S.; Chen, C.C. Understanding fintech continuance: Perspectives from self-efficacy and ect-is theories. *Ind. Manag. Data Syst.* **2020**, *120*, 1659–1689. [CrossRef]

29. Cheng, J.; Luo, J.; Du, Y.; Liu, Q. What kinds of entrepreneurial ecosystem can produce country-level female high entrepreneurial activity? *Stud. Sci. Sci.* **2021**, *39*, 695–702.

30. Ragin, C.C. Fuzzy sets: Calibration versus measurement. *Methodol. Vol. Oxf. Handb. Political Sci.* **2007**, *2*, 174–198.

31. Zhang, M.; Du, Y. Qualitative Comparative Analysis (QCA) in Management and Organization Research: Position, Tactics, and Directions. *Chin. J. Manag.* **2019**, *16*, 1312–1323.

32. Skaaning, S.-E. Assessing the robustness of crisp-set and fuzzy-set qca results. *Sociol. Methods Res.* **2011**, *40*, 391–408. [CrossRef]

33. Bell, R.G.; Filatotchev, I.; Aguilera, R.V. Corporate governance and investors' perceptions of foreign ipo value: An institutional perspective. *Acad. Manag. J.* **2014**, *57*, 301–320. [CrossRef]

34. Du, Y.; Kim, P.H. One size does not fit all: Strategy configurations, complex environments, and new venture performance in emerging economies-sciencedirect. *J. Bus. Res.* **2021**, *124*, 272–285. [CrossRef]

35. Fiss, P.C. Building better causal theories: A fuzzy set approach to typologies in organization research. *Acad. Manag. J.* **2011**, *54*, 393–420. [CrossRef]

36. Dwivedi, P.; Joshi, A.; Misangyi, V.F. Gender-inclusive gatekeeping: How (mostly male) predecessors influence the success of female ceos. *Acad. Manag. J.* **2018**, *61*, 379–404. [CrossRef]

37. Scott, L.; Flood, C.; Towey, B. Integrated Project Delivery for Construction. In Proceedings of the 49th ASC Annual International Construction Eduction Conference Proceedings, San Luis Obispo, CA, USA, 10–13 April 2013.
38. Levy, F. *BIM in Small-Scale Sustainable Design*; Wiley: Hoboken, NJ, USA, 2012.
39. Azhar, S.; Khalfan, M.; Maqsood, T. Building information modelling (BIM): Now and beyond. *Australas. J. Constr. Econ. Build.* **2012**, *12*, 15–28. [CrossRef]
40. Ashcraft, H. Transforming project delivery: Integrated project delivery. *Oxf. Rev. Econ. Policy* **2022**, *38*, 369–384. [CrossRef]
41. Azhar, N.; Kang, Y.; Ahmad, I.U. Factors influencing integrated project delivery in publicly owned construction projects: An information modelling perspective. *Procedia Eng.* **2014**, *77*, 213–221. [CrossRef]
42. Thomsen, C.; Darrington, J.; Dunne, D.; Lichtig, W. *Managing Integrated Project Delvery, Construction Management Association of America*; McLean: Virginia, VA, USA, 2009.
43. Viana, M.L.; Hadikusumo, B.H.; Mohammad, M.Z.; Kahvandi, Z. Integrated project delivery (IPD): An updated review and analysis case study. *J. Eng. Proj. Prod. Manag.* **2020**, *10*, 147–161.
44. El Asmar, M.; Hanna, A.S. Comparative analysis of integrated project delivery (IPD) cost and quality performance. *Proc. CIB W* **2012**, *78*, 2012.
45. Xiaoyu, W. Parametric Design of Pile Foundation Based on RevitAPI. Master's Thesis, Dalian University of Technology, Dalian, China, 2020.
46. Chenxi, T. Diffusion and Application of Building Information Model Technology. Master's Thesis, Xi'an University of Architecture &Technology, Xi'an, China, 2014.

 *buildings*

Article

Data-Driven Analysis for Facility Management in Higher Education Institution

Ashish Kumar Pampana [1], JungHo Jeon [2], Soojin Yoon [3,*], Theodore J. Weidner [4] and Makarand Hastak [5]

[1] Civil and Environmental Engineering, Oklahoma State University, 503 Engineering North, Stillwater, OK 74078, USA

[2] Department of Civil and Environmental Engineering and Engineering Mechanics, University of Dayton, Dayton, OH 45469, USA

[3] Division of Engineering Technology, Oklahoma State University, 511 Engineering North, Stillwater, OK 74078, USA

[4] Construction Engineering and Management, Purdue University, 550 Stadium Mall Drive, West Lafayette, IN 47907, USA

[5] Lyles School of Civil Engineering, Purdue University, 550 Stadium Mall Drive, West Lafayette, IN 47907, USA

* Correspondence: syoon@okstate.edu

Abstract: Planned Preventive Maintenance (PPM) and Unplanned Maintenance (UPM) are the most common types of facility maintenance. This paper analyzes current trends and status of Facility Management (FM) practice at higher education institutions by proposing a systematic data-driven methodology using Natural Language Process (NLP) approaches, statistical analysis, risk-profile analysis, and outlier analysis. This study utilizes a descriptive database entitled "Facility Management Unified Classification Database (FMUCD)" to conduct the systematic data-driven analyses. The 5-year data from 2015 to 2019 was collected from eight universities in North America. A preprocessing step included but was not limited to identifying common data attributes, cleaning noisy data, and removing unnecessary data. The outcomes of this study can facilitate the decision-making process by providing an understanding of various aspects of educational facility management trends and risks. The methodology developed gives decision makers of higher education institutions, including facility managers and institution administrators, effective strategies to establish long-term budgetary goals, which will lead to the enhancement of the asset value of the institutions.

Keywords: facility management; classification code; higher education institutions; planned preventive maintenance; unplanned maintenance; natural language processing; database; quantitative analysis

Citation: Pampana, A.K.; Jeon, J.; Yoon, S.; Weidner, T.J.; Hastak, M. Data-Driven Analysis for Facility Management in Higher Education Institution. *Buildings* **2022**, *12*, 2094. https://doi.org/10.3390/buildings 12122094

Academic Editor: Fani Antoniou

Received: 27 October 2022
Accepted: 23 November 2022
Published: 29 November 2022

Publisher's Note: MDPI stays neutral with regard to jurisdictional claims in published maps and institutional affiliations.

1. Introduction

"APPA—Leadership in Educational Facilities" defines maintenance as a combination of all the technical administrative actions taken during the service life of a building to retain its parts and functions [1]. Higher education institutions consist of different varieties of buildings in a large number compared to other organizations, which requires a more diverse approach in operational maintenance [2]. Planned Preventive Maintenance (PPM) is one of the maintenance strategies that aims to increase the reliability or lifespan of equipment as time-based or condition-based; it refers to a proactive approach to maintenance in which maintenance work is scheduled to take place regularly [3]. Unplanned Maintenance (UPM) occurs on a random basis as reactive or emergency maintenance. An unexpected component (or equipment) failure can cost a significant amount of money or time to restore, which results in uncertainty in budget allocation in the facility management [4,5]. A study published by APPA identified a major problem in the facilities management for university premises in North America; there is a lack of planning to adequately fund FM activities in the entire building life cycle [6]. Another study identified that $26 billion are needed to fix the accumulated deferred maintenance backlog (DeM) caused by the inability to fund capital renewal/replacement of building equipment, and \$5.7 billion are required to handle more urgent DeM [7]. As a result, insufficient facility maintenance, including DeM and UPM, have accelerated facility deterioration at most campus-sized institutions in the United States [7]. A study conducted in-depth interviews of 37 FM directors from Canada and identified that deferred maintenance of campus buildings resulted from declining financial aspects with growing institution size and concluded that there were insufficient funds, staff, and other resources to repair and maintain the built environment of campuses sufficiently [8].

Based on a case study [9], it was found that there is a need to improve communication between the university level facility maintenance and individual facility maintenance managers to track and implement programs, reduce redundancy, and strategically plan for the building as part of the overall campus. Unfortunately, the lack of a study exploring the status of PPM and UPM in the campus-scale higher education institutions is the primary barrier towards effective facility management. In addition, it remains unclear how the current standpoint can be analyzed based on quantitative and data-driven approaches. Therefore, there is a critical need to explore the current status of FM based on data-driven analyses.

In this context, this study analyzes the FM practices in the North American universities, with a particular focus on both PPM and UPM, based on the proposed systematic methodology. The objective was achieved via the following four steps. First, a survey was designed, distributed to facility managers at universities, and the results were analyzed to investigate the current status of PPM. Second, phone interviews were conducted to understand the overall FM practice. At this stage, natural language processing techniques (topic modeling and sentiment analysis) were used on the interview transcripts as an exploratory approach. Third, a database was developed based on the facility management data (e.g., work orders and labor hours) collected from eight universities. Fourth, three quantitative analyses (statistical comparison analysis, risk-profile analysis, and outlier analysis) were performed to analyze the database and identify critical information associated with PPM and UPM.

The results of this study are expected to facilitate the decision-making process of educational facilities by providing an understanding of various aspects of educational facility management trends and risks. It can allow administrators of higher education institutions (e.g., facility managers) to implement effective FM strategies systematically to establish long-term budgetary goals, which will lead to the enhancement of the asset value of the higher education institutions.

2. Background and Related Studies

2.1. Planned Preventive Maintenance (PPM) and Unplanned Maintenance (UPM)

PPM and UPM are two well-established approaches in the facility maintenance domain. A study proposed that PPM is carefully prepared in advance as it is done at scheduled times and is expected to be very efficient [3]. PPM is also defined as pro-active, where planning and execution of maintenance work are carried out in anticipation of the failure of facility [10]. Another study speculated in their case study that PPM can reduce the demand for correction [11]. Preventive Maintenance (PM) and planned maintenance are two primary components comprising PPM [1]. PM is a type of facility maintenance that increases the reliability or lifespan of a building and equipment is performed through periodic inspection, lubrication, and minor replacements [1,4,12]. Planned maintenance is a pre-determined job procedure that documents labor, materials, tools, and equipment to perform the task before implementing maintenance work [1].

In contrast to PPM, UPM is the work performed as the direct result of equipment failure. Since equipment failure occurs randomly, controlling UPM occurrences is a challenging task. A study reviewed maintenance definition for maintenance, repair, and replacement (MR&R) types, where UPM includes service calls, emergency responses, and unanticipated tasks [5]. UPM is also defined as reactive or emergency maintenance which leads to high maintenance costs [4]. Within UPM, reactive maintenance is a type of work done immediately after a failure to bring an asset back into operation [1]. Figure 1 illustrates the hierarchical structure of facility management as per operations & maintenance (O&M) defined by APPA [1].

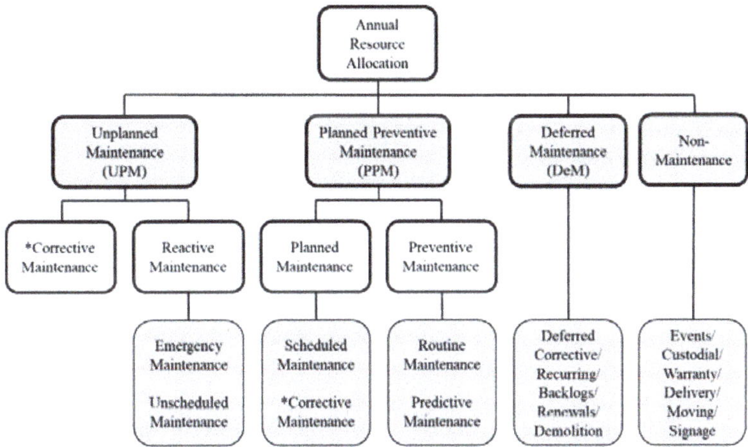

Figure 1. Hierarchical structure of facility management [1].

As shown in Figure 1, annual resource allocation for facility management can be divided into four major maintenance categories: PPM, UPM, deferred, and non-maintenance. PPM consists of planned and preventive maintenance. Planned maintenance refers to scheduled and corrective maintenance, while preventive maintenance reflects routine and predictive maintenance. The corrective maintenance can be categorized under both planned and unplanned maintenance based on APPA [1]. However, corrective maintenance has been used differently by institutions or facility managers [13–15], this study considered the corrective to be unplanned maintenance. Reactive maintenance refers to emergency and unscheduled maintenance. Deferred Maintenance can be divided into many maintenance types, such as deferred corrective, recurring, backlogs, renewals, demolition, etc. [1,7]. The non-maintenance work orders include events, custodial work, warranty work, delivery and transportation of equipment and supplies, signage, banners, etc. This study also excluded project-based work orders (i.e., renovations) which involve contractors outside of the facility management.

2.2. Current FM Guideline and FM Computerized Platform

A published guideline supported by APPA, "Maintenance Staffing Guidelines for Educational Facilities", focuses on determining the adequate maintenance staff size in managing educational facilities [16]. The guide also established baseline attribute standards for each maintenance level, which is now widely accepted as an industry standard. Another published guideline, "Operational Guidelines for Educational Facilities—Maintenance, second edition", introduced maintenance operations that offer best management practices for effective performance in each maintenance department along with the tools of determining staff levels with several case studies and statistical methods [17]. The staffing resources were calculated for a wide variety of campus sizes using the 'Aggregate method' in each case study. The full-time equivalent (FTE) calculation was performed by gathering all building-related data, determining staffing factors, selecting adjustment factors (e.g., campus age, varied facilities, DeM levels, campus missions, etc.), and applying a simple formula to get the FTE value. The formula used five adjustments ranging between −10%

and +10%. Adjustments were made to recognize economies of scale, condition of buildings, facility age, and campus mission. The adjustments were summed and used to increase or decrease the initial *FTE* estimate based on variations from the norm. Equation (1) shows the formula for *FTE* estimate.

$$Adjusted\ FTE = \left(1 + \sum factors\right) * Baseline\ FTE \tag{1}$$

A combination of computed FTE can support a work management system and provide an efficient organizational structure. APPA also introduced the "Facilities Performance Indicators (FPI)" program, which is based on a survey distributed to hundreds of North American universities, includes questions associated with facility condition index (FCI), current replacement value (CRV), energy cost, and age of buildings. The FPI report has been published every year and contains key information about the current trend and status of educational facilities. FPI aims to constantly improve the facilities by developing new tools in the field. Moreover, it provides insights on preventive maintenance programs, including reduced overtime needs, large-scale repairs, and customer service practices for improved facilities.

Additionally, the U.S. Department of Education, National Center for Education Statistics (NCES), and National Forum on Education Statistics (NFES) published a guideline, "Planning Guide for Maintaining School Facilities" [18], to develop, implement, and evaluate a facilities maintenance plan at the school district level. The guideline offers budgets, planning of school facilities maintenance, and facility audits. It also provides effective management of staff and contractors and training guidelines of school facilities for the hired staff.

Lastly, Whitestone Research published a cost reference guideline for facility maintenance and repair costs for over 1700 components and their associated maintenance tasks [19]. The components and tasks listed in Whitestone cost reference follow Uniformat II classification. The cost reference consists of various tools and critical information of the life of specific asset components, trade labor hours, historical inflation rate of maintenance and repair costs, and total cost required to maintain a facility over its service lifetime. The reference is a huge asset to the facility managers as it provides the estimates of 50-year maintenance cost profiles for 74 different models, which offers an advantage while creating budgets and cost estimates.

There are many computerized platforms available in the current market for facilities management. The platforms/variations of functionality that are applicable to this study [20] are as follows:

- IWMS: An integrated workshop management system (IWMS) is an all-in-one way to manage your facility. It includes from real estate portfolio management to floor plan. This is the most comprehensive tool in facility management;
- CMMS: A computerized maintenance management system (CMMS) focuses solely on handling facility maintenance requests. Once the MR&R is recognized, the CMMS coordinates from ticketing requests to delegating and performing the repair activity;
- CAFM: A computer-aided facility management (CAFM) is a platform to manage the actual workplace in facility management. The system handles floor plan creation, space utilization, and MR&R. This system is more effective for space management and accommodation of workers;
- EAM: An enterprise asset management (EAM) system focuses on asset management. This system tracks the number of computers and workstations, locations of the copiers, and printers. It helps facility managers update and manage the current asset and accounting.

For clarity and simplicity, all four platforms/variations are referred to as CMMS.

2.3. Facility Management Classifications

The advancement of equipment technologies and constantly evolving products in the facility domain have opened a new door towards the development of different classification systems; they categorize building elements and their related site work based on the functionality. The most widely used international classification systems in construction and facilities management include Uniformat II, OmniClass™, MasterFormat®, Uniformat™, Uniclass, UNSPSC, etc. These classification systems follow international standards, and facility managers at the universities rely on them to maintain their database, which records varying day-to-day activities. Table 1 illustrates current FM classification systems based on their origin, updated year, classification structure, hierarchy levels, a grouping of elements, and component details.

Table 1. Comparison of the current classification systems [21,22].

No	Classification System	Origin	Updated Year	Classification Structure	Levels	Elements Grouping	Component (Detailed/Neutral/Less Details/Not Detailed)
1	Uniformat II	North America	1999	Hierarchical	3	Functional	Not Detailed
2	MasterFormat®	North America	2020	Hierarchical	4	Mounted Elements	Less Detailed
3	Uniformat™	North America	2010	Hierarchical	5	Functional	Neutral
4	OmniClass™	North America	2015	Faceted	6	Functional	Less Detailed
5	Uniclass	United Kingdom	2015	Faceted	4 to 5	Functional	Less Detailed
6	UNSPSC	North America	2017	Hierarchical	5	Mounted Elements	Not Detailed

As can be seen from Table 1, Uniformat II was developed by ASTM (American Society of Testing and Materials) International [23]. It has a hierarchical structure with three standard levels: major group elements (e.g., substructure, shell, etc.), group elements (systems), and individual elements (subsystems). However, due to limited sub-elements in this system, different organizations can highly customize it by adding elements according to their requirements [22]. MasterFormat®, a product of 'Construction Specifications Institute' (CSI) and 'Construction Specifications Canada' (CSC), is solely based on mounted elements and has a hierarchical structure with four levels: divisions, sections, elements, and sub-elements [24]. Similar to MasterFormat®, Uniformat™ was developed by CSI & CSC, based on functional elements [21,25]. The structure of this classification system is hierarchical with five levels: categories, classes, two subclasses, and elements. Additionally, OmniClass Construction Classification System was developed by CSI & CSC [26]; this is similar to UK-based Uniclass [27] as both cover complete lifecycle classification of facility-built environment. The structure of OmniClass™ is faceted with six levels, which consists of work results from MasterFormat® and elements from Uniformat™ [21]. Another classification system used by the state of California, i.e., United Nations Standard Products and Services Code (UNSPSC) which is based on mounted elements and its structure is hierarchical with five levels [28]. The component details criteria, Detailed/Neutral/Less Details/Not Detailed, compared the specific details present based on component characteristics provided by Whitestone cost reference [19] such as units, trade, labor details, material costs, equipment type, task type, etc.

It was observed that the classification structure of four out of six systems were hierarchical and two were faceted or combinatory. A faceted structure is defined as the categorization of elements under a combination of facets [22]. All the aforementioned classification systems are used internationally, but most of them are specifically designed for the construction industry, not for facilities management. The available classification systems are either based on functionality or mounted elements with less or no details.

Therefore, there is a critical need to develop a classification system that includes component details based on both mounted elements and functionality, which can be suitable for diverse building types. This study introduces Facility Management Unified Classification Database (FMUCD) based on functionality and conduct data driven analysis to provide guidelines the facility management to make an appropriate decision in an uncertain situation at higher education institutions.

3. Methodology

The objective of this study was to explore the current status of FM practices by establishing Facility Management Unified Classification Database (FMUCD) and performing data-driven analysis for facility management in higher education institutions. Figure 2 illustrates the overall research framework. First, the survey questionnaires were distributed to the universities for data collection. Second, phone interviews were conducted, various questions about facility management practices were asked, and detailed work order history data from CMMS were collected from each university. At this stage, NLP analysis (topic modeling and sentiment analysis) was additionally conducted based on interview transcripts. Third, the database was developed using the collected raw data where all work orders were classified into different descriptive codes based on the Equipment Naming Convention; it was designed for this study by integrating the standard classification of major grouping elements of building Uniformat II with the elements published in Whitestone Cost References [19,23]. Lastly, further quantitative analyses were conducted: (1) statistical comparison analysis, (2) risk-profile analysis, and (3) outlier analysis.

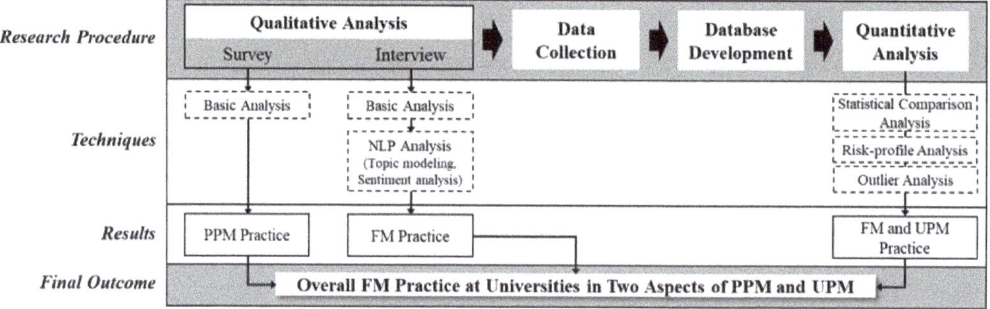

Figure 2. Overall research framework.

3.1. Qualitative Analysis

A survey was conducted to explore the current state of PPM at universities in North America. The survey, which consists of ten questions, was developed based on five aspects (process, cost plan, budget allocation, scheduling, and decision making) of PPM. For example, two survey questions were designed to investigate the current practice and workflow of PPM in universities. The survey was distributed to facility managers who were registered as a member of the APPA at twelve universities. When collecting responses from the universities, the responses with incomplete information were excluded. In addition, out of a total of ten questions, only five questions were analyzed and presented in this study because the remaining five questions were related to personal information, data availability, etc. Table 2 summarizes the five important questions, multiple answers provided for each question, and the corresponding number of responses.

Table 2. Survey result.

No	Question	Answers	Responses	
			n	**%**
1	How do you evaluate progress of the PPM work assignments?	Reports generated by CMMS	24	50%
		Key Performance Indicator—KPI	15	31%
		Paper reports	7	15%
		Other	2	4%
2	Where are PPM work orders prioritized among all work order types?	Above scheduled corrective	12	43%
		Dedicated crew for PPM work	8	29%
		Lowest priority	7	25%
		Above critical	1	3%
3	What is included in the PPM work order estimates?	Work	26	25%
		Set-up	21	20%
		Clean-up	20	19%
		Documentation	15	14%
		Travel (before the PPM)	13	12%
		Travel (after the PPM)	8	8%
		Others	3	3%
4	How do you estimate PPM worker time?	Prior experience	17	28%
		Multiple factors (e.g., guide, prior records)	16	26%
		Prior time records	13	21%
		Estimating guide (RS Means, Dodge, other)	9	15%
		Manufacturer recommendations	6	10%
5	Prioritize how you would improve PPM effectiveness	Prioritization strategy	19	76%
		Additional funding	4	16%
		Additional staff	2	8%

Analyzing the survey results led to the following three main observations: First, the progress of the PPM work assignments was mainly monitored based on the reports generated by CMMS (No. 1 in Table 2). This suggests that CMMS has been mainly adopted by at least half of the facility managers in universities in order to automatically monitor PPM work progress. Second, work, set-up, clean-up, and documentation were identified as the most significant four factors included in the PPM work order estimates; they accounted for 78% of the responses in question No. 3 in Table 2. Third, it was found that most of the university facility managers (76%) responded that the prioritization strategy is the most critical component to improve the effectiveness of the current PPM practice, as illustrated in question No. 5 in Table 2.

Phone interviews were conducted to understand the current status (e.g., types of management systems, maintenance components, and data recorded) of facility management and investigate practical issues in higher education institutions (i.e., universities) in North America. Compared to the survey analysis illustrated in the previous section, the focus of the interview was on exploring the overall FM practice, not being limited to the PPM. A flyer was created and distributed to facility managers who were registered as a member of APPA at thirty-five universities. As a result, twelve participants were recruited for a phone interview which was conducted from November 2019 to January 2020. A total of thirteen questions (three for planning and definition, six for data quality and variables, one for prioritization, and three for methodology) were developed and asked to respondents during the interview. (Additional survey questions can be developed in the future for a more comprehensive understanding of the current status of facility management practice at universities.) The phone interview took approximately 30 min, and each interview was recorded and transcribed digitally.

In this study, seven interview questions were excluded for further analysis since they were associated with definitions of terminologies, willingness to offer raw data, and personal information. As a result, responses to the remaining six important questions were analyzed and presented in Table 3.

Table 3. Phone interview result.

No	Question	Answers	Responses (N = 12)	
			n	%
1	Do you have an organized maintenance plan?	Schedule maintenance	5	42%
		PPM	5	41%
		System/Subsystem	2	17%
2	How do you classify the building systems and components?	Uniformat	5	42%
		MasterFormat	4	33%
		OmniClass	2	17%
		Other	1	8%
3	Is each maintenance task performed on an individual component tracked?	Yes	7	58%
		No	5	42%
4	When do you record work order information?	End of activity	7	58%
		End of shift	5	42%
		Mid-shift	0	0%
5	Who records work order?	Craft/Technician	10	83%
		Supervisor	2	17%
		Office clerk	0	0%
6	Where is the data recorded?	Electronically/CMMS	9	75%
		Both	2	17%
		Manually/Papers	1	8%

It was observed that scheduled maintenance (42%) and PPM (41%) were two major organized maintenance plans adopted in most universities. Within each university, building systems and components were classified based on Uniformat (42%) and MasterFormat (33%). The maintenance task was performed on an individual component tracked. Additionally, it was found that a work order was mostly recorded at the end of the activity (58%) by the technician (83%) using CMMS (75%). The result of the interviews is assumed to reflect the recommended practices of the operation perspective in the facility management at the referenced higher education institutions.

3.2. Data Driven Analysis for Qualitative Data

Two natural language processing (NLP) techniques (topic modeling and sentiment analysis) were applied to the collected interview transcriptions containing a significant amount of textual data (over 50,000 words) to reveal important latent information that was not able to be captured during the interview. NLP techniques have been increasingly used as a quantitative method to derive meaningful insights such as keywords [29], topics [30], and sentiment [31] from a set of textual data (e.g., transcripts) obtained from the interview. Previous studies have demonstrated the efficacy and potential of applying NLP techniques, addressing limitations (e.g., time-consuming, subjective, and error-prone) that reside in qualitative approaches such as interviews and surveys. In other words, conducting NLP analysis provides an opportunity to find unexpected observations or insights based on semantic and syntactic similarities that can be observed within textual data comprising interview transcriptions.

Raw data, interview transcriptions, from 12 universities were preprocessed through the following steps: removing stop words (e.g., "the", "am" and "a") and noises (e.g., blanks and punctuation), word stemming, and tokenization.

Latent Dirichlet Allocation (LDA)—one of the well-established topic modeling approaches—was adopted to identify keywords and prevalent topics in the interview. LDA allows for identifying patterns that can be observed within textual data without a tedious labeling process [32]. In general, LDA produces a couple of topic groups, each of which consists of corresponding keywords. Labeling topic group (naming) relies on human interpretation and judgment [33]. As a result, two topics were identified based on the semantic similarity of keywords in Table 4, which implies that the focus of respondents during the interview was on two aspects of PPM and the maintenance system. Another interesting observation was that Archibus, an integrated platform system for infrastructure and building management [34] frequently appeared during the interviews, which suggests that it was one of the most widely used software in the universities.

Table 4. Topics and keywords identified from the interview.

No	Topic	Keywords
1	PPM	Maintenance, plan, work, preventive, evaluate, worker, frequency, critical, year, order, asset, schedule, fix
2	Maintenance system	PMS, system, work, zone, order, equipment, time, Archibus, record, manager, worker

Sentiment analysis was further conducted to identify the facility managers' degree of positiveness or negativeness towards the use of PPM. Note that it was assumed that PPM was the main subject of the phone interview since it was identified as the main topic, as illustrated in Table 4. For the analysis, a large number of tokenized words derived from the previous LDA analysis were used as input to the well-established pre-trained Python module, Valence Aware Dictionary and sentiment Reasoner (VADER) [35]. VADER allows for quantitatively assessing the level of sentiments for the given texts. As a result, it provided a sentiment score between 0 and 1, where 0 indicates complete negative sentiment and 1 denotes complete positive sentiment. The criteria for positive (0.7~1.0), neutral (0.4~0.7), and negative (0.0~0.4) range was set based on the previous studies [36,37].

The results revealed that five universities (B, E, H, I, and L in Figure 3) responded that they were using the PPM (No. 1 in Table 3) showed positive sentiment scores. This finding supports that the universities are willing to adopt PPM with the effectiveness and advantages of the PPM.

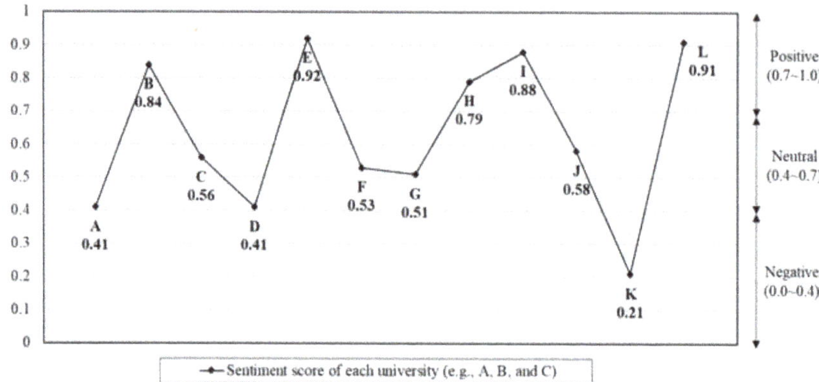

Figure 3. Sentiment analysis results based on phone interview transcriptions.

3.3. Facility Management Unified Classification Database (FMUCD)

Over the years, higher education institutions in North America have employed many classification systems (e.g., Uniformat II, Uniformat^{TM}, OmniClass^{TM}, and MasterFormat$^{®}$) to classify building systems, construction, and maintenance activities. As illustrated in Figure 4, Uniformat II [23] provides a more specific facility management structure with three levels (level 1-major group elements, level 2-systems, and level 3-subsystems). For example, in the figure, level 1 includes shell, interiors, services, etc. Regarding "Services" at level 1, it can be divided into HVAC, plumbing, electrical, conveying, and fire protection. For HVAC at level 3, it consists of heating, cooling, distribution systems, controls & instrumentation, terminal & package units, energy supply, etc.

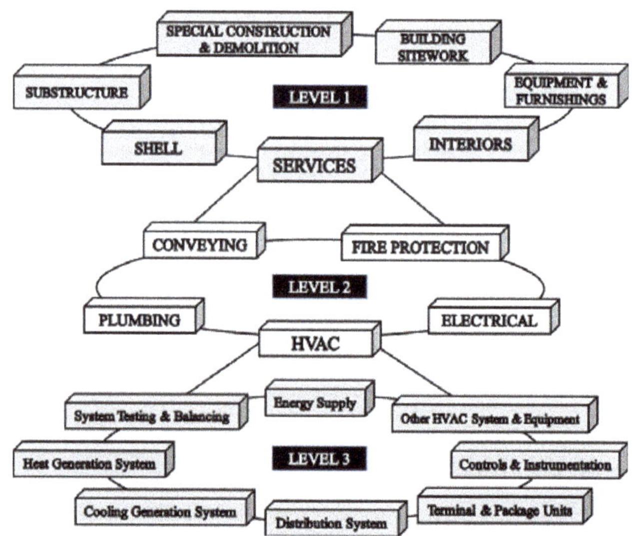

Figure 4. Uniformat II classification [23].

This study established a descriptive code entitled Facility Management Unified Classification Code (FMUCO) in the database. The purpose of the FMUCO is (1) to compile the current data from universities to create Mega data and (2) to conduct the data-driven analysis to explore the current status of the facility management in higher education institutions. The FMUCO code is created by combining Uniformat II with generic descriptions of building components from Whitestone cost reference [19] shown in Figure 5.

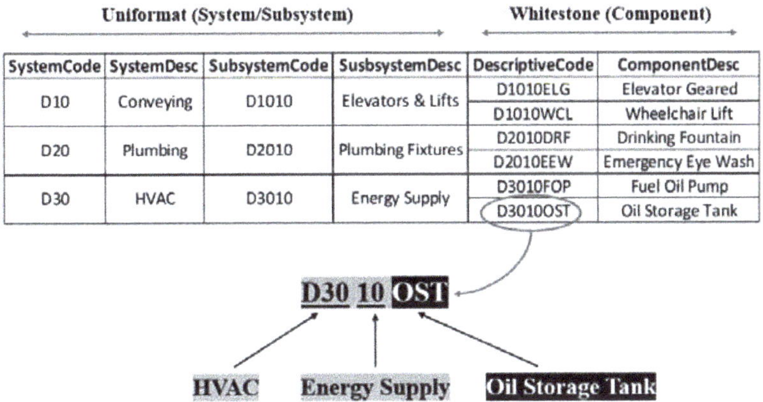

Figure 5. Facility Management Unified Classification Code (FMUCO).

As illustrated in Figure 5, the proposed descriptive code is composed of an 8-digit code; the first three digits describe the system code, the next two digits define the subsystem, and the last three digits are the abbreviation of the component description. The FMUCO has 543 descriptive codes, new elements can be added in the future. This classification method permitted the collected data for each university, which varied significantly in terms of data type, data points, and data attributes (e.g., work order description, cost information, labor hours, etc.), to be managed for the study. Data preprocessing was performed to develop a structured and organized database shown in Figure 6. This preprocessing step included but was not limited to identifying common data attributes, cleaning noisy data, and removing unnecessary data.

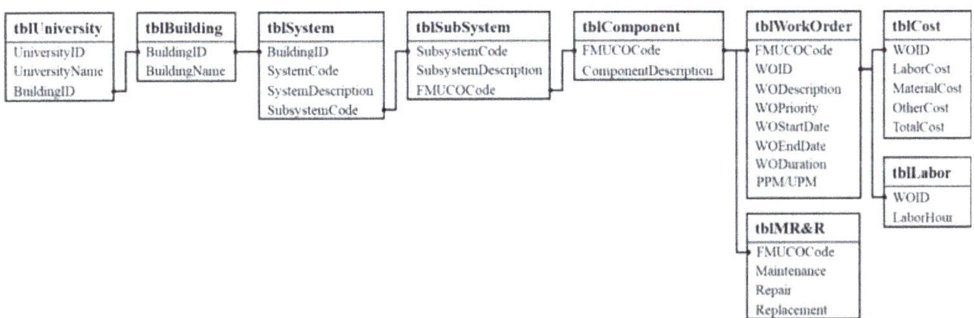

Figure 6. FMUCD structure.

3.4. Quantitative Analysis

The database developed for this study allowed identification of critical information and risks involved in the facilities management at the component level. Three types of data-driven approaches were adopted for quantitative analysis: (1) statistical comparison analysis, (2) risk-profile analysis, and (3) outlier analysis. Statistical comparison analysis (1) was conducted to explore the current trend of PPM and UPM for the referenced universities.

At this stage, the ten systems (e.g., HVAC, electrical, plumbing, conveying, fire protection, etc.) were compared to identify the highest number of work orders and labor hours associated with PPM and UPM at the system level. (2) Risk-profile analysis was conducted on the top three systems to distinguish the risks in the subsystem level of UPM. The risk profiles for top systems aimed to provide basic knowledge to the facility managers about the subsystems with a high probability of getting a UPM work order. The outlier analysis (3) was conducted to identify components with a high risk of generating UPM work orders.

3.5. Data Driven Analysis for Quantitative Data: Statistical Comparison Analysis

The statistical comparison analysis was performed on the developed database to explore the current trend in PPM and UPM shown Figures 7 and 8. As can be seen from Figures 7 and 8, the bar charts indicate the annual average numbers, i.e., the five-year trend of PPM and UPM with work order counts (WO) and labor hours (LH) per million square feet (MSF) at eight universities for the years 2015 to 2019.

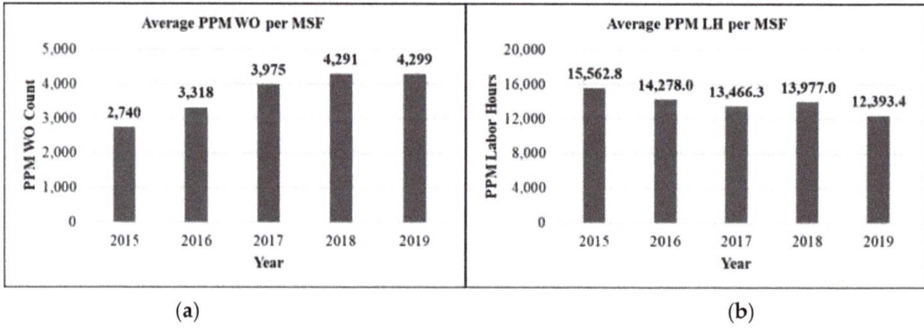

(a) (b)

Figure 7. Five-year trend of PPM: (**a**) Work Order Counts; (**b**) Labor Hours.

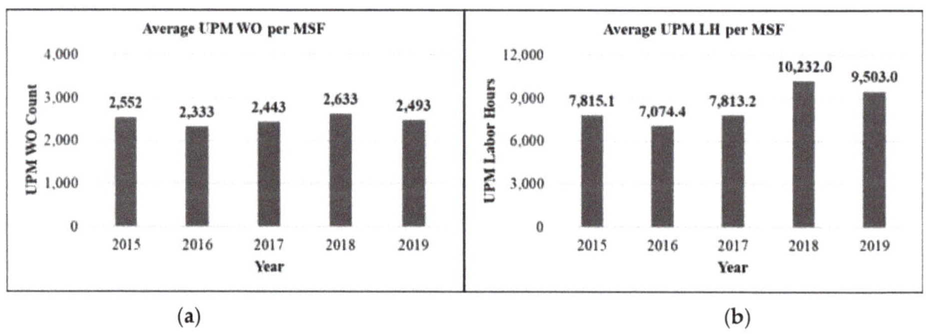

(a) (b)

Figure 8. Five-year trend of UPM: (**a**) Work Order Counts; (**b**) Labor Hours.

Comparing the five-year trend of PPM and UPM revealed that, the PPM recorded an average of 3725 work orders, while there was an average of 2491 UPM work orders during the given period. Similarly, the average PPM labor hours were 13,935.5 and the average UPM labor hours were 8487.5. As deterioration of buildings is considered, although the budget for PPM has been increased, it is revealed that the budget for UPM has remained consistent. Therefore, such a finding will be able to utilize as a guideline for facility managers or decision makers to allocate the budget for the PPM and UPM. Figures 9 and 10 illustrate the number of work orders and labor hours at the system level for the entire area maintained. The annual average work order count and labor hours of ten systems were investigated for PPM and UPM.

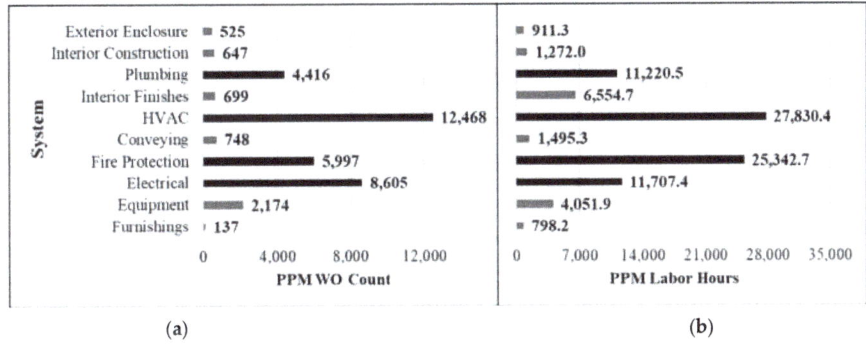

(a) (b)

Figure 9. System-level comparison of the PPM: (**a**) Work Order Counts; (**b**) Labor Hours.

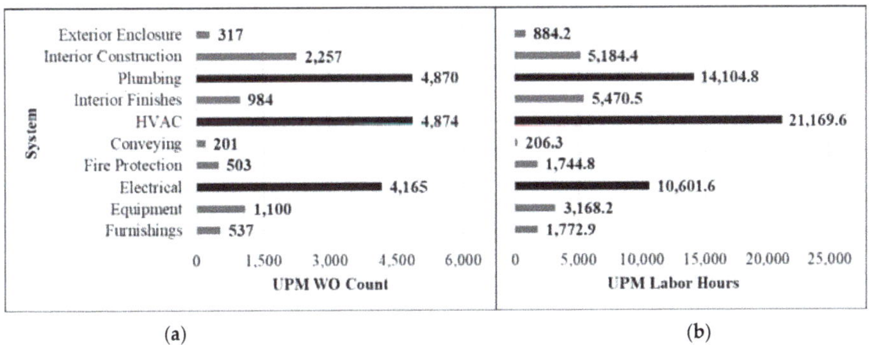

(a) (b)

Figure 10. System-level comparison of the UPM: (**a**) Work Order Counts; (**b**) Labor Hours.

As can be seen from Figures 9 and 10, HVAC was identified as the most significant system resulting in the highest number of work orders and labor hours every year, followed by electrical, fire protection, and plumbing systems in the PPM, while the HVAC system again was identified as the most critical system resulting in the highest number of work orders and labor hours, followed by plumbing, and electrical systems in the UPM. Although work order counts for the top two systems are similar, HVAC (4874) and plumbing (4870), HVAC consumed significantly higher number of labor hours in both PPM and UPM. Additionally, interior construction, interior finishes, and furnishings are also identified to be on the higher side compared to PPM whereas conveying systems and exterior enclosure generated lower UPM work orders.

3.6. Data Driven Analysis for Quantitative Data: Risk Profile Analysis

The risk-profile analysis in facilities management can be defined as the assessment of the inoperability of building equipments. A study conducted the severity analysis of Indian coal mine accidents with the historical data of 100 years with Weibull and Exponential distributions for evaluating hazard rate functions; whereas Poisson and Negative Binomial distributions for risk profiles of mine accidents [38]. To compare which distribution fits best to the data, a recent study analyzed the robustness of different methods of comparing fitted distributions such as AIC (Akaike Information Criterion), BIC (Bayesian Information Criterion), LRT (log-likelihood ration test), etc. [39]. AIC and BIC measure the performance of the models based on their complexity. AIC is a prediction error estimator which prevents overfitting of data whereas BIC penalizes the model more based on the number of parameters. While comparing the AIC and BIC, lower scores are preferred and both information criteria are used for appropriate model selection, and it can also be used for distribution selection [39]. The Negative Binomial (NB) distribution for a discrete random variable (X) can be calculated based on Equation (2) [40]:

$$P(X = x|r, p) = \binom{x-1}{r-1} p^r.(1-p)^{x-r} ,\qquad (2)$$

where $x = r, r + 1, \ldots , p$ refers to the independent Bernoulli trials, r is a fixed integer. From Equation (1), it can be said that X follows NB distribution at which rth success occurs. The parameters of NB fit are denoted by the number of successes (r) and event probability (p).

In this study, survival function risk profiles were developed to identify the high probability of getting a UPM work order at the subsystem level. Risk-profile consists of three steps: (1) Data mining, (2) Distribution fitting, and (3) Generation of the survival function risk-profiles. The data mining (1) is to select the appropriate data points from the raw data. The distribution fitting (2) is to find appropriate probability distributions by calculating AIC and BIC scores. The last step is generation of the survival function risk-profiles (3) where, the top three systems (HVAC, electrical, and plumbing from Figure 10) with their respective subsystems (e.g., heating, cooling, distribution, etc. for HVAC), identified to distinguish the risks in the UPM. As a result, Table 5 shows the comparison of distribution fits for the systems and subsystems based on AIC & BIC scores. The distribution fitting and comparisons were performed using R-programming.

Table 5. Goodness-of-fit of distributions for systems and subsystems.

System	Subsystem	Data Range	Poisson AIC	Poisson BIC	Negative Binomial AIC	Negative Binomial BIC	No. of Successes (r)	Event Prob. (p)
HVAC	Total HVAC	1 to 1023	51,684.54	51,689.29	9325.54	9335.03	1.2583	0.0130
	Heating	0 to 184	13,351.13	13,355.87	5510.13	5519.62	0.4350	0.0453
	Cooling	0 to 118	11,943.97	11,948.72	4509.96	4519.45	0.3058	0.0505
	Distributions	0 to 153	17,653.43	17,658.18	6981.43	6990.92	1.1479	0.0489
	Terminal Units	0 to 200	16,133.57	16,138.31	6121.44	6130.93	0.5614	0.0414
	Controls	0 to 440	26,742.70	26,747.44	7761.73	7771.22	0.9742	0.0264
Plumbing	Total Plumbing	0 to 566	47,664.31	47,669.06	9334.98	9344.47	1.6313	0.0157
	Fixtures	0 to 497	43,823.68	43,828.42	8801.60	8811.09	1.1198	0.0160
	Domestic	0 to 129	14,114.42	14,119.17	6627.55	6637.04	1.2877	0.0652
	Sanitary	0 to 62	8295.90	8300.64	5164.35	5173.84	1.0520	0.1273
	Rainwater	0 to 19	3023.30	3028.05	2351.56	2361.05	0.3733	0.2628
	Other Plumbing	0 to 52	5490.91	5495.66	2945.88	2955.37	0.2667	0.1169
Electrical	Total Electrical	1 to 541	54,697.38	54,702.13	9205.89	9215.38	1.1549	0.0132
	Service	0 to 81	8876.86	8881.61	4844.37	4853.86	0.6360	0.0967
	Lighting	1 to 393	43,445.45	43,450.20	8577.99	8587.48	0.9948	0.0169
	Communications	0 to 141	15,373.17	15,377.91	6454.95	6464.44	0.9544	0.0567
	Other Electrical	0 to 208	10,382.39	10,387.14	2933.79	2943.28	0.0483	0.0177

Table 5 shows that NB distribution fits the data best based on lower AIC & BIC scores. The table also shows the NB fit parameters (r and p) which were used to generate the risk profiles of the individual systems as well as their subsystems. Figure 11 illustrate the results of the survival function risk profiles for HVAC, plumbing, and electrical systems.

The risk-profiles are presented in Figure 11 where the x-axis represents the number of work order occurrences in a year for a building and y-axis represents the probability of inoperability. The probability of inoperability refers to all the occurrences which hindered the operation of the building elements. The probabilities for each occurrence were calculated for the x-axis ranging from 1 to 100. Each plot represents the probability of all major subsystems of a respective system with 850 data points of the top 25 buildings with most UPM work orders were identified for each of the eight universities for 2 to 5 years. As can be seen from Figure 11a, the controls & instrumentation resulted in highest inoperability probability as HVAC control panel, airflow and thermostat adjustment requests are very frequent in a building. Distribution systems resulted in the second most work order generating subsystem with repair requests as it is comprised of components like air handlers, fans, filters, ventilation, etc. Terminal & package units and heat generation systems were found to generate moderate number of MR&R requests with cooling generation systems being the lowest probability of generating UPM work orders. In Figure 11b, plumbing fixtures resulted in the highest probability of inoperability in plumbing systems. The key components in fixtures are sink, toilet, shower, bathtub, etc. Domestic water distribution being the second most prone subsystem followed by sanitary waste. Rain water drainage and other plumbing systems resulted in low inoperability probability. Additionally, Figure 11c illustrates that lighting and branch wiring subsystem dominated the system in terms of inoperability in electrical systems. Communications and security being moderate in terms of work order requests followed by electrical service and distributions. Other electrical system was found to be negligible in terms of UPM work order requests.

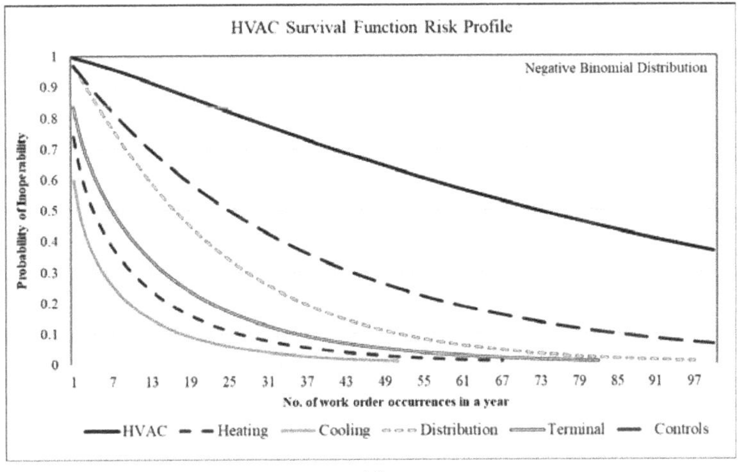

(a)

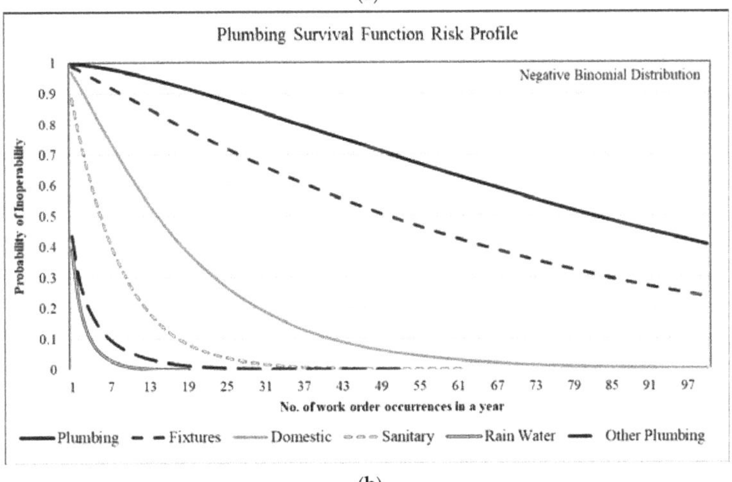

(b)

Figure 11. *Cont.*

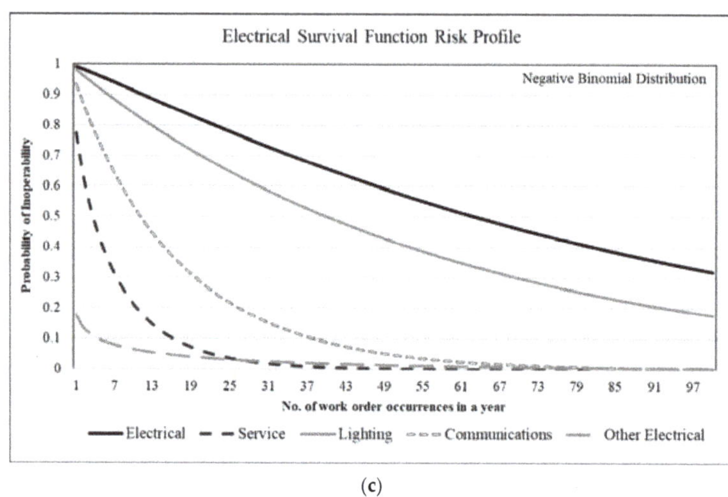

(c)

Figure 11. Survival Function Risk Profile for UPM Work Orders: (**a**) HVAC; (**b**) Plumbing; and (**c**) Electrical.

Interestingly enough, the HVAC work consisted of mostly controls and distribution systems work orders where controls and instrumentation having only 4 components (control panel, thermostat, digital controls, and meters) generated adjustment work orders in majority while distribution system generated more MR&R activities having more diverse components. On the other hand, plumbing work was dominated by plumbing fixtures and electrical work primarily consisted of lighting and branch wiring work orders. Considering the fact that universities spend a great deal of resources doing PPM work in fire protection which benefited the FM in reducing UPM work significantly but failed to do the same for other major systems. Therefore, the proposed diverse analyses, including a statistical analysis and a risk-profile analysis, are necessary to acknowledge the current status of the facility management from different angles.

Additionally, the outlier analysis allowed for understanding which building elements require careful consideration when planning PPM work. Out of the top 25 UPM buildings selected, the outliers from the HVAC system included the exhaust fan, air-conditioner, unit heater, fan, and thermostat (temperature issues). Similarly, the top components having the higher risk for generating electrical work orders involved the light fixtures, circuit breaker, smoke detector, and receptacle. The top outliers for plumbing systems were found to be toilet & stall, sink, urinals, floor drains, and shower. Table 6 presents the components recorded for over 100 number of occurrences generated for a building in a year.

As shown in Table 6, thermostat adjustments and issues recorded the highest number of workorder for a university in a year. This is one of the most requested facility operations in the buildings. For HVAC, air conditioners, air handlers, and radiators also generate high work order numbers. For electrical, light changing requests are frequent and changing of batteries in equipments seems more like routine requests. For Plumbing, sink and toilet repair requests are the most common request followed by the bathtub and shower enclosure. As a result, the outlier analysis helps facility managers (1) recognize the components registering more than 100 work orders in MR&R, and (2) to prepare budget allocation for facility management.

Table 6. Outlier components for UPM.

No	Top Component	No. of Occurrence	System	Subsystem	Descriptive Code
1.	Thermostat	390	HVAC	Controls & Instrumentation	D3060THE
2.	Fluorescent Light Fixture	345	Electrical	Lighting & Branch Wirings	D5020FLF
3.	Sink	232	Plumbing	Plumbing Fixtures	D2010SNK
4.	Battery	207	Electrical	Other Electrical Systems	D5090BAT
5.	Toilet & Wash Basin	180	Plumbing	Plumbing Fixtures	D2010TWB
6.	Air Conditioner	136	HVAC	Terminal & Package Units	D3050ACO
7.	Air Handler	127	HVAC	Distribution Systems	D3040AHD
8.	HVAC Control Panel	127	HVAC	Controls & Instrumentation	D3060HVA
9.	Bathtub & Shower Enclosure	116	Plumbing	Plumbing Fixtures	D2010BSE
10.	Radiator	103	HVAC	Heat Generation Systems	D3020RFW

4. Discussion and Conclusions

This study attempted to analyze the current trend and status of Facility Management (FM) practice at higher education institutions by proposing (1) the Facility Management Unified Classification Database (FMUCD), and (2) the systematic data-driven analyses: survey questionnaires and phone interviews, Natural Language Process (NLP) approaches, statistical analysis, risk-profile analysis, and outlier analysis.

The current trends and status of PPM at universities were mainly identified from the survey, phone interview, and statistical comparison analysis. The survey revealed that the progress of the PPM work was mostly monitored based on the Computer Maintenance Management System (CMMS) reports and four factors (work, set-up, clean-up, and documentation) were critical for the PPM estimates. Analyzing interview results suggested that schedule maintenance and PPM were two major organized maintenance plans at universities. At this stage, the application of NLP approaches found that the focus of the interview was on PPM, supported by the positive sentiment scores. From the statistical analysis, it was revealed that although PPM work order count increased over the years, UPM work orders remains consistent. Therefore, such a finding will be applied to be a guideline for facility managers or decision makers to allocate budgets for PPM and UPM; the budget of the UPM can be similar to the last year while, the budget of the PPM can be increased according to the budget flexibility. Additionally, HVAC was identified as the most significant system resulting in the highest number of work orders and labor hours every year in both PPM and UPM.

Findings related to UPM were mostly derived from risk-profile analysis and outlier analysis. At the system level, the main trades were HVAC, electrical, and plumbing which generated higher work orders and labor hours. Especially, while distribution systems and controls & instrumentation in HVAC were found to generate the maximum number of UPM work orders, lighting and branch wirings and communication & security for electrical, and plumbing fixtures in plumbing systems were identified as a major proportion of UPM work. Therefore, the proposed FMUCD and the results of the data-driven analyses will provide guidelines and best practices for the facility management to make an appropriate decision in an uncertain situation at higher education institutions. Moreover, the broader impact of this research is that it would help stakeholders of any campus-sized institution to develop, operate, maintain, upgrade, and disperse their assets in a cost-effective manner.

Author Contributions: Conceptualization, S.Y., T.J.W., M.H.; Methodology, A.K.P., S.Y., T.J.W. and J.J.; Software, A.K.P., J.J.; Validation, A.K.P., S.Y., J.J. and M.H.; Formal Analysis, A.K.P., S.Y., J.J. and T.J.W.; Investigation, T.J.W., S.Y., J.J. and M.H.; Resources, S.Y., T.J.W. and M.H.; Data Curation, S.Y. and T.J.W.; Writing—Original Draft Preparation, A.K.P., S.Y., T.J.W. and J.J.; Writing—Review and Editing, A.K.P., S.Y., T.J.W. and J.J.; Visualization, A.K.P., S.Y.; Supervision, S.Y., T.J.W. and M.H.;

Project Administration, S.Y., T.J.W. and M.H.; Funding Acquisition, S.Y., T.J.W. and M.H. All authors have read and agreed to the published version of the manuscript.

Funding: This research received no external funding.

Data Availability Statement: Some or all data, models, or code that support the findings of this study are available from the corresponding author upon reasonable request.

Conflicts of Interest: The authors declare that the research was conducted in the absence of any commercial or financial relationships that could be construed as a potential conflict of interest.

References

1. Association of Physical Plant Administrators (APPA) Facilities Maintenance and Operations. Available online: https://www.appa.org/bok/facilities-maintenance-and-operations/ (accessed on 12 September 2021).
2. Amaratunga, D.; Baldry, D. Assessment of Facilities Management Performance in Higher Education Properties. *Facilities* **2000**, *18*, 293–301. [CrossRef]
3. Doyen, L.; Gaudoin, O. Modeling and Assessment of Aging and Efficiency of Corrective and Planned Preventive Maintenance. *IEEE Trans. Reliab.* **2011**, *60*, 759–769. [CrossRef]
4. Rani, N.A.A.; Baharum, M.R.; Akbar, A.R.N.; Nawawi, A.H. Perception of Maintenance Management Strategy on Healthcare Facilities. *Procedia-Soc. Behav. Sci.* **2015**, *170*, 272–281. [CrossRef]
5. Yoon, S.J. *New Paradigm for Deferred Maintenance Budget Allocation 2018*; Purdue University: West Lafayette, IN, USA, 2018.
6. Rose, R. *Buildings—The Gifts That Keep on Taking: A Framework for Integrated Decision Making*; APPA CFaR Center for Facilities Research; APPA: Alexandria, VA, USA, 2007; ISBN 1890956384.
7. Yoon, S.; Weidner, T.; Hastak, M. Total-Package-Prioritization Mitigation Strategy for Deferred Maintenance of a Campus-Sized Institution. *J. Constr. Eng. Manag.* **2021**, *147*, 4020185. [CrossRef]
8. Wright, T.S.A.; Wilton, H. Facilities Management Directors' Conceptualizations of Sustainability in Higher Education. *J. Clean. Prod.* **2012**, *31*, 118–125. [CrossRef]
9. Lavy, S. Facility Management Practices in Higher Education Buildings. *J. Facil. Manag.* **2008**, *6*, 303–315. [CrossRef]
10. Oseghale, G.E.; Ikpo, I.J. An Evaluation of Industrial Facilities Defects in Selected Industrial Estates in Lagos State, Nigeria. *Civ. Eng. Dimens.* **2014**, *16*, 171–179. [CrossRef]
11. Che-Ani, A.I.; Ali, R. Facility Management Demand Theory. *J. Facil. Manag.* **2019**, *17*, 344–355. [CrossRef]
12. Fregonese, L.; Achille, C.; Adami, A.; Fassi, F.; Spezzoni, A.; Taffurelli, L. BIM: An Integrated Model for Planned and Preventive Maintenance of Architectural Heritage. In Proceedings of the 2015 Digital Heritage, Granada, Spain, 28 September–2 October 2015; Volume 2, pp. 77–80.
13. Trojan, F.; Marçal, R.F.M. Proposal of Maintenance-Types Classification to Clarify Maintenance Concepts in Production and Operations Management. *J. Bus. Econ.* **2017**, *8*, 560–572.
14. Stenström, C.; Norrbin, P.; Parida, A.; Kumar, U. Preventive and Corrective Maintenance-Cost Comparison and Cost-Benefit Analysis. *Struct. Infrastruct. Eng.* **2016**, *12*, 603–617. [CrossRef]
15. Wang, Y.; Deng, C.; Wu, J.; Wang, Y.; Xiong, Y. A Corrective Maintenance Scheme for Engineering Equipment. *Eng. Fail. Anal.* **2014**, *36*, 269–283. [CrossRef]
16. Weidner, T.; Adams, M. *Maintenance Staffing Guidelines for Educational Facilities*; APPA: Alexandria, VA, USA, 2002; ISBN 1-890956-23-6.
17. Bigger, A.S.; Becker, J.T. *Operational Guidelines for Educational Facilities-Maintenance*, 2nd ed.; APPA: Alexandria, VA, USA, 2011; Volume 2, ISBN 1-890956-67-8.
18. Szuba, T. *Planning Guide for Maintaining School Facilities*; DIANE Publishing: Darby, PA, USA, 2003; ISBN 1428925597.
19. Whitestone Research. *The Whitestone Facility Maintenance and Repair Cost Reference*, 18th ed.; Whitestone Research: Santa Barbara, CA, USA, 2014.
20. Sheynkman, A. What's the Difference: IWMS, CMMS, CAFM and EAM? Available online: https://spaceiq.com/blog/difference-between-iwms-cmms-cafm-eam/ (accessed on 27 July 2021).
21. Afsari, K.; Eastman, C.M. A Comparison of Construction Classification Systems Used for Classifying Building Product Models. In Proceedings of the 52nd ASC Annual International Conference Proceedings, Provo, UT, USA, 13–16 April 2016; pp. 1–8.
22. Kula, B.; Ergen, E.; Kula, B.; Ergen, E. Review of Classification Systems for Facilities Management. In Proceedings of the 13th International Congress on Advances in Civil Engineering, Izmir, Turkey, 12–14 September 2018; pp. 1–8.
23. Charette, R.P.; Marshall, H.E. *UNIFORMAT II Elemental Classification for Building Specifications, Cost Estimating, and Cost Analysis*; Department of Commerce, Technology Administration, National Institute of Standards and Technology: USA, 1999; ISBN NISTIR 6389.
24. MasterFormat®. Available online: https://www.csiresources.org/standards/masterformat (accessed on 21 July 2021).
25. *UniFormat^{TM}: A Uniform Classification of Construction Systems and Assemblies*; The Construction Specifications Institute: Alexandria, VA, USA, 2010; ISBN 978-0-9845357-1-2.
26. OmniClassTM. Available online: https://www.csiresources.org/standards/omniclass (accessed on 21 July 2021).

27. Gelder, J. The Principles of a Classification System for BIM: Uniclass 2015. In Proceedings of the 49th International Conference of the Architectural Science Association, Melbourne, VIC, Australia, 2–4 December 2015; Volume 1, pp. 287–297.
28. UNSPSC. Available online: https://www.unspsc.org/ (accessed on 21 July 2021).
29. Leh, A.; Köhler, J.; Gref, M.; Himmelmann, N.P. Speech Analytics in Research Based on Qualitative Interviews. *VIEW J. Eur. Telev. Hist. Cult.* **2018**, *7*, 138. [CrossRef]
30. Valenti, A.P.; Chita-Tegmark, M.; Tickle-Degnen, L.; Bock, A.W.; Scheutz, M.J. Using Topic Modeling to Infer the Emotional State of People Living with Parkinson's Disease. *Assist. Technol.* **2021**, *33*, 136–145. [CrossRef] [PubMed]
31. Parmar, M.; Maturi, B.; Dutt, J.M.; Phate, H. Sentiment Analysis on Interview Transcripts: An Application of NLP for Quantitative Analysis. In Proceedings of the 2018 International Conference on Advances in Computing, Communications and Informatics (ICACCI), Bangalore, India, 19–22 September 2018; pp. 1063–1068.
32. Hagen, L.; Uzuner, O.; Kotfila, C.; Harrison, T.M.; Lamanna, D. Understanding Citizens' Direct Policy Suggestions to the Federal Government: A Natural Language Processing and Topic Modeling Approach. In Proceedings of the 2015 48th Hawaii International Conference on System Sciences, Kauai, HI, USA, 5–8 January 2015; Volume 2015, pp. 2134–2143.
33. Lee, T.Y.; Smith, A.; Seppi, K.; Elmqvist, N.; Boyd-Graber, J.; Findlater, L. The Human Touch: How Non-Expert Users Perceive, Interpret, and Fix Topic Models. *Int. J. Hum. Comput. Stud.* **2017**, *105*, 28–42. [CrossRef]
34. Archibus Inc. Available online: https://archibus.com/about/ (accessed on 10 August 2021).
35. Hutto, C.; Gilbert, E. Vader: A Parsimonious Rule-Based Model for Sentiment Analysis of Social Media Text. In Proceedings of the International AAAI Conference on Web and Social Media, Ann Arbor, MI, USA, 1–4 July 2014; Volume 8, pp. 216–225.
36. Chatterjee, A.; Perrizo, W. Sentiment Trends and Classifying Stocks Using P-Trees. In *From Social Data Mining and Analysis to Prediction and Community Detection*; Springer: Berlin/Heidelberg, Germany, 2017; pp. 103–121.
37. Chatterjee, A.; Perrizo, W. Investor Classification and Sentiment Analysis. In Proceedings of the 2016 IEEE/ACM International Conference on Advances in Social Networks Analysis and Mining (ASONAM), San Francisco, CA, USA, 18–21 August 2016; pp. 1177–1180.
38. Maiti, J.; Khanzode, V.V.; Ray, P.K. Severity Analysis of Indian Coal Mine Accidents—A Retrospective Study for 100 Years. *Saf. Sci.* **2009**, *47*, 1033–1042. [CrossRef]
39. Chiew, E.; Cauthen, K.; Brown, N.; Nozick, L. Comparison of Distribution Selection Methods. *Commun. Stat.-Simul. Comput.* **2022**, *51*, 1982–2005. [CrossRef]
40. Washington, S.; Karlaftis, M.; Mannering, F.; Anastasopoulos, P. *Statistical and Econometric Methods for Transportation Data Analysis*; Chapman and Hall: London, UK; CRC: Boca Raton, FL, USA, 2020; ISBN 0429244010.

Article

Stakeholder Behavior Risk Evaluation of Hydropower Projects Based on Social Network Analysis—A Case Study from a Project

Min An [1,2], Weidong Xiao [1,3], Hui An [1,3,*] and Jin Huang [1,2,*]

[1] Hubei Key Laboratory of Construction and Management in Hydropower Engineering, China Three Gorges University, Yichang 443002, China
[2] College of Economics & Management, China Three Gorges University, Yichang 443002, China
[3] College of Hydraulic & Environmental Engineering, China Three Gorges University, Yichang 443002, China
* Correspondence: anhui@ctgu.edu.cn (H.A.); huangjin@ctgu.edu.cn (J.H.); Tel.: +86-155-7270-1791 (J.H.)

Abstract: Since construction involves many stakeholders and their behavioral risk interaction, which brings risks to the project construction, it is necessary to strengthen the research on the risk management of hydropower projects. This study comprehensively considers the characteristics of hydropower project construction and identifies relevant stakeholders to build and improve the stakeholder behavior risk evaluation index system. On this basis, the social network analysis method is used to build an evaluation model of stakeholders' behavioral risk transmission network, identify core factors and key relationships, analyze the path of behavioral risk transmission, take measures to cut off the transmission of core factors and key relationships, and test the effect of the risk network after control. The results show that: the evaluation model can effectively identify the core behavioral risk factors and key relationships in the construction process. Then, after taking targeted measures on the core behavioral risk factors and key relationships, hydropower projects are less affected by behavioral risk factors, and the risk transmission paths are reduced, which reduces the probability of behavioral risks arising from stakeholders and improves the behavioral governance efficiency of stakeholders. Applying this research model to the risk management of international hydropower projects can provide better guidance to the stakeholders and improve the accuracy and effectiveness of analyzing the behavioral risks of stakeholders in hydropower projects.

Keywords: hydropower projects; social network analysis; transmission path; stakeholder behavioral risk

Citation: An, M.; Xiao, W.; An, H.; Huang, J. Stakeholder Behavior Risk Evaluation of Hydropower Projects Based on Social Network Analysis—A Case Study from a Project. *Buildings* **2022**, *12*, 2064. https://doi.org/10.3390/buildings12122064

Academic Editor: Alban Kuriqi

Received: 26 October 2022
Accepted: 22 November 2022
Published: 24 November 2022

Publisher's Note: MDPI stays neutral with regard to jurisdictional claims in published maps and institutional affiliations.

1. Introduction

The gradual increase in environmental pollution and global warming has made the energy transition urgent [1,2]. Hydropower is valued as a clean energy source, and countries have increased the construction of hydropower projects, hoping that this could bring multiple benefits to them [3]. At the same time, hydropower projects are characterized by long construction cycles, complex construction procedures, and difficult environmental proofs, resulting in delays, increased costs, substandard quality, and environmental pollution risks [4,5]. As the number and complexity of construction projects increase, the extent of the risk for stakeholders is increasing, and the risk management of hydropower projects faces serious challenges.

Risk management in hydropower projects mainly includes objective and subjective risks [6,7]. Objective risk means that it does not depend on human consciousness and transcends human subjective consciousness, and one can only change the risk and occurrence conditions in a limited time and space to reduce its probability of occurrence, such as: natural environmental risks, technological risks, etc. [6]. With the progress of science and technology and the refinement of the social division of labor, the probability of objective risk in hydropower projects is gradually reduced [8]. Meanwhile, subjective risks are behavioral risks caused by human factors that can be avoided or controlled by people themselves, which now receive more attention [7]. Specifically, stakeholders are the most active productivity factors throughout the engineering projects, and the complexity and uncertainty of their behavior may trigger the occurrence of risk events [9,10]. However, most of the current studies control the occurrence of objective risk events, analyze the probability of the occurrence of risk events and the resulting losses [11], and have not yet conducted systematic research on stakeholder behavioral risks. Therefore, how to identify the behavioral risk factors and key relationships of hydropower project stakeholders is a worthy research problem for hydropower project risk management.

Stakeholders are not only the sources of behavioral risk but also the disseminators and receivers of behavioral risk [12]. The cooperative exchange of many stakeholders provides the path basis for behavioral risk transmission, which makes risk transmission complex and dependent [13]. In turn, the construction of hydropower projects brings together many stakeholders and their behavioral risk into a tight network structure, generating behavioral risk relationships and forming a complex behavioral risk transmission path. Therefore, the use of the social network analysis method in this study is conducive to restoring the process of influence between behavioral risks, analyzing the degree of influence between behavioral risks, and clarifying the transmission path between behavioral risks, which is of great significance for improving the efficiency of stakeholder behavior governance.

2. Literature Review

2.1. Project Risk Management

The risk management framework for engineering projects mainly includes risk identification, analysis, assessment, and control [14]. Risk evaluation studies on internal factors such as the construction schedule, cost, and quality of hydropower projects are currently conducted through this framework [15–17]. At the same time, the social environment external risk control is also one of the elements of successful hydropower project construction, including social risks and environmental risks, such as policy changes, natural disasters, etc. [18–20]. However, only considering the evaluation of risks such as the schedule, cost, quality, and social environment is not comprehensive enough, and scholars continue to explore the research on the risk management of hydropower projects in terms of organizational structure and construction safety [21,22]. From the above analysis, the existing hydropower project risk management mainly focuses on objective risk studies such as the schedule, cost, quality, and safety, neglecting the control and prevention of subjective risks caused by the behavioral risk factors of hydropower project stakeholders, etc.

Thevendran and Mawdesley [23] identify the human risk factor as the most influential construction risk and emphasize its necessity in project risk management. On this basis, some scholars have qualitatively analyzed the importance of human risk factors from internal stakeholders such as owners, contractors, and designers [24–26]. Despite the awareness of the importance and necessity of stakeholders' behavioral risk factors, it is far less popular and in-depth than the traditional risk factor research and lacks systematic behavioral risk factor identification and evaluation. Some scholars have even constructed their behavioral risk factor index systems from internal stakeholders such as owners, contractors, and designers to quantitatively analyze the impact of behavioral risk on project construction [27]. However, the behavioral risks of external stakeholders, such as migrants, governments, and environmental departments, are not considered, resulting in an imperfect behavioral risk indicator system. The analysis of the relationship between the behavioral risks of stakeholders is insufficient. Meanwhile, due to the mutual influence of behaviors among stakeholders, some scholars further study the synergy and partnership among stakeholders in project construction [28–31]. However, behavioral risk relationships are not integrated into the risk network to analyze the impact of overall relationships on the construction of engineering projects.

2.2. Social Network Analysis (SNA)

The method of social network analysis originated from sociology. With "people" as the core, it analyzes the subjective initiative of individuals in the network and the constraints of the social structure on people in the network [32]. In addition to the field of sociology, psychology, medicine, and finance also widely apply social network analysis to study the interaction between individual rational choices and collective constraints [33–35].

In recent years, many scholars [36] have introduced SNA into the field of engineering to study the relationship between individuals and collectives related to engineering projects. This method is also increasingly widely used in the field of engineering project management. Lin [37] used SNA to identify the core stakeholders of hydropower projects but did not comprehensively consider the influence relationship of risk factors among the stakeholders. On this basis, Herrera, et al. [38] used SNA to analyze the interaction influence relationship between designer groups and their members in a multidimensional manner but lacked the analysis of the influence of behavioral relationships between other stakeholders. Tang, et al. [39] used SNA to explore the cost–risk relationships of stakeholders in project construction to obtain key cost–risk relationships. Lu, et al. [40] used SNA to study the organizational structure risk relationship of stakeholders in construction projects and found that streamlining the organizational structure and improving the efficiency of personnel communication can reduce risk generation. In summary, the current use of SNA studies engineering project risk relationships from core stakeholders, stakeholder cost risks, and organizational structure risks, but it does not take measures to prevent the impact of the stakeholder behavior risk relationships generated, which makes it difficult to achieve the ultimate project risk management goals. Therefore, based on the analysis of risk relationships related to stakeholders, some scholars use SNA to combine stakeholder management with risk management, propose preventive measures and mechanisms to control the risk relationships generated, and provide a reference for preventing various types of risks [41,42]. However, this study only considers the advantages and disadvantages of risk relationships and takes measures to prevent the impact of risk. The risk relationship transmission path of stakeholders has not been studied, and there are problems such as the unclear influence process and the unclear degree of influence when describing the behavior risk relationship of stakeholders. Therefore, this study uses the SNA method to build a behavioral risk transmission network for stakeholders of hydropower projects, analyze the behavioral risks among stakeholders of the project, and determine the core behavioral risk factors and key relationships. Starting from the risk transmission path, take corresponding measures to cut off risk factors and relationship transmission and prevent behavior risk events.

In summary, different scholars have studied engineering project risk management in terms of objectivity, subjective behavioral risk, and the use of social network analysis. However, there are still two shortcomings: (1) In the current research on project risk management, the evaluation index system of stakeholders' behavioral risks is not improved. (2) In the existing research on engineering risk management using SNA, the analysis of the behavioral risk association relationship is lacking. Stakeholders are not embedded in the network structure to analyze the degree of influence between behavioral risks and the transmission paths.

Therefore, for the shortcomings of the existing studies, this study will be improved in the following aspects: (1) From the construction process of hydropower projects involving many stakeholders and frequent risky accidents. The stakeholders of hydropower projects and their behavioral risk factors are screened and identified using the literature, interviews, and questionnaires to improve the behavioral risk evaluation index system. (2) Use social network analysis to consider the relationship between specific behavioral risk factors, establish a behavioral risk transmission network, explain the key relationships and diffusion paths between risks, and take corresponding measures to prevent the occurrence of risks. The innovations are: (1) Building and improving the project construction stakeholders' behavior risk evaluation index system; (2) Using the social network analysis method to identify core factors and key relationships, analyze behavioral risk transmission paths, and take measures to cut off the propagation of core factors and key relationships. This study provides a reference for the related research of hydropower project risk management and promotes the sustainable development of hydropower project construction.

3. Research Approaches

In this study, the social network analysis method is applied to the project risk management theory, and an evaluation model of stakeholder behavior risk management for hydropower projects is proposed. This evaluation model uses social network analysis tools to visualize the behavioral risk assessment of stakeholders in hydropower projects and to quantitatively analyze the behavioral risk relationship and impact degree between individual networks and the overall network. At the same time, after taking measures to control the core behavioral risk factors and key relationships, social network analysis tools are used to test the behavioral risk response of stakeholders, and visual and quantitative analysis is conducted on the behavioral risk after testing. The combination of the social network analysis method and project risk management theory is conducive to the combination of the qualitative and quantitative analysis of project risk management to better realize risk identification, assessment, and response evaluation. Moreover, the behavioral risk relationship of stakeholders is visualized, which is conducive to the project risk management to clarify the behavioral risk relationship of stakeholders and the specific path of risk transmission.

The specific steps are as follows: First, identify and determine stakeholders and their behavioral risks by using a literature review, expert consultation, and other methods, and list the relevant stakeholders and behavioral risk factors. Second, the behavioral risk relationship of stakeholders is evaluated by a questionnaire and other methods to determine the behavioral risk relationship and transmission path of stakeholders. Third, the social network analysis method is used to evaluate the behavioral risk relationship and transmission path visually and quantitatively and determine the core behavioral risk factors and key behavioral relationships. Finally, targeted measures are taken to control the relationship between core behavioral risk factors and key behaviors, and social network analysis methods are used to respond to the controlled behavioral risk network, as shown in Figure 1.

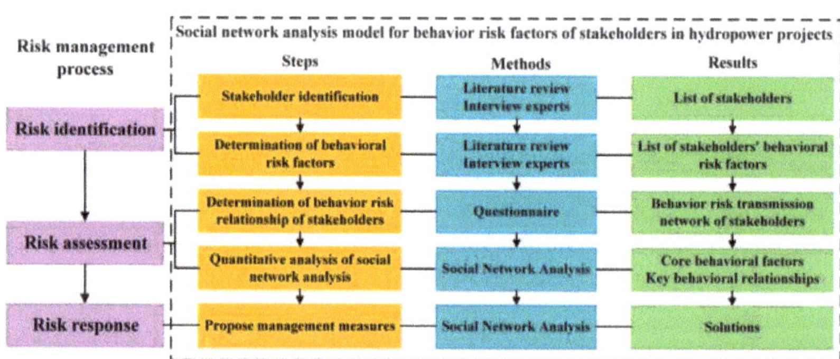

Figure 1. Research model framework.

3.1. Indicators System Construction

This study constructs an evaluation index system for stakeholders and their behavioral risk factors. The steps are as follows: First, preliminarily identify stakeholders and their behavioral risk factors through literature combing. Second, experts with rich engineering project management practice, scientific research experience, and research work in hydropower are invited to participate in the identification and classification of stakeholders and behavioral risk factors, and the information of experts is shown in Table 1.

Finally, the preliminary identification results are corrected and optimized by comprehensive expert suggestions to determine the final stakeholders and their behavioral risk factors evaluation index.

Table 1. Experts Information.

Expert	Work Unit	Position	Work Experience (Years)	Expert	Work Unit	Position	Work Experience (Years)
1	College	Professor	10	6	Supervisor	Director representative	8
2	Owners	Department manager	7	7	Designer	Engineer	7
3	Designer	Engineer	6	8	Contractors	Project manager	8
4	College	Associate professor	8	9	College	Lecturer	6
5	Contractors	Project manager	6	10	Owners	Department manager	8

Stakeholders are broadly defined as individuals, groups, or organizations that may be affected by the decisions, activities, or outcomes of a project. The purpose is to avoid the arbitrary or deliberate exclusion of certain stakeholders. Select the literature on the stakeholder risk assessment of large-scale engineering projects and hydropower projects. The stakeholders appearing in each work of literature are summarized and counted, and 14 stakeholders are initially identified, as shown in Table 2.

Table 2. Initially Identified Stakeholders.

Literature	Stakeholders													
	S1	S2	S3	S4	S5	S6	S7	S8	S9	S10	S11	S12	S13	S14
Lee, et al. [43]	√	√	√	—	√	—	—	—	√	—	—	—	√	—
Xia, et al. [44]	√	√	√	√	√	√	—	—	—	√	√	—	√	—
Zhang, et al. [45]	√	√	√	√	—	—	—	—	—	—	—	—	—	—
Ding, et al. [46]	√	√	—	—	√	—	—	—	√	—	—	√	—	√
Mok, et al. [47]	√	√	√	√	√	√	—	—	—	—	—	—	—	—
Sadkowska [48]	√	—	√	√	√	√	√	—	—	—	—	—	—	—
Amadi, et al. [49]	√	√	√	—	—	√	—	—	—	—	√	—	√	√
Daniel, et al. [50]	√	√	√	√	—	—	—	√	—	√	—	—	—	—
Luo, et al. [51]	√	√	√	√	—	√	√	—	—	—	—	—	—	—
Bahadorestani, et al. [52]	√	√	—	√	√	√	√	√	—	√	—	—	—	—
He, et al. [53]	√	√	—	√	√	—	—	—	—	—	—	—	√	—
Nguyen, et al. [54]	√	√	√	—	—	—	√	—	—	—	—	—	—	—
Jia, et al. [55]	√	√	—	√	√	—	—	—	—	—	—	—	—	—
Wen and Qiang [56]	√	√	—	√	√	—	√	—	—	—	√	—	—	—

S1–S14: Owners; Contractors; Subcontractor; Designer; Government; Supplier; Supervisor; Environmental Departments; Media; Financial Institution; Research Institutions; Operator; Natives; Consulting Company.

To identify the final stakeholders associated with the project construction of the hydropower project, experts were invited to conduct interviews to clearly explain the definition of each stakeholder in the initial identification list. The experts were also consulted on the following issues: whether the names and definitions of the initially identified stakeholders (*Si*) were appropriate and whether they needed to be modified; what important roles the above stakeholders (*Si*) usually played in a typical project; whether stakeholders (*Si*) play a role in the construction of the project and whether their actions will have an impact on the construction of the project; whether initially identified stakeholders (*Si*) need to be removed or added, and for what reasons. Based on the results of the above expert interviews, the stakeholders of the hydropower project construction were modified and improved, as shown in Figure 2. The stakeholders in the construction of hydropower projects are identified as the owners (the responsible body for hydropower project construction, which is responsible for project planning, financing, construction implementation, etc.), the designer (responsible for hydropower project design work), the contractors (responsible for hydropower project construction, transportation, labor, and other works), the supervisor (responsible for hydropower project supervision tasks), the material and equipment suppliers (responsible for providing hydropower project materials and equipment), the government (hydropower project location of government agencies), the immigrant (the masses affected by the construction of hydropower projects and involuntary relocation), and the environmental departments (in accordance with the relevant laws to implement the supervision and management of environmental protection law enforcement departments), respectively, with S1, S2, S3, S4, S5, S6, S7, and S8.

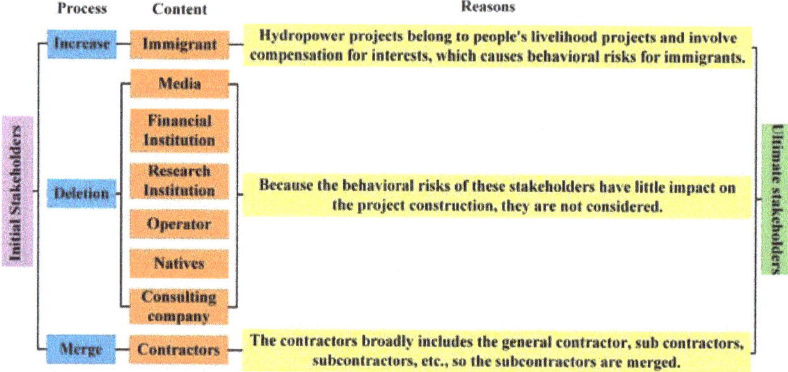

Figure 2. Stakeholder Identification Process.

Risk consists of two factors: the uncertainty of the occurrence of the event and the hazards arising after the occurrence. Therefore, hydropower project stakeholder behavioral risk is defined as the factors that are related to the subjective behavior of stakeholders in the construction of hydropower projects and have an uncertain impact on the successful achievement of project objectives. Identifying behavioral risk factors is not only a key step in establishing a behavioral risk evaluation model but is also a basis for effective risk management. Therefore, the literature studying the risk factors and impact categories of hydropower projects was selected, the behavioral risk factors appearing in each work of literature were summarized and counted, and 22 behavioral risk factors were initially identified, as shown in Table 3.

Table 3. Initially Identified Behavioral Risk Factors.

Factors	Literature							
	Yang and Zou [57]	Wu, et al. [58]	Yang, et al. [13]	Wang, et al. [27]	Xiang, et al. [26]	Xia, et al. [59]	Darvishi, et al. [60]	Barghi and Shadrokh Sikari [61]
R1	√	√	—	√	—	—	—	—
R2	—	√	—	—	√	—	—	—
R3	—	—	√	—	—	—	—	√
R4	√	—	—	√	—	√	—	√
R5	—	√	—	√	—	—	—	—
R6	—	—	√	—	√	—	—	—
R7	√	√	—	—	—	√	—	√
R8	—	—	√	√	√	—	—	√
R9	—	—	√	—	√	—	—	—
R10	—	—	√	√	√	—	—	—
R11	√	—	—	√	—	—	—	—
R12	√	—	—	—	—	—	√	—
R13	—	—	—	√	—	—	√	—
R14	—	—	—	—	—	—	√	√
R15	√	√	—	—	—	√	—	√
R16	√	—	—	—	—	—	—	—
R17	—	√	—	—	—	√	—	√
R18	—	√	√	—	—	√	—	√
R19	—	—	√	—	—	√	—	√
R20	—	—	—	√	—	√	—	√
R21	—	—	√	—	—	√	—	√
R22	—	—	√	—	—	√	√	—

R1–R22: Difficulty in paying funds; Proactive change request; Design changes; Project not delivered on schedule; The design does not consider ecological protection; Construction costs exceeded budget; Construction quality not up to standard; Inappropriate construction safety measures; Lack of timely implementation monitoring; Irregular supervision process; Substandard quality; Lack of timely supply of material and equipment; Supply price adjustment; Policy changes and adjustments; Decision approval and delay; Call off construction; Poor coordination skills; Create public opinion; Destruction of cultural customs; Poor quality of life; Disrupting construction sites; Not strictly enforcing environmental standards.

To finally determine the behavioral risk factors of stakeholders in hydropower project construction, experts were invited to conduct interviews to screen and classify the behavioral risk factors of stakeholders. The experts were also consulted on the following issues: whether the behavioral risk factors (*Rj*) are subjective behaviors of stakeholders (*Si*); whether the behavioral risk factors (*Rj*) of stakeholders (*Si*) themselves are independent; whether the behavioral risk factors (*Rj*) among stakeholders (*Si*) have a direct impact; whether the behavioral risk factors (*Rj*) are applicable to the project construction. Based on the results of the above expert interviews, the risk factors for stakeholder behavior in the construction process of hydropower projects are modified and refined, as shown in Figure 3.

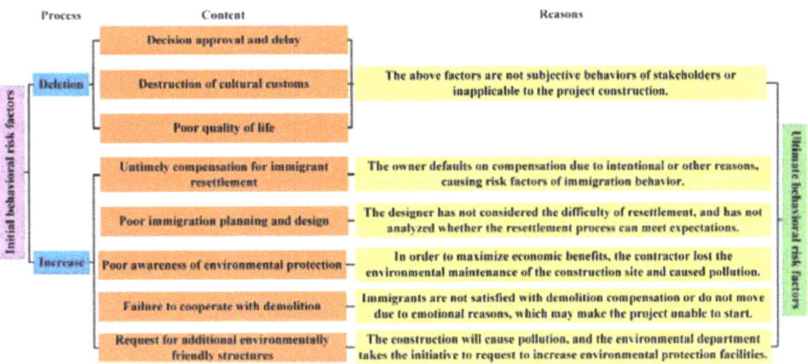

Figure 3. Behavioral Risk Factor Identification Process.

In summary, a total of eight stakeholders and their corresponding 24 behavioral risk factors were identified for the construction of hydropower projects, as shown in Table 4.

Table 4. Identification and classification of behavioral risk factors for hydropower projects.

Stakeholders	Behavior Risk Number	Behavioral Risk Factors	Behavioral Risk Factors Description
Owners (S1)	R1	Difficulty in paying funds	Owners are difficult to finance, lack of fund preparation, and lack of willingness to pay.
	R2	Proactive change request	May directly lead to a chain reaction in the construction process and increase the direct and indirect cost of the project and quality risk.
	R3	Untimely compensation for immigrant	Improper immigrants easily cause public resentment and anger, and there are problems such as insufficient investment in immigrants.
Designer (S2)	R4	Design Changes	Incomplete design drawings and lack of communication with the construction party and construction unit resulted in changes.
	R5	Poor immigration planning and design	The difficulty of immigrants has not been considered, as well as whether it can meet the expectation of immigration, resulting in backward work and other problems.
	R6	The design does not consider ecological protection	Designers lack environmental awareness or ignore environmental issues to save costs.
Contractors (S3)	R7	Project not delivered on schedule	Contractors have their own uncertain factors, as well as the actual construction and the planned progress of deviation.
	R8	Construction costs exceeded budget	The actual construction cost exceeds the planned cost due to mismanagement, malicious low bids, price fluctuation, and other reasons.
	R9	Construction quality not up to standard	Lax supervision of construction materials and shoddy phenomena occur, easily causing engineering quality and safety accidents.
	R10	Inappropriate construction safety measures	Without scientific safety production, standardization, and standardized management of the site, there are safety risks on the site.
	R11	Poor awareness of environmental protection	The environment of the construction site is not managed, resulting in air, water, and ecological pollution to the surrounding environment.
Supervisor (S4)	R12	Lack of timely implementation monitoring	Delayed supervision of the site and failure to rectify hidden dangers in time lead to project risks.
	R13	Irregular supervision process	The supervisor and the contractor conspire to pursue their own interests and lower the project quality standard.

Table 4. *Cont.*

Stakeholders	Behavior Risk Number	Behavioral Risk Factors	Behavioral Risk Factors Description
Material & Equipment Suppliers (S5)	R14	Substandard quality	Use sub-standard materials and equipment instead of quality standard equipment to provide maximum benefit to the contractor.
	R15	Lack of timely supply of material and equipment	Material shortage, suppliers do not perform their own responsibilities, and material supply is not timely.
	R16	Supply price adjustment	As market prices rise, suppliers take the initiative to increase the agreed supply price, resulting in disputes with contractors.
Government (S6)	R17	Policy changes and adjustments	Changes in national laws and regulations and other relevant documents cause local governments to issue the latest policies for governance.
	R18	Call off construction	As the project is not up to standard and is in violation of laws and regulations, the government directly stops the construction rectification.
	R19	Poor coordination skills	Due to the lack of capacity of government personnel, the coordination of various parties cannot be well completed.
Immigrant (S7)	R20	Create public opinion	Dissatisfied with the immigration plan, the media and other means are used to protect their rights and create relevant public opinion.
	R21	Disrupting construction sites	Dissatisfied with the immigration scheme, some immigrants may take relatively radical actions to disrupt the construction site.
	R22	Not cooperating with demolition	Not being satisfied with the compensation or emotional reasons for not moving may prevent the project from starting.
Environmental Departments (S8)	R23	Not strictly enforcing environmental standards	Some law enforcement officials conspire with contractors to pursue their own interests and lower environmental enforcement standards.
	R24	Request for additional environmentally friendly structures	The construction process lacks the relevant environmental protection facilities; the environmental departments require that it be increased.

3.2. SNA Model Construction

3.2.1. Overall Network Structure

1. Network density

Network density is a measure of node compactness in a risk network model. The higher the network density, the closer the connection between nodes, and the overall network structure presents a stable state. The calculation of bivariate directed network density is shown in Equation (1),the range of D is (0,1) [62].

$$D = \frac{L}{n(n-1)} \tag{1}$$

where L is the number of relationships between risk factors influencing each other; and n is the number of behavioral risk factors of stakeholders.

2. Block Model

The block model describes the relationship between actors, represents the relationship between the positions of each actor, reflects the influence relationship between each block, and makes the influence relationship of the whole behavioral risk network clearer. The behavioral risk factors of stakeholders in hydropower projects are divided into set discrete subgroups according to certain criteria, which are called "blocks", and each block is a subgroup of the whole risk network. Main steps of block model analysis: first, the behavioral risk factors of stakeholders in hydropower projects are classified by the *CONCOR* method (the iterative correlation convergence method, which iterates the correlation coefficients between each row or column in the matrix and eventually produces a correlation coefficient matrix consisting of only 1 and -1 to achieve a partitioning of the corresponding individual actors, thus simplifying the data), and each class is taken as a block to obtain the block matrix and density matrix. Second, the value of each block is determined according to certain criteria, i.e., 1-block or 0-block. The criteria used for relationships of different natures are different, and the most common is the α- Density index. Compare the density matrix with the density of the whole network. If the density is greater than the density of the whole network, take "1" in the image matrix and "0" in the image matrix to obtain the image matrix. Finally, core blocks are identified by obtaining the block matrix, density matrix, and image matrix and analyzing the location characteristics of blocks in the whole network [62].

3.   Clustering coefficient

The clustering coefficient reflects the closeness of the whole network. A larger value of the clustering coefficient indicates that the behavioral risk factors are more closely linked, the hidden risk is greater, and the likelihood of project failure is greater. Equation (2) is the calculation process [62].

$$C = \frac{1}{n}\sum_{i=1}^{n}\frac{E_i}{k_i(k_i-1)} \tag{2}$$

where n is the number of behavioral risk factors of stakeholders; E_i represents the total number of relationships between stakeholder i behavioral risk factors; k_i denotes the number of stakeholder i-related behavioral risk factors; and $k_i(k_i-1)$ denotes the total number of stakeholders point i behavioral risk factors.

4.   Intermediate central potential

The intermediate centrality potential is to measure the gap between the point with the highest centrality in the network and other points. It tests the ability of a specific point to control the entire network. The calculation is shown in Equation (3) [42].

$$C_B = \frac{\sum\limits_{i=1}^{n}(C_{\max}-C_i)}{n^3-4n^2+5n-2} \tag{3}$$

where n is the number of behavioral risk factors of stakeholders; C_i is the intermediate centrality of the stakeholder i; and $C_{\max}$ is the maximum of all C_i.

5.   Accessibility

Reachability refers to a kind of data transferability closed circle that is used to judge the degree of network connectivity; the higher the value indicates that its risk factors transfer more smoothly. The calculation is shown in Equations (4) and (5) [62].

$$(A+I) \neq (A+I)^2 \neq, \cdots, (A+I)^r = (A+I)^{r+1} = M, r < n-1 \tag{4}$$

$$C = 1 - \left(\frac{V}{N(N-1)/2}\right) \tag{5}$$

where M represents the reachable matrix; C represents the network reachable; A represents the adjacency matrix, I represents the identity matrix; V represents the number of unreachable point pairs in the network; and N represents the network scale.

3.2.2. Individual Network Structure

1.   Intermediary

An intermediary is defined as a person in the middle, regardless of whether he receives a reward. Its function is to group the whole network node and to study how different groups transmit through risk factors. To better clarify the definition of "intermediary", further understand how to transfer the risk factors of stakeholder behavior. The specific description is as follows: If the risk relationship transmission path is A→B→C, B is the intermediary. Specific roles are distributed according to the positions of the three, and circles of the same color represent the same group. If all three are in the same group, B is the coordinator; if A and C are in the same group, B is not in the group, and B is the consultant; if B and C are in the same group and A is not in the group, then B is the gatekeeper; if A and

B are in the same group and C is not in the group, then B is the representative; if A, B, and C are in different groups, B is the liaison [62], which is described in Figure 4. The results of this study selected 35% of the 24 behavioral risk factors as important risk factors [13].

2.   Point Median Center Degree

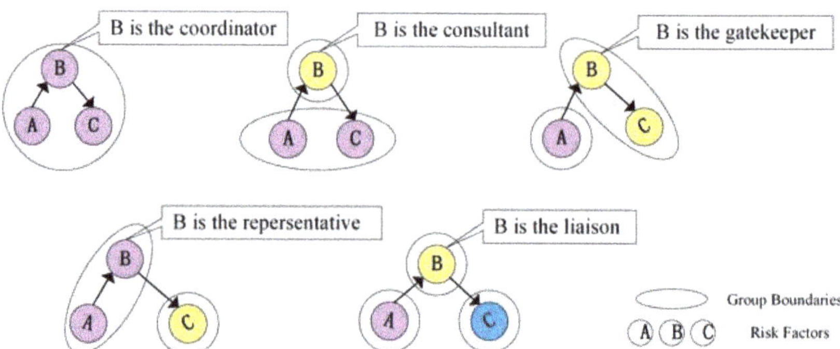

Figure 4. Description of the five types of intermediaries.

The intermediate centrality of a point refers to the fact that a point is in a critical position in the network if it is on multiple interaction paths. The larger the value of a point, the stronger its ability to control the conduction of other nodes and the more critical the network position. The calculation is shown in Equations (6) and (7) [63]. The results of this study selected 35% of the 24 behavioral risk factors as important risk factors [13].

$$C_i = \sum_j^n \sum_k^n b_{jk}(i), j \neq k \neq i, and, j < k \tag{6}$$

$$b_{jk}(i) = g_{jk}(i) / g_{jk} \tag{7}$$

where g_{jk} and g_{jk} (i) denote the number of paths and paths between stakeholders j and k, respectively; and b_{jk} (i) denotes the ability to interact between stakeholders j and k.

3.   Line Center Degree

The line center degree is the ability to transfer and control risk factors in the network. Measure the control degree of a line on information. The greater the value, the stronger the risk control transmission ability. It is calculated in Equation (8) [62]. In this study, 15% of the line intermediate centrality values higher than 0 risk relations were considered as critical relations [13].

$$C_{p \to q} = \sum_j^n \sum_k^n b_{jk}(p \to q), j \neq k \neq p \neq q, j < k \tag{8}$$

where b_{jk} (p→q) represents the ability of the control stakeholders j and k of relation p→q to communicate; and n denotes the total number of behavioral risk factors of stakeholders.

4. Empirical Analysis

4.1. Research Examples

This study takes the Chongqing JL Hydropower Project as the empirical object. This project is in Qi Jiang District, Chongqing, and involves many immigrants. It is a comprehensive large-scale power station hub project focusing on power generation and considering flood control, water supply, and shipping. The reservoir has a total storage capacity of 5.163 billion cubic meters, a regulated storage capacity of 900 million cubic meters, and a backwater length of 156.6 km.

This hydropower project has complex construction procedures, many risk factors, and many stakeholders with complex relationships, which are in line with the characteristics of most hydropower projects. It can provide a basis for verifying the rationality and feasibility of the stakeholder behavior risk transfer network evaluation model for hydropower project construction [64]. At the same time, the project is in the early decision-making stage, so it is necessary to carry out risk analysis and prevention in advance for its construction process in order to reduce the economic losses of various stakeholders and provide a reference for the construction of other hydropower projects [65].

4.2. Questionnaire Design and Statistical Analysis

The questionnaire content is mainly considered according to the behavior risk evaluation index system of stakeholders of hydropower projects. Proceeding from the self-interest and subjective behavior of various stakeholders, it is mainly based on the construction period, cost, quality, production capacity, and market. The project construction stakeholder behavioral risk factor inter-impact questionnaire was applied to the Chongqing *JL* hydropower project. The content of the questionnaire only requires the respondents to fill in the corresponding matrix and identify the possible influence relationship, which greatly reduces the number of judgments and improves the efficiency and quality of the questionnaire. The questionnaire involves a total of eight stakeholder corresponding behavioral risk factor influence matrices [66], and the instructions and requirements for filling them out are shown in Appendix A. (Due to the large number of questionnaires, this paper only takes the owners as an example, and the questionnaires of other stakeholders are similar.)

The questionnaire data are the basis for constructing the risk network of hydropower projects. Therefore, before the construction of the major *JL* hydropower project in Chongqing, 269 questionnaires were sent out in the form of email and paper, 249 were returned, and 240 were valid. The valid response rate was 89.2%. The background information of participants was shown in Table 5. From Table 5, we can see that: (1) The source of participant units includes stakeholders selected from the study, in which there are more sample data of owners, designers, contractors, and governments, while there are fewer sample data of supervisors, material and equipment suppliers, immigrants, and environmental departments. (2) A higher percentage of participants had a higher education and a longer working life. Among them, 93.72% have a bachelor's degree or above and 77.55% have a working life of more than 5 years. (3) A total of 88.72% of participants have experience in hydropower project construction, which indicates that most participants have rich experience in hydropower project construction.

Table 5. Background information for participants.

Unit Source	Owners 14.27%	Designer 15.73%	Contractors 28.64%	Supervisor 11.36%	Suppliers 10.56%	Government 12.44%	Immigrants 5.28%	Environmental Departments 1.72%
Education	Junior College 5.59%	Undergraduate 64.41%	Master 26.21%	Doctor 3.1%	Other 0.69%			
Work life	≤5 years 22.45%	6–10 years 40.5%	11–15 years 27.55%	16–20 years 8.45%	≥20 years 1.05%			
Participation in hydropower project construction	0 11.28%	1 38.72%	2 25.88%	3 13.97%	≥4 10.15%			

First, construct the "Filler–Fill Results" matrix through 240 valid questionnaires. Second, the consistency analysis of the "filler–finisher" matrix showed that the ratio of the first characteristic value to the second characteristic value was 3.046, which was greater than the general principal value of 3 [67]. This proved that the collected data had a single answer mode, and the questionnaire data met the requirements of social network analysis. Finally, because the relationship in the questionnaire belongs to the single answer mode, the questionnaire results are sorted according to the principle of the minority obeying the majority, and the adjacency binary directed matrix of the influence relationship between the behavioral risk factors of hydropower project stakeholders is constructed. "1" means that the behavioral risk factors of the row will have an impact on the behavioral risk of the column, and vice versa for "0". The original data used for analysis are shown in Appendix B.

4.3. Risk Network Visualization

A visualization of the risk network for stakeholder behavior in hydropower projects is shown in Figure 5. Among them, each node *SiRj* indicates *j* behavioral risk factors of *i* stakeholders, with a total of 24 nodes. The node color represents the stakeholder group. The directed arrow line between each node indicates the connection between the behavioral risk factors of each stakeholder, and the arrow represents the direction of the relationship between the nodes.

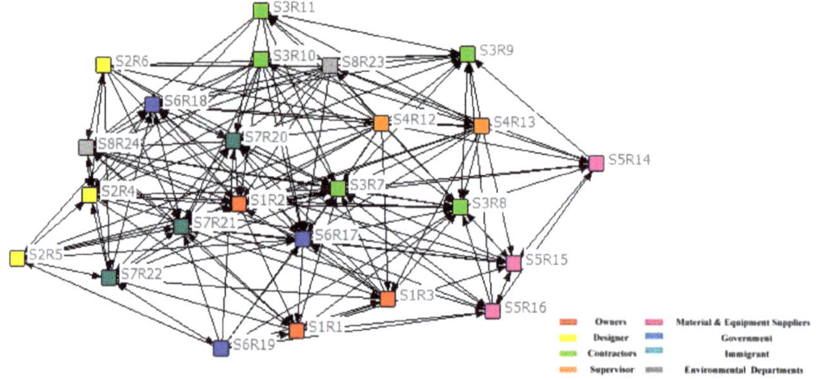

Figure 5. Behavioral risk network visualization for hydropower projects.

From Figure 5, the overall network is more densely connected, which indicates that the behavioral risk factors are interdependent and highly connected, reflecting the complexity of risk transmission in the construction process of the *JL* hydropower project.

There are S1R2 (Owners–Proactive change request), S3R7 (Contractors–Project not delivered on schedule), S6R17 (Government–Policy changes and adjustments), and S7R20 (Immigrant–Create public opinion) behavioral risk factors located in the center of the network, which may have a greater impact on the overall construction process.

On the contrary, behavioral risk factors located at the edges of the network, such as S2R5 (Designer–Poor immigration planning and design), S6R19 (Government–Poor coordination skills), S5R14 (Material & Equipment Suppliers–Substandard quality), S3R9 (Contractors–Construction quality not up to standard), etc., may have a smaller impact.

4.4. Risk Network Metrics Analysis

The analysis of risk network indicators focuses on the overall network and individual networks [63]. The overall network uses a block model to delineate the core risk subgroups in the network. Individual networks are analyzed in terms of intermediary analysis as well as intermediate centrality to identify the actors that are at the core of the network. Therefore, this study analyzes the overall network and individual network for the behavioral risk network of stakeholders in hydropower projects to identify the key behavioral risk factors and key relationships.

4.4.1. Overall Network Analysis

First, the network density in the behavioral risk network of hydropower projects is analyzed, and the value is 0.3351, which is between (0,1). This shows that the relationship between behavioral risks in the network is complex. Secondly, the adjacent binary directed matrix is iterated by the *CONCOR* method and finally divided into eight "blocks". Finally, the density matrix and the "like" matrix [68] are used to determine whether the "block" is in the core position of the network (both receiving and emitting relations; internal relations are close), and the core blocks are block 1 (S1R1 and S3R8), block 2 (S3R7, S3R9 and S5R15), block 3 (S1R2, S2R4, and S8R24), block 4 (S6R17, S6R18, S7R20, and S7R21), block 6 (S5R16 and S6R19), and block 8 (S4R12, S4R13, and S8R23).

4.4.2. Individual Network Analysis

First, the intermediary analysis is conducted on the behavioral risk network of hydropower projects, and the total number of five types of intermediaries accounted for 59.4% of all risk factors (540 in total). The behavioral risk results of the top eight who assumed the role of intermediaries are taken as important risk factors. Second, the intermediate centrality of their points is analyzed to obtain the intermediate centrality of each point, and the results of the top eight stakeholder behavioral risk factors are taken as important risk factors. Finally, the key risk factors for individual network analysis are obtained by taking the union of the important risk factors of both the intermediary and the point intermediate center degree as S1R1 (Owners–Difficulty in paying funds), S2R4 (Designer–Design Changes), S3R7 (Contractors–Project not delivered on schedule), S3R8 (Contractors–Construction costs exceeded budget), S4R12 (Supervisor–Lack of timely implementation monitoring), S6R17 (Government–Policy changes and adjustments), S6R18 (Government–Call off construction), S7R20 (Immigrant–Create public opinion), and S7R21 (Immigrant–Disrupting construction sites).

4.4.3. Key Relationships Analysis

The risk network was analyzed for line middle centrality, and it was found that there existed 185 groups of risk relationships with a line middle centrality greater than 0. However, the smaller the value, the smaller the impact between behavioral risk factors. Therefore, only the top 30 relationships were selected as key relationships, Table 6 shows the key relationships.

Table 6. Identification of key relationships in the top 30 risk network rankings.

Ranking	Risk Factors	Ranking	Risk Factors	Ranking	Risk Factors
1	S3R9→S6R18	11	S5R15→S6R17	21	S7R21→S6R18
2	S7R22→S6R19	12	S6R17→S6R18	22	S3R8→S3R9
3	S2R4→S2R6	13	S3R7→S5R16	23	S7R22→S6R18
4	S4R12→S8R23	14	S1R1→S7R22	24	S8R24→S6R18
5	S4R12→S4R13	15	S7R20→S6R18	25	S3R8→S5R15
6	S4R12→S3R11	16	S1R2→S1R1	26	S6R17→S7R22
6	S4R12→S3R10	17	S2R5→S6R18	27	S1R1→S7R21
8	S6R18→S4R12	18	S3R8→S5R14	28	S3R8→S1R1
9	S2R4→S2R5	19	S1R1→S1R3	29	S3R9→S7R20
10	S3R7→S7R20	20	S3R7→S3R9	30	S5R14→S3R9

4.5. Risk Network Control and Inspection

4.5.1. Core Risk Identification and Control

The core risk factors obtained from the individual network analysis are in the core block of the overall network analysis. Then, it is the core risk of the behavioral risk transmission network of the hydropower project [69], and its identification process is shown in Figure 6.

The key behavioral risk factors derived from the individual network analysis: S1R1 (Owners–Difficulty in paying funds) and S3R8 (Contractors–Construction costs exceeded budget) belong to block 1, S3R7 (Contractors–Project not delivered on schedule) belongs to block 2, S2R4 (Designer–Design Changes) belongs to block 3, S6R17 (Government–Policy changes and adjustments), S6R18 (Government–Call off construction), S7R20 (Immigrant–Create public opinion), and S7R21 (Immigrant–Disrupting construction sites) belong to block 4, and S4R12 (Supervisor–Lack of timely implementation monitoring) belongs to block 8. Since the overall network analysis yields block 1, block 2, block 3, block 4, and block 8 as the core blocks, the behavioral risk factors of the above hydropower project stakeholders are all core risk factors.

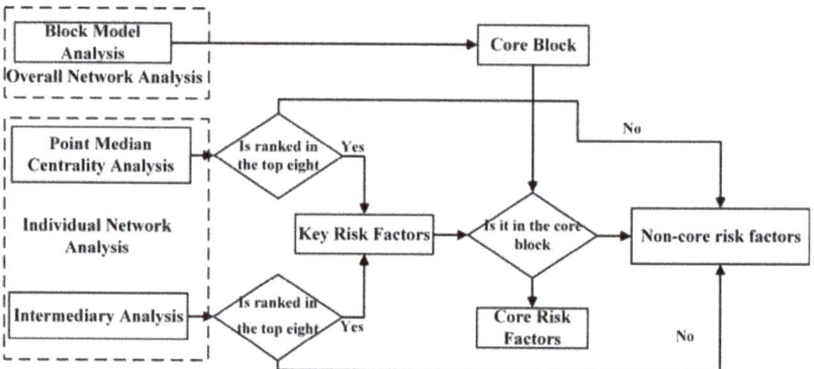

Figure 6. Core Risk Factor Identification Process.

Among the nine core behavioral risk factors, R1 (Difficulty in paying funds) belongs to the owners, R4 (Design Changes) belongs to the designer, R7 (Project not delivered on schedule) and R8 (Construction costs exceeded budget) belong to the contractors, R12 (Lack of timely implementation monitoring) belongs to the supervisor, R17 (Policy changes and adjustments) and R18 (Call off construction) belong to the government, and R20 (Create public opinion) and R21 (Disrupting construction sites) belong to the immigrant. The core behavioral risk factors involve six stakeholders who are all key participants in the construction of hydropower project projects, and project construction is also a key stage of the hydropower construction process. On the other hand, the nine key behavioral risk factors have nothing to do with material and equipment suppliers and environmental departments. It is related to the weak role of material and equipment suppliers in hydropower projects. Pay attention to the environmental problems in the construction of hydropower projects, enhance the environmental protection awareness of all stakeholders, and reduce the interest loss caused by the environment.

As can be seen from Table 6, except for the risk relationships ranked 2nd and 30th, all others are closely related to the nine core risk factors: R1, R4, R7, R8, R12, R17, R18, R20, and R21. Taking effective measures to control the key relationships is conducive to preventing some of the risks from being transmitted in the network [70–72]. Therefore, relevant measures are taken to control the relationship between core behavioral risk factors and key risks, as shown in Table 7.

Table 7. Risk response measures.

Type	Risk Factors	Stakeholders	Response
	R1	Owners	(1) Establish sound rules and regulations for fund management and standardize the basic accounting work of construction projects. (2) Financial supervision, financial management, and fund control must be integrated into the project establishment and feasibility study stage. (3) Strengthen the management and control of project price settlement to prevent the occurrence of over-estimation, over-calculation, and false claims.
	R4	Designer	(1) Fully understand the requirements of the owners and timely communication to ensure the feasibility and accuracy of the program. (2) The quality department guides the design department to sort out the workflow to ensure that design changes are at a controllable level.
Core Factors	R7,R8	Contractors	(1) Make a good construction organization and design plan, establish the target system of progress control, and clarify the personnel of progress control. (2) Conduct technical and economic analysis, determine the best construction plan, combine construction methods, and reduce material consumption costs. (3) Hold regular construction progress coordination meetings and adopt network planning techniques to implement the dynamic control of project progress.
	R12	Supervisor	(1) Supervisors should stick to their posts, conscientiously perform their supervisory duties, and not slacken their work.

Table 7. *Cont.*

Type	Risk Factors	Stakeholders	Response
	R17,R18	Government	(2) Strictly control the quality of construction, check the quality of raw materials and intermediate products, and do a good job of side stations and acceptance work. (1) Policy changes and adjustments should be in line with the actual situation and should not be changed casually. (2) Calling off construction cannot be a temporary notice; the site during the suspension of work needs to urge the contractor to rectify and implement the existing problems.
	R20,R21	Immigrant	(1) Actively express your demands with the relevant departments, exercise your rights legally and reasonably, and make efforts to cooperate with the relevant departments to do your duty. (2) Rational view of the project construction; shall not use force or false public opinion to defend rights; use the law to reasonably defend rights.
	Ranking	Key Relations	Response
Key Relationships	2	S7R22→S6R19	The government should actively coordinate with the immigrants, owners, and other parties involved in the project to solve the problem to achieve a balance of interests.
	30	S5R14→S3R9	Contractors should strictly control the quality of materials and equipment; the quality does not meet the standards to accept and use; prevent the construction quality that is not qualified.

4.5.2. Effectiveness Check

From the behavioral risk network of stakeholders in hydropower projects, after 9 core behavioral risk factors and 30 groups of key relationships were controlled, the risk factors were reduced from 24 to 15, and the risk relationships were reduced from 185 to 47. The risk network with the core behavioral risk factors and key relationships removed is shown in Figure 7. By visual comparison between Figures 5 and 7, the risk network becomes sparse, indicating that the risk factors in the network are less closely linked.

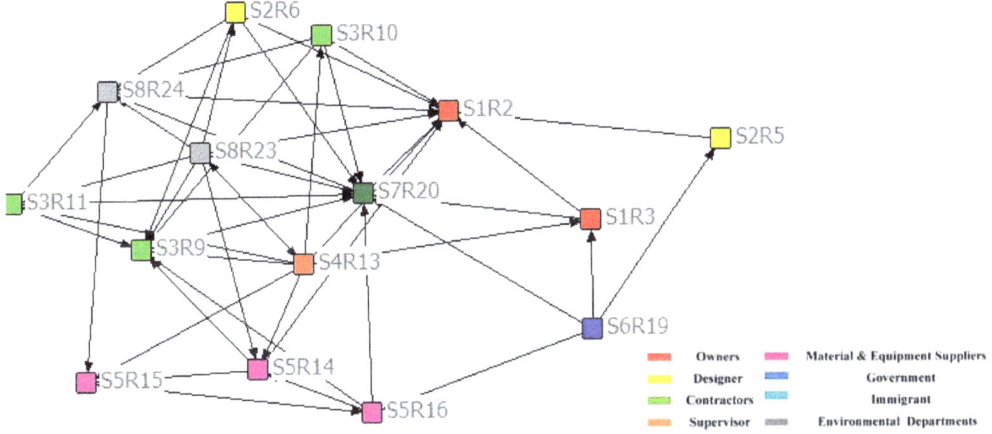

Figure 7. Stakeholder behavioral risk network after the elimination of core risks.

After the core behavioral risk factors and key relationships were controlled, the overall network effect was tested in terms of network density, clustering coefficient, intermediate central potential, and network accessibility, and the specific results were as follows.

(1) After the relationship between core risk factors and key risks is controlled, the overall network density is reduced from 0.3351 to 0.2238, which is a decrease of 33.2%. This shows that the close degree of risk relationship in the network is reduced, and the complexity of the network is effectively improved.

(2) The clustering coefficient is reduced from 0.387 to 0.241 after the core risk factors and key risk relationships are controlled, which is a 37.7% reduction. This shows that the diversity of risk transmission paths is reduced, which inhibits the frequency and activity of risk transmission.

(3) The intermediate central potential decreases from 0.1752 to 0.0904 after the core risk factors and key risk relationships are controlled, which is a decrease of 48.3%. This shows that the gap between risks is narrowed, and the control of risk factors over the whole network is weakened.

(4) The network accessibility value decreases from 0.1420 to 0.0444 after the core risk factors and key risk relationships are controlled, which is a decrease of 68.7%. This shows that the transmission path of multiple risk factors is cut off, and the connectivity of the network is effectively reduced.

5. Discussion

Under the framework of engineering project risk management, Wang, et al. [27] identify internal stakeholders such as designers and contractors and construct their corresponding system. Since internal stakeholders are not only limited to designers and contractors, there are also problems such as a lack of internal stakeholders and incomplete evaluation indexes. Based on this, this study adds internal stakeholders such as owners and supervisors and improves their corresponding behavioral risk evaluation indexes, such as untimely compensation for immigrants and unreasonable migration planning and design. Meanwhile, the existing studies focus less on the behavioral risks of external stakeholders such as governments and immigrants [37,40,42]. However, because the construction sites of hydropower projects are provided by the government, and the residents in the construction area need to actively cooperate with the project construction, external stakeholders are bound to be one of the conditions for the success of the project construction. Based on this, this study incorporates external stakeholder behavioral risk factors such as governments, immigrants, and environmental departments into the evaluation index system, such as: policy changes and adjustments and non-cooperation with demolition and relocation. The conclusion of this study shows that governments and immigrants can have an impact on the construction process of hydropower projects, and measures need to be taken to prevent this. Thus, the behavioral risk evaluation index system is constructed from the internal and external stakeholders involved in hydropower projects, which improves the behavioral risk indicators related to project construction and lays the foundation for a comprehensive analysis of the behavioral risk factors of hydropower project stakeholders.

Existing studies use engineering project risk management theory, which can effectively identify and quantify risk factors [22,28] but do not integrate multiple risk factors into risk networks to analyze their transmission paths. Therefore, this study combines social network analysis theory with engineering project risk management theory, integrates each behavioral risk factor into social network analysis, and constructs a stakeholder behavioral risk transmission network evaluation model. Meanwhile, some scholars also explain the importance and necessity of the behavioral risk of engineering project stakeholders and believe that behavioral risk has an influence on engineering project construction [19,25], without considering the influence relationship between behavioral risks. Based on this, this study uses the SNA method to analyze the impact of the overall relationship on the construction of hydropower projects and how to transfer the risk relationship. Targeted measures are taken to cut off the spread of core risks and key relationships. This highlights the ability to visualize risk network analysis among behavioral risk factors and enables hydropower project managers to clearly understand the core behavioral risk factors, key relationships, and risk transmission paths during the project construction.

This study proposes a conduction network model based on the SNA method to analyze risk relationships and control risks. Compared with traditional project risk analysis, this model breaks the limitations of the traditional project risk analysis framework and provides a risk management scheme for the construction phase of hydropower projects. It has the following practical risk management significance: (1) Dynamically understanding the impact between the behavioral risk factors of stakeholders in the construction process through the risk transmission path is conducive to taking measures to prevent the occurrence of risks and reduce the losses caused by risk events. (2) Provide project managers with new ideas for risk management, which should not only consider objective factors such as technology, investment, and safety management but also focus on the behavioral risks of stakeholders to promote the sustainable and healthy development of hydropower projects.

6. Conclusions

This study constructs and improves the stakeholder behavior risk evaluation system to study and evaluate the stakeholder behavior risk in its construction process. The stakeholder behavior risk network of hydropower projects is constructed through the SNA method, and the Chongqing JL hydropower project is selected as the research object to verify the rationality and feasibility of the evaluation model. To provide risk management experience for the construction of other hydropower projects in the future, the following conclusions are drawn from its study.

(1) The risk management of hydropower projects should not only pay attention to the behavioral risk factors of internal stakeholders but also pay more attention to the behavioral risk factors of external stakeholders.

(2) Attention should also be paid to the key relationships for the risk management of hydropower projects. According to the characteristics of the sending and receiving relationship of behavioral risk factors, specific countermeasures are proposed to block the transmission of core behavioral risks and cut off the transmission of key risk relationships.

(3) The methodology we put forward in this manuscript was an effective way to reduce the risks of hydropower projects management.

In this study, only the presence or absence of influence relationships between risk factors are considered when constructing the behavioral risk network of stakeholders in hydropower projects. However, there are strong and weak influence relationships between risk factors, and risk factors are induced only when the influence level exceeds a certain value. Therefore, this is the shortcoming of this study in the evaluation of behavioral risk networks, and it needs to be continued in the future.

Author Contributions: Study conception and design: M.A. and H.A.; analysis and interpretation of the results: M.A. and W.X.; draft manuscript preparation: W.X. and J.H.; writing—review & editing, supervision: H.A. All authors have read and agreed to the published version of the manuscript.

Funding: This research was funded by the National Science Foundation of China Youth Science Foundation Project (Grant NO. 72004116) "Precise Quantification of Water Pollution Loads and Strategic Optimization of Emission Permit Allocation in Three Gorges Reservoir Area under the Constraints of Economic and Environmental Regulations" and the Hubei Province Humanities and Social Sciences Key Research Base Project (Grant NO. 2021-SDSG-05) "An Empirical Study on the Evaluation of Urban–Rural Integration Development and Spatial Pattern Evolution in Chongqing Western Region".

Data Availability Statement: All of the data, models, and code generated or used during the study appear in the submitted article. Some or all of the data, models, or code that support the findings of this study are available from the corresponding author upon reasonable request. Informed consent was obtained from all subjects involved in the study.

Conflicts of Interest: The authors declare no conflict of interest.

Appendix A. Questionnaire on the Influence between Behavioral Risk Factors of Stakeholders in Hydropower Projects (Taking the Owner as an Example)

Dear Experts,

Hello! Thank you very much for taking time out of your busy schedule to do this questionnaire, we promise: the data obtained from this questionnaire will only be used for academic research, and absolutely no information provided by you will be disclosed without your permission.

In recent years, the state has advocated for the use of clean energy, and water resources power development as clean energy is vigorously promoted by the state; water resources power development needs to strengthen the infrastructure construction, such as: hydropower station construction, etc. When carrying out the infrastructure construction of large hydropower projects, involving the coordination of the interests of multiple participants, each participant takes some actions to maximize their own interests, their behavior may have an impact on other participants, and there are difficulties in achieving the quality, cost, schedule, and other goals of large hydropower projects.

The purpose of this research is to investigate the impact of risk factors on the behavior of stakeholders in large hydropower projects during the project construction, to analyze and identify key stakeholders and key behavior factors based on the survey data, and to provide help for the management of large hydropower projects during the project construction. We hope you can fill out the questionnaire truthfully and objectively, and we sincerely thank you for your cooperation.

Instructions for filling out:

1. The following content does not all need to be filled in; each survey object only needs to fill in a stakeholder behavior risk factors impact, such as: you in this project (mainly to the owners involved) only need to fill in Table 1.

2. Letter and number combinations appear in the questionnaire to explain; for example, S1R1 indicates that the stakeholder is the owner, and its corresponding behavioral risk factor is the difficulty of the payment of funds.

3. Explanation of the influence relationship between rows and columns of the table: the occurrence of behavioral risk factor S1R1 (column risk) can directly lead to (cause) the occurrence of or increase in S1R2 (row risk).

4. Please carefully judge one by one whether the following matrix of column risk factors has an impact on the row risk factors; if you think it will have a direct impact, click in the corresponding position of "□" to select, and click again to cancel the selection.

Influencing Factors \ Influenced Factors	S1R1 Difficulty in paying funds	S1R2 Proactive change request	S1R3 Untimely compensation for immigrant	S2R4 Design Changes	S2R5 Poor immigration planning and design	S2R6 The design does not consider ecological protection	S3R7 Project not delivered on schedule	S3R8 Construction costs exceeded budget
S1R1 Difficulty in paying funds	\	□	□	□	□	□	□	□
S1R2 Proactive change request	□	\	□	□	□	□	□	□
S1R3 Untimely compensation for immigrant	□	□	\	□	□	□	□	□

Influencing Factors	S3R9 Construction quality not up to standard	S3R10 Inappropriate construction safety measures	S3R11 Poor awareness of environmental protection	S4R12 Lack of timely implementation monitoring	S4R13 Supervision process is not standardized	S5R14 Material and equipment quality is not up to standard	S5R15 Lack of timely supply of material and equipment	S5R16 Market price adjustment
S1R1 Difficulty in paying funds	□	□	□	□	□	□	□	□
S1R2 Proactive change request	□	□	□	□	□	□	□	□
S1R3 Untimely compensation for immigrant	□	□	□	□	□	□	□	□

Influencing Factors	S6R17 Policy changes in adjustment	S6R18 Call off construction	S6R19 Poor coordination skills	S7R20 Creating Public Opinion	S7R21 Disrupting the construction site	S7R22 Not cooperating with demolition	S8R23 Not strictly enforcing environmental standards	S8R24 Request for additional environmentally friendly structures
S1R1 Difficulty in paying funds	□	□	□	□	□	□	□	□
S1R2 Proactive change request	□	□	□	□	□	□	□	□
S1R3 Untimely compensation for immigrant	□	□	□	□	□	□	□	□

S1—Owners; S2—Designer; S3—Contractors; S4—Supervisor; S5—Material and Equipment Suppliers; S6—Government; S7—Immigrant; S8—Environmental Departments.

Appendix B. Adjacency Matrix of Influence Relations of the Risk Network of Stakeholder Behavior in Hydropower Projects

	S1R1	S1R2	S1R3	S2R4	S2R5	S2R6	S3R7	S3R8	S3R9	S3R10	S3R11	S4R12	S4R13	S5R14	S5R15	S5R16	S6R17	S6R18	S6R19	S7R20	S7R21	S7R22	S8R23	S8R24
S1R1	0	0	1	0	0	0	1	0	0	0	0	0	0	0	1	0	0	0	0	0	1	1	0	0
S1R2	1	0	0	1	0	0	1	1	0	0	0	0	0	0	0	0	0	0	0	0	0	0	0	0
S1R3	0	1	0	0	0	0	1	1	0	0	0	0	0	0	0	0	1	1	0	1	1	1	0	0
S2R4	1	0	0	0	1	1	1	1	0	0	0	0	0	0	0	0	0	0	0	0	0	0	0	0
S2R5	0	1	0	1	0	0	1	0	0	0	0	0	0	0	0	0	0	1	0	0	1	1	0	0
S2R6	0	1	0	1	0	0	1	0	1	0	0	0	0	0	0	0	0	0	0	1	1	1	0	1
S3R7	0	0	0	0	0	0	0	1	1	0	0	0	0	0	0	1	0	0	0	1	0	0	0	0
S3R8	1	0	0	0	0	0	1	0	1	0	0	0	0	1	1	0	0	0	0	0	0	0	0	0
S3R9	0	0	0	0	0	0	1	1	0	0	0	0	0	0	0	0	0	1	0	1	0	0	0	0
S3R10	0	1	0	1	0	0	1	1	1	0	0	0	0	0	0	0	1	1	0	1	1	0	0	1
S3R11	0	0	0	1	0	0	1	0	1	0	0	0	0	0	0	0	1	0	0	1	0	0	0	1
S4R12	0	1	1	0	1	1	1	1	1	1	1	0	1	1	1	0	1	1	0	1	1	0	1	0
S4R13	0	1	1	0	0	0	1	1	1	1	1	1	0	1	1	0	1	1	0	0	1	0	1	0
S5R14	0	1	0	0	0	0	1	0	1	0	0	0	0	0	1	0	0	0	0	0	0	0	0	0
S5R15	0	0	0	1	0	0	1	1	0	0	0	0	0	0	0	1	1	0	0	0	0	0	0	0
S5R16	1	0	0	1	0	0	1	1	1	0	0	0	0	1	1	0	1	0	0	1	0	0	0	0
S6R17	1	0	0	0	1	0	1	1	0	0	0	0	0	0	0	1	0	1	0	0	1	1	0	1
S6R18	0	1	0	1	0	0	1	0	0	0	0	1	0	0	0	0	1	0	0	1	1	0	0	0
S6R19	1	0	1	0	1	0	1	1	0	0	0	0	0	0	0	1	1	0	0	1	1	1	0	0
S7R20	0	1	0	1	0	0	1	0	0	0	0	0	0	0	0	0	1	1	0	0	1	1	0	1
S7R21	0	0	1	0	0	0	1	0	0	0	0	0	0	0	1	0	1	1	0	1	0	1	0	1
S7R22	0	1	0	1	0	0	1	0	0	0	0	0	0	0	0	0	1	1	1	1	1	0	0	1
S8R23	0	1	0	1	0	1	0	1	1	0	0	0	1	1	0	0	1	1	0	1	1	0	0	1
S8R24	1	1	0	1	0	0	1	1	0	0	0	0	0	0	1	0	0	1	0	0	0	0	0	0

References

1. Grubler, A.; Wilson, C.; Bento, N.; Boza-Kiss, B.; Krey, V.; McCollum, D.L.; Rao, N.D.; Riahi, K.; Rogelj, J.; De Stercke, S.; et al. A low energy demand scenario for meeting the 1.5 °C target and sustainable development goals without negative emission technologies. *Nat. Energy* **2018**, *3*, 515–527. [CrossRef]
2. Gielen, D.; Boshell, F.; Saygin, D.; Bazilian, M.D.; Wagner, N.; Gorini, R. The role of renewable energy in the global energy transformation. *Energy Strategy Rev.* **2019**, *24*, 38–50. [CrossRef]
3. Ometto, J.P.; Cimbleris, A.C.P.; dos Santos, M.A.; Rosa, L.P.; Abe, D.; Tundisi, J.G.; Stech, J.L.; Barros, N.; Roland, F. Carbon emission as a function of energy generation in hydroelectric reservoirs in Brazilian dry tropical biome. *Energy Policy* **2013**, *58*, 109–116. [CrossRef]
4. Chang, X.; Liu, X.; Zhou, W. Hydropower in China at present and its further development. *Energy* **2010**, *35*, 4400–4406. [CrossRef]
5. Ma, L.; Wang, L.; Wu, K.-J.; Tseng, M.-L.; Chiu, A.S.F. Exploring the Decisive Risks of Green Development Projects by Adopting Social Network Analysis under Stakeholder Theory. *Sustainability* **2018**, *10*, 2104. [CrossRef]
6. Siraj, N.B.; Fayek, A.R. Risk Identification and Common Risks in Construction: Literature Review and Content Analysis. *J. Constr. Eng. Manag.* **2019**, *145*, 03119004. [CrossRef]
7. Ashkanani, S.; Franzoi, R. Gaps in megaproject management system literature: A systematic overview. *Eng. Constr. Arch. Manag.* **2022**. *ahead of print*. [CrossRef]
8. Liang, X.; Yu, T.; Guo, L. Understanding Stakeholders' Influence on Project Success with a New SNA Method: A Case Study of the Green Retrofit in China. *Sustainability* **2017**, *9*, 1927. [CrossRef]
9. Aladag, H.; Isik, Z. The Effect of Stakeholder-Associated Risks in Mega-Engineering Projects: A Case Study of a PPP Airport Project. *IEEE Trans. Eng. Manag.* **2020**, *67*, 174–186. [CrossRef]
10. Wang, Q.; Zuo, W.; Li, Q. Engineering Harmony under Multi-Constraint Objectives: The Perspective of Meta-Analysis. *J. Civ. Eng. Manag.* **2020**, *26*, 131–146. [CrossRef]
11. Wu, D.; Lambert, J.H. Engineering Systems and Risk Analytics. *Risk Anal.* **2020**, *40*, 1–7. [CrossRef] [PubMed]
12. Ma, L.; Zhang, X.; Wang, H.; Qi, C. Characteristics and Practices of Ecological Flow in Rivers with Flow Reductions Due to Water Storage and Hydropower Projects in China. *Water* **2018**, *10*, 1091. [CrossRef]
13. Yang, M.; Chen, H.; Xu, Y.; Gong, W. Stakeholder-Associated Risks and Their Interactions in PPP Projects: Social Network Analysis of a Water Purification and Sewage Treatment Project in China. *Adv. Civ. Eng.* **2020**, *2020*, 8897196. [CrossRef]
14. Keshk, A.M.; Maarouf, I.; Annany, Y. Special studies in management of construction project risks, risk concept, plan building, risk quantitative and qualitative analysis, risk response strategies. *Alex. Eng. J.* **2018**, *57*, 3179–3187. [CrossRef]
15. Balta, S.; Birgonul, M.T.; Dikmen, I. Buffer Sizing Model Incorporating Fuzzy Risk Assessment: Case Study on Concrete Gravity Dam and Hydroelectric Power Plant Projects. *ASCE-ASME J. Risk Uncertain. Eng. Syst. Part A Civ. Eng.* **2018**, *4*, 04017039. [CrossRef]
16. Pehlivan, S.; Oztemir, A.E. Integrated Risk of Progress-Based Costs and Schedule Delays in Construction Projects. *Eng. Manag. J.* **2018**, *30*, 108–116. [CrossRef]
17. Urgilés, P.; Claver, J.; Sebastián, M.A. Analysis of the Earned Value Management and Earned Schedule Techniques in Complex Hydroelectric Power Production Projects: Cost and Time Forecast. *Complexity* **2019**, *2019*, 3190830. [CrossRef]
18. Su, H.; Yang, M.; Kang, Y. Comprehensive Evaluation Model of Debris Flow Risk in Hydropower Projects. *Water Resour. Manag.* **2015**, *30*, 1151–1163. [CrossRef]
19. Gao, J.; Ren, H.; Cai, W. Risk assessment of construction projects in China under traditional and industrial production modes. *Eng. Constr. Archit. Manag.* **2019**, *26*, 2147–2168. [CrossRef]

20. Zheng, Q.; Shen, S.-L.; Zhou, A.; Lyu, H.-M. Inundation risk assessment based on G-DEMATEL-AHP and its application to Zhengzhou flooding disaster. *Sustain. Cities Soc.* **2022**, *86*, 104138. [CrossRef]
21. Zhou, J.-L.; Zhe-Hua, B.; Sun, Z.-Y. Safety Assessment of High-Risk Operations in Hydroelectric-Project Based on Accidents Analysis, SEM, and ANP. *Math. Probl. Eng.* **2013**, *2013*, 530198. [CrossRef]
22. Yu, Y.; Darko, A.; Chan, A.P.C.; Chen, C.; Bao, F. Evaluation and Ranking of Risk Factors in Transnational Public–Private Partnerships Projects: Case Study Based on the Intuitionistic Fuzzy Analytic Hierarchy Process. *J. Infrastruct. Syst.* **2018**, *24*, 04018028. [CrossRef]
23. Thevendran, V.; Mawdesley, M.J. Perception of human risk factors in construction projects: An exploratory study. *Int. J. Proj. Manag.* **2004**, *22*, 131–137. [CrossRef]
24. Chan, E.H.W.; Au, M.C.Y. Building contractors' behavioural pattern in pricing weather risks. *Int. J. Proj. Manag.* **2007**, *25*, 615–626. [CrossRef]
25. Shen, W.; Tang, W.; Yu, W.; Duffield, C.F.; Hui, F.K.P.; Wei, Y.; Fang, J. Causes of contractors' claims in international engineering-procurement-construction projects. *J. Civ. Eng. Manag.* **2017**, *23*, 727–739. [CrossRef]
26. Xiang, P.; Jia, F.; Li, X. Critical Behavioral Risk Factors among Principal Participants in the Chinese Construction Industry. *Sustainability* **2018**, *10*, 3158. [CrossRef]
27. Wang, T.; Gao, S.; Li, X.; Ning, X. A meta-network-based risk evaluation and control method for industrialized building construction projects. *J. Clean. Prod.* **2018**, *205*, 552–564. [CrossRef]
28. Wu, G.; Zuo, J.; Zhao, X. Incentive Model Based on Cooperative Relationship in Sustainable Construction Projects. *Sustainability* **2017**, *9*, 1191. [CrossRef]
29. Adeleke, A.Q.; Bahaudin, A.Y.; Kamaruddeen, A.M.; A Bamgbade, J.; Salimon, M.G.; Khan, M.W.A.; Sorooshian, S. The Influence of Organizational External Factors on Construction Risk Management among Nigerian Construction Companies. *Saf. Health Work* **2018**, *9*, 115–124. [CrossRef]
30. Feng, J.; Wang, Y.; Zhang, K. Evaluation of the Quality Supervision System for Construction Projects in China Considering the Quality Behavior Risk Transmission. *Symmetry* **2020**, *12*, 1660. [CrossRef]
31. Shin, N.; Yoo, J.-S.; Kwon, I.-W.G. Fostering Trust and Commitment in Complex Project Networks through Dedicated Investment in Partnership Management. *Sustainability* **2020**, *12*, 10397. [CrossRef]
32. Quatman, C.; Chelladurai, P. Social Network Theory and Analysis: A Complementary Lens for Inquiry. *J. Sport Manag.* **2008**, *22*, 338–360. [CrossRef]
33. Wrzus, C.; Hänel, M.; Wagner, J.; Neyer, F.J. Social network changes and life events across the life span: A meta-analysis. *Psychol. Bull.* **2013**, *139*, 53–80. [CrossRef]
34. Ghafouri, H.B.; Mohammadhassanzadeh, H.; Shokraneh, F.; Vakilian, M.; Farahmand, S. Social network analysis of Iranian researchers on emergency medicine: A sociogram analysis. *Emerg. Med. J.* **2014**, *31*, 619–624. [CrossRef]
35. Feng, P.; Sun, D.; Gong, Z. A Case Study of Pyramid Scheme Finance Flow Network Based on Social Network Analysis. *Sustainability* **2019**, *11*, 4370. [CrossRef]
36. Loosemore, M. Social network analysis: Using a quantitative tool within an interpretative context to explore the management of construction crises. *Eng. Constr. Arch. Manag.* **1998**, *5*, 315–326. [CrossRef]
37. Lin, S.-C. An Analysis for Construction Engineering Networks. *J. Constr. Eng. Manag.* **2015**, *141*, 04014096. [CrossRef]
38. Herrera, R.F.; Mourgues, C.; Alarcón, L.F.; Pellicer, E. Understanding Interactions between Design Team Members of Construction Projects Using Social Network Analysis. *J. Constr. Eng. Manag.* **2020**, *146*, 1–13. [CrossRef]
39. Tang, H.; Wang, G.; Miao, Y.; Zhang, P. Managing Cost-Based Risks in Construction Supply Chains: A Stakeholder-Based Dynamic Social Network Perspective. *Complexity* **2020**, *2020*, 8545839. [CrossRef]
40. Lu, W.; Xu, J.; Söderlund, J. Exploring the Effects of Building Information Modeling on Projects: Longitudinal Social Network Analysis. *J. Constr. Eng. Manag.* **2020**, *146*, 04020037. [CrossRef]
41. He, Z.; Huang, D.; Zhang, C.; Fang, J. Toward a Stakeholder Perspective on Social Stability Risk of Large Hydraulic Engineering Projects in China: A Social Network Analysis. *Sustainability* **2018**, *10*, 1223. [CrossRef]
42. Wang, Y.; Wang, Y.; Wu, X.; Li, J. Exploring the Risk Factors of Infrastructure PPP Projects for Sustainable Delivery: A Social Network Perspective. *Sustainability* **2020**, *12*, 4152. [CrossRef]
43. Lee, C.; Won, J.W.; Jang, W.; Jung, W.; Han, S.H.; Kwak, Y.H. Social conflict management framework for project viability: Case studies from Korean megaprojects. *Int. J. Proj. Manag.* **2017**, *35*, 1683–1696. [CrossRef]
44. Xia, N.; Zhong, R.; Wu, C.; Wang, X.; Wang, S. Assessment of Stakeholder-Related Risks in Construction Projects: Integrated Analyses of Risk Attributes and Stakeholder Influences. *J. Constr. Eng. Manag.* **2017**, *143*, 04017030. [CrossRef]
45. Zhang, S.; Pan, F.; Wang, C.; Sun, Y.; Wang, H. BIM-Based Collaboration Platform for the Management of EPC Projects in Hydropower Engineering. *J. Constr. Eng. Manag.* **2017**, *143*, 0001403. [CrossRef]
46. Ding, J.; Chen, C.; An, X.; Wang, N.; Zhai, W.; Jin, C. Study on Added-Value Sharing Ratio of Large EPC Hydropower Project Based on Target Cost Contract: A Perspective from China. *Sustainability* **2018**, *10*, 3362. [CrossRef]
47. Mok, K.Y.; Shen, G.Q.; Yang, R. Stakeholder complexity in large scale green building projects. *Eng. Constr. Arch. Manag.* **2018**, *25*, 1454–1474. [CrossRef]
48. Sadkowska, J. Difficulties in Building Relationships with External Stakeholders: A Family-Firm Perspective. *Sustainability* **2018**, *10*, 4557. [CrossRef]

49. Amadi, C.; Carrillo, P.; Tuuli, M. PPP projects: Improvements in stakeholder management. *Eng. Constr. Arch. Manag.* **2019**, *27*, 544–560. [CrossRef]
50. Daniel, E.I.; Pasquire, C.; Dickens, G. Development of Approach to Support Construction Stakeholders in Implementation of the Last Planner System. *J. Manag. Eng.* **2019**, *35*, 04019018. [CrossRef]
51. Luo, L.; Shen, Q.G.; Xu, G.; Liu, Y.; Wang, Y. Stakeholder-Associated Supply Chain Risks and Their Interactions in a Prefabricated Building Project in Hong Kong. *J. Manag. Eng.* **2019**, *35*, 04019018. [CrossRef]
52. Bahadorestani, A.; Karlsen, J.T.; Farimani, N.M. Novel Approach to Satisfying Stakeholders in Megaprojects: Balancing Mutual Values. *J. Manag. Eng.* **2020**, *36*, 04019047. [CrossRef]
53. He, Z.; Huang, D.; Fang, J.; Wang, B. Stakeholder Conflict Amplification of Large-Scale Engineering Projects in China: An Evolutionary Game Model on Complex Networks. *Complexity* **2020**, *2020*, 9243427. [CrossRef]
54. Nguyen, V.T.; Do, S.T.; Vo, N.M.; Nguyen, T.A.; Pham, S.V.H.; Vignali, V. An Analysis of Construction Failure Factors to Stakeholder Coordinating Performance in the Finishing Phase of High-Rise Building Projects. *Adv. Civ. Eng.* **2020**, *2020*, 6633958. [CrossRef]
55. Jia, L.; Qian, Q.K.; Meijer, F.; Visscher, H. Stakeholders' Risk Perception: A Perspective for Proactive Risk Management in Residential Building Energy Retrofits in China. *Sustainability* **2020**, *12*, 2832. [CrossRef]
56. Wen, S.; Qiang, G. Managing Stakeholder Concerns in Green Building Projects With a View Towards Achieving Social Sustainability: A Bayesian-Network Model. *Front. Environ. Sci.* **2022**, *10*, 290. [CrossRef]
57. Yang, R.J.; Zou, P.X.W. Stakeholder-associated risks and their interactions in complex green building projects: A social network model. *Build. Environ.* **2014**, *73*, 208–222. [CrossRef]
58. Wu, Z.; Nisar, T.; Kapletia, D.; Prabhakar, G. Risk factors for project success in the Chinese construction industry. *J. Manuf. Technol. Manag.* **2017**, *28*, 850–866. [CrossRef]
59. Xia, N.; Zou, P.X.W.; Liu, X.; Wang, X.; Zhu, R. A hybrid BN-HFACS model for predicting safety performance in construction projects. *Saf. Sci.* **2018**, *101*, 332–343. [CrossRef]
60. Darvishi, S.; Jozi, S.A.; Malmasi, S.; Rezaian, S. Environmental risk assessment of dams at constructional phase using VIKOR and EFMEA methods (Case study: Balarood Dam, Iran). *Hum. Ecol. Risk Assess. Int. J.* **2019**, *26*, 1087–1107. [CrossRef]
61. Barghi, B.; Sikari, S.S. Qualitative and quantitative project risk assessment using a hybrid PMBOK model developed under uncertainty conditions. *Heliyon* **2020**, *6*, e03097. [CrossRef] [PubMed]
62. Huang, N.; Bai, L.; Wang, H.; Du, Q.; Shao, L.; Li, J. Social Network Analysis of Factors Influencing Green Building Development in China. *Int. J. Environ. Res. Public Health* **2018**, *15*, 2684. [CrossRef] [PubMed]
63. Ganbat, T.; Chong, H.-Y.; Liao, P.-C.; Leroy, J. Identification of critical risks in international engineering procurement construction projects of Chinese contractors from the network perspective. *Can. J. Civ. Eng.* **2020**, *47*, 1359–1371. [CrossRef]
64. Zhou, J.-L.; Bai, Z.-H.; Sun, Z.-Y. A hybrid approach for safety assessment in high-risk hydropower-construction-project work systems. *Saf. Sci.* **2014**, *64*, 163–172. [CrossRef]
65. Agarwal, S.S.; Kansal, M.L. Risk based initial cost assessment while planning a hydropower project. *Energy Strat. Rev.* **2020**, *31*, 100517. [CrossRef]
66. Dixit, S.; Stefańska, A.; Musiuk, A.; Singh, P. Study of enabling factors affecting the adoption of ICT in the Indian built environment sector. *Ain Shams Eng. J.* **2021**, *12*, 2313–2319. [CrossRef]
67. Xie, L.; Han, T.; Skitmore, M. Governance of Relationship Risks in Megaprojects: A Social Network Analysis. *Adv. Civ. Eng.* **2019**, *2019*, 1426139. [CrossRef]
68. Yang, X.; Zhang, J.; Zhao, X. Factors Affecting Green Residential Building Development: Social Network Analysis. *Sustainability* **2018**, *10*, 1389. [CrossRef]
69. Li, C.Z.; Hong, J.; Xue, F.; Shen, G.Q.; Xu, X.; Mok, M.K. Schedule risks in prefabrication housing production in Hong Kong: A social network analysis. *J. Clean. Prod.* **2016**, *134*, 482–494. [CrossRef]
70. Valentin, V.; Naderpajouh, N.; Abraham, D.M. Integrating the Input of Stakeholders in Infrastructure Risk Assessment. *J. Manag. Eng.* **2018**, *34*, 04018042. [CrossRef]
71. Zhang, S.; Fang, Y. Research on Construction Project Organization Based on Social Network Analysis. *Wirel. Pers. Commun.* **2018**, *102*, 1867–1877. [CrossRef]
72. Yuan, M.; Li, Z.; Li, X.; Luo, X. Managing stakeholder-associated risks and their interactions in the life cycle of prefabricated building projects: A social network analysis approach. *J. Clean. Prod.* **2021**, *323*, 129102. [CrossRef]

 *sustainability*

Article

Risk Prioritization in a Natural Gas Compressor Station Construction Project Using the Analytical Hierarchy Process

Georgios K. Koulinas [1,*], **Olympia E. Demesouka** [1], **Gerasimos G. Bougelis** [2,3] and **Dimitrios E. Koulouriotis** [1]

[1] Department of Production and Management Engineering, Democritus University of Thrace, Vas. Sofias 12 St., 67132 Xanthi, Greece

[2] Department of Engineering Project Management, Faculty of Science and Technology, Hellenic Open University, Parodos Aristotelous 18 St., 26335 Patra, Greece

[3] TERNA S.A., 85, Mesogeion Ave., 11526 Athens, Greece

* Correspondence: gkoulina@pme.duth.gr

Abstract: Recently, the seamless construction and operation of natural gas pipelines has become even more critical, while the oil and gas industry's capability to operate effectively with acceptable risks and hazardous situations is mainly dependent on safety. As a result, it is very important to have a wide knowledge of effective management tactics for enhancing implementation of safety regulations and procedures. The problem of assuring workers' health and safety in the workplace is a crucial component in the endeavor to raise the productivity of labor and the level of competitiveness of building projects. To promote the health, safety, and well-being of workers, issues that are embedded within the concept of sustainability, we propose in this study a safety risk-assessment process that uses the analytical hierarchy process for assigning priorities to risks on construction worksites. This process uses a popular multicriteria method. The success of this strategy was shown by its application to the building of a natural gas compressor plant in Greece. The main contribution of this study is the application of a well-known multicriteria method for assessing risks in a natural gas compressor station construction project and prioritizing hazards to allocate budget for risk-mitigation measures.

Keywords: multicriteria analysis; risk management; assessment; natural gas pipeline

Citation: Koulinas, G.K.; Demesouka, O.E.; Bougelis, G.G.; Koulouriotis, D.E. Risk Prioritization in a Natural Gas Compressor Station Construction Project Using the Analytical Hierarchy Process. *Sustainability* **2022**, *14*, 13172. https://doi.org/10.3390/su142013172

Academic Editors: Albert P. C. Chan, Srinath Perera, Xiaohua Jin, Dilanthi Amaratunga, Makarand Hastak, Patrizia Lombardi, Sepani Senaratne and Anil Sawhney

Received: 5 September 2022
Accepted: 9 October 2022
Published: 14 October 2022

Publisher's Note: MDPI stays neutral with regard to jurisdictional claims in published maps and institutional affiliations.

1. Introduction

The most environmentally friendly form of hydrocarbon combustion is natural gas. It is available in large quantities, has a wide range of applications, helps fulfill the rising need for energy worldwide, and can work in tandem with other forms of renewable energy. In addition, natural gas is an essential resource for such industries as heating and power generation, manufacturing, and transportation, not only in Europe but all around the world.

Even though the combustion of natural gas releases greenhouse gases, it produces a far lower amount of carbon dioxide (CO_2) and other air pollutants than the vast majority of the fuels it is gradually replacing, particularly coal. The use of natural gas has increased dramatically over the last decade, making up over a third of the increase in total energy demand. This is higher than any other fossil fuel.

Natural gas is currently responsible for around a quarter of the world's electrical generation. It is anticipated that it will play a significant part in easing the transition to energy systems that produce zero net emissions over the medium range; however, its utility over the longer term remains unknown in a future where renewable energy sources predominate. Recently, natural gas pipelines have become vital for the functioning of every country, since natural gas that is transported via pipelines contributes to the economic expansion of cities and industries. It is anticipated that natural gas pipelines will continue to be significant for the global economy because they have the ability to transport hydrogen that is created from natural gas or electrolysis and has the potential to be a game changer in the transition to a cleaner source of energy [1]. Additionally, after goods derived from petroleum, natural gas is the principal energy resource in the Euro area while it is considered as the most significant energy source for the manufacturing industry.

During the last few decades, many studies contributed to the relative scientific field of risk management in natural gas infrastructures. Simonoff et al. [2] developed risk measures and scenarios to better understand how consequences of pipeline failures are linked to causes and other incident characteristics, and [3] proposed a model for quantitative risk assessment on metering stations and metering-regulation stations for natural gas with natural ventilation. In addition, ref. [4] described an application of a methodology for quantitative risk assessment that considered failure frequencies found in a public database, and consequences were computed as a function of pipe diameter and operating pressure for each network's section. Also, ref. [5] performed job-hazard assessment to predict hazards while executing nonroutine tasks in gas transmission stations, while [6] developed a model for accident classification in the natural gas sector according to possible fatalities, using rough set theory and decision rules. Recently, ref. [7] assessed the safety state of oil and gas activities and identified risk factors that cause hazards to people and to the environment using formal risk assessment and Bayesian networks.

Additionally, some studies used multicriteria decision analysis methods for oil and gas industry applications, such as [8], who presented a new methodology for identifying and assessing risks simultaneously by applying a multiattribute group decision-making technique. In the study of [9], the researchers proposed an approach for pipeline route selection based on SWOT analysis and the Delphi method for determining decision-makers' beliefs, and then the PROMETHEE model was used to integrate these beliefs with subjective judgments and identify the suitable pipeline route. Paradopoulou and Antoniou [10] performed REGIME multicriteria decision analysis to prioritize alternative LNG terminal locations on the island of Cyprus in the Mediterranean Sea, while Strantzali et al. [11] proposed a decision-support tool that embodies multicriteria analysis, using the PROMETHEE II method, for the evaluation of potential LNG export terminals in Greece. A comprehensive literature review and a framework for classification of decision-support methods used for technical, economic, social and environmental assessments within different energy sectors including upstream oil and gas, refining and distribution can be found in the study of [12].

In the recent studies of Marhavilas et al. [13–16], a combination of both the typical and fuzzy AHP and HAZOP method used for risk assessments in the sour crude oil industry.

The applications of AHP and fuzzy AHP in the health and safety research field include a wide range. In their study, [17] used AHP for measuring health and safety awareness in selecting a maintenance strategy within the Norwegian oil and gas industry, while [18] used a fuzzy extension of AHP with trapezoidal fuzzy numbers for safety evaluations in hot and humid workplaces. Additionally, [19] presented a framework for safety risk assessments in construction projects that was based on the cost of a safety model and the analytic hierarchy process, and [20] developed a methodology for safety device selection that used AHP and mechanical hazard classification. Podgórski [21] used typical AHP for evaluating how workplace safety and health management systems are working. In addition, [22] applied nonlinear fuzzy analytic hierarchy process and logarithmic fuzzy preference programming for performing safety evaluations within coal mines in China, and Xie et al. [23] developed a technique for evaluating the environmental quality of two commercial buildings. Janackovic et al. [24] ranked and selected occupational safety indicators using fuzzy AHP, and Kasap and Subasi [25] employed fuzzy AHP to quantify occupational risk in open pit mining. Additionally, Carpitella et al. [26] optimized system maintenance by combining reliability analysis with multicriteria techniques like fuzzy TOPSIS and AHP. Recently, [27] applied a combination of the Pythagorean fuzzy AHP and VIKOR method for health and safety risk assessment in dangerous workplaces, while Koulinas et al., [28] and Marhavilas et al., [29] used fuzzy AHP and real data to perform risk assessments in construction projects.

The literature review above is summarized in Table 1.

Table 1. Summary of the relative literature.

Reference	Natural Gas Infrastructure Risk Management				
	Quantitative Method	Qualitative Method			
Simonoff et al., 2010 [2]		X			
Bajcar et al., 2014 [3]	X				
Vianello and Maschio, 2014 [4]	X				
Li et al., 2016 [5]	X				
Cinelli et al., 2019 [6]	X	X			
Mrozowska, 2021 [7]	X	X			
Multicriteria methods for oil and gas industry					
	Quantitative method	Qualitative method	PROMETEE, PROMETEE II	REGIME	Group decision-making
Mojtahedi et al., 2010 [8]	X	X			X
Tavana et al., 2013 [9]	X	X	X		
Papadopoulou and Antoniou, 2014 [10]	X			X	
Strantzali et al., 2019 [11]	X		X		
AHP and FAHP applications for health and safety research field					
	AHP	FAHP			
Chandima Ratnayake and Markeset, 2010 [17]	X				
Zheng et al., 2012 [18]		X			
Aminbakhsh et al., 2013 [19]	X				
Caputo et al., 2013 [20]	X				
Podgórski, 2015 [21]	X				
Wang et al., 2016 [22]		X			
Xie et al., 2017 [23]	X				
Janackovic et al., 2017 [24]		X			
Kasap and Subasi, 2017 [25]		X			
Carpitella et al., 2018 [26]		X			
Gul, 2020 [27]		X			
Koulinas et al., 2019 [28]		X			
Marhavilas, Tegas, et al., 2020 [29]		X			

A great survey on risk analysis and assessment methodologies in the workplace can be found in the study of [30]. In addition, [31] provided a systematic literature review on the use of risk-acceptance criteria in occupational health and safety risk assessment.

The present approach intends to serve as a practical tool for knowledge and expertise transfer. The remaining five sections of the paper are: describing the analytical hierarchy process, presenting the compressor station, explaining the suggested framework, describing the application, and discussing the findings.

2. The Concept of the Analytical Hierarchy Process

The analytical hierarchy process (AHP) proposed by Saaty [32] is a well-known approach for evaluating many criteria in which the factors at hand are arranged in a hierarchical manner. It is founded not just on mathematics but also on human psychology, fusing together rational thought with emotional inclination. The ability to incorporate qualitative and quantitative criteria during the evaluation is one of the benefits of using this method. Another benefit is the ability to use the experience, knowledge, and intuition of the person making the decision when determining the weights of the elements. On the other hand, the subjective character of the modeling process is the fundamental flaw of this approach and, more generally, of similar multicriteria methods. This implies that the methodology cannot ensure that the judgments will be absolutely accurate.

The AHP approach allows for the multicriteria problem to be organized into a hierarchical structure. Following this, the local and global priorities for the problem's criteria and subcriteria may be defined using pairwise comparisons and weightings. During the process of conducting the pairwise comparisons, the AHP takes the judgments of the decision-maker regarding how important one criterion is in comparison to another as its input. As an output, the AHP generates a ranking according to the importance of each criterion and/or subcriterion of the analysis. A standard scale (Table 2) is used in order to convert the qualitative estimates of importance that the decision-maker has into numerical values.

Table 2. The basic scale of the AHP method [33].

Description	Level of Importance
Two factors are equally important	1
Factor *i* is moderately more important than factor *j*	3
Factor *i* is strongly more important than factor *j*	5
Factor *i* is very strongly more important than factor *j*	7
Factor *i* is extremely more important than factor *j*	9
Intermediate values	2, 4, 6, and 8

The fact that the approach examines the input judgments of the decision-maker for any possible instances of inconsistency is a feature that is highly significant to the method. The latter leads in an improvement in overall quality. In this particular study, we employ standard AHP in order to rate the risks identified for every task of the project.

3. Description of a Natural Gas Compressor Station

A compressor station is an essential component of a natural gas pipeline network, which transports natural gas from specific producing sites to the end customers. In this paper, the case of the Kipi Compressor Station of the Trans Adriatic Pipeline (TAP) is studied [34]. Distance, friction, and elevation variances inhibit the flow of natural gas via a pipeline and lower pressure. The compressor stations are ideally located throughout the collection and transportation pipeline network to assist maintain gas flow rate to the clients. Because the gas has a tendency to slow down as it passes through the pipeline network, engineers build compressor stations along the pipeline to maintain the gas flowing toward its destination.

During times of low demand, compressor stations are also able to deliver natural gas to storage sites in the surrounding region. In addition, the passage of the gas through the pipeline results in the formation of water droplets and various types of hydrocarbons inside the gas itself. Scrubbers, strainers, and filters are used in compressor stations to remove dirt and other contaminants from the flow of gas, in addition to separating the aforementioned objects.

3.1. The Natural Gas Compression Process

As described in Figure 1, and in [34], initially, the gas enters the station through the yard piping, which is the term given to the network of pipes that link the main gas pipeline to the compressor station.

The gas is routed through a number of filters and scrubbers in the yard by means of pipes, which eliminates any liquid or solid pollutants that may be present in the gas stream. After that, it goes back into the pipe at the compression station yard and enters a compressor unit. The compressor works to repressurize the gas so that it will flow steadily through the primary natural gas pipeline network. However, the process of increasing the pressure of the gas results in the generation of heat, which needs to be controlled. As a solution to this issue, the compressor station is equipped with a cooling system that is meant to remove the additional heat. This is often accomplished by employing a number of fans to assist in chilling the pipes as they reflect the heat away. Because of this cooling process, which also involves shifts in pressure and temperature, part of the liquid that was present in the gas condenses and separates itself from the primary flow of gas.

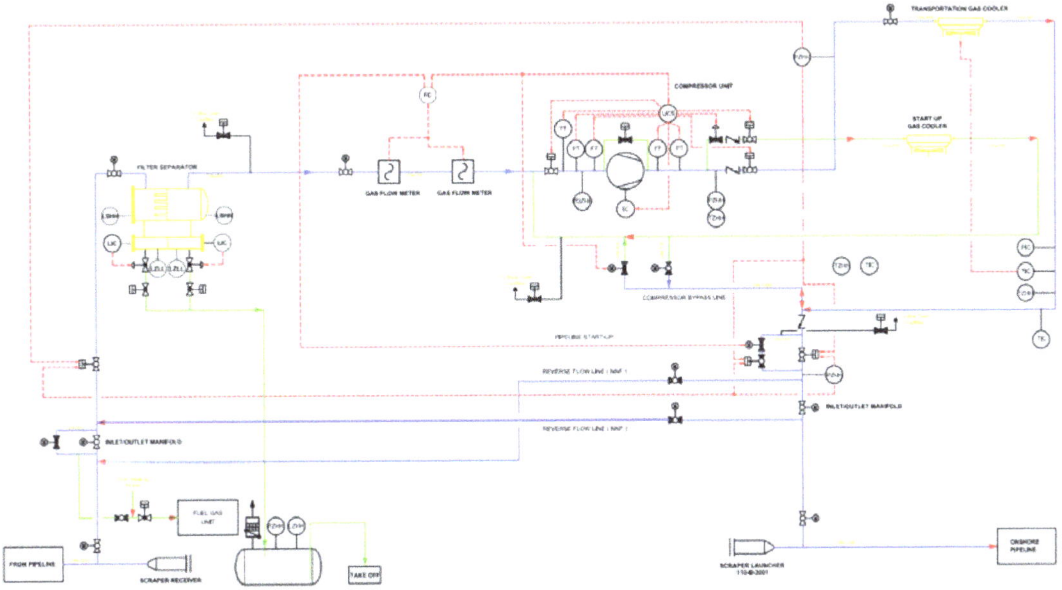

Figure 1. The flowchart of the natural gas compression process.

Other operations, such as the addition of mercaptan, the smell of which is sulfurous and indicates the existence of natural gas, may be a part of the process at the gas compressor station once the pressure of the gas has been reestablished.

A comprehensive system monitoring, gas pressure monitoring, and safety control apparatus are some of the other components that are often present in a gas compressor station. In the event that there is a disruption in the power supply, backup generators are an important component that plays a role in helping to maintain the natural gas pipeline running continuously and evenly.

Given that the compressor station is designed to filter, meter and compress natural gas for further transportation through the pipeline network, it mainly consists of the following:

- gas turbine compressors
- gas coolers, filtering, metering and piping Systems
- utilities (e.g., fuel gas, instrument air)
- electrical equipment
- I&C equipment
- civil structures
- one vent stack for station/piping depressurization

The gas is brought in by a scraper reception facility, which serves as the operational interface between the pipeline and the station. The station is where the gas is compressed. The natural gas that is being transported via the gas transmission pipeline is brought into the compressor station after it has gone through the scraper reception facilities. Before the gas can enter the metering and compression units, it must first be passed through an intake separator, which removes any solid particles and free water that may be present in the gas stream.

Two distinct steps of separation will make the separation process simpler. The droplets in the gas stream are subjected to gravity and/or centrifugal forces during the initial step of the separation process. After this initial stage, there is a second stage that is comprised of cartridges that use coalescing effects in order to produce liquid droplets that are of a larger size. Last but not least, the gas stream progresses via a demisting and vane step, which gets rid of the bigger droplets. The contaminants and liquids that have been separated will be collected in a sump located below the separator in the form of a horizontal pipe.

For reasons of custody, it is necessary to measure the volume of natural gas that is delivered to the gas transmission system located farther downstream. This will be accomplished by the utilization of ultrasonic flow meters (USM). In addition to this, the measured amount of gas flow will be utilized in the process of controlling the performance of the compressor.

A gas analyzing unit will perform an examination for the purpose of custody to determine the quality of the natural gas that is being transported from the upstream gas transmission system (GAU). The measurement will be carried out mechanically, either constantly or discontinuously, depending on how the relevant network code specifies it should be done.

The process gas chromatograph (PGC) is the primary component of the gas analyzing unit (GAU) system, which is designed to analyze at least the following parameters:

- C1 to C6 and CO_2 concentration
- hydrocarbon dew point
- water dew point
- sulfur concentration
- oxygen concentration

Because this is the primary gas entry point to the pipeline, the gas will be analyzed in more depth than it will be at the intermediate stations, which will merely monitor the concentration of C1 to C6 hydrocarbons and the hydrocarbon dew point. The numbers needed for the flow calculation, such as density and compressibility factor, are computed based on the results of measuring the composition of the gas. In addition, this composition provides the information necessary to construct indices such as the Net and Gross calorific value, as well as the Wobbe index. Additional quality-control methods are used for the purpose of monitoring the gas when it is introduced into the pipeline system.

The gas will enter the gas compression units once it has completed its journey via the gas metering unit. Depending on the capacity of the station, gas turbine-driven turbo compressors are anticipated to be utilized for the purpose of compressing the gas. The compressors are set up in a parallel configuration. Each compressor unit is built with unit shutoff valves, which may be used to separate the compressor unit on either the suction or discharge side. When a gas first enters the suction of a compressor, it is sent via a suction strainer on its way to the suction line. This serves as a protective measure against the formation of bigger deposits in the suction line. A flow meter is utilized on the gas supply before it is allowed to enter the compressor proper. After that, the gas will be compressed by a turbo compressor that has three different rotors, or impellers. A gas turbine will serve as the source of propulsion for the turbo compressor.

After exiting the compressor at the specified pressure, the gas then travels to the discharge header, where it is directed through the discharge check valve and the unit shutoff valve en route.

In the event of low flow, turbo compressors are prone to surging, which has the potential to cause the machine's destruction. A short recycling with a hot bypass valve (HBV) and a longer cooled recycle with an antisurge control valve (ASV) are both designed and put into the system so as to prevent surge operation from occurring. Controlling the machine at low flows is the antisurge valve (ASV), which prevents the machine from running too closely to the surge area. In the event that the antisurge valve does not respond quickly enough to rapid transients in the process, the hot bypass valve (HBV) will open entirely, which will cause the machine to trip. At each machine, the necessary process parameters are monitored. These include flow, suction and discharge pressure, and temperature. It is important to keep in mind that the antisurge cycle, namely, the cooler, is intended to be used with a single compressor unit. On the other hand, in the event that the units have to be run with a low flow, the station recycling valve will be used to accomplish this.

The same lines that are utilized for recycling will also be used for the purposes of starting up. The presence of these separate lines makes it possible for the compressor to begin functioning, even when other equipment is already in use. This starting line will be sent to the beginning of the startup header. The gas that will be used in the gas turbine will come from the suction header of the compressor. Nevertheless, in conformity with EN 12583, this line also features a separate shutoff valve that may be utilized if necessary. The fuel gas is being supplied by the fuel gas unit, and it will then be sent further to the gas turbine through a direct channel.

Another gas line is run all the way to the compressor seal gas panel from the side of the compressor that discharges the gas. This is necessary because the compressor needs a steady gas flow to the dry gas seals in order to function properly. Additionally, this gas flow is necessary even while the devices are in their pressurized stop position. As a result, it will be obtained from a position that is not directly associated with the shutoff valves that control the compressor unit. The gas will be extracted from the discharge side because the pressure has to be slightly greater than the suction or settle-out pressure. Purging of the tandem dry gas seal will be accomplished with the usage of the gas (primary and secondary seals). Air will be used to clean the tertiary compressor seal once it has been purged. Because there is always some quantity of seal gas that enters the process lines via the machine, the lines need to be depressurized during prolonged standstills (for example, to the suction line) in order to guarantee that there is adequate driving force for the flow of seal gas. It is impossible to prevent some of the seal gas from escaping through the vent lines of the dry gas seals, hence this is an inevitable aspect of the sealing system. In order to prevent the release of greenhouse gases, any air that escapes via the primary vent line (the connection between the primary and secondary seals) will be burnt in the boiler unit.

The gas will then be sent to a gas chiller when it has completed its journey through the compression unit. This cooler is necessary because a maximum temperature of 50 degrees Celsius must be maintained for the gas that is directed toward the pipeline. The cooler, also known as the transportation cooler, has a total of five compartments. Out of these five bays, four bays are required for duty, while the remaining bay serves as a standby bay. Each bay has the necessary number of one-pass heat exchanger bundles as well as two fans. A temperature measurement device located in the discharge header of the cooler is used, in conjunction with variable speed drives for the fans' motors, to maintain a consistent temperature at the cooler's output. In the event that the output temperature cannot be attained for whatever reason, the flow originating from the compressors will be lowered in the appropriate proportion. In the event that this preventive precaution is not enough, the compressors will be turned off. After that, the gas is transferred to the pipeline system using devices known as scraper launchers. Due to the fact that the design of the station is somewhat elevated above the design of the pipeline, it is anticipated that there will be a pressure shutoff valve at the station outlet, which will also serve as the station's shutoff valve. A startup cooler will also be provided, in addition to the transit cooler that was already mentioned. After being connected to the compressor's startup header, the cooler is then routed back to the suction header of the compressor. This refrigerator is constantly operational, and its entire capacity may be accessed at any time. A temperature measurement device located in the discharge header of the cooler is used, in conjunction with variable speed drives for the fans' motors, to maintain a consistent temperature at the cooler's output. The second reason for having this cooler is so that it can offer cooling capacity in the event that a compressor is working inside the antisurge area. In this scenario, gas is redirected from the compressor discharge line via the cooler and back to the compressor suction.

A scraper launcher facility serves as the operational interface between the pipeline and the compressor station. This is where the gas is launched once it has been compressed at the station.

Regarding the utility systems, these are described in the following sections.

3.2. Condensate Tank

The primary function of the condensate tank is to collect and store in a common condensate tank all liquids that have been separated in the various individual filters and separators until they are removed by a vacuum truck. A high-level alarm will sound whenever there is a dangerously high amount of liquid inside the tank [34].

Because of the potential for the environment within the tank to become explosive, a flame arrestor will be installed in it so that the station may continue to function in an appropriate and secure manner. In order to prevent any leaks into the earth, the tank will be constructed as a double-walled tank that will also have a leak detecting system.

To maintain a liquid temperature of at least +5 degrees Celsius even when the surrounding air temperature is at its coldest, the whole condensate tank will be electrically trace-heated and insulated.

3.3. Fuel Gas Unit

The gas turbine, the hot-water boiler unit, and the power generating unit are the three major users of the fuel gas unit, and thus the primary function of this unit is to condition the station incoming gas to meet their specific requirements [34]. The suction side of the compressor station is where the fuel gas is extracted from. Fuel gas treatment is designed to run in two separate trains with 100 percent capacity each. A filter is used to remove liquids and deposits from the pipeline before the gas is released into the atmosphere. After going through the filtration process, the gas is sent via a heat exchanger that is powered by hot water. This heat exchanger will preheat the gas in order to compensate for the temperature loss that will occur as a result of the Joule–Thompson effect, which will be accomplished by lowering the pressure. In this component of the system, a pressure relief valve will be provided in order to prevent an excessive buildup of pressure brought on by the heating of the gas in the event that the heat exchanger becomes clogged. The pressure of the gas will be lowered upstream of the heat exchanger until it reaches the desired pressure of 18 to 34 bar (depending on gas turbine supplier). In the event that the controller fails, there will be two medium-driven shutoff valves installed upstream of the pressure reduction valve. This will prevent the system from becoming overpressurized. On the low-pressure side of the system, a relief valve has been planned for installation, and its sole function will be to prevent the system from shutting down as a result of pressure peaks in the event that the redundant system is automatically switched on. While the fuel gas is being taken upstream of the metering system, the turbine flow meter that is meant to be suited for fiscal purposes will be measuring the fuel gas stream as it flows through the system. After this step, the gas is prepared for use in the gas turbine by being conditioned.

3.4. Hot-Water Boiler System

For both the radiator in the room- or building-heating system and the gas preheating in the fuel gas system, the heating medium, which is water that may be conditioned for heating purposes, will be given. This water will serve as the heating medium. In order to prevent the release of greenhouse gases, the seal gas that is produced by the compressor units will be burnt in the boiler units.

3.5. Vent and Blowdown System

In the event that an emergency depressurization is required, the station will be outfitted with a vent and blowdown system that has the capability of lowering the operating pressure to 6.9 bar in less than 15 min. The vented gas will be collected through one of three distinct headers at the end of the process. The suction area of the station is included in the first header, while the compressors are included in the second header and the discharge area is included in the third header. The blowdown system is constructed in such a way that it directs a consistent mass flow to the vent stack. This will be accomplished by the carefully orchestrated opening of the blowdown lines that are located in close proximity to the vent. After passing through the silencer, the gas is routed to the vent stack for final disposal. In addition to the emergency blowdown system, the blowdown system also has a number of manually operated vents that are connected to it. Venting for maintenance purposes requires the use of these manual vents.

3.6. Instrument Air System

The quality of the instrument air that is supplied to the compressor station shall be determined in accordance with DIN ISO 8573-1 [34]. In order to deliver the necessary quantity of air, three instrument air compressors have been installed simultaneously, and an additional unit has been set aside as a backup. Piston compressors have been chosen because they provide the appropriate degree of flexibility. A three-stage cleaning process is planned for the area downstream of the compressors. There is a stage dedicated to the removal of liquid droplets, followed by two stages dedicated to the removal of solids. After going through this cleaning process, the air is then sent through an adsorption drier in order to achieve the desired water dew point. One adsorption system will be on duty at all times, while the other will be in standby mode. Following an additional cleaning process, the air is then sent to the instrument air network through a buffer vessel so that it can handle peak demands. After the first stage of filtering, the liquids that have been removed will be sent to an oil/water separation stage before the effluent is disposed of.

4. The Proposed Framework

Due to the nature of construction sites as one of the most common locations for occurrences of accidents, conducting an evaluation of the project's safety risk is an essential component of effective construction project management. Figure 2 below depicts the suggested risk analysis and assessment framework based on the AHP application.

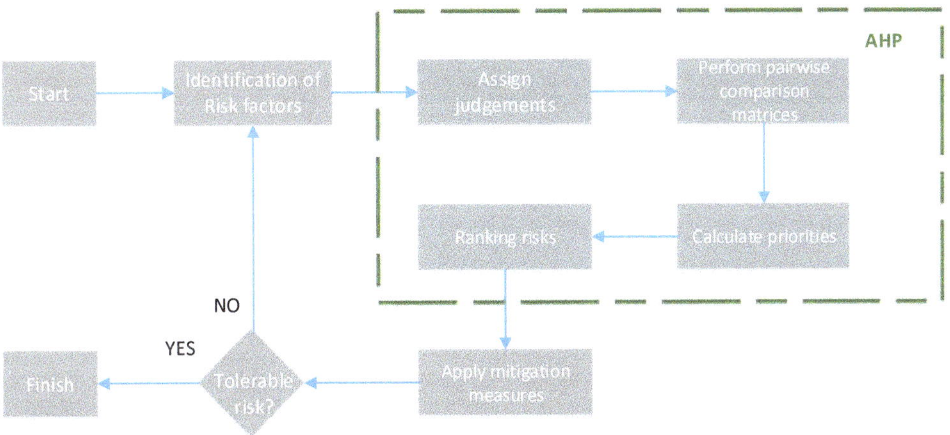

Figure 2. The flowchart of the proposed framework.

Firstly, the risks that might arise during the completion of each task are enumerated, and then the pairwise comparison matrices for those activities are filled up. The standard AHP procedure is used to obtain the weights for each risk, and the ranking is determined by ordering the risks' weights from highest to lowest. The risk manager is able to allot funding for risk-mitigation measures that may be tailored to the most significant risks associated with each individual activity after the most significant risk associated with each task has been recognized.

5. Application on a Natural Gas Compressor Station Construction Project

The suggested framework was used in a construction project in Greece of the Kipi compressor station of the Trans Adriatic Pipeline, with all the buildings that are foreseen in the relative area. Initially, an administration building that accommodates offices of engineering and management staff that support the station function is built, and it consists of a guard house, offices of engineering and management staff, conference and break rooms, kitchen, sanitary rooms for men, women, and people with disabilities, HVAC, server, electrical room etc. Next, a stores and workshop building accommodates electrical and mechanical workshops, small parts storage, male and female lockers, showers, sanitary rooms, meeting room and break room with kitchen, electrical and server room, archive, workshop office, mechanical workshop, storage area. Also, a utility building provides space for station vehicles as well as space for heating system, instrument and plant air, firefighting material and equipment storage, and lube oil storage. Finally, an electrical and control building, which is the main building that provides space for electricity supply facilities and the operation and control of compressor station, such as transformers (connection to public medium voltage grid), switch gears (medium voltage, low voltage), uninterruptable power supply with battery room, cathodic protection facilities, station control system, and HVAC rooms. The following Table 3 summarizes the tasks of the project and the corresponding risks of each task.

Table 3. The project activities and corresponding risks.

Activity ID	Activity	Risk ID	Risk
T1	Circulation	R1.1	Driving incident/accident
		R1.2	Circulation incident on construction site
		R1.3	Transport of the material
		R1.4	Weather condition
		R1.5	Presence of diesel fuel/carburant/lubricants
T2	Office work	R2.1	Bad ergonomic/physical stress
		R2.2	Climate exposition
		R2.3	Passive smoke
		R2.4	Bad hygiene condition
T3	Work in open space	R3.1	Bad condition of the ground and working zone
		R3.2	Presence of insects/wild animals
		R3.3	Extreme weather conditions
T4	Reaction to the emergence	R4.1	Unpreparedness of personnel
		R4.2	Impracticability of emergency ways and exits
T5	Coactivity	R5.1	Simultaneous operations in the same zone
		R5.2	Degraded situation in the proximity
T6	Work in night time	R6.1	Prolonged working time
		R6.2	Reduced visibility
T7	Manual work	R7.1	Bad ergonomic/physical stress
		R7.2	Torquing
		R7.3	Fall/impact of equipment and material on the personnel
		R7.4	Injury by manual tools
T8	Lifting operations	R8.1	Failure of crane
		R8.2	Fall of load
		R8.3	Failure of the lifting
		R8.4	Persons, equipment and structure in the proximity
		R8.5	Lifting with construction machinery

Table 3. *Cont.*

Activity ID	Activity	Risk ID	Risk
T9	Excavation and groundwork	R9.1	Collapsing of soil
		R9.2	Use of excavator
		R9.3	Presence of network/cables underground
		R9.4	Unexploded ordnance
		R9.5	Open holes and trenches on worksite
		R9.6	Unfavorable work zone
T10	Confined space	R10.1	Unfavorable work zone
		R10.2	Presence of toxic substances
		R10.3	Presence of energized sources
T11	Working at height	R11.1	Fall of personnel
		R11.2	Fall of objects
		R11.3	Improper use of portable ladder
T12	Scaffolding and PEMP	R12.1	Work on MEWP
T13	Concrete pouring	R13.1	Use of heavy machinery for pouring
		R13.2	Use of rotating machine for mixing concrete
		R13.3	Noise
T14	Welding and cutting	R14.1	Presence of naked flames/sparks
		R14.2	Use of rotating and electrical tools
		R14.3	Optical radiation
		R14.4	Noise
T15	Torch cutting	R15.1	Presence of naked flames/sparks
		R15.2	Presence of gas cylinders
T16	Abrasive blasting	R16.1	Abrasive projection
		R16.2	Asphyxia
		R16.3	Environmental pollution
		R16.4	Noise
T17	Painting activity	R17.1	Use of paints and chemicals
		R17.2	Fire ignition
T18	Use of chemicals	R18.1	Exposition to chemical substances
		R18.2	Storage of chemicals products
T19	Use of site engines	R19.1	Equipment with internal combustion (compressors, power generator, etc.)
		R19.2	Rotating engine parts
		R19.3	Environmental pollution
		R19.4	Noise
		R19.5	Use of pneumatic material (grinders, pneumatic hammers, vibrators, etc.)
		R19.6	High-pressure cleaning
T20	Electrical works	R20.1	Electrocution
		R20.2	Use of electrical tools and cables
T21	Ionizing radiation	R21.1	Mobilization of radioactive source on site
		R21.2	Ionizing radiation
		R21.3	Incident affecting the source
T22	Pressure test	R22.1	Equipment under pressure
		R22.2	Overpressure
		R22.3	Presence of nitrogen
		R22.4	Environmental pollution
T23	Work on energized equipment	R23.1	Failure of insulation procedure
		R23.2	Asphyxia

The decision-maker, responsible for making the decisions needed by the multicriteria approaches, was the engineer serving as the risk manager of the project. This technique gives the risk managers a choice mechanism for effectively prioritizing hazards and subsequently leads to efficient allocation restricted budget for expenditures in accident prevention.

The present case study consists of a separate hierarchy for every single task, given that the AHP is applied for the risks of each project activity. For example, the hierarchy for the task "Working at height" consists of three risks (Fall of personnel, Fall of objects, Improper use of portable ladder), which will be assessed using the multicriteria method (Figure 3).

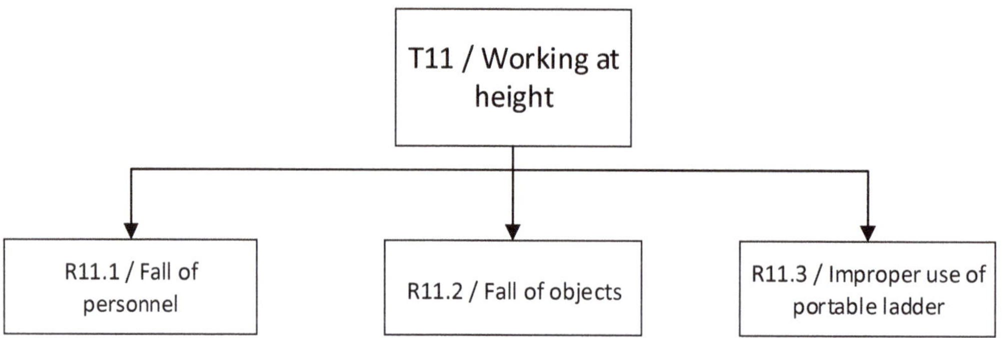

Figure 3. An example hierarchy for T11/Working at height task.

The expert risk manager has to apply evaluations and fill pairwise comparison matrices for the risks of every project activity.

Table 3 shows the risks of each task and the expert's choices. The influence of each risk on the overall level of safety in the workplace while carrying out each activity is established through the use of pairwise comparisons. The typical AHP technique generates the consistency ratios (CRs) in order to measure and assure that the judgments made by the decision-maker are consistent with one another. The appropriate local priorities that were computed using the typical AHP are outlined in Table 4 below. It is important to point out that every judgment turned out to be accurate, as evidenced by the fact that the CR for each matrix was less than 10%.

Table 4. The risk list, judgments, and results for each activity of the project.

Task ID	Risk ID	Pairwise Comparison Matrix					Score	Ranking
T1		R1.1	R1.2	R1.3	R1.4	R1.5		
	R1.1	1.00	0.17	0.50	3.00	5.00	13.36%	3
	R1.2	6.00	1.00	3.00	7.00	9.00	53.37%	1
	R1.3	2.00	0.33	1.00	5.00	7.00	23.59%	2
	R1.4	0.33	0.14	0.20	1.00	3.00	6.35%	4
	R1.5	0.20	0.11	0.14	0.33	1.00	3.34%	5
T2		R2.1	R2.2	R2.3	R2.4			
	R2.1	1.00	9.00	7.00	4.00		62.88%	1
	R2.2	0.11	1.00	0.33	0.14		4.28%	4
	R2.3	0.14	3.00	1.00	0.33		9.40%	3
	R2.4	0.25	7.00	3.00	1.00		23.44%	2
T3		R3.1	R3.2	R3.3				
	R3.1	1.00	5.00	3.00			63.70%	1
	R3.2	0.20	1.00	0.33			10.47%	3
	R3.3	0.33	3.00	1.00			25.83%	2

Table 4. *Cont.*

Task ID	Risk ID	Pairwise Comparison Matrix						Score	Ranking
T4		R4.1	R4.2						
	R4.1	1.00	2.00					66.67%	1
	R4.2	0.50	1.00					33.33%	2
T5		R5.1	R5.2						
	R5.1	1.00	3.00					75.00%	1
	R5.2	0.33	1.00					25.00%	2
T6		R6.1	R6.2						
	R6.1	1.00	0.50					33.33%	2
	R6.2	2.00	1.00					66.67%	1
T7		R7.1	R7.2	R7.3	R7.4				
	R7.1	1.00	3.00	0.20	0.33			12.22%	3
	R7.2	0.33	1.00	0.14	0.20			5.70%	4
	R7.3	5.00	7.00	1.00	2.00			52.32%	1
	R7.4	3.00	5.00	0.50	1.00			29.76%	2
T8		R8.1	R8.2	R8.3	R8.4	R8.5			
	R8.1	1.00	0.50	5.00	3.00	0.25		16.27%	3
	R8.2	2.00	1.00	6.00	4.00	0.50		26.48%	2
	R8.3	0.20	0.17	1.00	0.50	0.14		4.30%	5
	R8.4	0.33	0.25	2.00	1.00	0.17		6.89%	4
	R8.5	4.00	2.00	7.00	6.00	1.00		46.06%	1
T9		R9.1	R9.2	R9.3	R9.4	R9.5	R9.6		
	R9.1	1.00	0.50	7.00	5.00	3.00	9.00	30.61%	2
	R9.2	2.00	1.00	7.00	5.00	4.00	8.00	40.50%	1
	R9.3	0.14	0.14	1.00	0.50	0.25	2.00	4.52%	5
	R9.4	0.20	0.20	2.00	1.00	0.50	4.00	8.05%	4
	R9.5	0.33	0.25	4.00	2.00	1.00	5.00	13.36%	3
	R9.6	0.11	0.13	0.50	0.25	0.20	1.00	2.97%	6
T10		R10.1	R10.2	R10.3					
	R10.1	1.00	0.20	0.33				10.47%	3
	R10.2	5.00	1.00	3.00				63.70%	1
	R10.3	3.00	0.33	1.00				25.83%	2
T11		R11.1	R11.2	R11.3					
	R11.1	1.00	3.00	5.00				63.70%	1
	R11.2	0.33	1.00	3.00				25.83%	2
	R11.3	0.20	0.33	1.00				10.47%	3
T12	R12.1							100%	1
T13		R13.1	R13.2	R13.3					
	R13.1	1.00	2.00	4.00				55.84%	1
	R13.2	0.50	1.00	3.00				31.96%	2
	R13.3	0.25	0.33	1.00				12.20%	3
T14		R14.1	R14.2	R14.3	R14.4				
	R14.1	1.00	0.50	3.00	5.00			33.36%	2
	R14.2	2.00	1.00	3.00	4.00			45.05%	1
	R14.3	0.33	0.33	1.00	2.00			13.60%	3
	R14.4	0.20	0.25	0.50	1.00			7.99%	4
T15		R15.1	R15.2						
	R11.1	1.00	2.00					66.67%	1
	R11.2	0.50	1.00					33.33%	2

Table 4. *Cont.*

Task ID	Risk ID		Pairwise Comparison Matrix					Score	Ranking
T16		R16.1	R16.2	R16.3	R16.4				
	R16.1	1.00	2.00	5.00	4.00			50.68%	1
	R16.2	0.50	1.00	3.00	2.00			26.41%	2
	R16.3	0.20	0.33	1.00	0.50			8.63%	4
	R16.4	0.25	0.50	2.00	1.00			14.28%	3
T17		R17.1	R17.2						
	R17.1	1.00	2.00					66.67%	1
	R17.2	0.50	1.00					33.33%	2
T18		R18.1	R18.2						
	R18.1	1.00	3.00					75.00%	1
	R18.2	0.33	1.00					25.00%	2
T19		R19.1	R19.2	R19.3	R19.4	R19.5	R19.6		
	R14.1	1.00	0.25	6.00	4.00	0.50	2	15.30%	3
	R14.2	4.00	1.00	9.00	7.00	2.00	5	42.35%	1
	R14.3	0.17	0.11	1.00	0.50	0.14	0.25	3.03%	6
	R14.4	0.25	0.14	2.00	1.00	0.20	0.5	4.94%	5
	R19.5	2.00	0.50	7.00	5.00	1.00	4	25.73%	2
	R19.6	0.50	0.20	4.00	2.00	0.25	1	8.66%	4
T20		R20.1	R20.2						
	R20.1	1.00	0.50					33.33%	2
	R20.2	2.00	1.00					66.67%	1
T21		R21.1	R21.2	R21.3					
	R21.1	1.00	0.50	2.00				28.57%	2
	R21.2	2.00	1.00	4.00				57.14%	1
	R21.3	0.50	0.25	1.00				14.29%	3
T22		R22.1	R22.2	R22.3	R22.4				
	R22.1	1.00	5.00	3.00	6.00			57.67%	1
	R22.2	0.20	1.00	0.50	2.00			12.51%	3
	R22.3	0.33	2.00	1.00	3.00			22.16%	2
	R22.4	0.17	0.50	0.33	1.00			7.66%	4
T23		R23.1	R23.2						
	R23.1	1.00	2.00					66.67%	1
	R23.2	0.50	1.00					33.33%	2

Applying the AHP method allows the risk manager to extract more accurate information regarding the importance of each risk for every activity. More specifically, in the last two columns of Table 3, the score of each risk and the corresponding ranking are listed. Thus, the manager identifies the most influential risk factor and can allocate budget resources to reduce as much as possible the total project risk.

In addition, it is worth mentioning that three classes of activities are considered regarding the results. Firstly, there are some activities observed that the first ranked risk is more important than the others. In this analysis, we considered that risk is much more important than the others if it has a score of far more significance than 60% (namely, the sum of scores for the rest of the risks is by far smaller than 40%). In this class belong activities such as T2, T3, T4, T5, T6, T10, T11, T12, T15, T17, T18, T20, and T23. Next, we observed that for some tasks, there is a more critical risk (the one ranked as first), but its score is between 50% and 60%, namely, it is marginally responsible for the majority of the task's risk. In this group, we classified activities such as T1, T7, T13, T16, T21, and T22.

Finally, we considered a group of tasks for which there is a predominant risk factor, but the majority of the risk of the activity is not due to it. In this class belong activities such as T8, T9, T14, and T19. Figure 4 presents the weight of the first ranked risk for each task of the project.

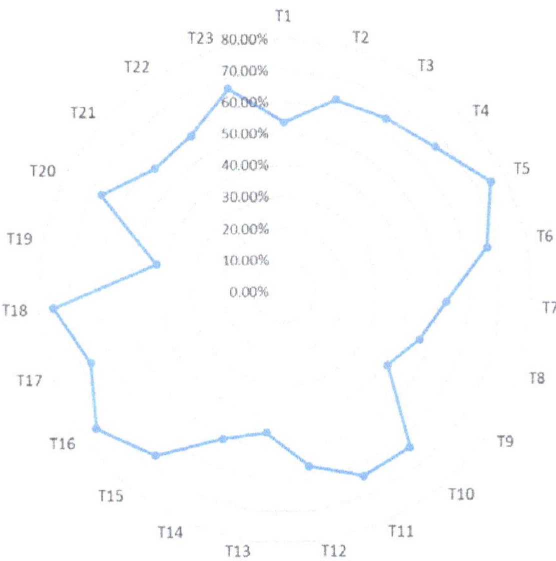

Figure 4. The weights of the first ranked risk for every task.

These findings provide evidence that supports the hypothesis that was initially proposed for this research: that it is essential to make use of a multicriteria analysis method in order to determine the significance of risk factors for project activities. This is because there are activities for which the risk manager needs to revise the given judgments or allot a budget in order to ensure that other risk factors besides the predominant ones are taken into consideration.

6. Conclusions

The technique of assigning priorities to various aspects of risk may unquestionably be of assistance to managers in devising strategies to reduce or eliminate the most significant risk factors via the utilization of preventive measures. A more efficient allocation of a limited budget may reduce costs associated with assistance and mortgages, and in general makes it possible for managers to have the budget available that can be used to reduce project risks to a greater extent. In addition to this, an effective allocation of a limited budget may reduce expenses associated with assistance and mortgages. The key contribution provided by this study is the application of a well-known multicriteria technique to ranking and prioritizing risks. In this case, the AHP was used to express judgments based on the decision-maker's experience and value system as it related to the analysis of risk factors for each activity in the construction of a natural gas compressor station.

This framework may be used as a guide to help prioritize the implementation of safety measures and the allocation of scarce resources in order to reduce the likelihood of as many accidents as possible. Not only might it be utilized as a teaching tool but it could also be used to help managers with less expertise make better judgments. It might also be used as a template for training newcomers and transferring knowledge from seasoned professionals to others with less expertise. The proposed approach benefits from ability to use the experience, knowledge, and intuition of the person making the decision when determining the weights of the elements. On the other hand, the subjective character of the modeling process is the fundamental flaw of this approach and, more generally, of similar multicriteria methods. This implies that the methodology cannot ensure that the judgments will be absolutely accurate.

Although the proposed method was successful, it might be enhanced by doing a sensitivity analysis on the risk manager's assessments of the second and third set of risks, i.e., those tasks in which the principal risk is associated with a relatively low overall score. As a result, the proposed framework may be honed to better fit the specific circumstances under investigation.

Author Contributions: Conceptualization, G.K.K., O.E.D., G.G.B. and D.E.K.; data curation, G.K.K., O.E.D. and G.G.B.; formal analysis, G.K.K., O.E.D., G.G.B. and D.E.K.; investigation, G.K.K. and G.G.B.; methodology, G.K.K., O.E.D., G.G.B. and D.E.K.; software, G.K.K., O.E.D., G.G.B.; supervision, G.K.K. and D.E.K.; validation, G.K.K., O.E.D., G.G.B. and D.E.K.; visualization, G.K.K., O.E.D.; writing—original draft, G.K.K., O.E.D. and G.G.B.; Writing—review and editing, G.K.K., O.E.D. and D.E.K. All authors have read and agreed to the published version of the manuscript.

Funding: This research received no external funding.

Data Availability Statement: Some or all data, models, or codes that support the findings of this study are available from the corresponding author upon reasonable request.

Conflicts of Interest: The authors declare no conflict of interest.

References

1. Neumann, A. *The Role of Natural Gas in Europe towards 2050*; Edmonds, J., Emberson, D., Gabriel, S.A., Holz, F., Karstad, P.I., Klöckner, C.A., Nord, L.O., Rúa, J., Pollet, B.G., Rasmussen, P., et al., Eds.; NETI Policy Report 01/2021; NTNU: Trondheim, Norway, 2021.
2. Simonoff, J.S.; Restrepo, C.E.; Zimmerman, R. Risk management of cost consequences in natural gas transmission and distribution infrastructures. *J. Loss Prev. Process Ind.* **2010**, *23*, 269–279. [CrossRef]
3. Bajcar, T.; Cimerman, F.; Širok, B. Model for quantitative risk assessment on naturally ventilated metering-regulation stations for natural gas. *Saf. Sci.* **2014**, *64*, 50–59. [CrossRef]
4. Vianello, C.; Maschio, G. Quantitative risk assessment of the Italian gas distribution network. *J. Loss Prev. Process Ind.* **2014**, *32*, 5–17. [CrossRef]
5. Li, W.; Zhang, L.; Liang, W. Job hazard dynamic assessment for non-routine tasks in gas transmission station. *J. Loss Prev. Process Ind.* **2016**, *44*, 459–464. [CrossRef]
6. Cinelli, M.; Spada, M.; Kadziński, M.; Miebs, G.; Burgherr, P. Advancing Hazard Assessment of Energy Accidents in the Natural Gas Sector with Rough Set Theory and Decision Rules. *Energies* **2019**, *12*, 4178. [CrossRef]
7. Mrozowska, A. Formal Risk Assessment of the risk of major accidents affecting natural environment and human life, occurring as a result of offshore drilling and production operations based on the provisions of Directive 2013/30/EU. *Saf. Sci.* **2021**, *134*, 105007. [CrossRef]
8. Mojtahedi, S.M.H.; Mousavi, S.M.; Makui, A. Project risk identification and assessment simultaneously using multi-attribute group decision making technique. *Saf. Sci.* **2010**, *48*, 499–507. [CrossRef]
9. Tavana, M.; Behzadian, M.; Pirdashti, M.; Pirdashti, H. A PROMETHEE-GDSS for oil and gas pipeline planning in the Caspian Sea basin. *Energy Econ.* **2013**, *36*, 716–728. [CrossRef]
10. Papadopoulou, M.P.; Antoniou, C. Environmental impact assessment methodological framework for liquefied natural gas terminal and transport network planning. *Energy Policy* **2014**, *68*, 306–319. [CrossRef]
11. Strantzali, E.; Aravossis, K.; Livanos, G.A.; Nikoloudis, C. A decision support approach for evaluating liquefied natural gas supply options: Implementation on Greek case study. *J. Clean. Prod.* **2019**, *222*, 414–423. [CrossRef]
12. Shafiee, M.; Animah, I.; Alkali, B.; Baglee, D. Decision support methods and applications in the upstream oil and gas sector. *J. Pet. Sci. Eng.* **2018**, *173*, 1173–1186. [CrossRef]
13. Marhavilas, P.K.; Filippidis, M.; Koulinas, G.K.; Koulouriotis, D.E. The integration of HAZOP study with risk-matrix and the analytical-hierarchy process for identifying critical control-points and prioritizing risks in industry—A case study. *J. Loss Prev. Process Ind.* **2019**, *62*, 103981. [CrossRef]
14. Marhavilas, P.K.; Filippidis, M.; Koulinas, G.K.; Koulouriotis, D.E. An expanded HAZOP-study with fuzzy-AHP (XPA-HAZOP technique): Application in a sour crude-oil processing plant. *Saf. Sci.* **2020**, *124*, 104590. [CrossRef]
15. Marhavilas, P.K.; Filippidis, M.; Koulinas, G.K.; Koulouriotis, D.E. A HAZOP with MCDM Based Risk-Assessment Approach: Focusing on the Deviations with Economic/Health/Environmental Impacts in a Process Industry. *Sustainability* **2020**, *12*, 993. [CrossRef]
16. Marhavilas, P.; Filippidis, M.; Koulinas, G.; Koulouriotis, D. Safety Considerations by Synergy of HAZOP/DMRA with Safety Color Maps—Applications on: A Crude-Oil Processing Industry/a Gas Transportation System. *Processes* **2021**, *9*, 1299. [CrossRef]

17. Ratnayake, R.C.; Markeset, T. Technical integrity management: Measuring HSE awareness using AHP in selecting a maintenance strategy. *J. Qual. Maint. Eng.* **2010**, *16*, 44–63. [CrossRef]

18. Zheng, G.; Zhu, N.; Tian, Z.; Chen, Y.; Sun, B. Application of a trapezoidal fuzzy AHP method for work safety evaluation and early warning rating of hot and humid environments. *Saf. Sci.* **2012**, *50*, 228–239. [CrossRef]

19. Aminbakhsh, S.; Gunduz, M.; Sonmez, R. Safety risk assessment using analytic hierarchy process (AHP) during planning and budgeting of construction projects. *J. Saf. Res.* **2013**, *46*, 99–105. [CrossRef]

20. Caputo, A.C.; Pelagagge, P.M.; Salini, P. AHP-based methodology for selecting safety devices of industrial machinery. *Saf. Sci.* **2013**, *53*, 202–218. [CrossRef]

21. Podgórski, D. Measuring operational performance of OSH management system—A demonstration of AHP-based selection of leading key performance indicators. *Saf. Sci.* **2015**, *73*, 146–166. [CrossRef]

22. Wang, Q.; Wang, H.; Qi, Z. An application of nonlinear fuzzy analytic hierarchy process in safety evaluation of coal mine. *Saf. Sci.* **2016**, *86*, 78–87. [CrossRef]

23. Xie, Q.; Ni, J.-Q.; Su, Z. Fuzzy comprehensive evaluation of multiple environmental factors for swine building assessment and control. *J. Hazard. Mater.* **2017**, *340*, 463–471. [CrossRef] [PubMed]

24. Janackovic, G.; Stojiljkovic, E.; Grozdanovic, M. Selection of key indicators for the improvement of occupational safety system in electricity distribution companies. *Saf. Sci.* **2017**, *125*, 103654. [CrossRef]

25. Kasap, Y.; Subaşı, E. Risk assessment of occupational groups working in open pit mining: Analytic Hierarchy Process. *J. Sustain. Min.* **2017**, *16*, 38–46. [CrossRef]

26. Carpitella, S.; Certa, A.; Izquierdo, J.; La Fata, C.M. A combined multi-criteria approach to support FMECA analyses: A real-world case. *Reliab. Eng. Syst. Saf.* **2018**, *169*, 394–402. [CrossRef]

27. Gul, M. Application of Pythagorean fuzzy AHP and VIKOR methods in occupational health and safety risk assessment: The case of a gun and rifle barrel external surface oxidation and colouring unit. *Int. J. Occup. Saf. Ergon.* **2018**, *26*, 705–718. [CrossRef]

28. Koulinas, G.; Marhavilas, P.; Demesouka, O.; Vavatsikos, A.; Koulouriotis, D. Risk analysis and assessment in the worksites using the fuzzy-analytical hierarchy process and a quantitative technique—A case study for the Greek construction sector. *Saf. Sci.* **2018**, *112*, 96–104. [CrossRef]

29. Marhavilas, P.K.; Tegas, M.G.; Koulinas, G.K.; Koulouriotis, D.E. A Joint Stochastic/Deterministic Process with Multi-Objective Decision Making Risk-Assessment Framework for Sustainable Constructions Engineering Projects—A Case Study. *Sustainability* **2020**, *12*, 4280. [CrossRef]

30. Marhavilas, P.K.; Koulouriotis, D.; Gemeni, V. Risk analysis and assessment methodologies in the work sites: On a review, classification and comparative study of the scientific literature of the period 2000–2009. *J. Loss Prev. Process Ind.* **2011**, *24*, 477–523. [CrossRef]

31. Marhavilas, P.K.; Koulouriotis, D.E. Risk-Acceptance Criteria in Occupational Health and Safety Risk-Assessment—The State-of-the-Art through a Systematic Literature Review. *Safety* **2021**, *7*, 77. [CrossRef]

32. Saaty, T.L. *The Analytic Hierarchy Process*; McGraw-Hill: London, UK, 1980.

33. Saaty, T.L. How to make a decision: The analytic hierarchy process. *Eur. J. Oper. Res.* **1990**, *48*, 9–26. [CrossRef]

34. Bougelis, G. Risk Assessment Using Decision Matrix Risk Assessement Technique and Multicriteria Decision Making Methods FEAHP and FTOPSIS—Case Study on a Natural Gas Construction Site. Master's Thesis, Hellenic Open University, Patra, Greece, 2021.

 *sustainability*

Article

Scientometric Analysis and AHP for Hierarchizing Criteria Affecting Construction Equipment Operators' Performance

Kleopatra Petroutsatou *, Ilias Ladopoulos and Konstantina Tsakelidou

Faculty of Engineering, Department of Civil Engineering, Aristotle University of Thessaloniki, 54124 Thessaloniki, Greece; iladopou@civil.auth.gr (I.L.); ktsakelid@civil.auth.gr (K.T.)
* Correspondence: kpetrout@civil.auth.gr

Abstract: The construction sector constitutes a significant indicator of a country's economic growth. Construction equipment is an integral part of every construction project, and its contribution during construction determines any project's completion. It also represents a significant capital investment for companies in this sector. A major strategic goal for such companies is the increase in the equipment's productivity, which is affected mostly by its operators. The aim of this research is to recognize and prioritize the criteria affecting the performance of construction equipment operators. Scientometric analysis, using VOSViewer software, was implemented for the formation of different kinds of bibliometric networks, proposing a holistic approach to this research field. Those networks delineated the field with regard to construction equipment operators and revealed the correlations between the network's items, which were formed because of previous research, and finally, conclusions were drawn. An extensive literature review in conjunction with structured interviews with experts and operators determined the factors affecting the operators' performance, with a view to creating a hybrid decision model based on the Analytical Hierarchy Process (AHP), as implemented by the Transparent Choice tool. Many experts evaluated the criteria affecting the operators' performance, leading to remarkable conclusions. Moreover, a few pointers for future research are provided.

Keywords: construction equipment; analytic hierarchy process (AHP); scientometric analysis; productivity; operator

Citation: Petroutsatou, K.; Ladopoulos, I.; Tsakelidou, K. Scientometric Analysis and AHP for Hierarchizing Criteria Affecting Construction Equipment Operators' Performance. *Sustainability* **2022**, *14*, 6836. https://doi.org/10.3390/su14116836

Academic Editors: Srinath Perera, Albert P. C. Chan, Xiaohua Jin, Dilanthi Amaratunga, Makarand Hastak, Patrizia Lombardi, Sepani Senaratne and Anil Sawhney

Received: 14 April 2022
Accepted: 27 May 2022
Published: 2 June 2022

Publisher's Note: MDPI stays neutral with regard to jurisdictional claims in published maps and institutional affiliations.

1. Introduction

Construction projects are currently prevailing in every aspect of human life, with the goal of improving the quality of people's lives. As clearly is defined in the European Commission (2012) Road Transport Report, their standards are strictly specified so that they will eventually correspond to the demanding reality. Their successful completion relies on successful project management, which must strongly emphasize the efficient utilization of labor, material, and equipment in order to deliver a successful project on time, within the budget, and as per the defined quality standards [1]. Under this framework, the productivity of construction projects was always an issue worth examining [2,3]. Productivity is used to denote a relationship between output and its associated input used in the production system [2]. It depends on a variety of factors, such as construction equipment, which represents a significant capital investment for companies in this sector [3]. Efforts to improve productivity have been made in recent decades, focusing on the most influential factors.

A project's productivity is directly affected by fleet management, which concerns the selection of suitable construction equipment for each task according to its requirements [4]. The fleet and asset management function is responsible for strategic decisions regarding fleet composition, fleet average age, capital expenditure, finance, tax, and return on investment. It uses the data developed in other functions, interfaces with the company strategic planning process, and develops the rates, estimates, budgets, benchmarks, and standards needed to manage the whole process [4]. Nowadays, construction companies are facing multiple difficulties with how to properly and effectively manage their fleet of construction equipment. Fleet management is a feature that allows companies to avoid or minimize the risks associated with investing in equipment, efficiency, productivity, overall transportation costs, and impartial compliance in legislation [5]. On the other hand, low productivity means inefficiency of resources with the inevitable results of cost and time overruns [6,7].

Previous research on construction project productivity primarily focused on the efficiency of construction project delivery and focused on tangible input-output schema within the construction process [8]. Liberda et al. [9] managed to identify the most critical aspects in terms of human, external, and management issues that affect construction productivity. Ghoddousi and Hosseini [10] conducted a survey of the factors affecting the productivity of construction projects in Iran and concluded that the most important grounds affecting sub-contractor productivity include, in descending order: materials/tools, construction technology and method, planning, supervision system, reworks, weather, and jobsite condition. Hasan et al. [11] identified more than 46 articles from different sources concerning the factors affecting construction productivity within the last 30 years. They finally concluded that despite noticeable differences in the socio-economic conditions across both developed and developing countries, an overall reasonable consensus exists on a few of the significant factors impeding productivity.

As Hedman et al. [12] certify, the equipment operators are a crucial factor influencing the duration of the time loss, which refers to planned downtime, setup time, measurement and adjustment, equipment failure, etc. This perspective is strengthened by He et al. [13]; they studied a construction project's resilience (CPR) by measuring specific systemic indicators from the perspective of employee behavior, such as operators.

Moreover, the construction equipment operators' performance is related to their safety preconditions during earthwork, which rely upon the synergy of the work and their interactions with each other and with their supervisors [14].

The basis of this study is set on the criteria affecting the construction equipment operators' performance. Skills and aptitude are also significant factors that are considered to be critical for the performance of earthmoving equipment operators, but they co-exist with more quantitative factors, which are examined in this study. Several studies have highlighted the relationship between aptitude and employee performance. Aptitude is the potential to demonstrate the ability to perform a certain kind of work at a certain level [15]. This research contributes to the body of knowledge by combining those two abilities with other, still untapped, factors. It is agreed that the operators' performance is a mixture of tangible and intangible factors. It is described as their ability to complete their work, fulfilling certain standards, based on the goals or objectives set by their employers [12,13].

In an effort to highlight the effect of the power of equipment operators on the construction project productivity, this study dives deep into the human factors to extract the tangible (or subjective) and intangible (or objective) criteria related to the construction equipment operators' performance in the field. The definition of worker productivity is widely examined. Tangen [16] examined the ways in which the concepts of "productivity" and "performance" are dealt with in the literature, demonstrating that the terms used within these fields are often vaguely defined and poorly understood.

However, performance entails more. It includes their willingness and ability to communicate or collaborate, their promptness, and their demeanor at work. Consequently, the abovementioned factors have a significant impact on the overall performance of the construction project. The expectations and standards set by their supervisors can shape the operators' experience, can affect performance, and can certainly have an impact on their productivity and, ultimately, the project's success. In a nutshell, productivity concentrates on the output, i.e., what is produced, whereas performance is often activity-based and is quantitative or qualitative [17]. Maqsoom et al. [17] realized through their research that worker productivity is critical within construction projects as it is the measure of the rate at which work is performed, and more importantly, it helps to build knowledge on how to motivate the workers to perform at high levels. Much earlier, Navon [18] measured indirect productivity parameters and converted them into sought indicators in order to comprehensively point out the importance of the operator's performance to the project's productivity.

In order to quantify the earthmoving equipment operators' performance factors, this research focuses on identifying and hierarchizing those factors. The necessary data concerning the performance criteria were investigated through: (i) scientometric analysis, (ii) structured interviews with construction equipment experts, and (iii) structured interviews with construction equipment operators. The findings of this research will be beneficial for contractors, project managers, and equipment operators as they reveal the key issues regarding the attitudes and behaviors that play an integral role in enhancing productivity in construction projects [19].

2. Literature Review

This paper conducts a two-step literature review by adopting an interpretivist philosophical approach and inductive reasoning to generate new theories on the phenomena under investigation. In the first step, a scientometric analysis was conducted, as described in Section 2.1, to reveal the necessity of connecting the operator's performance with the construction equipment's productivity. This analysis involves the application of the "science mapping" method, which acts as both a descriptive and a diagnostic tool for research policy purposes, processing immense reservoirs of bibliometric data [20–24].

The second step justifies the criteria selection by looking into the relevant past studies that were extracted by the previous step (Section 2.2). It collects a great amount of related literature from the place where the criteria concerning operator performance are extracted and presented in a comprehensive list. Most importantly, in this section, each selected factor has been scrutinized, with a view to justifying every sub-criterion.

The above process is deemed as necessary in order to form a final criteria and sub-criteria list, as key constituents for the AHP decision tree, presented in Section 3.2.

2.1. Scientometric Analysis

This study goes deep into the published literature to reveal the void regarding the research made on the criteria that affect the construction equipment operators' performance. A scientometric analysis is used to objectively map the scientific knowledge on this specific field and to identify the research themes and the corresponding challenges based on the scientometric results, with the use of the VOSviewer application [20]. In order to create those scientometric networks, a four-step process was followed, as described in Figure 1.

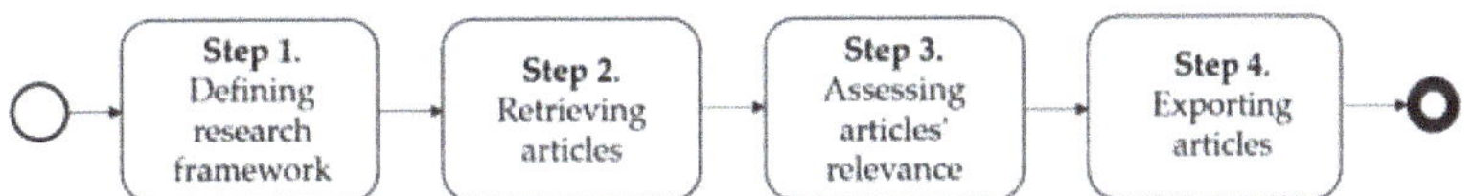

Figure 1. Flowchart of map creation in VOSViewer.

In step one, the research framework is defined, with the intention of recognizing and setting the desired goals. At this point, an initial investigation is conducted to seek the necessary research components by separating the relevant from the irrelevant.

During step two, the articles were retrieved which were closely related to the examined topic. Those articles were extracted by well-recognized bibliographic databases, such as Web of Science and Scopus, covering a period from 2001 to 2021. To identify the relevant publications, search terms were used (Table 1). Figure 2 illustrates the evolution of the research made from 2001, where an increase from 2016 and onwards has been observed. Step three includes a comprehensive relevance assessment of the extracted documents in order to finalize the publications to be inserted for scientometric mapping into VOSViewer and to comment upon the extracted maps.

Table 1. Search Terms in Web of Science and Scopus.

Boolean Operator	Terms	Description
	construction equipment	The term that describes the main topic and the core search rule
OR	machinery	Used for searching all machinery- and equipment-based
OR	equipment	publications, in order to exclude the irrelevant
AND	operator *	Term that specifies the distinctive topic, concerning operators
OR	product *	Term that specifies the distinctive topic, concerning productivity
AND	AHP	The applied Multi-Criteria Decision Analysis (MCDA) method
OR	Analytic * Hierarchy Process	Used to include references for AHP as Analytic or Analytical Hierarchy Process
NOT	medic *	All the terms concerning medical, health, and
NOT	Health	pharmaceutical issues
NOT	pharma *	

The asterisk (*) suggests that it can be replaced by any word or phrase.

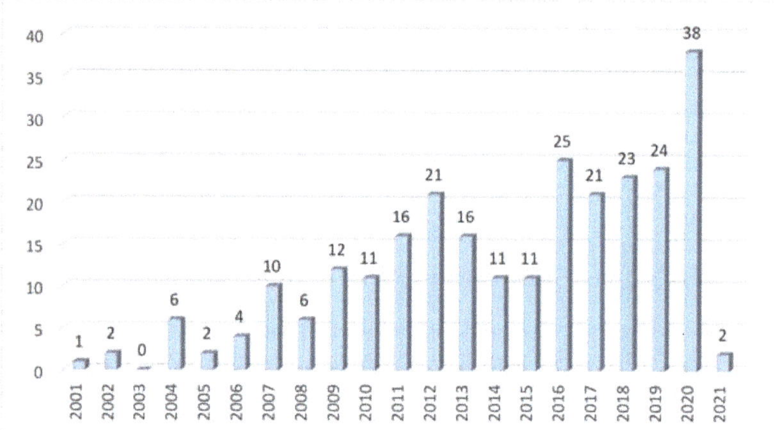

Figure 2. Total Number of Publications Related to Operator Productivity and AHP.

The fourth step of the scientometric mapping process includes the extraction of the selected literature in a recognized form for processing by the VOSViewer application.

Its final product is the production of a comprehensive network comprising the terms which coexist inside the overall publications, where their linkage strength, their appearances, and their relativity are visible, weighted, and clustered. Different clusters are represented by different automatically assigned colors and each color designates a specific research area. The terms inside each cluster are represented by circles, and their size reflects the number of publications in which they were found. The spacing between those circles indicates their relatedness, and their degree of relativity is indicated by the thickness of the curved lines connecting them. The degrees of relatedness between words are indicated by the curved lines.

This paper presents two types of visualization of terms by the VOSViewer network: (a) text data co-occurrence among the titles and their abstracts and (b) keyword co-occurrence. Their visualization networks are presented in Figures 3 and 4, and the produced clusters by subject are in Tables 2 and 3, respectively.

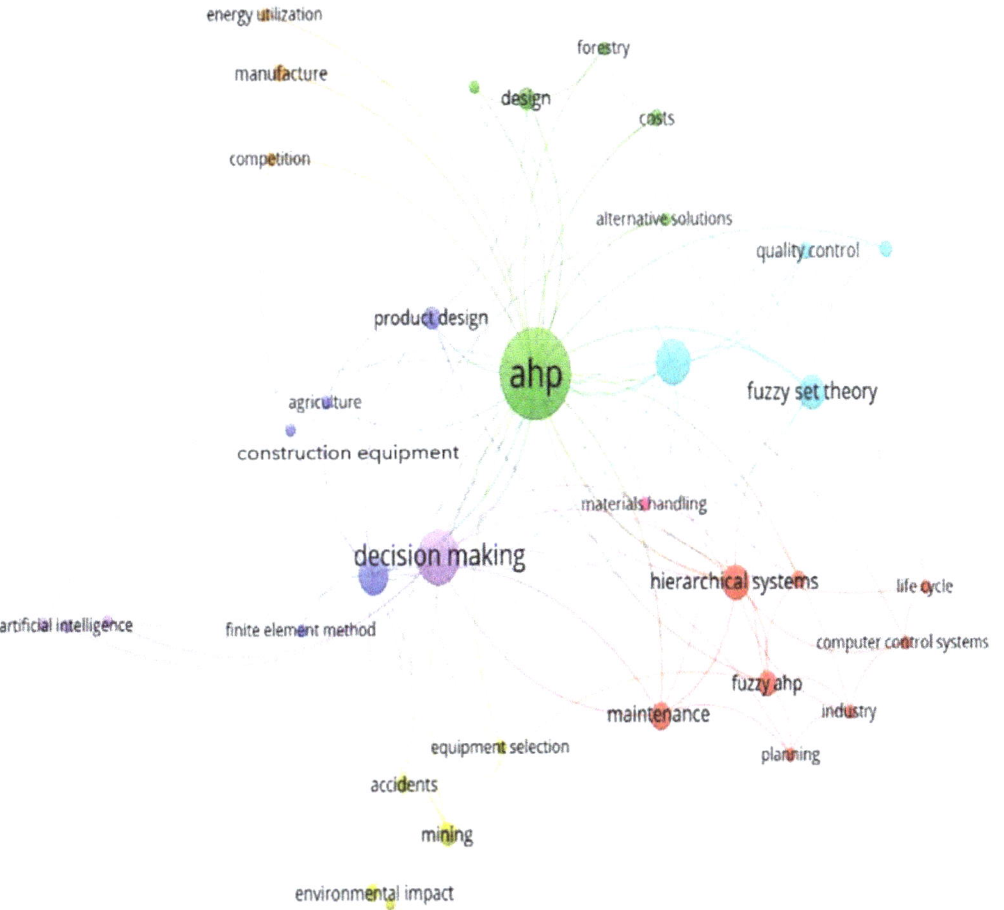

Figure 3. VOSViewer map based on title and abstract text data.

Table 2. Text Data Co-occurrence Clustering.

Cluster Number	Main Subject	Color	Terms Included
1	Hierarchization methods	Red	8
2	AHP	Green	6
3	Equipment	Dark Blue	5
4	Ore mining	Light Green	5
5	Decision Making	Purple	4
6	Construction Equipment	Sky Blue	4
7	Industry	Orange	3
8	Material Handling	Pink	1

Table 3. Keyword Co-occurrence Clustering.

Cluster Number	Main Subject	Color	Terms Included
1	Production	Red	10
2	Decision Making	Green	10
3	AHP Applications	Dark Blue	7
4	Industry	Yellow	7
5	Maintenance	Purple	6
6	Strategy and Indexes	Sky Blue	4
7	Productivity	Orange	3

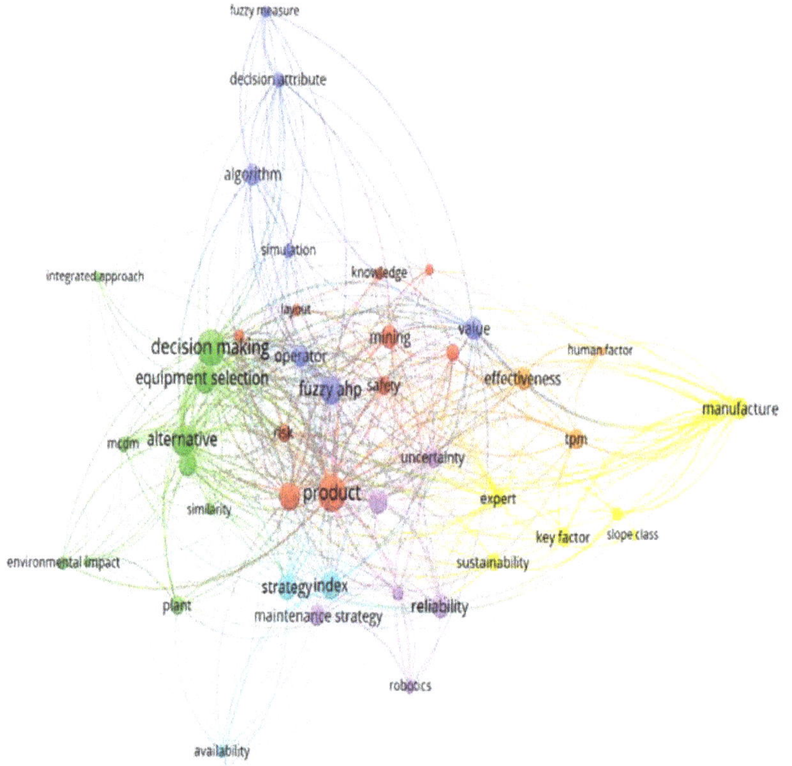

Figure 4. VOSViewer map based on keywords (network visualization).

2.1.1. Text Data Co-Occurrence among the Titles and Their Abstracts

In this scientometric network, "ahp" constitutes a heavily weighted subject in the scientific community, presenting a significant proximity with the "decision making" term, as the AHP is a specific decision-making method. A strong proximity also exists between the "construction equipment" and the "decision making" terms, a fact that supports the application of the MCDA methods to a variety of the utilization aspects of construction equipment. Nevertheless, the "operator" or "productivity" terms are absent inside the network, while terms such as "maintenance" and "equipment selection" are orbiting and directly linked with the main reference terms of the AHP and decision making. This approach indicates a void in the literature with regard to the discussed topic.

Further scrutinization of the map leads to further implied conclusions:

- The AHP is very popular among several MCDA methods with regard to the use of equipment. The term "equipment" includes construction equipment and general equipment (in industry, agriculture, manufacturing, etc.)
- By the way that the term "decision making" is linked with the other terms, it is related to issues such as maintenance, design, equipment selection, material handling, quality control, etc.
- The absence of the terms "production", "productivity", and "operator" can be explained by the fact that these terms are not defining the titles and the abstracts of the selected publications, which does not necessarily mean that they do not exist inside the rest of these documents.
- The fact that the last four clusters have fewer terms highlights the void inside the literature when it comes to relating construction equipment productivity with decision making in the industry sector and in material handling.

2.1.2. Keyword Co-Occurrence

The analysis based on keywords indicates that the network delineates a strong link between the terms "decision making", "equipment selection", and "mcdm" by including them in the same cluster, indicating that methods such as the AHP are often used for decision making. The terms "operator", "simulation", "decision attribute", and "fuzzy ahp" belong to the same cluster, indicating a sort of correlation. There is also a noticeable proximity between the terms "decision making" and "operator", indicating their strong linkage, even though they do not belong to the same cluster. The term "operator" is also close to "safety", "reliability", "knowledge", and "experience", which are significant factors affecting factors the operator's performance.

Some further implied conclusions from the network visualization analysis are the following:

- The average linkage weight (denser network) among the keywords is much stronger than among the titles and the text data of the abstracts; this is caused by the fact that the keywords are more or less used as common "de facto" terms.
- The clustering terms (visualized by different colors), in this case, are more distinct as their amount is greater, and they are used commonly.
- The term "decision making" is located at a close distance to "equipment selection" and "operator", which clearly indicates the importance of the operator when it comes to selecting the proper equipment for certain projects. Equipment selection includes purchasing and fleet management.
- Additionally, the "operator" is related with "safety", "reliability", "knowledge", and "experience", which are crucial factors for the operator's performance and efficiency.
- The "decision making" cluster (green) lies very close to the "production" (red) and "AHP applications" (dark blue) clusters, indicating their strong relatedness.
- The "productivity" cluster (orange) also includes the "human factor", a term which refers to the operators.
- The "industry" cluster (yellow) is the most distant; however, it includes heavily weighted terms as the construction industry is an essential part of the general term.

2.2. Criteria Selection

According to Atkinson [25], human factors have a clear causal link with machine productivity rates. He also concludes that the production performance of machinery is largely reliant on the operator's skill and competence. Construction equipment operators are frequently called upon to handle difficult and demanding situations, which impacts their performance. The examined literature indicates the main criteria affecting an operator's performance, as depicted in Table 4.

Table 4. Criteria Sourcing.

Criterion	Source
Operator's Competence	
Knowledge/Experience	Holt and Edwards, 2015 [26]
	Yang, Edwards, and Love, 2004 [27]
	Dumitrescu and Delsenicu, 2018 [28]
	Du, Dorneich, and Steward, 2016 [29]
	Langer et al., 2012 [30]
Training/Preparation	Du, Dorneich, and Steward, 2016 [29]
	Naskoudakis and Petroutsatou, 2016 [31]
	Dumitrescu and Delsenicu, 2018 [28]
Motive/Earnings	Yang, Edwards, and Love, 2004 [27]
	Holt and Edwards, 2015 [26]
	Dumitrescu and Delsenicu, 2018 [28]
Stress/Fatigue	Yang, Edwards, and Love, 2004 [27]
	Haggag and Elnahas, 2013 [32]
Relationships and Interaction	
Between employees	Dumitrescu and Delsenicu, 2018 [28]
Between employees and employer	Dumitrescu and Delsenicu, 2018 [28]
Disagreement resolution	Dumitrescu and Delsenicu, 2018 [28]
On-site communication	Beleiu, Crisan, and Nistor, 2015 [33]
Construction Equipment	
Use complexity	Dumitrescu and Delsenicu, 2018 [28]
	Yang, Edwards, and Love, 2004 [27]
Maintenance adequacy	Cheuk, Leung, and Tse, 2005 [34]
	Naskoudakis and Petroutsatou, 2016 [31]
Fleet availability	Naskoudakis and Petroutsatou, 2016 [31]
	Bahnassi and Hammad, 2012 [35]
	Lee et al., 2012 [36]
Innovation/New technologies	Naskoudakis and Petroutsatou, 2016 [31]
	Barati and Shen, 2018 [37]
	Albrektsson and Aslund, 2019 [38]
Task	
Complexity	Dumitrescu and Delsenicu, 2018 [28]
Project demands	Dumitrescu and Delsenicu, 2018 [28]
Timetable	Naskoudakis and Petroutsatou, 2016 [31]
	Dumitrescu and Delsenicu, 2018 [28]
Daily workload	Yang, Edwards, and Love, 2004 [27]
	Haggag and Elnahas, 2013 [32]
Natural/Environmental Factors	
Exposure to dust and emissions	Langer et al., 2012 [30]
	Naskoudakis and Petroutsatou, 2016 [31]
	Dumitrescu and Delsenicu, 2018 [28]
	Kokot and Ogierman, 2019 [39]
Weather conditions	Du, Dorneich, and Steward, 2016 [29]
	Dumitrescu and Delsenicu, 2018 [28]
Soil properties	Du, Dorneich, and Steward, 2016 [29]
	Barati and Shen, 2018 [37]
	Langer et al., 2012 [30]
Safety conditions	Naskoudakis and Petroutsatou, 2016 [31]
	Dumitrescu and Delsenicu, 2018 [28]
	Kokot and Ogierman, 2019 [39]
	Petroutsatou and Giannoulis, 2020 [40]
Light conditions and noise levels	Bahnassi and Hammad, 2012 [35]
	Naskoudakis and Petroutsatou, 2016 [31]
	Dumitrescu and Delsenicu, 2018 [28]

2.2.1. Operator's Competence

According to Holt and Edwards [26], the operator's competence is the operator's ability to effectively and efficiently apply the machine to the work task. It depends on the knowledge/experience [26] and the preparation/training that an operator has [27]. Motives can also be an additional factor in an employee's productivity levels; this factor is usually linked to earnings and insurance type [28]. Finally, stress and fatigue have been recognized by Haggag and Elnahas [32] as common conditions for operators, drastically affecting their competence.

2.2.2. Relationships and Interactions

The risk related to the labor system may be generated by human resource errors, an inadequate job description, dangerous equipment, improper social relationships between employees, and/or physical/environmental factors [28]. Focusing mostly on the manufacturing technologies, they realized that occupational stress is being enhanced by the new constraints which employees are now obliged to cope with and has also generated the need for organizations to redesign the work environment in order to counteract both the traditional and the emergent risks. According to Beleiu et al. [33], the relationships and interactions between employees, and between employees and their employers, are critical performance factors. In addition, the way a disagreement is resolved affects their performance, but it also depends on how fast it is resolved. They also stressed the importance of on-site communication as a determinant factor in the project's success, so that its efficiency can be supportive to construction equipment operators.

2.2.3. Construction Equipment

The research conducted by Dumitrescu and Delsenicu [28] identified that the equipment's complexity and maintenance adequacy can affect an operator's performance. Naskoudakis and Petroutsatou [31] emphasized the fact that, with regard to equipment development, new methods and designs are implemented to enhance reliability, machine control, comfort, and safety and to reduce the costs derived from failures and breakdowns, signifying the importance of the equipment's innovation as an influential factor. Their literature review also highlighted the importance of fleet management and construction equipment deployment. According to Vorster [4], fleet management should fulfill three overriding and critically important goals directly linked to human factors: (i) the equipment must be in the right place at the right time, (ii) the equipment must achieve the stated levels of reliability and uptime, and (iii) the total owning and operating costs must be kept to a competitive minimum.

2.2.4. Task

Dumitrescu and Delsenicu [28] define the task as the complexity of activities which are undertaken by an individual as part of a working process, the timeframe for activity completion, the job requirements, etc. Thus, each task can affect the operator's performance levels. They acknowledged that natural and environmental issues appearing in the field directly affect the operator's performance. Machines operate in an abrasive environment, where the operators' security is one of the most crucial issues in the construction sector [27,32]. Designing a work environment to meet the needs of the employees and the job requirements is a fundamental factor for enhancing productivity. Moreover, lighting conditions and noise levels, exposure to dust and emissions, soil properties, and weather conditions, especially in earthmoving work and road construction projects, are the main environmental factors, with a significant impact on their productivity levels.

3. Materials and Methods

3.1. Research Process

This research adopts a seven-step approach for identifying and hierarchizing the criteria that have the most effect on the construction equipment operator's performance.

The first step is to identify the relevant literature concerning the equipment operator's performance during a construction project. The review also focuses on investigating any criteria referred to by previous authors. The second step is to supplement those criteria with others mentioned by construction equipment experts and operators, through structured on-site interviews. Several oral interviews with construction management experts and construction equipment operators highlighted the importance of the criteria affecting the performance of the operators. In addition, construction sites were visited, and many interviews were gathered, leading to the consolidation of a final criteria list, as shown in Table 4. The third step is to classify those criteria into two main categories in order to distinguish between those which are based on personal experience and those which are verifiable facts. The formed opinion or viewpoint helps to distinguish the objective from the subjective criteria. Most commonly, subjective means something based on the personal perspective or preferences of an operator, meaning the subject who is observing, and this often implies that it comes with personal biases. In contrast, objective is the attempt to be unbiased, and this means that it is not influenced by an individual's personal viewpoint.

The fourth step of the process introduces the final development of the decision tree, as shown in Figure 5. This hierarchy model is vital when it comes to utilizing the AHP for weighting those influential criteria. It also constitutes the basis for forming a specialized questionnaire (fifth step), which is applicable for weighting the criteria through Saaty's scale [41] (as described in Section 3.2). The application of the AHP (sixth step) in the current research is reviewed in the following sections. The results coming from the AHP method formulate the final scoring of the criteria weighting, leading to the seventh and last step of the process. The overall process is visualized in Figure 5.

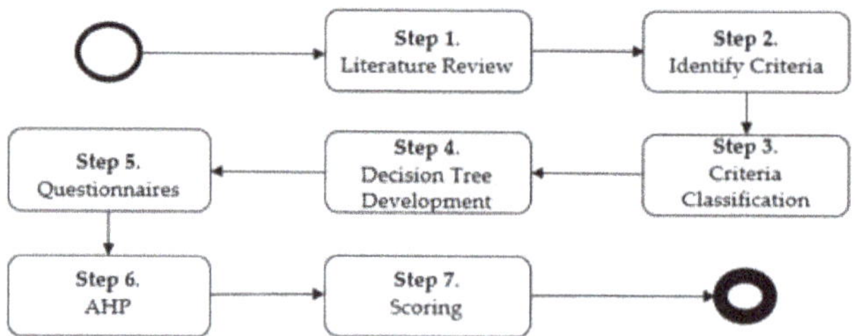

Figure 5. Methodology milestones.

3.2. Application of AHP

The AHP was presented by Thomas L. Saaty as a new approach to dealing with complex economic, technological, and sociopolitical problems, which often involve a great deal of uncertainty [42]. It is a structured technique for analyzing Multi-Criteria Decision Analysis (MCDM) problems according to a pairwise comparison scale [41]. To deal with complexity, our mind must model it by creating a structure and providing observations, measurements, and judgements and hopefully, of course, rigorous analysis to study the influences of the various factors included in the model [43,44].

The AHP is an MCDA method, used by Nassar et al. [45] to measure the relative importance among a set of criteria, and it is suitable for this research due to its ability to compare tangible (subjective) and intangible (objective) factors.

In this study, the Transparent Choice tool for the AHP [46] was employed as it concentrates all the procedure into one tool. It also incorporates a function to produce and disseminate the questionnaires, according to the imported criteria. The operators' performance criteria were classified into levels and sublevels, forming the hierarchy model in Figure 6. Even though the main function of the AHP is to guide the final decision to a certain alternative option, this research takes advantage of the AHP's criteria-weighting function to hierarchize them based on their final weighting score. Based on the same procedure utilized by Petroutsatou et al. [47], this process leads to one alternative, which is the best scoring criterion.

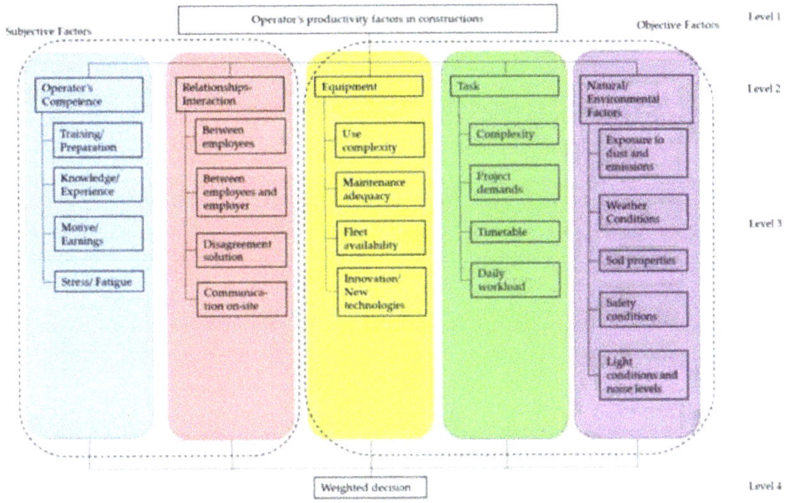

Figure 6. AHP decision tree model.

The model was imported to the Transparent Choice AHP platform; the questionnaires were created and distributed among the experts, who evaluated each criterion according to the AHP's fundamental evaluation scale. The number of evaluators who participated was 13, with different kinds of expertise in the construction sector, as allocated in Table 5. Special emphasis was given to the quality of the evaluators; this was based mostly on their expertise and not on their quantity. This fact does not affect the quality of the results as the AHP is a method with no specific statistical sample but is one that relies explicitly on the Consistency Ratio (CR), because in making paired comparisons, just as in thinking, people do not have the intrinsic logical ability to always be consistent [48]. Furthermore, this study does not constitute a polling exercise, as conducted by Tsafarakis et al. [49], where they exploited the capabilities of the AHP to investigate the preferences of individuals on public transport innovations using the Maximum Difference Scaling method.

Table 5. Evaluators' profile.

	Expertise	Quantity
1	Academia	6
2	Project Managers	2
3	Construction Equipment Operators	3
4	Construction Equipment Owners	2
	Total	13

The academia group includes professors of construction equipment disciplines with a vast experience in field engineering operations. Two experienced project managers were selected, due to their extended field work. Their perspective is based mostly on the project's performance indicators, which are directly affected by the performance of the operators they supervise. Construction equipment operators were chosen based on their experience in operating heavy earthwork machinery. Finally, representatives of OEMs, such as Caterpillar and JCB, were selected, representing the group of construction equipment owners.

The questionnaires were distributed to the above evaluators with the use of the Transparent Choice AHP Software, through its online survey application. The aggregated AHP results are presented in Table 6.

Table 6. Transparent Choice aggregated criteria weights.

#	Criteria	Weight	
		Local	Global
1.	**Equipment**	**24%**	**24%**
1.1	Fleet availability	37%	9%
1.2	Maintenance adequacy	42%	10%
1.3	Innovation/New technologies	12%	3%
1.4	Use complexity	9%	2%
2.	**Operator's competence**	**41%**	**41%**
2.1	Stress/Fatigue	15%	6%
2.2	Knowledge/Experience	52%	21%
2.3	Training/Preparation	19%	8%
2.4	Motive/Earnings	14%	6%
3.	**Task**	**15%**	**15%**
3.1	Project demands	34%	5%
3.2	Daily workload	18%	3%
3.3	Complexity	14%	2%
3.4	Timetable	34%	5%
4.	**Natural/Environmental Factors**	**10%**	**10%**
4.1	Exposure to dust and emissions	9%	1%
4.2	Soil properties	26%	2%
4.3	Weather conditions	21%	2%
4.4	Safety conditions	34%	3%
4.5	Light conditions and noise levels	10%	1%
5.	**Relationships—Interaction**	**11%**	**11%**
5.1	Disagreement solution	24%	3%
5.2	Between employees	19%	2%
5.3	Between employees and employer	26%	3%
5.4	On-site communication on site	32%	3%

The "local" column illustrates the sub-criteria (level 3) weighting in the context of each main (level 2) criterion. The "global" column illustrates each criterion or sub-criterion weighting in the context of the overall decision (level 1). The rankings for each group of evaluators are presented in Table 7.

Table 7. Transparent Choice evaluators weighting results comparison.

#	Criteria	Academia (6)		Project Managers (2)		Operators (3)		Owners (2)	
		Weight		Weight		Weight		Weight	
		Local	Global	Local	Global	Local	Global	Local	Global
1.	**Equipment**	**20%**	**20**	**25%**	**25%**	**28%**	**28%**	**24%**	**24**
1.1	Fleet availability	28%	6%	43%	11%	43%	12%	42%	10
1.2	Maintenance adequacy	48%	10%	42%	10%	40%	11%	25%	6
1.3	Innovation/New technologies	11%	2%	9%	2%	10%	3%	24%	6
1.4	Use complexity	12%	2%	5%	1%	6%	2%	9%	2
2.	**Operator's competence**	**41%**	**41%**	**26%**	**26%**	**51%**	**51%**	**38%**	**38**
2.1	Stress/Fatigue	20%	8%	10%	3%	10%	5%	15%	6
2.2	Knowledge/Experience	44%	18%	58%	15%	61%	31%	54%	21

Table 7. *Cont.*

#	Criteria	Academia (6)		Project Managers (2)		Operators (3)		Owners (2)	
		Weight		Weight		Weight		Weight	
		Local	Global	Local	Global	Local	Global	Local	Global
2.3	Training/ Preparation	25%	10%	15%	4%	13%	7%	16%	6
2.4	Motive/Earnings	11%	5%	18%	5%	17%	8%	16%	6
3.	**Task**	**18%**	**18%**	**24%**	**24%**	**8%**	**8%**	**11%**	**11**
3.1	Project demands	26%	5%	24%	5%	44%	4%	48%	5
3.2	Daily workload	22%	4%	11%	3%	18%	1%	12%	1
3.3	Complexity	13%	2%	9%	2%	11%	1%	26%	3
3.4	Timetable	39%	7%	56%	14%	27%	2%	15%	2
4.	**Natural/ Environmental Factors**	**9%**	**9%**	**15%**	**15%**	**7%**	**7%**	**8%**	**8**
4.1	Exposure to dust and emissions	11%	1%	14%	2%	5%	0%	9%	1
4.2	Soil properties	27%	3%	10%	2%	40%	3%	21%	2
4.3	Weather conditions	20%	2%	18%	3%	21%	1%	24%	2
4.4	Safety conditions	31%	3%	47%	7%	26%	2%	37%	3
4.5	Light conditions and noise levels	11%	1%	10%	1%	9%	1%	9%	1
5.	**Relationships— Interaction**	**12%**	**12%**	**11%**	**11%**	**6%**	**6%**	**18%**	**18**
5.1	Disagreement solution	23%	3%	29%	3%	30%	2%	14%	3
5.2	Between employees	22%	3%	25%	3%	16%	1%	9%	2
5.3	Between employees and employer	21%	3%	25%	3%	30%	2%	30%	6
5.4	On-site communication	34%	4%	22%	2%	24%	1%	47%	9

An additional examination was conducted to identify each evaluator's profile with regard to the main criteria ranking coming from their perspective. Table 8 illustrates the total scores for each main criterion in order to give a comparable form and to help extract further results.

Table 8. Main criteria ranking comparison.

Evaluators	Equipment	Operator's Competence	Task	Natural/ Environmental Factors	Relationships— Interaction
Academia	20%	41%	18%	9%	12%
Project Managers	25%	26%	24%	15%	11%
Operators	28%	51%	8%	7%	6%
Owners	24%	38%	11%	8%	18%

4. Results

4.1. Cumulative Evaluation

The questionnaires were answered with a view to prioritizing the criteria affecting the construction equipment operators' performance. Figure 7 illustrates the cumulative results by percentage. The operator's competence is the most influencing factor, with an overall score of 41%, while construction equipment and task follow with a score of 24% and 15%, respectively. According to the above results, construction companies or contractors should carefully select experienced and trained personnel in order to efficiently complete any construction project. Investing in further training for their operators could also be an option to leverage the overall productivity of their construction projects.

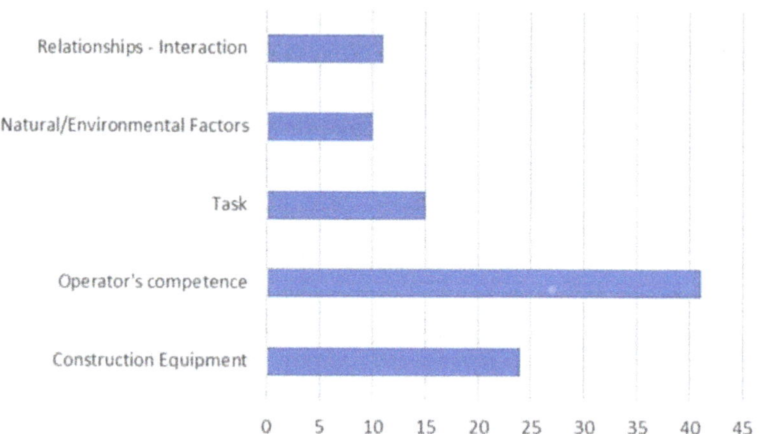

Figure 7. Cumulative results.

Each group of evaluators presented a different perspective, resulting in a different scoring for each criterion, as shown in Figure 8.

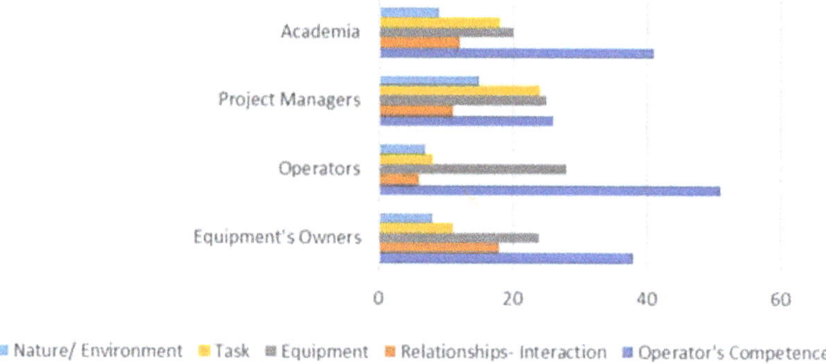

Figure 8. Criteria scoring for each group of evaluators.

The academia group presented similar results to the cumulative ones. The operator's competence was the most important criterion for all the groups of evaluators, with a different percentage in every group. The academia group and the construction equipment operators, for example, formed similar profiles, evaluating "operator's competence" as the most important criteria, with a total score of 41% and 51%, respectively. Construction equipment owners gave a total score of 38% for the operator's competence, 24% for construction equipment, and 18% for relationships—interaction. Project managers, on the other hand, evaluated the operator's competence with a total score of 26% and equipment and task with a total score of 25% and 24%, respectively.

The above analysis allows the formulation of each evaluator's different approach when dealing with earthwork operations. The operators consider the equipment as an extension of themselves, one which is totally dependent on their own skills and operating attitude. Consequently, their ability to efficiently handle the equipment improves the project's progression and the equipment's productivity. Thus, their competence is the dominant factor, with a direct effect on their performance. The academia group agrees too. The project managers score "relationships—interaction" at the lowest level among the criteria. The equipment owners rated the equipment operator's competence with the highest score.

4.2. Sub-Criteria Evaluation

According to Saaty [41], to make a decision we need to know the problem, the need and purpose of the decision, the criteria of the decision, the sub-criteria, the stakeholders and groups affected, and the alternative actions to take. In this study, where the AHP is used to hierarchize the criteria by their weighting score, the sub-criteria are used to expand the pairwise comparisons at a more in-depth level. In that way, the analysis gets to the root of the decision-making problem and becomes more precise and understandable. Based on Table 4, these sub-criteria comparisons are visualized and analyzed in the following sections.

4.2.1. Equipment

Wood and Gidado [50] suggested that the definition of a complex project should refer to the interaction, interdependencies, and interrelationships between the parts of a project and that a great deal of complexity lies within the organizational aspects of a project. The dynamics of innovation are based upon a wide spectrum of possibilities within the system, including incremental innovation at one extreme and breakthrough innovation at the other. Innovation is a process whereby the learning experience and the technology-adoption life cycle contribute to the creative thinking behind underlying motivational forces, whether technology- or market-driven [51]. When it comes to maintenance adequacy, the main objective is to provide maintenance capacity (resources) to meet the random maintenance workload, in order to achieve several objectives that include maximizing the system availability, safety, and the utilization of limited resources [52]. The area of asset management is gaining significance, especially in the availability contracts [53]. Maintaining a proper fleet of equipment can be of strategic importance to a company in cases where the award of a contract is also based upon the condition and availability of the equipment [54]. Furthermore, any unavailability of the proper equipment could cause its overturning, causing damage to property and personnel injury or even fatality, as Edwards [55] highlights.

According to the above literature review framework concerning equipment-related factors that affect the operator's productivity, the AHP analysis revealed the trends depicted in Figure 9.

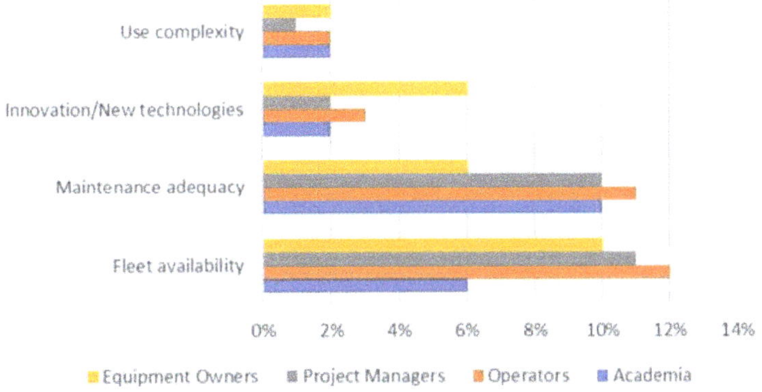

Figure 9. Equipment sub-criteria scoring.

Fleet availability and maintenance adequacy were of great interest for most of the evaluators, and they presented similar weighting results. Specifically, equipment operators granted 12% and 11% to fleet availability and maintenance adequacy, respectively, declaring those two factors as the most influential ones when it comes to construction equipment.

4.2.2. Operator's Competence

The research suggests that personal motivation is a critical internal driving force that, if harnessed, can significantly improve an operator's productivity rate when working mobile plant and machinery [56]. Edwards et al. [56] concluded that the operators' personal motivation can best be encouraged by paying attention to "personal satisfiers" and "security" aspects, with particular emphasis being given to work flexibility and variety, a safe work environment, and appropriate operator remuneration.

In terms of reducing fuel consumption, unit emissions and cost, Jukic and Carmichael [57] revealed that, compared to the baseline values, trained drivers saw a reduction in their fuel consumption by an average of 8.5 percent, reducing to 7.7 percent in the several weeks following training.

Regarding the operator's knowledge/experience, Edwards [58] indicated that the more competent (a mix of qualification and experience) an excavator operator is, the more efficiently (i.e., productively) they can employ the machine and vice versa.

Fatigue is one of the factors leading to reduction in productivity, poor quality of work, and increased risk of accidents in construction [59]. Handling heavy construction equipment is considered as a hazardous occupation and requires personnel to maintain high levels of work situational awareness (WSA). In an analysis made by Sneddon et al. [60], it was found that higher levels of stress, sleep disruption, and fatigue were significantly associated with lower levels of WSA.

The AHP results highlighted the aforementioned factors against the operator's competence among evaluators, as shown in Figure 10.

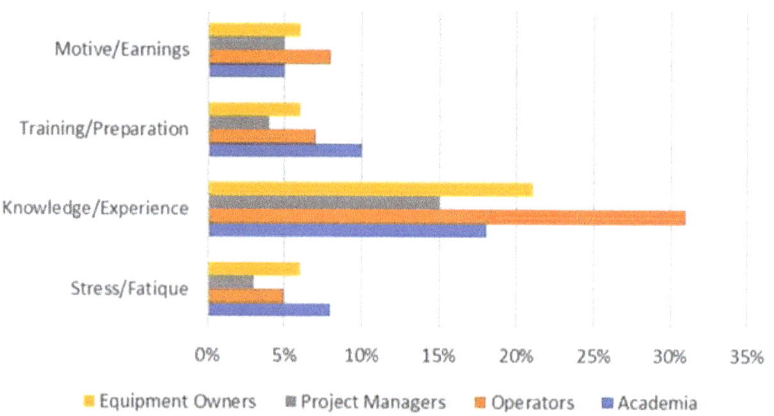

Figure 10. Operator's competence sub-criteria scoring.

The AHP weighting results point out the knowledge/experience criterion as being the most influential on the operator's competence, with total scores of 31% and 21% for the operator and the equipment owner evaluation groups, respectively.

4.2.3. Task

According to Dinakar [61], a clean and efficient planning mechanism, which clearly specifies the work and timetable to be used, can prevent delays in construction projects. Particularly in the European Union, it is a common practice to execute most of the public works through co-founded financial projects. Those projects are characterized by tight budgets and strict timetables [62]. Such timetables could be stressful for the earthwork equipment operators, causing their productivity degradation.

Wood and Gidado [50] tried to provide a greater understanding of the science of complexity in construction. Their research results suggested that the definition of a complex project should refer to the interaction, the interdependencies, and the interrelationships between parts of a project and that the largest amount of complexity lies within the organizational aspects of a project.

Izetbegović and Nahod [63] examined the relationship between the workload, the time pressure, and the work productivity of a construction project. Their findings showed a significant productivity reduction in the case of an additional workload, no matter whether the additional work was required or was a consequence of prior poor performance.

Choi et al. [64] examined the relationship between the construction worker's occupational safety and the application of wearable devices for localization. Their research was motivated by the increasingly demanding and hazardous construction environment. Additionally, Barlow [65] raised concerns about the poor performance of the construction industry, in the UK and elsewhere, caused by increasingly demanding customers and construction project complexity.

The above factors related to the project's tasks were weighted in relation to the construction equipment operator's performance, and the AHP results are presented in Figure 11.

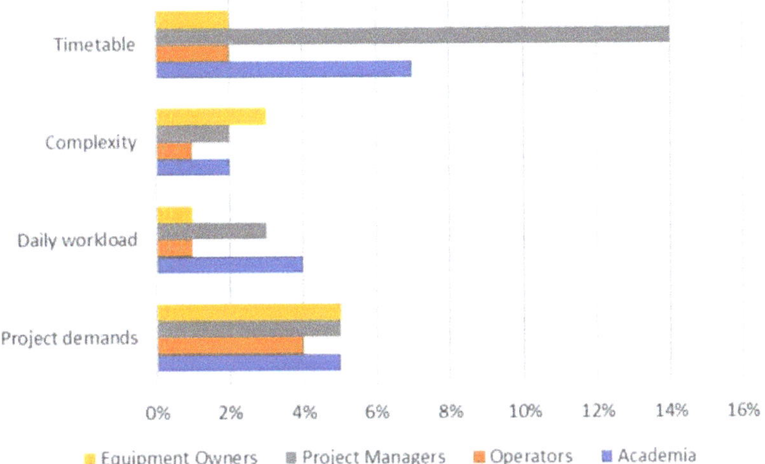

Figure 11. Task sub-criteria scoring.

The AHP results highlight the importance of the projects timetable from the project managers and the academia perspective by giving a weighting score of 14% and 7%, respectively. The weighting scores of the equipment owner and the operators point out that the project's timetable is of less importance (2% each), while their attention falls onto the project's demanding conditions (5% and 4%, respectively).

4.2.4. Natural/Environmental Factors

According to the World Health Organization (WHO), hearing loss is one of the top 10 most serious health problems worldwide, and noise-induced hearing loss (NIHL) is the leading occupational disease [66,67]. Duffy et al. [68] determined the factors associated with sun exposure behaviors among Operating Engineers (heavy equipment operators), highlighting their high risk for skin cancer due to high rates of exposure to ultraviolet light and low rates of sunblock use. Additionally, Eger et al. [69] highlighted the importance of light conditions and the operator's line of sight during construction works.

The unsafe behavior that is seen everywhere on construction sites is the biggest challenge for further improvement of construction safety performance. Focusing on the "human" related issues in construction safety, Fang et al. [70] reviewed the research and practices of safety management and came up with three key elements to look at, namely safety leadership, safety culture, and safety behavior. It is also notable that the subject of construction safety in general is widely referred to in the global literature.

Elazouni and Basha [71] managed to link problems with the operating construction equipment with low productivity and noted that weather conditions are one of the main factors that are unanticipated prior to the inception of the work and adversely affect productivity.

In order to highlight the importance of soil properties during construction, Parsakho et al. [72] investigated the effects of moisture, porosity, and soil bulk density during a forest road construction. Furthermore, Devi and Palaniappan [73] presented the influence of technological, operational, and site-related parameters, such as soil properties, on the performance of earthmoving operations.

Earthwork constructions emit a large amount of dust into the environment, which causes serious health hazards to construction workers. To reveal the characteristics of the health risks to workers caused by the dust generated during the earthwork construction phase, to polish the evaluation system of health damage in construction projects, and to improve the occupational health of workers, Luo et al. [74] and Chen et al. [75] established a health-risk evaluation system, which revealed the negative effect of dust exposure to the equipment operators' performance. Additionally, Ahn and Lee [76] presented a methodology for incorporating the analysis of operational efficiency into quantifying the amount of exhaust emission from construction operations and thus pointing out the effects of those emissions on construction projects productivity.

The above factors related to natural and environmental effects were weighted in relation to the construction equipment operators' performance, and the AHP results are presented in Figure 12.

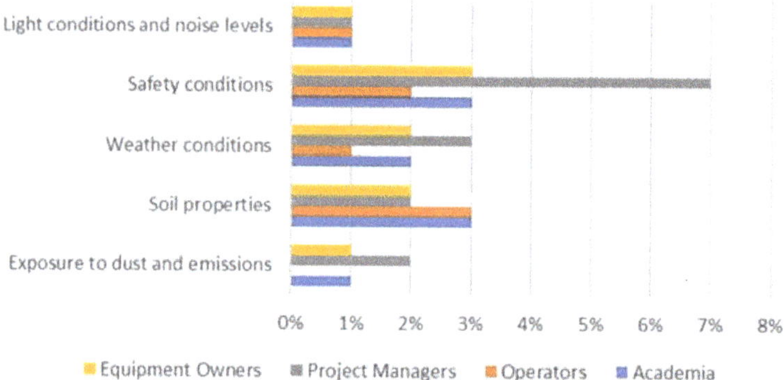

Figure 12. Natural/Environmental factor sub-criteria scoring.

All groups of evaluators agreed that safety conditions during construction and earthworks are of greater importance when it comes to the operator's performance. The highest score came from project managers (7%), as a result of it being their main obligation to ensure construction safety during construction works. Equipment owners (3%), academia (3%), and operators (2%) followed. On the other hand, the highest ranking given by the equipment operators was the soil properties (3%). The operators also considered that their exposure to dust and emissions had no effect on their performance.

4.2.5. Relationships—Interaction

The communication channels and the relationships developed between an employer and an employee are analyzed. The manager will be considered either as an agent of the employer or as an individual actor defending his or her own interests and with the ability to intervene between the three actors [77].

In order to identify the necessary factors for a safe construction site, Mohamed [78] conducted research in which he corroborates the importance of the role of management commitment, communication, workers' involvement, attitudes, and competence, as well as supportive and supervisory environments, in achieving a positive safety climate.

Additionally, investigations have been carried out which suggest that the motivation of employees in all industries is affected by the environment or culture in which they work [79]. Their research concluded that the environment of a construction site does affect demotivation levels of site personnel. Specifically, several variables were significantly linked to this result, including long hours, chaos, non-recognition for work done, and colleagues' aggressive management style.

This study incorporates the above research to investigate the influencing weight of those relationships—interaction factors on the performance of construction equipment operators and presents them in Figure 13.

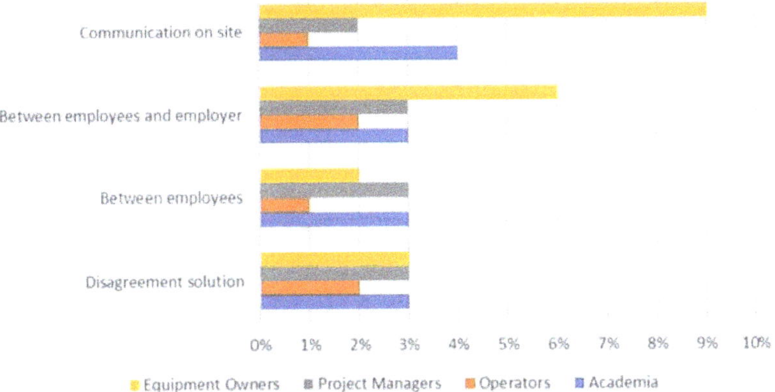

Figure 13. Relationships—Interaction.

The AHP results indicated the significance of the on-site communication, especially for the equipment owners, the academia group, and the equipment operators (9%, 4%, and 2%, respectively). On the other hand, what was more important for the equipment operators was the relationship between the employees and employers (2%), but also the ability to come to a solution to the problem when there is a disagreement in the field (2%). The importance of the relationship between employees and employers is also highlighted by the equipment owners (6%). The above diversity could be explained in terms of working mentality. Employees, such as the equipment operators, are the task receivers and those who are directly affected by the employer's decisions and management attitudes. The way they interact with superintendents and the way they reach a solution to a disagreement affects their psychological condition, their level of motivation and, of course, their will and temper for more productive work.

5. Discussion

It is generally accepted that construction equipment is an integral part of every project in the construction sector, and it represents a significant capital investment for the companies that own it. The efficient utilization of this resource makes the project successful [33]. This research started with the objective of identifying and hierarchizing the factors affecting the construction equipment operator's performance. This is a topic that is frequently and widely discussed in the construction industry sector, but not comprehensively examined and quantified, as was found through the literature review.

In previous research, many scholars have utilized different methods to exploit the results of other related publications, mostly by examining a project's productivity in general or in relation to other factors. However, no research has been found to systematically summarize those publications and provide a holistic approach on their interdependencies and, more specifically, to feature the linkage between equipment operator performance and the project's productivity. Two hundred and sixty-three topic-related publications were examined and visualized through the VOSViewer application. The statistical analysis of those articles revealed that the researchers' interest in construction equipment and operator productivity has been increasing over the past 5 years. Technological evolution seems to radically affect the construction sector, and therefore, it was examined as an influential factor on the construction equipment operators' performance.

The objective of this research was to identify the criteria and sub-criteria with a great effect on the equipment operator's performance. The structured interviews with experts in the field, combined with the conducted literature review led to the development of the decision tree model (Figure 6), with five main criteria affecting the operator's performance: (i) operator's competence, (ii) relationships—interaction, (iii) equipment (iv) task, and (v) natural/environmental factors. Furthermore, each criterion has been evaluated in relation to a total of twenty-one dependent sub-criteria.

The operators' competence was the most important criterion among all the groups of evaluators (i.e., academia, project managers, operators, and equipment's owners). This result supports that the idea that to ensure successful projects, an experienced and trained personnel is an important success factor. Furthermore, the availability of the equipment on the sites and the maintenance adequacy of the fleet are among the most prevailing factors between the equipment and the operator's productivity. Notably, prescriptive maintenance and the deployment of equipment are of the utmost importance according to our analysis. Regarding the task that should be delivered and to what extent this can affect the operator's performance, the academia group and the project managers highlight the importance of the project schedule, whereas the operators and equipment owners focus their attention on the project's demands. This stance discloses that the projects owners and academia demonstrate a holistic approach to the issue and not an activity-based concern, as the operators and equipment owners do. Regarding environmental factors, safety conditions are ranked first for ensuring the operator's enhanced productivity for the project managers, whereas, for the operators, soil properties are the determinant factor in their ensured work effectiveness. Lack of effective communication channels and conflicts among the construction teams are criteria that are ranked high as causal factors for an operator's poor productivity.

Based on the aforementioned research and our presented analysis, the practical applications of this study principally relate to helping stakeholders in plant and machinery to better understand the interrelationships between the factors investigated that affect the equipment operators' performance. More specifically, it offers practitioners valuable indicators to: (i) identify causal situations for the operators' inefficiencies, (ii) reinforce their fleet management, and (iii) thus ensure the project's success.

6. Conclusions

Enhanced productivity is an overarching goal in the construction sector as it integrates the effectiveness and efficiency of project's resources while guaranteeing the quality of the work. This paper explores, for the first time, qualitative evidence for the interdependencies between the equipment operator's performance and the construction equipment's productivity. Through an extensive literature review and interviews with experts, this paper was challenged to provide an annotative approach and pave the way for further constructive thinking on the examined topic.

This research objective was to: (i) recognize the factors affecting operators' performance levels that are closely related to a project's productivity and (ii) prioritize those factors by attributing total scores with Transparent Choice's tool for the AHP. The AHP was selected as the most suitable method for this research by utilizing its ability to weight and hierarchize the criteria, without the need to specify alternative attributes. The factors were divided into two groups, subjective and objective, and each group included two and three categories, respectively. On level three of the decision tree model twenty-one factors were investigated and shortlisted using the AHP. The decision tree model was evaluated by different types of evaluators, such as academia, equipment owners, operators, and project managers. Each group of evaluators formed a different profile by attributing different total scores to each criterion. The academia group was the group of evaluators that presented similar results to the cumulative ones and similar profiles to the operators' group. The operator's competence was considered by all groups of evaluators as the most important factor; in particular, "knowledge" and "experience" ranked first, followed by "training" and "on-site preparation" and contributed radically to the construction equipment operators' performance.

The limitations of this research relate to the fact that this was the first holistic approach to relating the equipment operators' performance with the tangible and intangible factors identified in the literature and expert interviews. More experts could surely enhance the robustness of our results. Moreover, in future research, the qualitative approach presented here could be expressed in mathematical equations in order to quantify the sensitiveness of the factors analyzed in relation to the equipment's productivity.

Author Contributions: Conceptualization, K.P.; Data curation, K.T.; Formal analysis, I.L.; Investigation, K.T.; Methodology, K.P. and I.L.; Project administration, K.P.; Software, I.L. and K.T.; Supervision, K.P.; Writing—original draft, I.L. and K.T.; Writing—review & editing, K.P. All authors have read and agreed to the published version of the manuscript.

Funding: This research received no external funding.

Acknowledgments: The authors would like to acknowledge the significant contribution made by the evaluators, without whose valuable input and support the quality of this work could not be ensured.

Conflicts of Interest: The authors declare no conflict of interest.

References

1. Gunduz, M.; Abu-Hijleh, A. Assessment of Human Productivity Drivers for Construction Labor through Importance Rating and Risk Mapping. *Sustainability* **2020**, *12*, 8614. [CrossRef]
2. Yi, W.; Chan, A. Critical review of labor productivity research in construction journals. *J. Manag. Eng.* **2014**, *30*, 214–225. [CrossRef]
3. Petroutsatou, K.; Marinelli, M. *Construction Equipment, Operational Analysis and Economics of Civil. Engineering Projects*, 2nd ed.; KRITIKI SA: Athens, Greece, 2018.
4. Vorster, M. *Construction Equipment Economics*, 1st ed.; Pen Publications: Columbus, IN, USA, 2009.
5. Petroutsatou, K.; Ladopoulos, I.; Vlachokostas, G. Comparative Evaluation of Fleet Management Software in the Greek Construction Industry. *IOP Conf. Ser. Mater. Sci. Eng.* **2022**, 1218. [CrossRef]
6. Petroutsatou, K.; Sifniadis, A. Exploring the consequences of human multitasking in industrial automation projects: A tool to mitigate impacts-Part II. *Organ. Technol. Manag. Constr.* **2018**, *10*, 1770–1777. [CrossRef]
7. Marinelli, M.; Petroutsatou, K.; Fragkakis, N.; Lambropoulos, S. Rethinking new public infrastructure value for money in recession times: The Greek case. *Int. J. Constr. Manag.* **2018**, *18*, 331–342. [CrossRef]
8. Antunes, R.; Gonzalez, V.; Walsh, K.; Rojas, O. Dynamics of project-driven production systems in construction: Productivity function. *J. Comput. Civil. Eng.* **2017**, *31*, 4017053. [CrossRef]
9. Liberda, M.; Ruwanpura, J.; Jergeas, G. Construction Productivity Improvement: A Study of Human, Management and External Issues. In Proceedings of the Construction Research Congress, Honolulu, HI, USA, 19–21 March 2003. [CrossRef]
10. Ghoddousi, P.; Hosseini, M.R. A survey of the factors affecting the productivity of construction projects in Iran. *Technol. Econ. Dev. Econ.* **2012**, *18*, 99–116. [CrossRef]
11. Hasan, A.; Baroudi, B.; Elmualim, A.; Rammeezdeen, R. Factors affecting construction productivity: A 30 year systematic review. *Eng. Constr. Archit. Manag.* **2018**, *25*, 916–937. [CrossRef]
12. Hedman, R.; Subramaniyan, M.; Almstrom, P. Analysis of Critical Factors for Automatic Measurement of OEE. *Procedia CIRP* **2016**, *57*, 128–133. [CrossRef]
13. He, Z.; Wang, G.; Chen, H.; Zou, Z.; Yan, H.; Liu, L. Measuring the Construction Project Resilience. *Buildings* **2022**, *12*, 56. [CrossRef]
14. Zhu, Z.; Yuan, J.; Shao, Q.; Zhang, L.; Wang, G.; Li, X. Developing Key Safety Management Factors for Construction Projects in China: A Resilience Perspective. *Int. J. Environ. Res. Public Health* **2020**, *17*, 6167. [CrossRef]
15. Johari, S.; Jha, K.N. How the Aptitude of Workers Affects Construction Labor Productivity. *J. Manag. Eng.* **2020**, *36*, 04020055. [CrossRef]
16. Tangen, S. Demystifying productivity and performance. *Int. J. Product. Perform. Manag.* **2005**, *54*, 34–46. [CrossRef]

17. Maqsoom, A.; Mubbasit, H.; Alqurashi, M.; Shaheen, I.; Alaloul, W.S.; Musarat, M.A.; Salman, A.; Aslam, B.; Zerouali, B.; Hussein, E.E. Intrinsic Workforce Diversity and Construction Worker Productivity in Pakistan: Impact of Employee Age and Industry Experience. *Sustainability* **2022**, *14*, 232. [CrossRef]
18. Navon, R. Automated project performance control of construction projects. *Autom. Constr.* **2005**, *14*, 467–476. [CrossRef]
19. Ofori, G.; Zhang, Z.; Ling, F. Key barriers to increase construction productivity: The Singapore case. *Int. J. Constr. Manag.* **2020**. [CrossRef]
20. Tijssen, R.; Van Raan, T. Mapping Changes in Science and Technology: Bibliometric Co-Occurrence Analysis of the R&D Literature. *Eval. Rev.* **1994**, *18*, 98–115. [CrossRef]
21. Cobo, M.; López-Herrera, A.G.; Herrera-Viedma, E.; Herrera, F. Science Mapping Software Tools: Review, Analysis, and Cooperative Study Among Tools. *J. Am. Soc. Inf. Sci. Technol.* **2011**, *62*, 1382–1402. [CrossRef]
22. Jan van Eck, N.; Waltman, L. Software survey: VOSviewer, a computer program for bibliometric mapping. *Scientometrics* **2010**, *84*, 523–538. [CrossRef]
23. Zhong, B.; Wu, H.; Li, H.; Sepasgozar, S.; Luo, H.; He, L. A scientometric analysis and critical review of construction related ontology research. *Autom. Constr.* **2019**, *101*, 17–31. [CrossRef]
24. Van Eck, N.; Waltman, L. VOSViewer Manual 2018. Available online: https://www.vosviewer.com/documentation/Manual_VOSviewer_1.6.8.pdf (accessed on 22 December 2021).
25. Atkinson, R. Project management: Cost, time and quality, two best guesses and a phenomenon, its time to accept other success criteria. *Int. J. Proj. Manag.* **1999**, *17*, 337–342. [CrossRef]
26. Holt, G.; Edwards, D. Analysis of interrelationships among excavator productivity modifying factors. *Int. J. Product. Perform. Manag.* **2015**, *64*, 853–869. [CrossRef]
27. Yang, J.; Edwards, D.; Love, P.E.D. Measuring the impact of daily workload upon plant operator production performance using Artificial Neural Networks. *Civ. Eng. Environ. Syst.* **2004**, *21*, 279–293. [CrossRef]
28. Dumitrescu, A.; Delsenicu, D. Risk assessment in manufacturing SMEs' labor system. *Procedia Manuf.* **2018**, *22*, 912–915. [CrossRef]
29. Du, Y.; Dorneich, M.; Steward, B.L. Virtual operator modeling method for excavator trenching. *Autom. Constr.* **2016**, *70*, 14–25. [CrossRef]
30. Langer, T.H.; Iversen, T.; Andersen, N.; Mouritsen, O.; Hansen, M. Reducing whole-body vibration exposure in backhoe loaders by education of operators. *Int. J. Ind. Ergon.* **2012**, *42*, 304–311. [CrossRef]
31. Naskoudakis, I.; Petroutsatou, K. A Thematic Review of Main Researches on Construction Equipment over the Recent Years. *Procedia Eng.* **2016**, *164*, 206–213. [CrossRef]
32. Haggag, S.; Elnahas, S. Event-based detection of the digging operation states of a wheel loader earth moving equipment. *Int. J. Heavy Veh. Syst.* **2013**, *20*, 157–173. [CrossRef]
33. Beleiu, I.; Crisan, E.; Nistor, R. Main factors Influencing Project Success. *Int. Manag. Res.* **2015**, *11*, 59–72.
34. Cheuk, A.N.; Leung, J.T.; Tse, P.W. Effective Architecture for Web-Based Maintenance System and Its Security. In Proceedings of the ASME 2005 International Design Engineering Technical Conferences and Computers and Information in Engineering Conference: 20th Biennial Conference on Mechanical Vibration and Noise, Long Beach, CA, USA, 24–28 September 2005; Volume 1, pp. 601–606. [CrossRef]
35. Bahnassi, H.; Hammad, A. Near Real-Time Motion Planning and Simulation of Cranes in Construction: Framework and System Architecture. *J. Comput. Civil. Eng.* **2012**, *26*, 54–63. [CrossRef]
36. Lee, G.; Cho, J.; Ham, S.; Lee, T.; Lee, G.; Yun, S.-H.; Yang, H.-J. A BIM- and sensor-based tower crane navigation system for blind lifts. *Autom. Constr.* **2012**, *26*, 1–10. [CrossRef]
37. Barati, K.; Shen, X. Modeling and optimizing fuel usage of on-road construction equipment. In Proceedings of the Construction Research Congress 2018: Sustainable Design and Construction and Education—Selected Papers from the Construction Research Congress, New Orleans, LA, USA, 2–4 April 2018; pp. 198–207. [CrossRef]
38. Albrektsson, J.; Aslund, J. Fuel Optimal Control of an Articulated Hauler Utilising a Human Machine Interface. In Proceedings of the 2018 IEEE International Conference on Industrial Technology (ICIT), Lyon, France, 19–22 February 2018; pp. 175–180. [CrossRef]
39. Kokot, G.; Ogierman, W. The numerical simulation of FOPS and ROPS tests using LS-DYNA. *Mechanika* **2019**, *25*, 383–390. [CrossRef]
40. Petroutsatou, K.; Giannoulis, P. Analysis of construction machinery market: The case of Greece. *Int. J. Constr. Manag.* **2020**. [CrossRef]
41. Saaty, T. T. Decision making with the analytic hierarchy process. *Int. J. Serv. Sci.* **2008**, *1*, 83–98. [CrossRef]
42. Shapira, A.; Goldenberg, M. AHP-based equipment selection model for construction projects. *J. Constr. Eng. Manag.* **2005**, *131*, 1263–1273. [CrossRef]
43. Jato-Espino, D.; Castillo-Lopez, E.; Rodriguez-Hernandez, J.; Canteras-Jordana, J.C. A review of application of multi-criteria decision making methods in construction. *Autom. Constr.* **2014**, *45*, 151–162. [CrossRef]
44. Saaty, T.; Ozdemir, M. The unknown in decision making. What to do about it. *Eur. J. Oper. Res.* **2006**, *174*, 349–359. [CrossRef]
45. Nassar, K.; Thabet, W.; Beliveau, Y. A procedure for multi-criteria selection of building assemblies. *Autom. Constr.* **2003**, *12*, 543–560. [CrossRef]

46. Transparent Choice AHP Software. Available online: https://www.transparentchoice.com/ahp-software (accessed on 8 April 2022).

47. Petroutsatou, K.; Ladopoulos, I.; Nalmpantis, D. Hierarchizing the Criteria of Construction Equipment Procurement Decision Using the AHP Method. *IEEE Trans. Eng. Manag.* **2021**, 1–12. [CrossRef]

48. Saaty, T.L. How to Make a Decision: The Analytic Hierarchy Process. *Interfaces* **1994**, *24*, 19–43. [CrossRef]

49. Tsafarakis, S.; Gkorezis, P.; Nalmpantis, D.; Genitsaris, E.; Andronikidis, A.; Altsitsiadis, E. Investigating the preferences of individuals on public transport innovations using the Maximum Difference Scaling Method. *Eur. Transp. Res. Rev.* **2019**, *11*, 3. [CrossRef]

50. Wood, H.; Gidado, K. Project Complexity in Construction. In *The International Construction Conference, Royal Institute of Chartered Surveyors, RICS COBRA*; RICS Foundation UK: Dublin, Ireland, 2008.

51. Tangkar, M.; Arditi, D. Innovation in the Construction Industry. *Civ. Eng. Dimens.* **2000**, *2*, 96–103. [CrossRef]

52. Duffuaa, S.; Alfares, H. Methods and Approaches for Maintenance Capacity Planning. In Proceedings of the 2015 International Conference on Industrial Engineering and Operations Management, Dubai, United Arab Emirates, 3–5 March 2015.

53. Alabdulkarim, A.; Ball, P.; Tiwari, A. Rapid modeling of field maintenance using discrete event simulation. In Proceedings of the Winter Simulation Conference, Phoenix, AZ, USA, 11–14 December 2011. [CrossRef]

54. Nepal, M.P.; Park, M. Downtime model development for construction equipment management. *Eng. Constr. Archit. Manag.* **2004**, *11*, 199–210. [CrossRef]

55. Edwards, D.; Parn, E.; Sing, M.; Thwala, W.D. Risk of excavators overturning. Determining horizontal centrifugal force when slewing freely suspended loads. *Eng. Constr. Archit. Manag.* **2019**, *26*, 479–498. [CrossRef]

56. Edwards, D.; Yang, J.; Wright, B.C. Establishing the link between plant operator performance and personal motivation. *J. Eng. Des. Technol.* **2007**, *5*, 173–187. [CrossRef]

57. Jukic, D.; Carmichael, G. Emission and cost effects of training for construction equipment operators. *Smart Sustain. Built Environ.* **2016**, *5*, 96–110. [CrossRef]

58. Edwards, D. An artificial intelligence approach for improving plant operator maintenance proficiency. *J. Qual. Maint. Eng.* **2002**, *8*, 239–252. [CrossRef]

59. Aryal, A.; Ghahramani, A.; Becerik-Gerber, B. Monitoring fatigue in construction workers using physiological measurements. *Autom. Constr.* **2017**, *82*, 154–165. [CrossRef]

60. Sneddon, A.; Mearns, K.; Flin, R. Stress, fatigue, situation awareness and safety in offshore drilling crews. *Saf. Sci.* **2013**, *56*, 80–88. [CrossRef]

61. Dinakar, A. Delay Analysis in Construction Project. *Int. J. Emerg. Technol. Adv. Eng.* **2014**, *4*, 784–788.

62. Petroutsatou, K.; Ladopoulos, I. Integrated Prescriptive Maintenance System (PREMSYS) for Construction Equipment Based on Productivity. *IOP Conf. Ser. Mater. Sci. Eng.* **2020**, *1218*, 012006. [CrossRef]

63. Izetbegović, J.; Nahod, M.-M. The impact of the additional workload on the productivity in construction projects. In Proceedings of the Creative Construction Conference, Prague, Czech, 21 June 2014.

64. Choi, B.; Hwangb, S.; Leec, S.H. What drives construction workers' acceptance of wearable technologies in the workplace?: Indoor localization and wearable health devices for occupational safety and health. *Autom. Constr.* **2017**, *84*, 31–41. [CrossRef]

65. Barlow, J. Innovation and learning in complex offshore construction projects. *Res. Policy* **2000**, *29*, 973–989. [CrossRef]

66. World Health Organization 2015. Available online: https://apps.who.int/iris/bitstream/handle/10665/154589/9789241508513_eng.pdf (accessed on 8 April 2022).

67. Neitzel, R.; Daniell, W.; Sheppard, L.; Davies, H.; Seixas, N. Comparison of Perceived and Quantitative Measures of Occupational Noise Exposure. *Ann. Occup. Hyg.* **2008**, *53*, 41–54. [CrossRef]

68. Duffy, S.; Choi, S.H.; Hollern, R.; Ronis, D. Factors Associated with Risky Sun Exposure Behaviors Among Operating Engineers. *Am. J. Ind. Med.* **2012**, *55*, 786–792. [CrossRef]

69. Eger, T.; Salmoni, A.; Whissell, R. Factors influencing load–haul–dump operator line of sight in underground mining. *Appl. Ergon.* **2004**, *35*, 93–103. [CrossRef]

70. Fang, D.; Huang, Y.; Guo, H.; Lim, H.W. LCB approach for construction safety. *Saf. Sci.* **2020**, *128*, 104761. [CrossRef]

71. Elazouni, A.; Basha, I. Evaluating the performance of construction equipment operators in Egypt. *J. Constr. Eng. Manag.* **1996**, *122*, 109–114. [CrossRef]

72. Parsakho, A.; Hosseini, S.A.; Jalilvand, H.; Lotfalian, M. Physical soil properties and slope treatments effects on hydraulic excavator productivity for forest road construction. *Pak. J. Biol. Sci.* **2008**, *11*, 1422–1428. [CrossRef]

73. Devi, P.; Palaniappan, S. A study on energy use for excavation and transport of soil during building construction. *J. Clean. Prod.* **2017**, *164*, 543–556. [CrossRef]

74. Luo, Q.; Huang, L.; Xue, X.; Chen, Z.; Zhou, F.; Wei, L.; Hue, J. Occupational health risk assessment based on dust exposure during earthwork construction. *J. Build. Eng.* **2021**, *44*, 103186. [CrossRef]

75. Chen, X.; Guo, C.; Song, J.; Wang, X.; Cheng, J. Occupational health risk assessment based on actual dust exposure in a tunnel construction adopting roadheader in Chongqing, China. *Build. Environ.* **2019**, *165*, 106415. [CrossRef]

76. Ahn, C.; Lee, S.H. Importance of Operational Efficiency to Achieve Energy Efficiency and Exhaust Emission Reduction of Construction Operations. *J. Constr. Eng. Manag.* **2013**, *139*, 404–413. [CrossRef]

77. Havard, C.; Rorive, B.; Sobczak, A. Client, Employer and Employee: Mapping a Complex Triangulation. *Eur. J. Ind. Relat.* **2009**, *15*, 257–276. [CrossRef]
78. Mohamed, S. Safety Climate in Construction Site Environments. *J. Constr. Eng. Manag.* **2002**, *128*, 375–384. [CrossRef]
79. Smithers, G.; Walker, D. The effect of the workplace on motivation and demotivation of construction professionals. *Constr. Manag. Econ.* **2000**, *18*, 833–841. [CrossRef]

 *buildings*

Article

What Are the Readability Issues in Sub-Contracting's Tender Documents?

Ahmed Yousry Akal

Civil Engineering Department, Higher Institute of Engineering and Technology, Kafrelsheikh, Kafrelsheikh City 33511, Egypt; a_yousry85@yahoo.com

Abstract: Readability is an important aspect that each sub-contracting's tender documentation should have in order to ensure commonality in the interpretation of its terms by the general contractor and sub-contractor. Otherwise, their contractual relationship is fueled by conflict. Previous studies indicated that the documents provided to the sub-contractors in practice are often not easy to read; the reason behind this problem has not been explored yet. This paper bridges this gap by defining 14 readability issues, following a systematic content analysis of real documents of 34 tenders of the sub-contracting arrangement. Further, it introduces a framework of the anti-measures of the specified issues through examining the readability-associated literature. The research's chief finding is that 8 out of the 14 readability issues are responsible for 73.1184% of the ease-of-reading problems in the sub-contracting's tender documentation. These readability issues are as follows: poor presentation of the format of the tender documentation, sentences and clauses are too long and complicated, spelling and grammatical errors, abstractness or vagueness of words or sentences, using controversial phrases, repetition of provisions or clauses, poor illustration of procedure or process, and listing of irrelevant conditions to the tender scope. The study also, while discussing the readability issues, categorizes them into four pivots, including structural and presentation-related problems, lengthening and repetition-related problems, text-related problems, and terminology-related problems. To date, it is believed that such classification has not been realized in any of the prior literature. These results have implications that can benefit drafters by enabling them to know the possible dimensions of the readability problems and their countermeasures concerning the sub-contracting's tender documents for up-skilling their drafting style when formulating such documentation in the future.

Keywords: readability; sub-contractors; sub-contracting; tender documents; construction

1. Introduction

Sub-contracting is a contractual process where a firm or an individual adheres to its responsibilities and duties on behalf of another [1]. Nowadays, sub-contracting has gained worldwide prominence in the construction community [2,3]. According to Ulubeyli et al. [2], it has become an essential practice in any construction project to find that the project's general contractor is more focused on planning, organizing, and monitoring his/her project activities. Yet, the majority of the project's actual production works is implemented via the sub-contracting arrangement. In accordance with Hinze and Tracey [4], the volume of the works done by the sub-contractors in a project may represent, in many cases, 80–90% of the whole project's scope. This high percentage is owing to the technical and strategic functions that the sub-contractor can present to the general contractor. Technically, given the general contractor's lack of experience in executing the project's specialized trades and services such as painting, insulation, plumbing, etc., the necessity to hire the specialist sub-contractors for implementing these works is imperative [3,5]. This, indeed, does not only enable the general contractor to adequately finish his/her project's specialized trades and services, but further contribute to realizing them at lower costs more quickly [6]. Strategically, on the other hand, the general contractor's gains from the sub-contracting

Citation: Akal, A.Y. What Are the Readability Issues in Sub-Contracting's Tender Documents? *Buildings* **2022**, *12*, 839. https://doi.org/10.3390/buildings12060839

Academic Editors: Srinath Perera, Albert P. C. Chan, Dilanthi Amaratunga, Makarand Hastak, Patrizia Lombardi, Sepani Senaratne, Xiaohua Jin and Anil Sawhney

Received: 12 April 2022
Accepted: 13 May 2022
Published: 16 June 2022

Publisher's Note: MDPI stays neutral with regard to jurisdictional claims in published maps and institutional affiliations.

practice are sharing the project risks with the sub-contractor, easing his/her cash flow and financing related problems, and reducing his/her overhead adherence, such as office staff, accommodations, etc. [3,7].

Emphatically, all of the aforementioned functions highlight that the general contractor's capability to deliver his/her project within quality, schedule, and cost objectives depends significantly on receiving the sub-contractors services [7]. This, in turn, indicates that the sub-contractors are key pillar in executing the construction industry's pertinent projects and realizing their success. Generally, the prime contractor can obtain the sub-contractors services relying upon the tendering approach, including any one of the forms of negotiated tendering, open tendering, selective tendering, pre-registered tendering, and annual tendering [8]. Definitely, utilizing any of these forms by the general contractor to sublet a part of his/her project is associated with providing the sub-contractor(s) with the tender documents. The tender documents are a package of documentation, encompassing: an invitation letter to the tender, instructions for the bidders, tender form and appendices, contract conditions, specifications, design drawings, and bill of quantities and schedule of rates [9]. Building on the clarity and consistency of these documents, the tenderer can be equipped with sufficient information, including the financial, contractual, legal, administrative, and technical aspects regarding the tender scope. This information, in turn, enables the tenderer to perfectly study the tender, know his/her contractual obligations and rights, and price its schedule of rates easily and accurately [1,10]. Hence, the more clarity and consistency the tender documentation has, the more certainty the tenderer has when interpreting his/her assigned responsibilities and rights. Conversely, the less clarity and consistency the tender documents have, the more ambiguous the understanding of their terms becomes.

According to Youssef et al. [11], the clarity- and consistency-related issues of the words, sentences, paragraphs, and clauses in a contract's textual documentation are known as the readability issues. The severity of this issue lies in that when the readability of a text in a contract document is low, its possibility for being interpreted in terms of low commonality degrees by the contracting parties is high [12]. This, unfortunately, makes the consistency between the contract parties on their duties and rights unattainable. Consequently, their contractual relationship is fueled by disputes [13]. Focusing on the readability issue in the sub-contracting's tender documentation, the sub-contractors confirm that the documents provided to them in practice are often not plain and consistent [1,14]. Regrettably, since the construction community has been plagued by this problem, and hitherto, there has not been a sufficient answer for the next question: "what are the readability issues in the sub-contracting's tender documents?". The reason is completely comprehensible, as the sub-contractors associated studies always receive unfair interest from the construction industry researchers [14]. Therefore, it is not surprising to examine the literature on the factors affecting the construction documents' readability, which is really very limited [15], to find that scant investigations, if any, have been conducted on the sub-contracting documentation. This gap can negatively influence the success of applying the disputes-avoiding mechanisms (DAMs) of the sub-contracting arrangement. This is because the analysts of the DAMs (e.g., [16]) clearly reported that providing easy-to-read documents without ambiguity or contradictions in their interpretation is among the top-ranked effectual ways for avoiding the disputes in the sub-contracting arrangement. As Chong and Zin [13] explained, this mechanism is a proactive-based dispute preventing approach, and accordingly, its achievement depends on a previous knowledge of the sources of unclarity and inconsistency in the contract documentation for being eliminated. Thence, the lack of specifying these sources impedes the shaping of a disputes-free contractual relationship between the general contractor and the sub-contractor.

Against this backdrop, this research intends to draw the answers of the two most frequently raised questions in the construction community: (1) "what are the readability issues in the sub-contracting's tender documents?" and (2) "what are the measures for enhancing the readability in the sub-contracting's tender documents?". Based on the answers to these questions, the consequences of the present paper are twofold. First, it acquaints the drafters of the sub-contracting's tender documents with the agents responsible for the readability issues in these documents and their anti-measures. This is a highly desirable knowledge because, upon its basis, the drafters can remove the sources of the unclarity and inconsistency from the sub-contracting's tender documentation. Accordingly, the interpretation of these documents' content becomes clearer and more comprehensible, fostering the agreement between the general contractor and the sub-contractor on their contractual responsibilities and rights. This, in turn, establishes a harmonious framework free of the lesion of disputes between the general contractor and the sub-contractor. Drawing on this implication, the second contribution of the research is realizing a proactive anti-dispute strategy in the sub-contracting practice by providing easy-to-read documents for its contracting parties, without fuzziness or inconsistency in their explanation. These contributions will be realized by examining real documents of 34 tenders of the sub-contracting arrangement in Egypt, employing the Content Analysis Approach (CAA). This is one of the first recognized endeavors to define the readability issues in the sub-contracting's tender documentation from the documents submitted to the sub-contractors in practice. This is for both the international construction community in general and the developing construction markets like that of Egypt in particular.

This study chose to consider the case of Egypt's construction sector as the research context, given the greater expansion of the Egyptian government than ever before in terms of executing several mega national projects for serving its economic growth. According to a recent report on Egypt, the values of the contracts awarded in 2020 and those underway in Egypt are nearly USD 14.9 billion and USD 435.9 billion, respectively, positioning the country as the third-biggest project market in the Middle East and North Africa [17]. Undoubtedly, this expansion cannot be realized without the effective cooperation between the Egyptian prime contractors and their sub-contractors. Certainly, the success of this cooperation requires the contractual relationship between the general contractor and his/her sub-contractors to be free of the troubles of conflicts, disputes, litigations, and legal proceedings for running their construction project smoothly. As a proactive management strategy against these troubles [15], the tender documentation of the sub-contracting should be written clearly to ensure commonality in the interpretation of their terms by the general contractor and sub-contractor. Unfortunately, the literature in Egypt on the issues pertinent to the readability problem and their countermeasures with respect to the sub-contracting's tender documents is silent similar to their counterparts in the developing and developed countries. This gap, in turn, portends severe consequences on the effectiveness of the cooperation between the general contractor and the sub-contractor in specific and the construction project's work progress in general, either in the Egyptian construction market or any construction sector elsewhere. Hence, this research, by addressing the readability issues in the sub-contracting's tender documents in Egypt, bridges a significant gap in the construction tender management literature. More significantly, it serves as a pioneering study for directing the scholars in other economies to take a step forward towards examining their sub-contracting's tender documentation for assessing and improving their readability and consistency.

The remainder of this paper reviews, in Section 2, the research methods of the prior works concerning scrutinizing the readability issues in the construction documentation and their countermeasures. Further, it outlines the gaps un-approached by these works. Section 3 involves the methodology adopted to extract the readability issues from the assembled sub-contracting's tender documents and to define their anti-measures. Section 4 analyzes the findings and compares them with those of the found peer researches of the developing economies to generalize the implications of the study towards these countries. Section 5 discusses the findings and their implications. Finally, Section 6 sums up the study and introduces its limitations, along with the future research directions.

2. Literature Review

Generally, text readability is described as the measure of reflecting the ease of reading of a written textual document and comprehending its content [12]. For embedding this measure in the construction documentation, it is necessary to provide the industry's drafters with the factors obstructing the comfortability in reading and apprehending these documents, along with their corresponding countermeasures. Disappointingly, the responses of the construction industry researchers to these necessities are countable. More critically, most of the scholars' efforts have been focused on one type of construction documentation, i.e., the contracts (e.g., [13]). In accordance with Youssef et al. [11] and Koc and Gurgun [15], the works associated with exploring the readability issue in the contracts have been based upon: (a) comparative-based case study, (b) text analysis algorithms, (c) interview, and (d) questionnaire survey. In the course of the comparative-based case study, Broome and Hayes [18] concentrated on investigating the drafting style of the New Engineering Contract (NEC), comparing it to that of the FIDIC contracts. Building on interviews with 81 personnel from the organizations of the employers, contractors, and sub-contractors, the study denoted that the NEC conditions are clearer and more understandable than those of the FIDIC contracts. This has been ascribed to the improper drafting of the FIDIC conditions in terms of having too-long sentences, several redundant legal expressions, and poor layout. By Lam and Javed [19], another comparison has been fulfilled between the practitioners in the United Kingdom and Australia to recognize the probable pitfalls in the output specifications of the contracts interrelated with the public–private partnership/private finance initiative. Referring to many cross-referencing to other documents has been highlighted as an influential readability issue in emerging the pitfalls in the output specifications.

Using the text analysis algorithms, the second literature strand in the body of knowledge of the readability issue has been emerged. Rameezdeen and Rajapakse [12] measured the readability in the NEC 1993 and FIDIC 1999 New Red Book, utilizing the Flesch Reading Ease Score (FRES) algorithm of the text analysis. This algorithm employs the average sentence length along with the average figures of syllables per word to denote the reading degree of a text. Further, its standard range is from 0 to 100, where the closer the FRES is to 100, the higher a text's ease-of-reading becomes. Based on this algorithm, the FRES values of the NEC 1993 and FIDIC 1999 are 40.70 and 29.70, respectively, indicating the high readability of the NEC 1993. Six years afterward, Rameezdeen and Rodrigo [20] utilized the algorithms of the FRES, Average Sentence Length (ASL), and Average Packet Length (APL) to quantify the readability of the clauses pertinent to the FIDIC Red Book versions: 1969, 1977, 1987, and 1999. The independent variables in the ASL and APL are the number of words, sentences, and packets in the clause. Moreover, the lower the scores of the ASL and APL are, the higher the clause's readability is. According to these algorithms, FIDIC 1999 has been termed as the easiest readable edition because it has the highest FRES with the lowest ASL and APL, in comparison with the three other editions. A year later, the FRES algorithm has been called up again by Rameezdeen and Rodrigo [21] to study the impact of modifying the standard forms-based contracts on the readability. Using 281 amended clauses from 12 infrastructure projects executed in Sri Lanka against their original counterparts in FIDIC 1987 and 1999, the researchers concluded that amending the originally drafted clauses makes their clarity and readability too difficult process.

In another line of efforts, the research strategies of the questionnaire survey and interview shaped the mainstream trend in discussing the features of the readability issue, especially in the developing economies. In Malaysia, Chong and Zin [13] administered a questionnaire-based survey of 11 problems related to the clarity of the standard form-based contracts utilized by the public sector. Based on the responses of 30 Malaysian experts, lengthening the wording of the contract clauses' sentences has been graded as the top-ranked cause behind contract unclarity. Menches and Dorn [22], additionally, surveyed 26 students of a construction management course to scrutinize their emotional reactions towards drafting the contract clauses in both positive and negative styles of language. The findings illustrated that formulating the contract clauses in a positive manner of language raises the reader's positive emotional reactions, and vice versa. Three years later, Chong and Oon [23] carried out a two-round Delphi survey to explore the feasibility of using plain language in elucidating the legal formulating in Malaysia's construction contracts. All of the 12 participants in the survey unanimously affirmed that formulating the contract clauses in plain language serves as a line of defense against many readability issues, encompassing the sentences' length as well as their presentation in passive voice and negative manner of writing. In the same vein, Masfar [24] reaffirmed that simplifying the language style of the public works contract within Saudi Arabia by using plain language is essential to avert the readability problems of the length, complexity, and ambiguity of the contract clauses.

In another investigation, additionally, following the semi-structured interview research approach, Besaiso et al. [25] analyzed the perspectives of 12 Palestinian professionals concerning the readability, clarity, interpretation, and understanding of the clauses associated with FIDIC 1999 Red Book. In this respect, the experts criticized the readability and lucidity of the FIDIC clauses, given the extensive use of cross-referencing, the length of the sentences, and the presence of phrases with uncertain/double meanings. Most recently, through a comprehensive review and face-to-face group interview with three experts, Koc and Gurgun [15] presented 18 risks influencing the construction contracts' readability. The identified risks were then included in a questionnaire survey that used the Fuzzy Visekriterijumska Optimizacija I Kompromisno Resenje approach to assess their consequences on the readability of the contracts. The replies of 18 experts indicated that the unnecessary complexity in utilizing nouns and the inappropriate employing of referents is the most significant risk contributing to rise in readability issues in contract documents. Far from the few realized studies concerning exploring readability problems in contracts, fewer researches have been accomplished by Ali and Wilkinson [26], Chong and Zin [13], and Chong and Oon [23] to determine their countermeasures. These works have been achieved relying upon two methodologies. First, reviewing the related archival literature, either to compile a list [13] or develop a guideline [23,26] of the measures that can be followed to confront the readability issues. Second, surveying the compiled list of the measures among the practitioners to investigate the extent to which the presented measures are influential to boost the contracts' readability [13]. Drawing on these efforts, the scholarly-based knowledge has been provided with an important guideline of several measures for improving the contracts' readability. More details of these measures can be found in the aforementioned studies.

On the basis of the foregoing discourse, the prior works can be characterized by four noteworthy features. First, the studies in the area of the readability of construction documents are too limited, emphasizing on the contracts. Second, the research approaches of the interview and questionnaire survey have been broadly used in the methodologies of the readability works. Although utilizing these methods captures evidence from the extensive expertise of the parties involved in the contracts, the evidence is anecdotal [27]. More critically, usually the contributions provided in accordance with these methods are influenced by the number of the participants in the study. This, in turn, adds a major limitation to the extracted findings in terms of their generalization and representation [15]. Third, concerning the other literature on the readability, in which their approaches have been built on text-analysis algorithms, their outcomes are not sufficient to be relied upon for reflecting the contract's readability risks. This is completely understandable, as the independent variables of these algorithms do not consider the grammatical structure or the language style of the evaluated contract clause. These algorithms, however, appraise the readability of the contract clause in terms of the number of its words, sentences, and packets. Fourth, neither the researches associated with the questionnaire survey and interview nor those related to the text-analysis algorithms have been interested in touching on the readability issues with respect to the sub-contracting's tender documents.

Aggregating the aforementioned features together, the result is that there is an urgent need to perform a systematic examination of the sub-contracting's tender documents to obtain a deep and realistic comprehension of the readability issues in these documents. Hence, a better allocation for the anti-measures of these issues can be realized. Consequently, in a more clear and consistent manner, the sub-contracting's tender documentation can be drafted in the future for boosting the commonality in the explanation of their terms by the general contractor and sub-contractor.

3. Research Methodology

To objectively answer the study questions, the author adopted a scientifically sound and broadly utilized methodology consisting of 5 steps. Figure 1 summarizes the target, outcome, and sequence of these steps. Additionally, each step will be illustrated in detail within the subsequent sections.

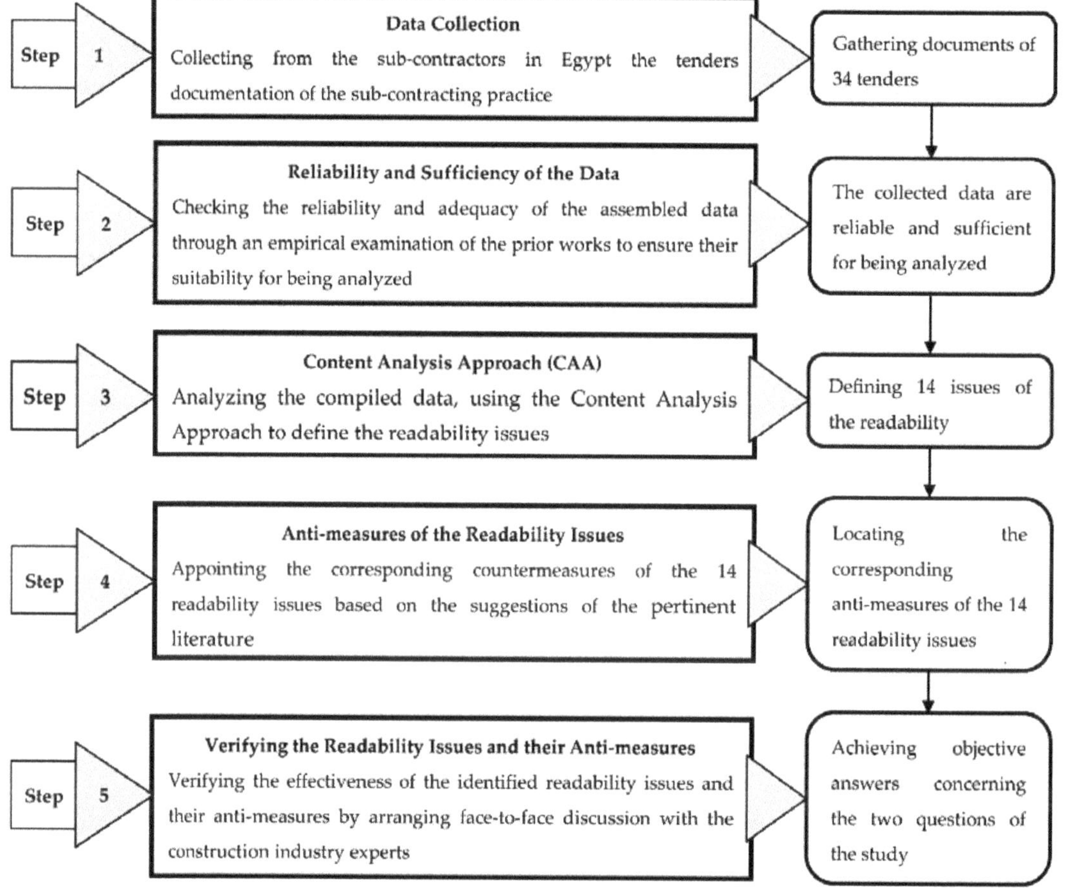

Figure 1. Research methodology steps.

3.1. Data Collection

In this research, the source of the data is the documents submitted to the sub-contractors in practice during their tender process with the general contractors. Obtaining data in studies pertinent to construction management has several methods, such as a questionnaire survey and interviewing. However, extracting the data from real documentation or contracts presents direct and factual information with respect to the issue being investigated. More importantly, it handles the shortcomings of the data gathered relying upon the questionnaire survey and interviewing in terms of the potential recall and bias of the participants in the survey or the interview [27]. To this end, two Egyptian sub-contractors based on the author's personal acquaintances have been contacted to provide the sub-contracting's tender documents. The firms of these two sub-contractors have been established in 1998 and 2017. Moreover, they have the last grade (i.e., seven) according to the classification system of the Egyptian Federation for Construction and Building Contractors (EFCBC), which is responsible for grading the construction companies in Egypt based on their capitals, employee numbers, and assets. Depending on these two Egyptian sub-contractors, the documents of 34 tenders have been compiled to form the data of this paper.

Table 1 shows the characteristics of the collected tenders in terms of their issued year, number of pages, and scopes. It appears by examining the tender documents that they are released from one of the leading construction companies in Egypt. This firm's class, in accordance with the classification system of the EFCBC, is a first-grade company. Owing to its participation in many mega national projects, which involve a lot of the specialized works, it always depends on receiving the sub-contractors services for accomplishing its contracted projects. A deep examination of the tender documents, additionally, informs that their common contents are a simple invitation letter to the tender, bill of quantities and schedule of rates, specific and general conditions, and requirements related to the occupational safety and health. Yet, the specifications and design drawings have been found in a little of the tender documentation. These tenders are: 06, 07, 11, 13, 18, 22, 23, 24, 27, 29, 31, and 33. It is worth mentioning that all the tender documents have been written in Arabic, since Egypt's first language is Arabic. Nevertheless, English has been used to describe some terms, mainly in the bill of quantities and schedule of rates as well as the design drawings.

Another observation concerning the tender documents is that the sub-contractors have been invited to the tenders and received their documentation via their e-mail accounts. However, if the sub-contractors want to participate in the tenders, they have to deliver their documents in hand to the sub-contractor department of the prime contractor. The last column of Table 1 includes the scopes of the tenders sent to the sub-contractors. As this column presents, various trades relevant to the civil, architectural, electrical, and mechanical engineering disciplines have been mentioned. In the civil engineering field, the main activities are: plain and reinforced concrete; excavation and dewatering; joint sealing; compaction and paving; road signs and surface markings; fencing and gates; insulation; laying curbs and interlocking tiles; and building using stones. Yet, the architectural trades encompass the works of aluminum and glass doors and windows, floor covering, and finishing. As for the electrical and mechanical specializations, the associated trades are: installing and commissioning of an electrical and mechanical filtration system for pools, establishing high-density polyethylene pipelines for drainage and cable protection, and electrical installation and commissioning of a fire alarm system. Certainly, the diversity in the tenders' scopes means that the drafting style of their documentation is different from one tender to another. This diversity, in turn, affords an excellent opportunity for the current study for illustrating several factors of the readability issue in the sub-contracting's tender documents.

Table 1. Tender documents characteristics.

Tender No.	Release Data	No. of Pages	Scope of Sub-Contracting Package
01	2017	10	Reinforced concrete, including shuttering, fabrication and erection of steel rebar, and pouring.
02	2017	9	Manual excavation, dewatering, and transferring of the excavation output.
03	2017	9	Sealing of joints in concrete slabs.
04	2017	12	Installing and commissioning of an electrical and mechanical filtration system for pools.
05	2017	9	Establishing the base-course layer in a highway.
06 *	2017	31	Road signs and surface markings.
07 *	2017	18	Fencing.
08	2018	9	Manual excavation.
09	2018	7	Mechanical drilling.
10	2018	9	Laying and leveling of concrete floors.
11 *	2018	13	Fencing and gates.
12	2018	9	Aluminum doors and windows.
13 *	2018	11	Road signs.
14	2018	13	Finishing works.
15	2018	9	Insulation.
16	2018	20	Earthworks, plain and reinforced concrete, and finishing works.
17	2018	9	Floor covering using ceramic, porcelain, and marble.
18 *	2018	19	Finishing works.
19	2018	10	Laying interlocking tiles and sealing of expansion joints.
20	2018	9	Laying curbs and interlocking tiles.
21	2018	10	Paving.
22 *	2018	25	Finishing works.
23 *	2018	35	Finishing works.
24 *	2018	11	Glass and glazing of doors.
25	2018	10	Plain and reinforced concrete, including shuttering, fabrication and erection of steel rebars, and pouring.
26	2018	10	Repairing and insulating concrete surfaces against water leakage.
27 *	2019	6	Insulation.
28	2020	7	High-density polyethylene piping for drainage.
29 *	2020	12	Electrical installation and commissioning of a fire alarm system.
30	2020	7	High-density polyethylene piping for protecting cables.
31 *	2020	9	Road signs.
32	2020	7	Building using riprap stones.
33 *	2020	29	Rail information and directional signboards.
34	2021	18	Finishing works.

* means the tender documents contain the specifications or the design drawings.

3.2. Reliability and Sufficiency of the Data

In view of the compiled documentation of the tenders, reliable and objective outputs from their analysis can be drawn. This is related to two reasons. First, since the documents of the assembled tenders reflect real-life cases from the construction community and they will be subjected to the CAA, they are precious for presenting reliable findings to the construction management literature. This has been assured by Li et al. [28] that analyzing real documented construction data using the CAA affords more trustworthy results than those relevant to the questionnaire survey and interview. Second, several studies—in which the data sources are real construction contracts, documents, and reports as well as the CAA are their major analytical tool—have been conducted based on a smaller sample of the construction documentation than that collected in the current research. For instance, the common causes of claims in Canada have been defined relying upon the data of 24 construction claim reports [29]. In addition, in the United States, Nguyen et al. [27] investigated the allocation of the risks in the public–private partnership (PPP) scheme on the basis of the content analysis of 21 contracts pertinent to the PPP projects. Undoubtedly, this empirical examination of the prior works indicates that the obtained data (i.e., 34 tenders) represents an acceptable sample for performing the CAA, and thus, it can be deemed as a firm foundation to afford objective results.

3.3. Content Analysis Approach

The CAA has been adopted to analyze the tender documents, so as to have a precise answer concerning the first question of the research: "what are the readability issues in the sub-contracting's tender documents?". The CAA is an observation-based research technique that is employed to systematically analyze the content of all the forms associated with the recorded communications [30]. Furthermore, it can be utilized with either the qualitative or quantitative information and in an inductive or deductive manner [31]. Owing to these features, the CAA has been employed extensively by the construction industry researchers to assist them to draw real data from the construction documentation, including reports, contracts, and news reports. This has been noted in the context of several important branches of the construction management researches, such as claims, PPP schemes, and prefabricated buildings (e.g., [27–29]).

To study the tender documents, a protocol of a three-step content analysis has been set. In the first step, the intention is to form an initial framework of the factors behind the readability issue in the sub-contracting's tender documentation. In this regard, the checklists of Chong and Zin [13] and Koc and Gurgun [15] have been relied upon as a guideline for exploring the readability issues in the assembled document packages. The registers of Chong and Zin [13] and Koc and Gurgun [15] have 15 and 18 risks influencing the construction contracts' readability, respectively. Moreover, they have 12 common risks, as has been mentioned in Koc and Gurgun [15]. More details pertinent to these two lists can be found in Chong and Zin [13] and Koc and Gurgun [15]. The reason for choosing these two checklists is that they have been developed following an accurate methodology, encompassing a comprehensive review of the relevant literature and validation with subject matter experts. Moreover, in addition to the lists of the readability issues of Chong and Zin [13] and Koc and Gurgun [15], to the best of the authors' knowledge, there is no other list, other than that of Chong and Oon [23]. However, the checklist of Chong and Oon [23] is completely similar to the checklist of Chong and Zin [13]. Therefore, the lists of Chong and Zin [13] and Koc and Gurgun [15] have been considered appropriate to direct the author when scrutinizing the tender documents.

Similar to the suggestion of Nguyen et al. [27], round one of the content analysis process has been based upon an initial set of the tender documentation. These documents belong to the tenders from 1–10 (see Table 1). This preliminary investigation is a very significant stage in the CAA to refine the checklists of Chong and Zin [13] and Koc and Gurgun [15], as they do not exemplify the readability issues of the sub-contracting's tender documentation. They, however, represent the construction contracts' readability risks. Appreciating this importance, the documents of each tender have been read in detail several times. According to Arshad et al. [32], this can help in realizing an objective understanding of the documentation content and preventing the author's subjectivity while extracting the result. At the end of studying the first 10 tender documents, 14 readability issues have been drawn. Table 2 presents these issues, illustrating that while 10 of the readability issues have been stated in the checklists of Chong and Zin [13] and Koc and Gurgun [15], the other 4 ones have been derived from analyzing the 10 tender documents. As a further refinement of the compiled list of the readability issues, the second step of the content analysis process has been started to examine the rest of the documentation (i.e., tenders from 11–34). Consequently, the possibility of adding any new unlisted issue is available. As has been followed in the prior step, the 24 documents packages have been carefully read multiple times. The finding indicated that the list of the readability issues of Table 2 is sufficient and no new issue has emerged.

In the third step of the content analysis, all the tender documents have been rechecked to reaffirm that no factor has been missed during the first and second rounds of scrutinizing the documents packages. Similar to the first and second steps of the content analysis procedure, the 34 tender documents have been accurately checked. The result of this stage affirmed that no new issue has been found, other than those mentioned in Table 2. This affirmation may be due to the precise investigation of the tender documentation during the first and second rounds of the content analysis process. These two rounds lasted for approximately 28 working hours over 2 weeks to extract the readability issues from the documents packages. Building on the finding of this step, all the found issues can be shown in Table 2, encompassing their negative impacts on the readability of the sub-contracting's tender documentation. In addition, it includes their sources, either from the relevant literature or the content analysis of the tender documents. It is worth mentioning that, in this step, for each readability issue, its Frequency of Appearance (FA), Relative Frequency of Appearance (RFA), and Ranking (R) have been defined for the statistical analysis. The FA of each readability issue has been determined by figuring up the number of times it appears in the tender documents. As for the RFA of each issue, it has been calculated by dividing its RA by the grand total of the RA of all the readability issues. Yet, for defining R, the issues have been ranked in a descending order of their RFA values, where the issue of the highest RFA receives the first rank. The FA of the readability issues can be found in Table 3, whereas their RFA and R appear in Table 4.

Table 2. Readability issues in the tender documents.

ID	Readability Issue	Negative Consequence on the Readability	Source		
			A	B	C
RI$_1$	Poor presentation of the format of the tender documentation (e.g., figures, tables, font, indentation, line spacing).	Adversely impacting the lucidity of the tender scope for the sub-contractor.		•	•
RI$_2$	Sentences and clauses are too long and complicated.	Reducing the willingness of the sub-contractors to read the tender documentation precisely; accordingly, overlooking matters that could be crucial in defining their obligations and rights.	•	•	•
RI$_3$	Spelling and grammatical errors (e.g., missing letters, nouns, and verbs, as well as poor sentence formation).	Impacting the sub-contractor to understand the tender documents' provisions and clauses correctly.	•	•	•
RI$_4$	Abstractness or vagueness of words or sentences.	Causing more than one meaning or misunderstanding for the sub-contractor.	•	•	•
RI$_5$	Using controversial phrases.	Resulting in interpreting the tender documents' provisions and clauses in a different sense than what the general contractor intends to tell.	•	•	•
RI$_6$	Using specific vocabulary, legal terms, and legal jargon.	Causing the clarity and readability problems owing to the presence of incomprehensible legal terminology for the sub-contractor.	•	•	•
RI$_7$	Referring to engineering terminology, code, or specification that are not frequent to all disciplines.	Causing the clarity and readability problems due to the presence of incomprehensible engineering terminology for the sub-contractor.		•	•

Table 2. *Cont.*

ID	Readability Issue	Negative Consequence on the Readability	Source		
			A	B	C
RI_8	Repetition of provisions or clauses.	Increasing the size of the tender documentation package; consequently, distracting the sub-contractor from the main provisions and clauses of the tender scope.	•	•	•
RI_9	Poor illustration of procedure or process.	Adversely impacting the information flow of the tender scope for the sub-contractor.	•		•
RI_{10}	Lack of/poor visual representations (e.g., drawings).	Adversely impacting the visual representation of the tender scope for the sub-contractor.		•	•
RI_{11}	Using abbreviations without illustrating their definitions.	Causing the clarity and readability problems as a result of the presence of incomprehensible acronyms for the sub-contractor.			•
RI_{12}	Listing conditions that are not related to the tender scope.	Increasing the size of the tender documentation package; consequently, distracting the sub-contractor from the main conditions of the tender.			•
RI_{13}	Inconsistencies among the tender clauses.	Resulting in divergent interpretations of the same clause.			•
RI_{14}	Transliteration of English words/idioms into Arabic.	Causing the clarity and readability problems given the presence of incomprehensible idioms for the sub-contractor.			•

A: [13]; B: [15]; C: content analysis of the tender documents.

Table 3. Frequency of appearance of the readability issues in the tender documents.

Tender No.	RI_1	RI_2	RI_3	RI_4	RI_5	RI_6	RI_7	RI_8	RI_9	RI_{10}	RI_{11}	RI_{12}	RI_{13}	RI_{14}
01	•	•	•	•	•	•		•	•	•		•		
02	•	•	•	•	•	•		•	•			•		
03	•	•	•	•	•	•		•	•	•		•		
04	•	•	•	•	•	•	•	•	•			•		•
05	•	•	•	•	•	•		•	•	•		•		
06	•	•	•	•	•	•	•	•	•	•	•	•		
07	•	•	•	•	•	•	•	•	•	•		•	•	•
08	•	•	•	•	•	•		•	•			•		
09	•	•	•	•	•	•		•	•	•		•		
10	•	•	•	•	•	•		•	•			•		
11	•	•	•	•	•	•		•	•	•		•	•	•
12	•	•	•	•	•	•		•	•	•		•		
13	•	•	•	•	•	•	•	•	•	•		•		•
14	•	•	•	•	•	•		•	•	•	•	•		•
15	•	•	•	•	•	•	•	•	•	•		•		
16	•	•	•	•	•	•	•	•	•	•	•	•		•
17	•	•	•	•	•	•		•	•	•		•		
18	•	•	•	•	•	•		•	•	•		•		
19	•	•	•	•	•	•	•	•	•	•		•		•
20	•	•	•	•	•	•		•	•			•	•	
21	•	•	•	•	•	•	•	•	•	•		•	•	
22	•	•	•	•	•	•	•	•	•	•	•	•		•

Table 3. *Cont.*

Tender No.	RI$_1$	RI$_2$	RI$_3$	RI$_4$	RI$_5$	RI$_6$	RI$_7$	RI$_8$	RI$_9$	RI$_{10}$	RI$_{11}$	RI$_{12}$	RI$_{13}$	RI$_{14}$
23	•	•	•	•	•	•	•	•	•	•	•	•		•
24	•	•	•	•	•	•		•	•	•	•	•		
25	•	•	•	•	•	•		•	•	•		•		
26	•	•	•	•	•	•		•	•	•		•		•
27	•	•	•	•	•		•	•	•			•		
28	•	•	•	•	•			•	•			•	•	•
29	•	•	•	•	•		•	•	•	•	•	•		•
30	•	•	•	•	•		•	•	•		•	•	•	
31	•	•	•	•	•		•	•	•	•		•	•	
32	•	•	•	•	•			•	•	•		•		
33	•	•	•	•	•		•	•	•	•	•	•	•	
34	•	•	•	•	•		•	•	•	•	•	•	•	•
FA	34	34	34	34	34	26	16	34	34	26	10	34	9	13

Table 4. Relative frequency of appearance and ranking of the readability issues.

ID	FA	RFA (%)	Ranking (R)
RI$_1$	34	9.1398	1
RI$_2$	34	9.1398	1
RI$_3$	34	9.1398	1
RI$_4$	34	9.1398	1
RI$_5$	34	9.1398	1
RI$_6$	26	6.9892	9
RI$_7$	16	4.3011	11
RI$_8$	34	9.1398	1
RI$_9$	34	9.1398	1
RI$_{10}$	26	6.9892	9
RI$_{11}$	10	2.6882	13
RI$_{12}$	34	9.1398	1
RI$_{13}$	9	2.4194	14
RI$_{14}$	13	3.4946	12
Grand Total	372	100%	

3.4. Anti-Measures of the Readability Issues

For avoiding the 14 specified readability issues, and consequently, improving the clarity of reading and understanding the sub-contracting's tender documentation, their anti-measures should be determined. This purpose is the scope of the second question of this paper: "what are the measures for enhancing the readability in the sub-contracting's tender documents?". To answer this question, the research associated with discussing the readability issues of the contracts (e.g., [15,25]) and their countermeasures (e.g., [13,23,26]) have been reviewed. Indeed, these studies do not include the anti-measures of all the 14 readability issues; they include the countermeasures of the readability issues RI$_1$, RI$_2$, RI$_6$, RI$_8$, and RI$_9$ and parts of those pertinent to RI$_4$ and RI$_7$. However, the deep scrutinizing of these researches guided the author to suggest the anti-measures of the rest of the readability issues. This is on the basis of the concept identified by these studies regarding the role of a countermeasure with respect to a readability issue. This notion is that the function of an anti-measure of a readability issue is minimizing its consequence or preventing its occurrence for making the reading easier, supporting the comprehension, and avoiding the misinterpretation risk. Building on this concept, the author has been enabled to derive the corresponding countermeasures of the rest of the readability issues. Table 5 elaborates the anti-measures of all the readability issues, together with their sources, either from the relevant literature or the author's suggestion. Further, it shows how these anti-measures can improve the readability of the sub-contracting's tender documentation.

Table 5. Anti-measures of the readability issues.

ID	Corresponding Anti-Measure	Positive Consequence on the Readability	Source				
			A	B	C	D	E
RI$_1$	Preparing adequate format for the tender documentation in terms of font size and type, indentation, line spacing, tables, and figures.	Improving the lucidity of the tender scope for the sub-contractor.	•				
RI$_2$	Reduce the number of words per sentence to be within 20 words.	Enabling the sub-contractor to easily read and comprehend the tender scope.	•	•	•		
RI$_3$	Reviewing the spelling and grammar of the tender documentation before being released.	Improving the readability and avoiding the misunderstanding risk.					•
RI$_4$	Draft the scope of the tender in an informative and understandable manner; Employ the words of the unique meaning, rather than those with multiple interpretations.	Supporting the clarity of the tender scope; Improving the readability and avoiding the misinterpretation risk.	•				•
RI$_5$	Avoiding the usage of the controversial phrases.	Avoiding the misinterpretation risk.					•
RI$_6$	Utilize everyday words; Abandoning the usage of legal language.	Increasing the clarity, readability, and understanding of the tender scope.		•	•	•	
RI$_7$	Employ engineering terminology frequent to all disciplinarians wherever possible; Attaching the necessary clauses of the referred code or specification with the tender's documentation package.	Enhancing the clarity, readability, and understanding of the tender scope.		•			•
RI$_8$	Eliminating the redundancy or repetition of words.	Reducing the size of the tender documentation package, leading to optimizing the concentration of the sub-contractor towards the tender scope.		•	•		
RI$_9$	Supporting the procedures/processes with flow chart or illustrative examples.	Enhancing the understanding of the sub-contractor in terms of the data of the tender scope; accordingly, avoiding the misunderstanding risk.		•	•		
RI$_{10}$	Attaching a clear presentation of all the related drawings with the tender documentation package.	Improving the visual representation of the tender scope for the sub-contractor.					•
RI$_{11}$	Mentioning the definitions of the utilized acronyms.	Increasing the clarity, readability, and understanding of the tender scope.					•
RI$_{12}$	Omitting the irrelevant conditions to the tender scope by eliminating the usage of the standard templates of the tender documentation.	Reducing the size of the tender documentation package, leading the sub-contractor to be more focused on the tender-relevant conditions.					•
RI$_{13}$	Checking the consistency among the tender clauses before releasing the tender documentation.	Avoiding the risks of misinterpretation and misunderstanding.					•
RI$_{14}$	Translating the English words/idioms into understandable Arabic phrases.	Improving the clarity, readability, and understanding of the tender scope.					•

A: [26]; B: [13]; C: [23]; D: [25]; E: author's suggestion.

3.5. Verifying the Readability Issues and their Anti-Measures

Although the assembled data have been discussed to be enough for undertaking the CAA and scientific sound steps have been followed to determine the readability issues and their anti-measures, the effectiveness of these factors needs to be verified. This is because, on the basis of the analysis conducted by the author on the compiled tender documents, by using the CAA, the readability issues of Table 2 have been revealed. Further, some of the countermeasures of Table 5 have been defined relying upon the author's suggestion. Hence, the subjectivity in outlining the elements of Tables 2 and 5 may exist. To check the soundness of the outcomes of Tables 2 and 5 as the factors responsible for causing the readability issues and controlling their consequences concerning the sub-contracting's tender documentation, interviews with the construction industry experts have been performed. The interviews have been arranged, employing face-to-face discussions with 3 experts. The number of experts is similar to the sample utilized by Koc and Gurgun [15] for verifying the suitability of their readability risks. Importantly, the experts' bio-data paid the author to appoint them from his personal network for conducting the interviews. In terms of their educational background, 2 of the experts hold Ph.D. in structural engineering, whereas the other has a bachelor's degree in civil engineering. As for their expertise within the construction field, it is lengthy, ranging from 16 to 18 years, with broad knowledge of the tendering procedures and their documents. This has been known from the top administrative positions which they occupy in their firms. While 2 of them are the owners of construction companies with grades of 6 and 7, according to the classification system of the EFCBC, the other expert is one of the project managers of a contracting firm with a grade of 1. Moreover, their companies have several contributions in the Egyptian construction sector, either as sub-contractors or general contractors.

To conduct the interviews, a package in a hard copy, encompassing a sample of the tender documentation, the readability issues of Table 2, and the anti-measures of Table 5 have been printed. Subsequently, each expert has been interviewed to discuss the sources of the readability issues as the author found in the sample of the tender documents. Moreover, at the interview, the expert has been asked to examine whether the factors of Table 2 cover the readability issues of the sub-contracting's tender documentation, or if some missing factors have to be involved. In the same vein, the countermeasures of Table 5 have been checked. All the interviewed experts unanimously highlighted that the elements of Table 2 reflect the relevant factors of the readability issues in the sub-contracting's tender documents and the anti-measures of Table 5 are sufficient to avoid their happening. This consensus, in turn, implies that the findings of this study are objective. Consequently, they can be introduced to the drafters of the sub-contracting's tender documents as effective solutions to formulate highly readable and consistent documents.

4. Analysis and Comparison of the Results

In this study, as Table 2 comprises, 14 issues, together with their negative consequences on the readability of the sub-contracting's tender documents, have been determined utilizing the CAA. Table 3 counts the FA of these readability issues as has been found while analyzing the documentation of the 34 sub-contracting tenders. In accordance with Table 3, 8 of the readability issues have been present in all the tender documents. They are RI_1, RI_2, RI_3, RI_4, RI_5, RI_8, RI_9, and RI_{12}. Yet, the other 6 issues, encompassing RI_6, RI_7, RI_{10}, RI_{11}, RI_{13}, and RI_{14} have appeared in some of the tender documents, with a FA ranging from 9 to 26. Based on the FA of the readability issues, their RFA and R have been computed. Table 4 includes these statistics. As this table presents, given the existence of RI_1, RI_2, RI_3, RI_4, RI_5, RI_8, RI_9, and RI_{12} in all the tender documents, they have the highest RFA of 9.1398%. As a result, they have been awarded the first ranking, and therefore, they are the most-frequent readability issues in the documentation of the sub-contracting tenders. Another observation from the analysis of these eight issues is that the summation of their RFA values is 73.1184%. This consequence, in turn, indicates that 73.1184% of the problems affecting the clarity of reading and understanding the sub-contracting's tender documents are associated with these 8 issues. Building on this finding, the consequence is that the more the focus on avoiding the occurrence of these issues is, the higher the possibility becomes for providing easy-to-read and comprehensible documentation of the sub-contracting tenders.

As can be extracted from Table 4, additionally, regarding the other six issues of the readability in terms of their RFA and R is that two of them, comprising RI_6 and RI_{10} have been ranked ninth, with an RFA of 6.9892%. Yet, RI_7 (RFA = 4.3011%), RI_{14} (RFA = 3.4946%), RI_{11} (RFA = 2.6882%), and RI_{13} (RFA = 2.4194%) have the positions of eleventh, twelfth, thirteenth, and fourteenth, respectively. With a deep insight into these six issues together, it can be summarized that they represent 26.8816% of the sources of the unclarity and inconsistency in the tender documents of the sub-contracting practice. Certainly, this small percentage can describe these six issues as factors with limited consequences with respect to the theme being discussed, especially when it is compared to the proportion relevant to the top-eight frequent issues of the readability. Nevertheless, neglecting their avoidance implies that the documents of the sub-contracting tenders are not perfectly functional for being understood without different interpretations or misunderstanding of their clauses. Hence, it is advised that, for drafting the sub-contracting's tender documents in a compatible and understandable manner, the readability issues of both those of the highest and lowest RFA in the tender documentation have to be addressed. Table 5 supports this end by identifying for each readability issue its corresponding anti-measure, along with its possible positive impact on improving the readability of the sub-contracting's tender documentation, regardless of its RFA.

The prior analysis of the readability issues is beneficial, whether for the drafters of the sub-contracting's tender documents or Egypt's construction sector, as this study has been performed with respect to these contexts. Nevertheless, associating the reached findings with those of the relevant literature can afford further consequences from the conducted analysis for being directed to a wider context. In this regard, the top-eight frequent issues of the readability have been compared with the outcomes of Chong and Zin [13] and Koc and Gurgun [15]. These works have been considered because they are the only ones that are concerned with grading the readability issues in descending order of their impact on grasping the construction documentation. Hence, their findings have been deemed appropriate for being compared with the outputs of the current study. As Table 6 illustrates, the context of the present paper is Egypt. In addition, the work of Chong and Zin [13] has been conducted in Malaysia for rating 11 readability issues. Yet, the study of Koc and Gurgun [15] is believed to be associated with Turkey's construction industry for sorting 18 readability risks. These features, in terms of the countries of these studies, indicate that the results of the comparison will be useful to the developing construction markets only.

Table 6. Rankings of the top-eight frequent issues of the readability in the developing countries.

Study Characteristics	Study	This Study	Chong and Zin [13]	Koc and Gurgun [15]
	Country	Egypt	Malaysia	Turkey
	Scope	Sub-Contracting's Tender Documents	Contracts	Contracts
	No. of Issues/Study	14	11	18
Ranking of the Readability Issues	RI_1	1st	-	12th
	RI_2	1st	1st	3rd
	RI_3	1st	9th	6th
	RI_4	1st	6th	2nd
	RI_5	1st	11th	5th
	RI_8	1st	3rd	9th
	RI_9	1st	10th	-
	RI_{12}	1st	-	-

-: means the readability issue has not been mentioned in that study.

According to Table 6, 3 out of 8 of the top-frequent readability issues of the present research have been assessed as highly ranked risks in Malaysia. These issues are RI_2, RI_4, and RI_8, having the first, sixth, and third places, respectively. On the other hand, 4 out of 8 of the most frequent issues of readability, including RI_2, RI_3, RI_4, and RI_5, have been marked with high scores in Turkey. Their associated ranks are third, sixth, second, and fifth, respectively. These two facts together mean that while RI_2 and RI_4 are readability issues, having a full occurrence of 100% in all the investigated countries, RI_3, RI_5, and RI_8, have a rate of frequency of 50%. In the same vein, the rest of the top-eight frequent issues of the readability in the sub-contracting's tender documents, comprising RI_1, RI_9, and RI_{12} are with an occurrence proportion of 0%. These statistics, in turn, classify the highly ranked readability issues in the construction documentation of the developing countries into 3 groups, as follows:

1. Group one: consists of RI_2 and RI_4 and it is the most critical group, since its associated issues have been graded across all the investigated countries as severe issues with respect to the readability of the construction documents;
2. Group two: includes RI_3, RI_5, and RI_8 and it is the second most critical group, as its related issues have appeared in 50% of the surveyed countries as issues with serious consequences on the clarity of interpreting the construction documentation;
3. Group three: involves RI_1, RI_9, and RI_{12} and it is the least critical group because its relevant issues have not been mentioned in the studied countries as issues with extreme impacts on the construction documents' readability.

Certainly, the aforementioned classification enriches the drafters of the construction documentation and the scholars in the developing countries with a prioritized plan to better comprehend the issues pertinent to their documents' readability. Accordingly, their efforts can be optimized to manage the effects of those issues; particularly this study affords them with the anti-measures of these issues, as Table 5 comprises. Another significant conclusion from Table 6 is that the researches of Chong and Zin [13] and Koc and Gurgun [15] have focused on the same type of construction documents, i.e., contracts. However, the ranks of their readability issues are somewhat different. For instance, in Chong and Zin [13], RI_5 and RI_8 have the positions of eleventh and third, respectively. Yet, in Koc and Gurgun [15], their associated ranks are fifth and ninth, respectively. These differences, in turn, denote that the ranks of the readability issues are context-bound, varying from country to country. Hence, the top-ranked issues of readability, with respect to the same type of construction document, can differ greatly relying upon the context of the country.

5. Discussion and Implications of the Results

This research highlights the readability issues in the sub-contracting's tender documents in Egypt. In light of reviewing the literature of the construction documents' readability risks, this investigation seems to be the first known contribution in this respect, either in Egypt or internationally. This supports the value of this study towards the knowledge account because it reveals the characteristics of the factors obstructing the comfortability in reading and apprehending the sub-contracting's tender documentation. This contribution has been achieved, using the CAA to analyze the documents of 34 tenders of the sub-contracting arrangement. As a result, 14 readability issues have been defined, along with their RF, RFA, and R for the statistical analysis. Of these, as Table 3 illustrates, 10 issues, including RI_1 to RI_{10} have been present in the prior works of the readability of contracts. Yet, four issues, from RI_{11} to RI_{14} have been noticed as distinctive factors regarding the sub-contracting's tender documents. This is a vital implication, because it adds 4 new elements to the limited existing risk checklists of construction documentation readability, particularly in the tender documents-related field. More significantly, it means that although the majority of the readability issues may be similar in different construction documents' types, each type of documentation has its relevant issues. Accordingly, it can be deduced that the construction documents' readability risks are documents-distinct factors. Koc and Gurgun [15] also agree with this significant conclusion that the readability issues may differ depending on the contract type. Based on this consensus, realizing additional researches in the future for scrutinizing each particular type of the construction documentation in terms of its readability issues is warranted. Hence, more inclusive theories and practices can be developed, supporting improving the wording of the construction documents.

By analyzing the RF, RFA, and R of the 14 readability issues, it has been shown that "poor presentation of the format of the tender documentation" (RI_1), "sentences and clauses are too long and complicated" (RI_2), "spelling and grammatical errors" (RI_3), "abstractness or vagueness of words or sentences" (RI_4), "using controversial phrases" (RI_5), "repetition of provisions or clauses" (RI_8), "poor illustration of procedure or process" (RI_9), and "listing conditions that are not related to the tender scope" (RI_{12}) are the top-eight frequent issues of the readability in the sub-contracting's tender documents. Each of which is with an occurrence in all of the tender documents, accounting for 9.1398% of the grand total of the RFA of the 14 readability issues. Thus, in total, they represent 73.1184% of the problems encountered by the sub-contractors regarding the ease of interpreting the documentation of the tenders to which they are invited to. This result is a crucial message for the drafters of the sub-contracting's tender documentation, making them aware of the major recurrent mistakes that they are responsible for when preparing these documents. Another significant message for those drafters in this regard is that they can recognize the other 6 readability issues, which have been mentioned in some of the tender documents. These issues together exemplify 26.8816% of the whole summation of the RFA of the readability issues. They are, in descending order of their RFA percentages: "using specific vocabulary, legal terms, and legal jargon" (RI_6), "lack of/poor visual representations" (RI_{10}), "referring to engineering terminology, code, or specification that are not frequent to all disciplines" (RI_7), "transliteration of English words/idioms into Arabic" (RI_{14}), "using abbreviations without illustrating their definitions" (RI_{11}), and "inconsistencies among the tender clauses" (RI_{13}).

By taking a closer look into these factors, the characteristics of the readability issues in the sub-contracting's tender documents can be summarized in four pivots: (a) structural and presentation-related problems, (b) lengthening and repetition-related problems, (c) text-related problems, and (d) terminology-related problems. The structural and presentation-related problems appear in RI_1, RI_9, and RI_{10}. This pivot highlights that the poorer the quality level on which the tender documentation is formatted and produced, the lower the visual representation and the information flow of the tender scope for the sub-contractor. This fact stems from the case that, when the sub-contractor is unable to know and see all the detailed data of the requested work consistently, avoiding the risk of misunderstanding becomes extremely low [15]. The consequence of this relation may extend further to discourage the sub-contractor to read the tender documentation and negatively impact on his/her decision towards participating in the tender. So, it is exceedingly recommended that releasing the tender documentation should be in a proper presentation, whether in the format or the content of its structure, data, and drawings. This is a highly necessary feature that each document should have for comprehensively and clearly providing the sub-contractor with the tender scope. This recommendation can easily be achieved by following the corresponding anti-measures of the issues of this pivot (see Table 5). This is another implication of this study, as it not only contributes to determining, analyzing, and ranking the readability issues in the sub-contracting's tender documents, but also introduces a framework of the countermeasures of the identified issues. Relying upon Table 5, RI_1 can be avoided, employing suitable font size and type, indentation, and line spacing for enhancing the documents' general format and their readability for the reader in particular [26]. Further, by supporting the tender procedures with a flow chart or illustrative examples and attaching all the detailed drawings adequately with the tender documentation package, RI_9 and RI_{10} can be eliminated, respectively. Notably, the consideration of these anti-measures has multiple benefits for the sub-contractor, including enabling him to see, read, and understand the tender documents more clearly; reducing his/her misunderstanding risk; and consequently, encouraging him to participate in the tender.

RI_2, RI_8, and RI_{12} represent the second dimension of the readability issues in the sub-contracting's tender documents. As these issues point, they contribute to lengthening the tender documents' sentences and clauses and increasing the size of the tender documentation package. According to Koc and Gurgun [15], the negative consequence of RI_2 on the readability, with respect to the contracts, encompasses reducing the willingness of the readers to read them precisely. Consequently, they can overlook matters that could be crucial in defining their obligations and rights. As for RI_8 and RI_{12}, they cause the contract documents to be voluminous, resulting in the complexity of extracting the information. As a result, the attention of the reader can be distracted from the main relevant conditions of the contract. Combining these impacts together, the possible result is that exposing the reader to the problems with ease-of-reading. As the author noticed when analyzing the 34 tender documents, three causative factors may be behind the occurrence of RI_2, RI_8, and RI_{12}. First, the drafters are not sufficiently skilled to formulate the tender documentation's sentences and clauses in a shorter and informative manner. Second, given their utilization of a standardized template for producing any tender documentation package, regardless the scope it reflects, the documents include repetitive and unnecessary clauses. Third, the documentation package has been issued without an accurate revision, either from the drafters or their managers. Although this analysis reveals the root causes of the lengthening and repetition-related issues of the sub-contracting's tender documents, it has two significant implications for controlling them. First, the drafters' skills need to be honed to master how a sentence or clause in a document can be written shortly in an informative way. This is achievable by involving the drafters in training courses to learn from the expertise of the academics and practitioners in this field. Second, the managers of the drafters should set a precise multi-step system for revising the documentation before being released. The steps of this system can incorporate a senior drafter to review the works of his/her junior drafting team, followed by the approval of the manager of the tenders' preparation department.

Table 5 provides additional recommendations for addressing the issues of RI_2, RI_8, and RI_{12}. In terms of RI_2, this table indicates that the words number per sentence should be within 20 words. This is an important feature that each sentence should have since long sentences have been highlighted by many scholars and practitioners as a major source of the lack of clarity and misinterpretation [13]. As for RI_8 and RI_{12}, it can be informed that the size of the tender documentation package must be as simple as possible by eliminating the repeated provisions, clauses, or the irrelevant conditions to the tender scope. This makes the reading of the sub-contracting's tender documentation easier and increases the attention of the sub-contractor on the pertinent terms of the tender.

Pivot three of the readability issues is concerned with the text-related problems. Its relevant issues are RI_3, RI_4, RI_5, and RI_{13}. As Table 2 pinpoints, the explanations of these issues reflect that the text-related problems are responsible for causing the tender documentation's sentences and clauses to have a poor language structure and be inconsistent, unclear, and incomprehensible. The consequences of these issues are that they cause the sub-contractor to interpret the tender documents' provisions and clauses in a different sense than what the general contractor intends to tell. Consequently, the chance of interpreting the tender documentation's provisions and clauses with a high degree of commonality by the sub-contractor and the prime contractor becomes low [12]. Hence, the agreement between these two parties on their duties and rights being elusive, leading to the risk of disputes [13]. Rameezdeen and Rodrigo [21] also support this analysis, that the lower the readability of a construction document is, the higher is the disputes between the contracting parties. In the same vein, Koc and Gurgun [33] confirmed that if the construction documents are not understandable because of the inconsistency and ambiguity in their clauses, the failure of the contractual relationship between the involved parties is inevitable. For avoiding such consequences, Table 5 suggests that first, the seniors of the drafting teams and the managers of the tenders' preparation departments should adopt the above-proposed revising system of the tender documentation. This system can assist in refining the tender documents' sentences and clauses in terms of their language structure, so as to enhance their readability. More importantly, it allows them to check the consistency among the tender clauses for assuring that they are consistent with each other having the same meaning for the sub-contractor. Second, they advise to employ the words of the unique meaning, rather than those with multiple interpretations, and avoiding using the controversial phrases. This is a valuable recommendation because it enables the sub-contractors to know their responsibilities and rights without the risks of misinterpretation or ambiguity.

RI_6, RI_7, RI_{11}, and RI_{14} signify the terminology-related problems. Referring to the descriptions of these issues in Table 2, they result in the presence of incomprehensible terminology for the sub-contractor, encompassing specialized legal and engineering terms, abbreviations, and literally translated words/idioms from English into Arabic. Unfortunately, finding the intended meaning of such specific terms could be a time-consuming and too-difficult process for the sub-contractor [34], resulting in the unclarity and readability risks [13,15]. The reason behind the existence of these problems is that the drafter considers the sub-contractors are familiar with all the terminology and abbreviations that he/she writes or translates. According to Besaiso et al. [25], this belief is incorrect since the readers of an engineering document of a contract or tender are almost engineers not schooled in law to understand the legalistic language of the contract or tender. Further, although they are engineers, it is ordinary to be unacquainted with the technical terminology, codes, specifications, and abbreviations of all engineering disciplines. More critically, Egypt's engineers and its sub-contractors are native Arabic speakers. Hence, including English words/idioms in the tender documentation or literally translating them into Arabic will make the documents inapprehensible to them. In consistence with this analysis, Besaiso et al. [25] justified the FIDIC clauses' unclarity, because they have been written utilizing very legalistic language. Additionally, Koc and Gurgun [15] revealed that employing infrequent engineering terminology to all disciplines and too many abbreviations are among the readability risks of the contracts, causing disparity between the contracting parties. This analysis informs the drafters of the sub-contracting's tender documents and their managers of a significant fact: not every term or abbreviation they add to the tender documentation provides ease-of-reading for the sub-contractor. This can, however, increase his/her fuzziness and incomprehension risks.

For addressing the terminology-related issues, Table 5 highlights that utilizing everyday words and abandoning employing legal language by the drafters are warranted to limit the presence of legalistic terms in the tender documents. Further, when it is essential to point to an engineering term in the tender documents, it should be frequent to all disciplinarians wherever possible. Similarly, the necessary clauses of the referred-to code or specification and the definitions of the utilized abbreviations must be attached with the tender documentation package. Moreover, any English words/idioms have to be translated into understandable Arabic phrases. Indeed, all of these anti-measures contribute to providing the sub-contractor with comprehensible terminology, supporting the highly needed aspects in any construction documentation, comprising clarity, readability, and understanding.

The above-mentioned analysis and discussion bring a detailed insight about the readability issues in the sub-contracting's tender documents by categorizing them into structural and presentation-related problems, lengthening and repetition-related problems, text-related problems, and terminology-related problems. The accuracy of this classification stems from involving the issues of the similar nature under the same group, depending on their descriptions and impacts on the readability for the reader. To date, it is believed that such framework has not been realized in any of the prior literature. This classification provides a significant implication for enhancing the drafters' and academics' knowledge to obtain an accurate description regarding the pivotal sources of the readability problems in a construction document. This study, additionally, in view of the top-eight frequent issues of the readability in the sub-contracting's tender documents, introduces another classification to benefit the developing countries generally. Relying upon investigating whether these eight issues are highly ranked risks in the found peer researches of Malaysia and Turkey, a hierarchy of three levels has been developed. The top of the hierarchy comprises RI_2 and RI_4, representing the issues impacting the readability in all the developing countries. Level two points to the issues present in 50% of the developing construction markets, including RI_3, RI_5, and RI_8. Level three is the least critical one because its issues, i.e., RI_1, RI_9, and RI_{12}, have 0% in terms of their occurrence as critical readability problems in Malaysia and Turkey. This hierarchy contributes to afford an initial classified checklist of the issues obstructing the construction documentation's readability in the developing economies, serving as the bedrock for helping the drafters and academics in those countries to define their associated readability problems.

6. Conclusions

This study contributes to answering two questions raised frequently in the construction community: "what are the readability issues in the sub-contracting's tender documents?" and "what are the measures for enhancing the readability in the sub-contracting's tender documents?". Building on applying the CAA to real documentation of 34 tenders of the sub-contracting arrangement in Egypt, 14 readability issues have been extracted. Further, through examining the prior works of readability, the corresponding anti-measures of the specified issues have been allocated. Subsequently, the soundness of the reached results has been confirmed by arranging face-to-face discussions with three experts. By determining the FA of the readability issues within the tender documents, "poor presentation of the format of the tender documentation", "sentences and clauses are too long and complicated", "spelling and grammatical errors", "abstractness or vagueness of words or sentences", "using controversial phrases", "repetition of provisions or clauses", "poor illustration of procedure or process", and "listing conditions that are not related to the tender scope" have been specified as the top-eight most frequent issues in the sub-contracting's tender documentation. These eight issues have then been compared with the outcomes of the found peer researches of Malaysia and Turkey. The findings of the comparison highlight that "sentences and clauses are too long and complicated" and "abstractness or vagueness of words or sentences" are severe issues obstructing the ease-of-reading and understanding of the construction documents in the developing countries. Relying upon discussing the identified readability issues, they have been categorized into four pivots, including "structural and presentation-related problems", "lengthening and repetition-related problems", "text-related problems", and "terminology-related problems". This classification, along with the other outputs of this paper, benefits the drafters and academics to obtain an accurate description regarding the possible pivotal sources of the readability problems in a construction document.

As in all studies, this research has limitations. First, since the readability issues have been drawn from the documents of 34 tenders of the sub-contracting practice in Egypt, replicating this research in the future by increasing the sample of the tender documents is recommended. Second, given the readability issues have been simply ranked in terms of their FA, relying upon applying the CAA to the documentation of the assembled tenders, the findings derived from these ranks should be viewed with caution until verifying these ranks. This can be realized in future research streams by involving the readability issues in a questionnaire and exploring the experts' perspectives regarding their frequency, severity, and criticality. Third, as the findings of the paper have been verified by three experts, surveying more practitioners in the future can enhance their reliability. Fourth, within the context of Egypt, which is a developing country, the study outcomes have been realized. Accordingly, its results, particularly in terms of the ranks of the readability issues, are limited to Egypt only. This is owing to the conclusion derived from the current paper that the most important issues of the readability are contingent upon the context of the country. Fifth, in this study, the readability issues have been classified into four dimensions by involving the issues of the similar nature under the same dimension, depending on their descriptions and impacts on the readability for the reader. Thus, validating this classification in the upcoming research directions, utilizing the analytical techniques of the Exploratory Factor Analysis, the Principal Component Analysis, or the Cluster Analysis, is important to refine its precision.

Funding: This research received no external funding.

Institutional Review Board Statement: Not applicable.

Informed Consent Statement: Not applicable.

Data Availability Statement: All referenced data exists within the paper.

Conflicts of Interest: The author declares no conflict of interest.

References

1. Laryea, S.; Lubbock, A. Tender pricing environment of subcontractors in the United Kingdom. *J. Constr. Eng. Manag.* **2014**, *140*, 04013029. [CrossRef]
2. Ulubeyli, S.; Manisali, E.; Kazaz, A. Subcontractor selection practices in international construction projects. *J. Civ.Eng. Manag.* **2010**, *16*, 47–56. [CrossRef]
3. Choudhry, R.; Hinze, J.; Arshad, M.; Gabriel, H. Subcontracting practices in the construction industry of pakistan. *J. Constr. Eng. Manag.* **2010**, *138*, 1353–1359. [CrossRef]
4. Hinze, J.; Tracey, A. The contractor-subcontractor relationship: The subcontractor's view. *J. Constr. Eng. Manag.* **1994**, *120*, 274–287. [CrossRef]
5. El-Mashaleh, M. A construction subcontractor selection model. *Jordan J. Civ. Eng.* **2009**, *3*, 375–383.
6. Arditi, D.; Chotibhongs, R. Issues in subcontracting practice. *J. Constr. Eng. Manag.* **2005**, *131*, 866–876. [CrossRef]
7. Mbachu, J. Conceptual framework for the assessment of subcontractors' eligibility and performance in the construction industry. *Constr. Manag. Econ.* **2008**, *26*, 471–484. [CrossRef]
8. Ajayi, O.; Ayanleye, A.; Achi, F.; Johnson, O. Criteria for Selection of Subcontractors and Suppliers in a Building project in Lagos State, Nigeria. In Proceedings of the 5th Built Environment Conference, Durban, South Africa, 18 July 2010.
9. Code Committee. *Egyptian Code for Managing Construction Projects*, 1st ed.; Housing and Building National Research Center: Cairo, Egypt, 2009.
10. Laryea, S. Quality of tender documents: Case studies from the UK. *Constr. Manag. Econ.* **2011**, *29*, 275–286. [CrossRef]
11. Youssef, A.; Osman, H.; Georgy, M.; Yehia, N. Semantic risk assessment for Ad Hoc and amended standard forms of construction contracts. *J. Leg. Aff. Disput. Resolut. Eng. Constr.* **2018**, *10*, 04518002. [CrossRef]
12. Rameezdeen, R.; Rajapakse, C. Contract interpretation: The impact of readability. *Constr. Manag. Econ.* **2007**, *25*, 729–737. [CrossRef]
13. Chong, H.; Zin, R. A case study into the language structure of construction standard form in Malaysia. *Int. J. Proj. Manag.* **2010**, *28*, 601–608. [CrossRef]
14. Shash, A. Bidding practices of subcontractors in Colorado. *J. Constr. Eng. Manag.* **1998**, *124*, 219–225. [CrossRef]
15. Koc, K.; Gurgun, A. Assessment of readability risks in contracts causing conflicts in construction projects. *J. Constr. Eng. Manag.* **2021**, *147*, 04021041. [CrossRef]
16. Shivanthi, B.; Devapriya, K.; Pandithawatta, T. Disputes between Main Contractor and Subcontractor: Causes and Preventions. In Proceedings of the 8th World Construction Symposium, Colombo, Sri Lanka, 8–10 November 2019.
17. *Ministry of Housing, Utilities, and Urban Communities*; 2021 in Partnership with Meed Projects; Egypt Project Market Outlook: Cairo, Egypt, 2021.
18. Broome, J.; Hayes, R. A comparison of the clarity of traditional construction contracts and of the new engineering contract. *Int. J. Proj. Manag.* **1997**, *15*, 255–261. [CrossRef]
19. Lam, P.; Javed, A. Comparative study on the use of output specifications for Australian and U.K. PPP/PFI Projects. *J. Perform. Constr. Fac.* **2015**, *29*, 04014061. [CrossRef]
20. Rameezdeen, R.; Rodrigo, A. Textual complexity of standard conditions used in the construction industry. *Australas. J. Constr. Econ. Build.* **2013**, *13*, 1–12. [CrossRef]
21. Rameezdeen, R.; Rodrigo, A. Modifications to standard forms of contract: The impact on readability. *Australas. J. Constr. Econ. Build.* **2014**, *14*, 31–40. [CrossRef]
22. Menches, C.; Dorn, L. Emotional reactions to variations in contract language. *J. Legal Affairs Dispute Resolut. Eng. Constr.* **2013**, *5*, 97–105. [CrossRef]
23. Chong, H.; Oon, C. A practical approach in clarifying legal drafting: Delphi and case study in Malaysia. *Eng. Constr. Archit. Manag.* **2016**, *23*, 610–621. [CrossRef]
24. Masfar, Z. Towards a Saudi plain language standard construction contract. *Int. J. Constr. Eng. Manag.* **2017**, *6*, 168–179.
25. Besaiso, H.; Fenn, P.; Emsley, M.; Wright, D. A Comparison of the suitability of FIDIC and NEC conditions of contract in Palestine: A perspective from the industry. *Eng. Constr. Archit. Manag.* **2018**, *25*, 241–256. [CrossRef]

26. Ali, N.; Wilkinson, S. Modernising Construction Contracts Drafting—A Plea for Good Sense. In Proceedings of the W113—Special Track 18th CIB World Building Congress, Salford, UK, 10–13 May 2010.
27. Nguyen, D.; Garvin, M.; Gonzalez, E. Risk allocation in U.S. public-private partnership highway project contracts. *J. Constr. Eng. Manag.* **2018**, *144*, 04018017. [CrossRef]
28. Li, Z.; Zhang, S.; Meng, Q.; Hu, X. Barriers to the development of prefabricated buildings in China: A news coverage analysis. *Eng. Constr. Archit. Manag.* **2021**, *28*, 2884–2903. [CrossRef]
29. Semple, C.; Hartman, F.; Jergeas, G. Construction claims and disputes: Causes and cost/time overruns. *J. Constr. Eng. Manag.* **1994**, *120*, 785–795. [CrossRef]
30. Kolbe, R.; Burnett, M. Content-analysis research: An examination of applications with directives for improving research reliability and objectivity. *J. Consu. Res.* **1991**, *18*, 243–250. [CrossRef]
31. Elo, S.; Kyngäs, H. The qualitative content analysis process. *J. Adv. Nurs.* **2008**, *62*, 107–115. [CrossRef]
32. Arshad, M.; Thaheem, M.; Nasir, A.; Malik, M. Contractual risks of building information modeling: Toward a standardized legal framework for design-bid-build projects. *J. Constr. Eng. Manag.* **2019**, *145*, 04019010. [CrossRef]
33. Koc, K.; Gurgun, A. The role of contract incompleteness factors in project disputes: A hybrid fuzzy multi-criteria decision approach. *Eng. Constr. Archit. Manag.* **2022**, *in press*. [CrossRef]
34. Koc, K.; Gurgun, A. Ambiguity factors in construction contracts entailing conflicts. *Eng. Constr. Archit. Manag.* **2021**, *29*, 1946–1964. [CrossRef]

Application Issues of Impacted As-Planned Schedule for Delay Analysis

Kyong Ju Kim [1], Bonghee Han [2,*], Min Seo Park [1], Kyoungmin Kim [3] and Eu Wang Kim [1]

[1] Department of Smart Cities, Chung-Ang University, Seoul 06974, Korea
[2] Department of Civil Engineering, Chung-Ang University, Seoul 06974, Korea
[3] School of Civil & Environmental Engineering, Urban Design and Study, Chung-Ang University, Seoul 06974, Korea
* Correspondence: bbongbal@cau.ac.kr

Citation: Kim, K.J.; Han, B.; Park, M.S.; Kim, K.; Kim, E.W. Application Issues of Impacted As-Planned Schedule for Delay Analysis. *Buildings* **2022**, *12*, 1442. https://doi.org/10.3390/buildings12091442

Academic Editors: Srinath Perera, Albert P. C. Chan, Dilanthi Amaratunga, Makarand Hastak, Patrizia Lombardi, Sepani Senaratne, Xiaohua Jin and Anil Sawhney

Received: 21 July 2022
Accepted: 1 September 2022
Published: 13 September 2022

Publisher's Note: MDPI stays neutral with regard to jurisdictional claims in published maps and institutional affiliations.

Abstract: Most construction projects are delayed, and many are subject to claims or disputes. Therefore, delay analysis is a critical component of any construction project to determine who is responsible for delays. This research examines four different techniques for estimating delay impacts using the impacted as-planned (IAP) method. A sample network was introduced as an example to discuss several concerns. The advantages and limitations of each approach were identified, and recommendations were given for each approach. When inserting an activity or activities representing delay events in IAP, it is necessary to use both constraints and logical relations among delay events, their logical predecessors, and successors. Constraints representing the actual date of delay events are the simplest and easiest. However, constraints should not be used in "single insertion" and "inserting only owner- or contractor-caused delay" approach. In addition, in the case of using constraints, it is critical to ensure that the impact of delay events is less than the duration of those delay events. Constraints should be avoided in this scenario, and delay events should be logically connected to their logical predecessors and successors without constraints. This study also identified through an example that inserting delay events only by logic can cause wrong analysis results. The results of this study will be helpful for delay analysts in identifying what kinds of problems occur in IAP methods and how to prevent those problems.

Keywords: delay analysis; delay impact; impacted as-planned; claims; dispute resolution

1. Introduction

The main objective of this study is to identify practical issues in applying impacted as-planned (IAP) delay analysis method and to suggest improved approaches to them. Currently there are some disputes on how to apply the IAP delay analysis method in practice. So, this study reviews what kind of different IAP approaches are being used, and investigates what their advantages and limitations are through a sample project network, and suggests how to improve those problems. Most construction projects are delayed [1–5]. Many of them are the subject of allegations or engage in legal battles. As a result, almost every construction project includes construction claims [5,6]. Delay analysis is a critical component of every construction project since it determines who is accountable for delays. There are various techniques for analyzing delays such as as-planned vs. as-built, impacted as-planned (IAP), as-planned but for, collapsed as-built, window analysis, time impact analysis, etc. Each technique has pros and cons in analyzing the delay.

Several researchers have investigated those techniques. By assessing the research related to delay analyses from 1982 to 11 February 2021, two primary research areas were detected in this field, namely, improving the delay analysis methods and resolving the disputes before they occur [7]. Bubshait and Cunningham [2] evaluated three delay measurement techniques (as-planned schedule method, as-built schedule method, and modified as-built method (time impact analysis)) and found that each delay analysis technique may have different results of delay. Stumpf [8] looked at schedule delay analysis and showed how results differed. Kao and Yang [9] compared four windows-based delay analysis methods to identify their advantages and limitations, in terms of the perspectives of use prerequisite, functional capability, analytical procedure, and accuracy of analysis results. Kim et al. [10] suggested that the result of delay analysis was differently produced depending on the type of baseline schedule, update program using as-built data or not, and approach to dealing with concurrent delay and acceleration measures. Based on a case study, Braimah [11] analyzed delay investigation techniques and identified problematic issues and their improvement needs. Arditi and Pattanakitchamroon [12] evaluated the benefits and drawbacks of commonly used delay analysis methods, as well as the impact of several factors on the selection of a delay analysis method. Although much research has been directed toward improving the use of these delay analysis approaches, the continuing difficulties associated with such claims suggest the need for additional empirical analysis into the extent of use of the approaches, their success rates in dealing with delay claims resolutions, and the obstacles affecting their appropriate implementation in practice [13].

Certain considerations must be addressed while selecting a proper delay analysis method because each delay claim is unique, and its characteristics will define which delay analysis method to implement [14]. When selecting analysis methods, it is important to consider the timing of delay analysis and the quality of available data and documents because the delay effect must be quantified using reliable data [15]. The availability of contemporaneous documentation, the quality of available schedules, and the existence of updated schedules are important factors in selecting a delay analysis technique.

There are so many cases where impacted as-planned (IAP) schedule method is the only possible option to analyze delays and who are responsible for the delays. The IAP is not recommended for dispute resolution in courts. However, many construction projects still have only baseline schedule without as-built schedules, which are routinely evaluated, monitored, and updated as a project progresses. In this instance, the sole alternative is to use the IAP. In applying IAP, however, different approaches in their details are being utilized in practice. Each approach has advantages and limitations. This research recognizes and examines some problems in applying IAP to dispute resolution and suggests how to solve them. This research outcome is projected to be helpful for delay analysts in recognizing what kinds of problems occur in IAP methods.

To achieve the research objectives, this study identifies and categorizes four different techniques which are being applied in delay analysis as the name of IAP, and introduces a sample network schedule which is simplified to analytically compare and assess four techniques. With the sample network, a scenario was given about how works have progressed. Based on the sample network and the scenario, delay analyses by four techniques were performed. In discussion, the results of the analyses, the pros and cons of each technique were compared, and the recommendations were given.

2. Impacted As-Planned Method

SCL (Society of Construction Law) [16] defines the IAP method as introducing delay event sub-networks into a logic-linked baseline programme, and its recalculation using CPM (Critical Path Method) programming software to determine the prospective influence these events have on the anticipated contract completion dates revealed within the baseline programme. The numerous delays are structured as activities in IAP and added to the as-planned CPM schedule in chronological order, revealing how the project is being delayed by one delay at a time [17]. For delay analysis, this method simply employs a baseline (as-planned) schedule. The number of delayed days is equivalent the variance between the completion dates of the schedules before and after the impacts [11]. This method does not measure the effect of actual work performed but heavily relies on the validity of a baseline schedule [18]. The basic steps to implement IAP are as follows [14,19]:

1.   Identify "un-impacted schedule" for delay analysis.

2. Insert or add an activity or activities (fragnet) into the "un-impacted schedule" to represent the impact of the delay(s).
3. Calculate the new schedule created by the "un-impacted schedule" with the fragnet or activity inserted. The resultant network is considered the "impacted schedule".
4. Compare the project completion date of the impacted and un-impacted schedules to determine the impact of the inserted fragnet(s).

3. Four Approaches

The IAP method is used to calculate the hypothetical impact of those inserted activities on a network by adding or inserting activities that represent delays or changes into a network. AACE (Association for the Advancement of Cost Engineering) International [14] recommends global or stepped insertion. This research focuses on chronological, stepped insertion since it is more generally used. In practice, delay analysis professionals use various approaches in adding or inserting the fragnet (an activity or activities representing the delay(s)) into an as-planned schedule. They can be divided into four categories. The first method involves inserting a single delay event into an as-planned schedule without additional delay events. The second approach is to insert delay events into a network in a chronological order based on the logical relationship with their predecessors and successors, without regard for the delayed events' actual dates. The third approach is to insert delay events into the network in a chronological order based on logical relation and using constraints based on the actual date of delay events. The fourth approach is to insert delay events caused by only the owner or only the contractor into the as-planned schedule. This research reviews the issues in applying these approaches and identifies improvement needs.

3.1. Sample Network

To analytically assess and compare these techniques, a sample network is introduced. Since a real project network cannot be introduced in this paper, the problems which were identified in the middle of analyzing real big complex projects having up to 5000 activities were dealt through the sample network. A sample network was used for simplification and the instinctive understanding; The schedule is as revealed in Figure 1. It is made of activities A, B, C, D, E, F, and G. Its total project duration is 40 days. Its critical activities are A, B, C, D, and G. Activities E and F have 5 days of float.

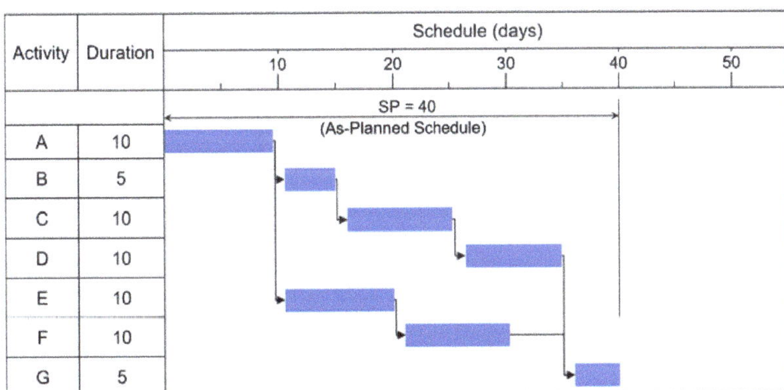

Figure 1. As-planned schedule of a sample project.

The project started as planned, but in the middle of the progress, three delayed events happened. "Delay event 1" was affected by the owner delaying the authorization on the shop drawing for "activity E". It hindered the start of "activity E" for 10 days from the 11th to the 20th date. Furthermore, the contractor had a labor problem ("delay event 2"), which hindered the start of activity C for 5 days from the 16th to the 20th date. In the progress, the contractor completed "activity F" 3 days earlier than planned. After the finish of activity F, the owner ordered a change ("delay event 3"), which happened for 5 days from the 38th day to the 42nd day, impacting the start of "activity G". Figure 2 revealed the as-built schedule. The project was 7 days late, as it turned out.

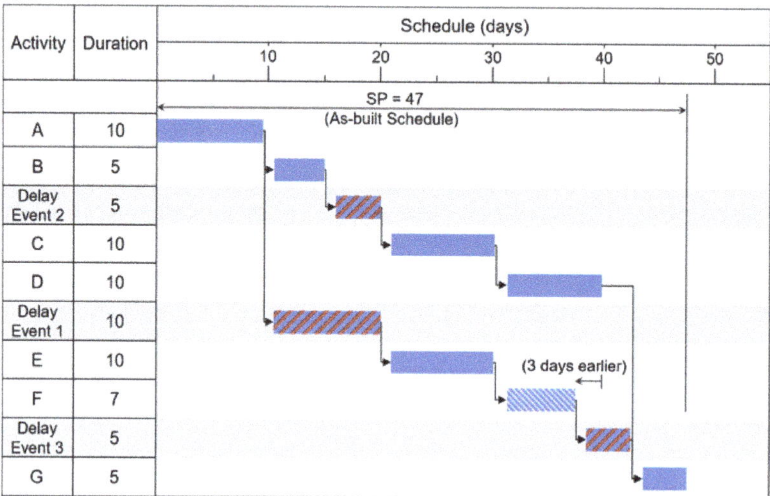

Figure 2. As-built schedule of a sample project.

3.2. Inserting a Single Delay Event

To quantify the impact of a single delay event, several researchers and delay analysts argue that it should be put into an as-planned schedule without any other delay events [20]. There are always a lot of delays in real projects. Therefore, inserting all delay events into an as-planned schedule chronologically and analyzing the impact of each delay event sequentially is difficult and time-intensive. For simplicity and convenience, this simplified approach is utilized in practice.

The delay event or the fragnet is connected to logically relevant predecessors and successors when only a single fragnet is inserted into an as-planned schedule. At each evaluation, the impact of each delay event is calculated. The effect of all delay events can be analyzed by repeating this process. The total estimated delay of a project is obtained by adding all delay effects together. The ratio of owner- and contractor-caused delay can be calculated. This approach assumes that the estimated total delay should be equal to the actual delay of the project. Therefore, the summation of the owner-responsible delay is evaluated by multiplying the ratio of owner-caused delay to the actual delay of the project. The contractor-caused delay is also calculated through the same process.

The general process for determining the impact of a delay is as follows: Firstly, only "delay event 1" is added to the as-planned schedule. Figure 3 shows the outcome. It takes 45 days to complete the job. The completion of the project is delayed as much as 5 days. The owner is responsible for the delay, which is classified as the "excusable and compensable (EC)" delay.

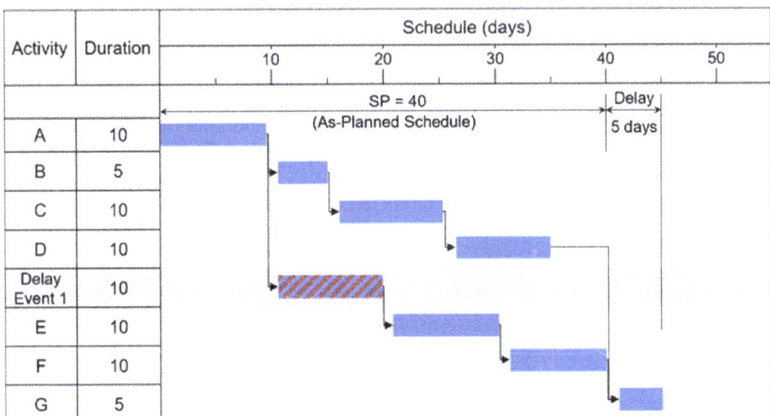

Figure 3. Impacted as-planned schedule by delay event 1 (owner-caused delay).

Next, as shown in Figure 4, "delay event 2" alone is added. Its total project duration becomes 45 days. Its estimated completion date is delayed 5 days. It is analyzed that "delay event 2" has 5 days of impact. The contractor is liable for the delay, which is classified as "non-excusable and non-compensable (NN)".

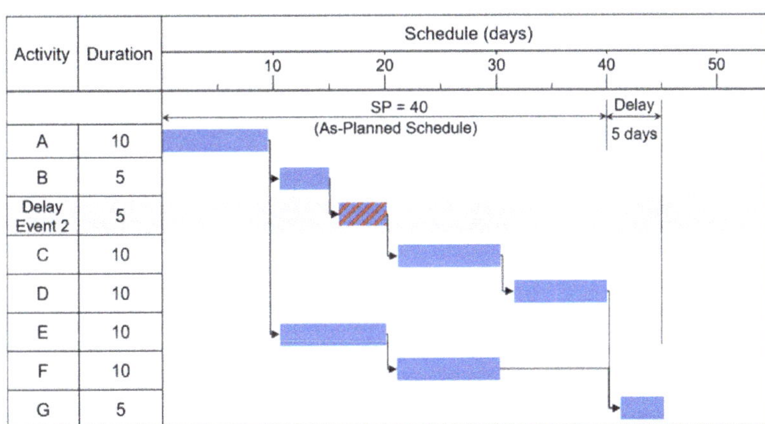

Figure 4. Impacted as-planned schedule by delay event 2 (contractor-caused delay).

"Activity F" was completed 3 days earlier than as-planned. However, it is not reflected in IAP; only delayed events are added to the as-planned schedule. As shown in Figure 5, the IAP schedule by only "delay event 3", its total project duration is 40 days. There is no delay because there is no change in the project duration according to the analysis.

In this analysis, (owner-caused) delay event 1 and (contractor-caused) delay event 2 have 5 days of impact respectively when a single delay event is inserted without any other delay events. All of the delays are added up to 10 days. Owner-responsible delay (EC) is calculated by multiplying actual delay by owner-caused delays divided by the sum of all delays caused by the owner and contractors. The same method is used to compute the contractor-responsible delay (NN).

$$\text{Owner-responsible delay (EC)} = 7 \times 5/10 = 3.5 \text{ days} \tag{1}$$

$$\text{Contractor-responsible delay (NN)} = 7 \times 5/10 = 3.5 \text{ days} \tag{2}$$

This approach is simple and easy to implement but may have some problems. This approach would count the impact of each delay event. As a result, it cannot reflect concurrent delays. This approach would also fail to account for the cumulated impact of previous delay events or any changes. As a result, this approach assumes that the sum of all delays is equal to actual delay of the project and that there is no acceleration.

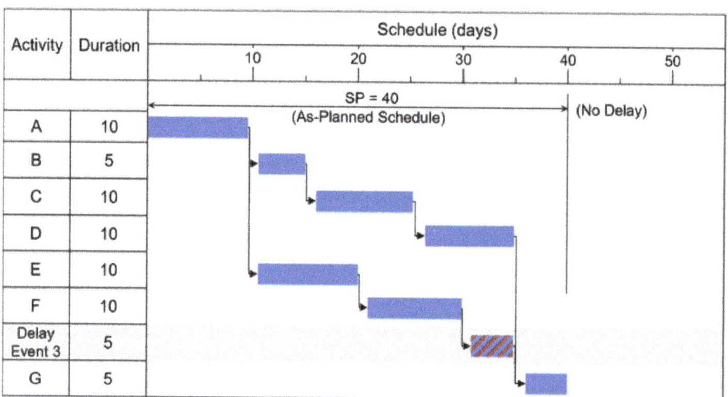

Figure 5. Impacted as-planned schedule by delay event 3 (owner-caused delay).

3.3. Stepwise Insertion without Using Constraints

All delay events are incorporated step-by-step and chronologically into an as-planned schedule using the approach. A delayed event or a fragnet should only be placed into the as-planned schedule by connecting them to logically linked predecessor and successor activities according to Kim and Kwon [21], and constraints should not be used. As shown in Figure 6, they are pointing out that if the delay event (the activity X) is added into the network by using constraints based on its actual start and finish date, the delay of the project completion could be overestimated.

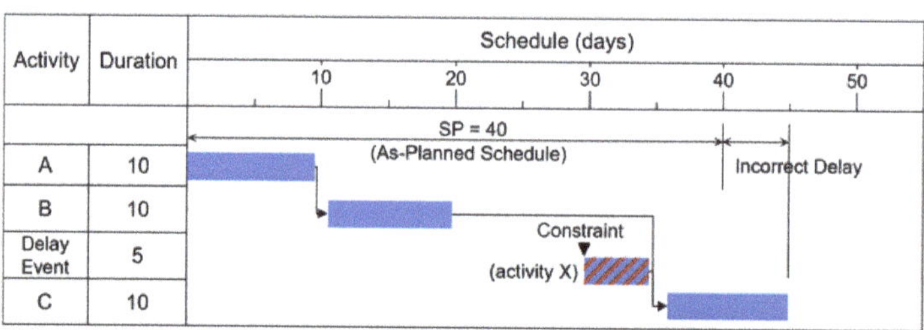

Figure 6. Overestimation of a delayed impact.

This approach requires adding the delay event with its duration into the network and connecting it to its logical predecessor and successor without changing anything of the as-planned schedule and without considering the actual start and finish date of the delay event. As the constraints reflect the actual date of each delay, but the as-planned schedule does not reflect updated actual progress, this approach needs only the duration of the delay event and logical relation should be employed and that constraint on the delay event should not be given.

The process to analyze delay impact in stepwise insertion without constraints is as follows: The IAP schedule by "delay event 1" is the same as Figure 3. The project is delayed as much as 5 days. The owner is liable for the delay, which is classified as an "EC" type. Figure 7 shows the IAP schedule by "delay event 1" and "delay event 2" on the planned schedule. As delay event 1 modified the critical path, there is no extra delay. Therefore, delay event 2 does not have an extra impact on the duration of the project. Here, delay events 1 and 2 are identified as concurrent delays, which is the type of "excusable and non-compensable (EN)" delay.

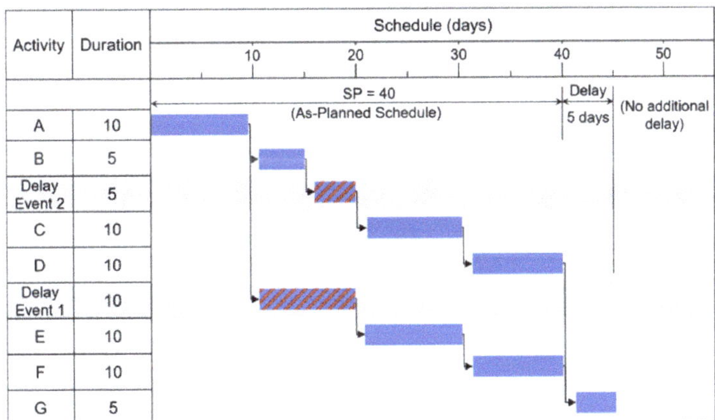

Figure 7. Impact of delay events 1 and 2.

"Delay event 3" happened for 5 days, and it had an impact on the start of activity G (Figure 8). To calculate the impact on the duration of the project, the delay event should be added to the as-planned schedule. The delay event 3 happened after "activity F" and had an impact on the start of "activity G". Activity F was completed 3 days earlier than planned. However, the IAP does not consider the actual performance of the as-planned schedule, as well as the actual start and finish date of the delayed event [17]. Therefore, the delay event 3 connected just to activity F as its logical predecessor and activity G as its logical successor.

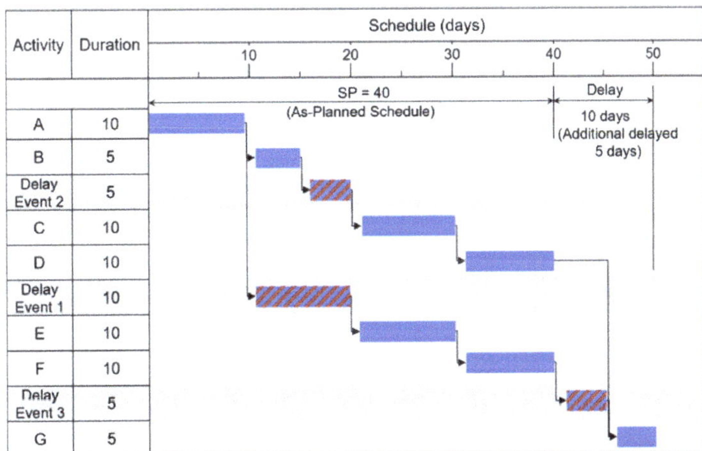

Figure 8. Impact of delay events 1, 2, and 3.

Figure 8 shows the IAP schedule by delay event 3. In addition, the project is 5 days behind schedule. The owner is responsible for the delay, which is classified as an "EC" type. It is analyzed as the total delay of this project is 10 days: 5 days are concurrent delays and 5 days are owner-caused delays.

3.4. Stepwise Insertion with Constraints

This approach is to insert an activity or activities reflecting delay events into an as-planned schedule in chronological order by using constraints assigning the date that the delay events happened. As shown in Figures 9 and 10, in step-by-step insertion using constraints, the as-planned schedule is impacted by delay events 1 and 2. The project is delayed 5 days. They are concurrent, which is the type of "ENs" delay.

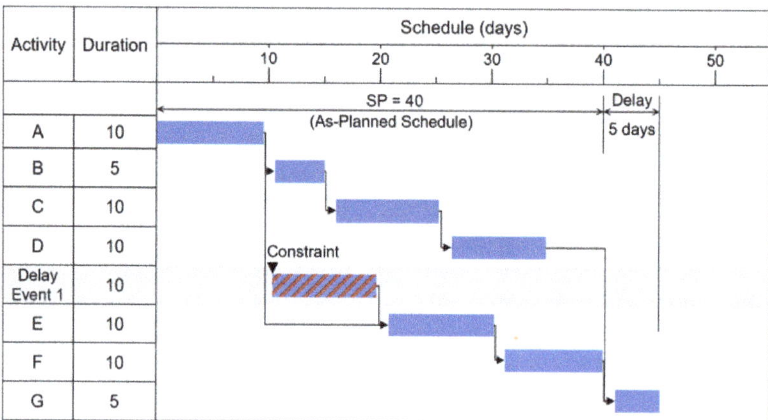

Figure 9. Adding delay event 1 by constraints.

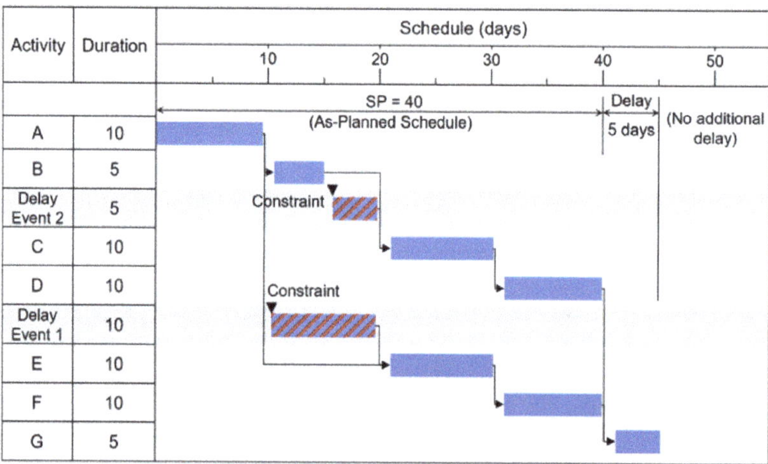

Figure 10. Adding delay events 1 and 2 by constraints.

"Delay event 3" caused by the owner's change order happened from the 38th day to the 42nd day of the project. The delay event was inserted by using constraints reflecting the actual start and finish date. Therefore, Figure 11 shows the IAP schedule. Delay event 3 has 2 days of impact on the duration of the project. The owner is responsible for the delay, which is the type of "EC" delay. Therefore, the total delay of the project is 7 days. It turns out that concurrent delay is 5 days and owner-caused delay is 2 days. The result is the same as its real progress.

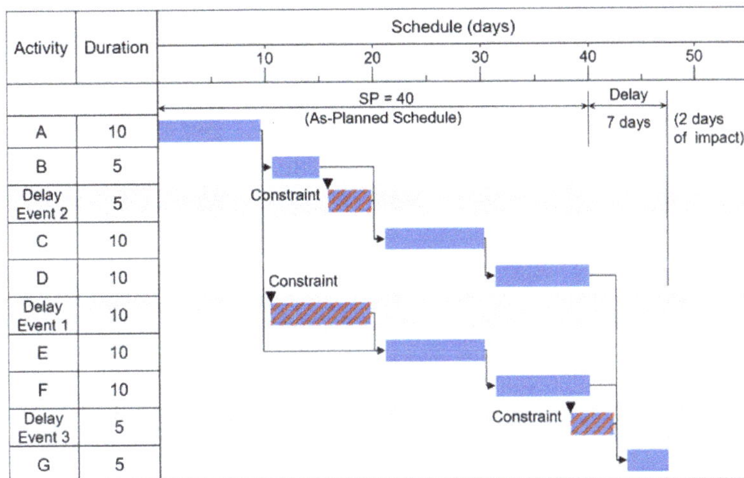

Figure 11. Impacted as-planned using constraint.

The as-planned schedule in step-by-step insertion contains most delay events and changes that occurred before any single delay event to be analyzed. Even if the as-planned schedule was updated with changes or delays, there is still a potential that the as-planned schedule does not incorporate all the network's earlier delays or changes. That is, the as-planned schedule may differ from the most recent as-built schedule. As shown in Figure 6, as pointed out by Kim and Kwon [21], inserting delay events with constraints could lead to an overestimate of the impact of a delay event. When the delaying events are added to the network by using constraints based on its actual start and finish date, the delay of the project completion could be overestimated. This type of overestimation, on the other hand, can only occur when the completion delay is greater than the duration of the delay event. This kind of problem can be prevented by making sure that the delay of the completion is less than the duration of the delay event, and that if not, the delay event should be connected to one of its reasonable predecessor activities in the network.

When delay events are added without constraints, it can also be examined as if there were delays even if there were none in a project. If "delay event 3" is related to its predecessor (activity F) and successor (activity G) as FS0 without taking into account its actual start and finish date as shown in Figure 8, the analysis' conclusion overestimates its influence and differs from its true progress. If the "delay event 3" is connected to its predecessor (activity F) as a finish-to-start relation with lag time 3 reflecting its real progress and to its successor (activity G) with FS0, its result would be the same as its real progress. Here, FS0 means finish-to-start relation with lag time 0. This outcome would be the same as the constraints. This means that connecting the delay event to its logical predecessors rather than using constraints that represent the actual date of the delay event provides no real benefit. In addition, the approach employing constraints is simple and easy to implement and can reduce disputes about the logical, soft connection between the delay event and its predecessor activities.

3.5. Adding Only Owner-Caused (or Contractor-Caused) Delays

Another approach is for the scheduler to simply take the as-planned schedule and add additional activities that represent delays (usually induced by the opposing party) to show why the project was completed later than expected [22]. This technique is used in practice to simply prove the other party's responsibility for the delay of the project. However, this approach can distort the delay each party is responsible for and may lead to the wrong decision.

Figure 12 shows the IAP schedule, which solely includes imposed by the owner. It shows that the owner is responsible for 10 days of project delays.

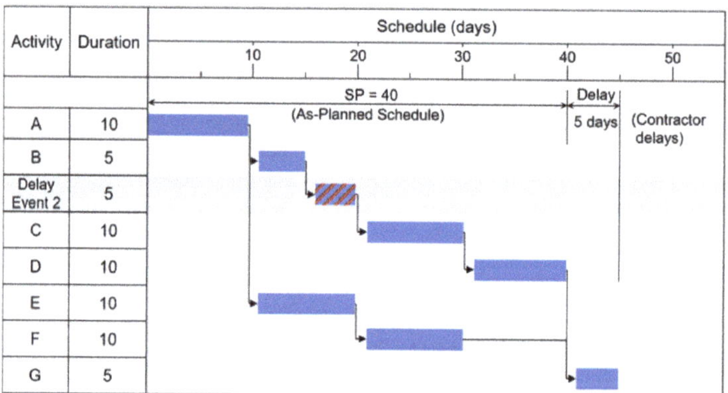

Figure 12. Owner-caused delay.

The IAP schedule including only contractor-caused delays is shown in Figure 13. It shows that contractor delays 5 days of project duration.

Figure 13. Contractor-caused delay.

The project's planned duration is 40 days, and the project's actual duration is 47. As a result, the total delay is 7 days. The owner-caused delay is 10 days, while the contractor-caused delay is 5 days, according to the analysis.

Owner-responsible delay (EC) is computed by multiplying the actual delay by owner-caused delays divided by the summation of all delays caused by the owner and contractors. Contractor-responsible delay (NN) is also calculated in the same way.

$$\text{Owner-responsible delay (EC)} = 7 \times 10/15 = 4.67 \text{ days} \tag{3}$$

$$\text{Contractor-responsible delay (NN)} = 7 \times 5/15 = 2.33 \text{ days} \tag{4}$$

This calculation assumes that the project's estimated total delay is equal to the project's actual delay. This approach does not allow for the analysis of concurrent delays caused by both the owner and the contractor. Therefore, in practice, this technique should only be utilized for simple pre-study.

4. Discussion

Table 1 summarizes the results of delay analysis by the four various techniques.

Table 1. Summary of the results of delay analysis.

Alternative	Delay (Days)		
	EC	EN	NN
Actual	2	5	-
Single insertion	3.5	-	3.5
Stepwise insertion without constraints	5	5	-
Stepwise insertion with logic and constraints	2	5	-
Inserting only owner- (or contractor-) caused delays	4.67	-	2.33

In a single insertion approach, a single delay event is inserted into an as-planned schedule without any additional delay events. This approach is currently being utilized by many delay analysts [23]. As shown in Table 1, however, it turns out that the types of delays are analyzed as different from actual progress. This approach considers the impact of each delay event at a time. As a result, it cannot analyze true concurrent delay. In addition, in the case of applying this approach, it is not recommended to utilize constraints based on actual dates of each delay event because an as-planned schedule does not reflect previous delays or changes that happened before each delay event in the network. Constraints with actual dates can result in a significant overestimate error when evaluating the impact of each delay event. This approach assumes that the sum of all delay impacts is equal to the project's actual delay for simplification [20]. Therefore, it does not consider acceleration in the project and has limitations in terms of acceleration analysis.

In a stepwise insertion without constraints, Kim and Kwon [21] proclaims that delay events should be inserted into an as-planned schedule without using constraints, and delay events should be connected logically to activities in the as-planned network. However, there is a good likelihood that the delay events in the connection do not have a logical predecessor (activity) in the network. When an owner orders a design change in the middle of a project, for example, there may be no predecessor activities to the design change. Furthermore, when soft decisions are required, connecting delay events to logical predecessors may be difficult, and can cause controversy in delay analysis. Here, the soft decision means connection requiring the manager's decisions that are related to optimizing site situations or minimizing a given impact [24]. They could be challenged in court since any change in the connection could have a significant influence on the outcomes of delay analysis [25]. Furthermore, as shown in Table 1, it turns out that the use of only logical relations excluding constraints can also lead to an overestimate in delay impact analysis.

In addition, two activities, predecessor activities being logically connected to the delay events and successor activities being impacted by them, could have long-time intervals. For example, the selection of a subcontractor for interior work can be scheduled right after the start of the project but interior works may not be started right after the selection of the subcontractor. The interior works are usually set to begin when the detailed work plan, shop drawing development, and all other previous works have been completed. Therefore, the activities "Selection of interior subcontractor" and "Interior works" could take a long-time interval. It can sometimes be longer than 1 year. As the activity "Selection of interior subcontractor" is near to the time of schedule development, an un-updated, as-planned schedule may reflect genuine progress of the activity. However, it may not reflect the real progress of activity "Interior works", which is far from the time of schedule development. In this case, there is little chance that only using a logical relationship among delay events, predecessor, and successor activities without using constraints provide more accurate results than that of stepwise insertion with constraints where constraints are added into the network based on their actual dates and as-planned schedule includes all previous delay events and changes.

The third approach based on stepwise insertion with constraints is simple and efficient to implement. The result of delay analysis by this approach in this example is the same as that of its real progress. As Kim and Kwon [21] argues, this approach can lead to overestimation. To avoid overestimation, however, it is just necessary to make sure that the impact on the project's duration is less than the duration of each delay event.

Approach inserting an only owner- (or contractor-) caused delay is simple and easy to implement. However, as illustrated in Table 1, it turns out that the types of delays are analyzed as different from actual progress. The outcome could be unfavorable to one party. In addition, constraints should not be used in inserting delay events into as-planned schedule. When only owner or, contractor-caused delays are added in the network, it does not reflect all changes in the schedule. However, constraints are added into the network based on their actual dates that the delay events actually happened. This can easily distort the output. This approach also implies that the sum of all estimated delay effects equals the project's actual delay. Therefore, it cannot examine acceleration. In addition, concurrent delays cannot be estimated because it only analyzes delays caused by one party.

5. Conclusions

This research examines various techniques to estimate delay impacts using the IAP technique. A prototype network was introduced as an example to discuss several concerns. The results of delay analysis by each technique were compared with actual impacts. The advantages and limitations of each approach were identified. Recommendations were given for each approach.

Stepwise insertion with logic and constraints is most recommended. When inserting a fragnet (an activity or activities) in IAP to reflect delay events, both constraints and logical relationships between delay events, their logical predecessors, and successors can be used. Constraints based on the actual date of delay events are the simplest and easiest to use. However, constraints should not be used in "single insertion" and "inserting only owner- or contractor-caused delay" approach. In addition, in the case of using constraints, it is essential to check if the impact of delay events is larger than the duration of those delay events. In that case, delay events should be logically connected to their logical predecessor and successors without constraints. This study also identified through an example that inserting delay events by logic also can cause wrong analysis results. The findings of this research will aid delay analysts in identifying what kinds of issues arise in IAP methods and how to avoid them.

The problems dealt in this paper were identified by investigating real big projects having almost 5000 activities. However, real network could not be introduced in this paper. A sample network was used for simplification. This paper has limitations in that it cannot show the complexity and dynamics of real projects. AACE [14] and SCL [16] provides recommended protocols for delay analysis. However, there are still a lot of controversy in delay analysis. The area of delay analysis still does not have a standard practice. More issues in analyzing delay should be investigated and standardized in the future.

Author Contributions: Conceptualization, K.J.K. and B.H.; Methodology, K.J.K., B.H. and K.K.; Validation, K.J.K. and K.K.; Formal Analysis, B.H. and M.S.P.; Investigation, M.S.P. and E.W.K.; Writing—Original Draft Preparation, K.J.K.; Writing—Review and Editing, B.H. and K.K.; Visualization, M.S.P. and E.W.K.; Supervision, K.J.K. and K.K.; Project Administration, K.J.K. All authors have read and agreed to the published version of the manuscript.

Funding: This research was supported by the Basic Science Research Program through the National Research Foundation of Korea (NRF) funded by the Ministry of Science and ICT (No. 2019R1A2C2010794) and the Chung-Ang University Research Scholarship Grants in 2021.

Institutional Review Board Statement: Not applicable.

Informed Consent Statement: Not applicable.

Data Availability Statement: The data presented in this study are available on request from the corresponding author.

Acknowledgments: The authors acknowledge the contribution of all respondents and facilitators who helped in the study.

Conflicts of Interest: The authors declare no conflict of interest.

References

1. Sepasgozar, S.M.; Karimi, R.; Shirowzhan, S.; Mojtahedi, M.; Ebrahimzadeh, S.; McCarthy, D. Delay Causes and Emerging Digital Tools: A Novel Model of Delay Analysis, Including Integrated Project Delivery and PMBOK. *Buildings* **2019**, *9*, 191. [CrossRef]
2. Bubshait, A.A.; Cunningham, M.J. Comparison of Delay Analysis Methodologies. *J. Constr. Eng. Manag.* **1998**, *124*, 315–322. [CrossRef]
3. Salim, W.N.; Sujana, C.M. Project delay analysis of highrise building project in Jakarta. *IOP Conf. Ser. Earth Environ. Sci.* **2020**, *426*, 012051. [CrossRef]
4. Alkass, S.; Mazerolle, M.; Harris, F. Construction delay analysis techniques. *Constr. Manag. Econ.* **1996**, *14*, 375–394. [CrossRef]
5. Zaneldin, E.K. Investigating the types, causes and severity of claims in construction projects in the UAE. *Int. J. Constr. Manag.* **2020**, *20*, 385–401. [CrossRef]
6. Ho, S.P.; Liu, L.Y. Analytical model for analyzing construction claims and opportunistic bidding. *J. Constr. Eng. Manag.* **2004**, *130*, 94–104. [CrossRef]
7. Çevikbaş, M.; Işık, Z. An overarching review on delay analyses in construction projects. *Buildings* **2021**, *11*, 109. [CrossRef]
8. Stumpf, G.R. Schedule delay analysis. *Cost Eng.* **2000**, *42*, 32–43.
9. Kao, C.K.; Yang, J.B. Comparison of windows-based delay analysis methods. *Int. J. Proj. Manag.* **2009**, *27*, 408–418. [CrossRef]
10. Kim, Y.J.; Lee, W.C.; Hong, J.S.; Shin, D.W. A Case study on delay analysis methods in the construction projects. *Korean J. Constr. Eng. Manag.* **2004**, *5*, 129–137.
11. Braimah, N. construction delay analysis techniques—A review of application issues and improvement needs. *Buildings* **2013**, *3*, 506–531. [CrossRef]
12. Arditi, D.; Pattanakitchamroon, T. Analysis methods in time-based claims. *J. Constr. Eng. Manag.* **2008**, *134*, 242–252. [CrossRef]
13. Braimah, N. Approaches to delay claims assessment employed in the uk construction industry. *Buildings* **2013**, *3*, 598–620. [CrossRef]
14. AACE International. *Forensic Schedule Analysis, Recommended Practice No. 29R-03, TCM Framework: 6.4–Forensic Performance Assessment*; AACE: Fairmont, WV, USA, 2011.
15. Kim, B.S.; Seong, G.G.; Bae, I.H.; Bang, H.S.; Choi, H.J.; Kim, O.K. An design of analyzing process by construction extension of time. *Korean J. Constr. Eng. Manag.* **2019**, *20*, 104–113. [CrossRef]
16. Society of Construction Law (SCL). *Delay and Disruption Protocol*, 2nd ed.; SCL: Hinckley, UK, 2017; ISBN 978-0-9543831-2-1.
17. Trauner, J.T. *Construction Delays: Documenting Causes, Wining Claims, Recovering Costs*; R.S. Means Company Inc.: Kingston, MA, USA, 1990; ISBN 978-0876291740.
18. Arditi, D.; Pattanakitchamroon, T. Selecting a delay analysis method in resolving construction claims. *Int. J. Project Manag.* **2006**, *24*, 145–155. [CrossRef]
19. Sargent, A. Delay Analysis Methods: Examining the Impacted As-Planned Method. Available online: https://vertexeng.com/insights/delay-analysis-methods-examining-the-impacted-as-planned-method/ (accessed on 31 March 2022).
20. Lee, G. *Report on Delay Analysis on MBC (Munhwa Broadcasting Corporation) New Home Office Construction Project*; Seoul Southern District Court: Seoul, Korea, 2015.
21. Kim, S.G.; Kwon, S. Case Analysis on application of project delay analysis method in domestic construction project. *Korean J. Constr. Eng. Manag.* **2019**, *20*, 98–106. [CrossRef]
22. Zack, J.G. But-for schedules-analysis and defense. *Cost Eng.* **2001**, *43*, 13–17.
23. Shin, H.C. An Analysis of Problems and Survey on the Application Method for Delay Analysis. Master's Thesis, Chung-Ang University, Seoul, Korea, 2019. Available online: http://dcollection.cau.ac.kr/common/orgView/000000229553 (accessed on 19 August 2022).
24. Popescu, C. Selecting As-planned base in project dispute *Trans. Am. Assoc. Cost. Eng.* **1991**, C2.1–C2.4.
25. Architectural Institute of Korea (AIK). *Final Report on Delay Analysis on MBC (Munhwa Broadcasting Corporation) New Home Office Construction Project*; Architectural Institute of Korea: Seoul, Korea, 2019.

Article

A Project Scheduling Game Equilibrium Problem Based on Dynamic Resource Supply

Cuiying Feng [1,2,3], Shengsheng Hu [1,*], Yanfang Ma [4] and Zongmin Li [2]

1   School of Management, Zhejiang University of Technology, Hangzhou 310023, China
2   Business School, Sichuan University, Chengdu 610064, China
3   Zhejiang Keyou Information Engineering Co., Ltd., Yiwu 322001, China
4   School of Economics and Management, Hebei University of Technology, Tianjin 300401, China
*   Correspondence: zjuthss@163.com

Abstract: In a resource-constrained project scheduling problem, most studies ignore that resource supply is a separate optimization problem, which is not in line with the actual situation. In this study, the project scheduling problem and the resource supply problem are regarded as a dynamic game system, with interactive influences and constraints. This study proposes a Stackelberg dynamic game model based on the engineering supply chain perspective. In this model, the inherent conflicts and complex interactions between the Multi-mode Resource-Constrained Project Scheduling Problem (MRCPSP) and the Multi-Period Supply Chain Problem (MPSCP) are studied to determine the optimal equilibrium strategy. A two-level multi-objective programming method is used to solve the problem. The MRCPSP is the upper-level planning used to optimize project scheduling and activity mode selection to minimize project cost and duration; MPSCP is a lower-level planning method that seeks to make resource transportation decisions at a lower cost. A two-layer hybrid algorithm, consisting of Genetic Algorithm (GA) and Particle Swarm Optimization (PSO), is proposed to determine the optimal equilibrium strategy. Finally, the applicability and effectiveness of the proposed optimization method are evaluated through a case study of a large hydropower construction project, and management suggestions for related departments are provided.

Keywords: multi-cycle supply chain; project scheduling; Stackelberg dynamic game; two-level multi-objective programming; GA with double strings; particle swarm optimization

Citation: Feng, C.; Hu, S.; Ma, Y.; Li, Z. A Project Scheduling Game Equilibrium Problem Based on Dynamic Resource Supply. *Appl. Sci.* **2022**, *12*, 9062. https://doi.org/10.3390/app12189062

Academic Editor: Antonio J. Nebro

Received: 20 August 2022
Accepted: 6 September 2022
Published: 9 September 2022

Publisher's Note: MDPI stays neutral with regard to jurisdictional claims in published maps and institutional affiliations.

1. Introduction

Managing work in the form of projects has become a common practice to improve work efficiency. Currently, approximately 20% of the world's economic activity is in the form of projects, generating an annual economic value of roughly $12 trillion [1]. Project scheduling refers to the scientific and reasonable arrangement of the beginning and execution times of each activity in a project in order to achieve the established goal [2]. The Resource-Constrained Project Scheduling Problem (RCPSP) is a form of planning based on constraining the resources required by project activities. The Classic RCPSP scheduling decision must satisfy the temporal and resource constraints, and its solution is a scheduling plan that optimizes the management objective under these constraints. [3]

Many scholars have studied the RCPSP, and extension problems have been developed. Liu et al. [4] designed an RCPSP model based on the time window delay from the perspective of owner-contractor interaction. Kim et al. [5] considered the delay penalty on the basis of minimizing the total project time. Cheng et al. [6] considered the problem of night shifts in construction projects and minimized the project duration, cost, and utilization of night shifts while meeting the constraints of operational logic and labor availability. In the study of Demeulemeester and Herroelen [7] as well as Debels and Vanhoucke [8], an activity can be interrupted after every integer unit of its activity time. Muritala Adebayo

Isah and Byung-Soo Kim [9] presented a stochastic multiskilled resource scheduling model for RCPSP, which considers the impacts of risk and uncertainty on activity durations. The standard RCPSP assumes that an activity can only be executed in one mode, with a fixed duration and resource requirements. On this basis, Elmaghraby [10] proposed a new concept; in practice, management departments can flexibly arrange appropriate execution modes for project activities to achieve corresponding goals, and each mode has different durations and resource demands, i.e., the Multi-mode Resource-Constrained Project Scheduling Problem (MRCPSP). Varma et al. [11] discussed a multi-mode problem without the use of non-renewable resources. Zhu et al. [12] considered the MRCPSP with generalized resource constraints. Bellenguez and Emmanuel [13] discussed a special case: in an MRCPSP, each activity requires specific skills, while resources are employees with fixed skills, and employees must be selected according to their skills when arranging activities.

MRCPSP is a critical issue in engineering supply chain management, especially in large-scale engineering construction projects. The resource supply is complex and changeable, and the resource transportation policy is updated according to the different ordering schedules of project scheduling [14]. At this point, a Multi-Period Supply Chain Problem (MPSCP) arises, directly affecting both the cost and schedule of the project. If the project schedule is made without considering the constraints of upstream resource supply capacity, the supply delay or interruption of suppliers will delay the construction period and increase both the project cost and risk, among other factors. Similarly, resource supply driven by non-engineering schedule planning will lead to a lower resource utilization rate and a higher inventory cost. In this case, resource constraint is not only a constraint condition of MRCPSP, but also an optimization problem closely related to MRCPSP with the characteristics of a dynamic game. However, in most studies, project scheduling and resource supply are considered as two independent optimization problems, ignoring the interaction and conflict between them, possibly leading to a suboptimal solution for resource supply and project delay. Therefore, it is more realistic to consider project scheduling and resource supply as an integrated system for dynamic game optimization.

Relevant research by Sarker [15] demonstrated that the simultaneous optimization of project scheduling and resource supply can improve the efficiency of project scheduling and reduce the overall cost. Xie et al. [16] took the project duration and cost as the optimization objectives, considered the variable resource availability and expressed it by interval variables, and established a dual-objective optimization model of the MRCPSP under the constraint of variable resource availability. Lv et al. [17] further expanded renewable resources into flexible resources with capacity differences, and established a problem model considering capacity differences in which the capacity level affects activity duration. Schwindt and Trautmann [18] considered the time-dependent resource capacity and divided the aggregate demand of intermediate and final products into batches in the batch production mode. Shu-Shun Liu et al. [19] proposed a two-stage optimization model based on constrained programming to address the bridge maintenance scheduling problem.

Many scholars have proposed rich algorithms to solve the integrated system optimization problem of project scheduling and resource supply chains. Asta et al. [20] designed a hybrid algorithm that combines Monte Carlo and hyper-heuristic methods to solve this problem. Xie et al. [21] studied MRCPSP under the condition of uncertain activity duration and designed an approximate dynamic programming algorithm based on the rollout to solve it. Peteghem et al. [22] studied MRCPSP with resource preemption characteristics, introduced an extended serial scheduling generation scheme to improve mode selection, and designed a two-population genetic algorithm. Furthermore, many studies have proven that GA and PSO are more effective and have different advantages in solving such problems [23–27].

GA was first proposed by J. Holland in 1975. It is a random search algorithm that draws on natural selection and genetic mechanisms in the biological world and follows the principle of "survival of the fittest" [28]. Its basic idea is to imitate the natural evolution process through genetic manipulation of individuals with certain structural forms in the population, so as to generate a new population and gradually approach the optimal solution. PSO was proposed by J. Kennedy and R. C. Eberhart in 1995 [29]. It is a random search algorithm based on group cooperation, developed by simulating the foraging behavior of birds. It finds the global optimum by following the currently searched optimum.

The existing research has made important achievements in project scheduling problems and algorithm designs. However, when constructing the model, the interaction between decision makers is ignored. Secondly, the project scheduling problem from the perspective of the engineering supply chain is a multi-objective and multi-stage complex decision problem; previous studies [26,27] have shown that using the bi-level programming method can generate better results.

The innovation of this paper is that the project scheduling problem and resource supply problem are regarded as an integrated system of a dynamic game, involving interactive influences and constraints. Moreover, a two-level multi-objective programming method is adopted, which organizes the whole process of "objective—modeling—algorithm—optimization—decision." A large hydropower construction project is taken as an example to prove the scientificity and feasibility of the method.

The rest of the paper is organized as follows: Section 2 gives the key problem statement of MRCPSP-MPSCP integrated system and research methods; Section 3 details the modeling method and hypothesis of establishing the two-level dynamic game model; Section 4 proposes the two-level GADS/DIWPSO hybrid algorithm to solve the established model; Section 5 gives a practical case to emphasize the practicability and effectiveness of the optimization method, and proposes forward management suggestions to related departments; and finally, Section 6 provides conclusions and future research directions.

2. Research Overview

2.1. Problem Description

Project scheduling has always been considered the core of engineering supply chains, as the construction and operation of the supply chain are driven by the development of the project schedule. The engineering construction department first defines the resource demand of each demand point in each time period by forming a project schedule plan; next, the resource supplier attempts to meet the resource demand. However, as the resource supplier is also a decision-making subject with its own constraints, it optimizes its cost and time goals by formulating a resource transportation strategy and sends this information back to the engineering construction department, thus affecting the formulation of the project's schedule. Conflict and cooperation coexist in the engineering supply chain. The engineering construction department has higher decision-making power (i.e., the leader), whereas the resource supplier is subordinate (i.e., the follower). This "leader–follower" behavior is, in its essence, a Stackelberg game, with the characteristics of multi-periodicity in practice. Therefore, project scheduling and resource supply comprise an inseparable integrated system, which is the game analysis and dynamic coordination problem of the integrated system of "project scheduling–resource supply" from the perspective of the engineering supply chain. The successful operation of this system helps reduce project costs, shorten construction periods, and improve project quality and resource utilization.

The research object of this project is a large hydropower construction project located in southeast China. A concrete double-curvature arch dam is the main project, with many construction activities with priority relationships and shared resources; each activity has several alternative modes, and each mode has a certain duration and resource demand. To meet the requirements of shared resources, it is necessary to specify the ordering time and quantity in each time period when making the project scheduling scheme, and the resource supplier further formulates the resource transportation strategy. These constitute the dynamic game decision-making system of the MRCPSP-MPSCP integrated system, and the structural model is shown in Figure 1.

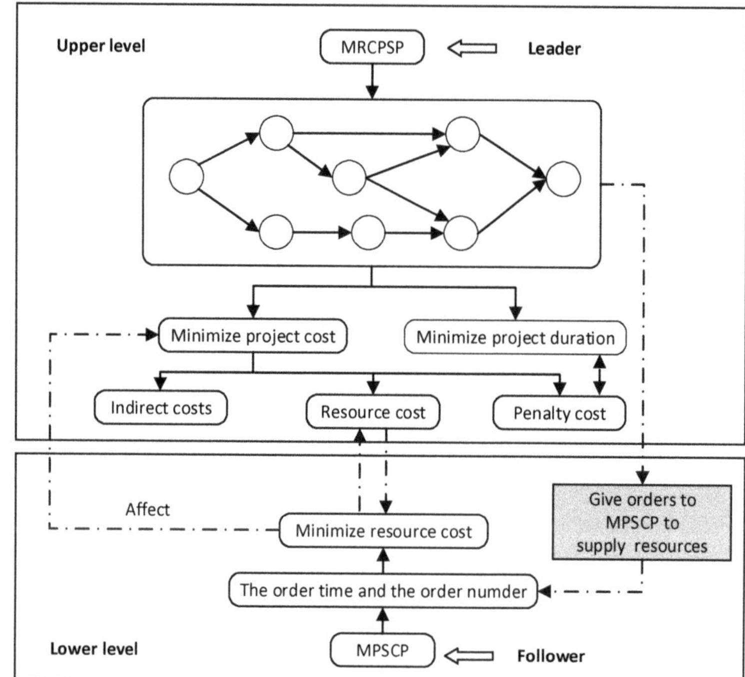

Figure 1. Structural model of the MRCPSP-MPSCP integrated system.

2.2. Research Methodology

In this paper, we adopt a two-level multi-objective mode of programming which informs the whole process of "objective—modeling—algorithm—optimization—decision." According to the characteristics of the dynamic game of this problem in the engineering supply chain, we adopt a two-level modeling method to express the interaction between MRCPSP and MPSCP. To determine the optimal equilibrium strategy of the model, a two-layer hybrid algorithm, composed of a GA with double strings and an improved PSO, is proposed. Considering the existence of many uncertainties in the engineering supply chain, for example, the project activity time is a typical uncertain variable; Bidot et al. [30] considered a project schedule with a random activity duration. In addition, factors such as weather conditions, labor efficiency, and transportation environment make the decision-making process more complicated. Therefore, random variables are used in this study to describe various variables in an uncertain environment. Finally, the applicability and effectiveness of the proposed optimization method are evaluated through a case of a large hydropower construction project.

3. Model Establishment

To properly express the dynamic game characteristics of the MRCPSP-MPSCP integrated system, a two-level multi-objective programming model is established, which includes the upper and lower models.

3.1. Symbols and Assumptions

3.1.1. Indicators

j: Project activity index, $j \in J = \{1, 2, \ldots, J\}$
k: Material type index, $k \in K = \{1, 2, \ldots, K\}$
t: Time period index, $t \in \{1, 2, \ldots\}$
m: Activity mode index, $m \in M = \{1, 2, \ldots, M\}$

i: Activity mode index, $i \in I = \{1, 2, \ldots, I\}$

s: Demand point index, $s \in S = \{1, 2, \ldots, S\}$

3.1.2. Parameters Related to Project Scheduling

B: Total available budget

D: Project planning cycle

IC_s: Inventory capacity at the demand point s

P_j: Set of predecessors of activity j

c_{jm}: Direct cost of activity j in mode m

d_{jm}: Operation time of activity j in m mode

r_{jmk}: The demand of m mode of activity j for resource k in each time period

r_k: Maximum supply capacity of resource k in each time period

c_0: Overhead cost per time period

EF_j: The earliest completion time of activity j

LF_j: The latest completion time of activity j

R_s: The set of activities for which demand point s is responsible

Pc_k: Unit purchase cost of resource k

Oc_k: Each order cost of resource k

Ic_k: Storage cost of resource k in each time period

3.1.3. Parameters Related to Resource Supply

$T(t)$: Delivery date of time period t

R_k: Maximum amount of resource k transported each time

P_{ik}: Supply capacity of resource k at supply point i

c_{isk}: Unit transportation cost of resource k on the transportation path (i, s)

t_{isk}: Unit transportation time of resource k on the transportation path (i, s)

3.1.4. Decision Variables

$v_{isk}(t)$: The allocation of resource k on transportation path (i, s) in time period t

$$x_{jmt} = \begin{cases} 1, & \text{If activity } j \text{ executes mode } m \text{ in time period } t \\ 0, & \text{Otherwise} \end{cases}, \text{ represents the mode}$$

selection of activity j

$$z_{kt} = \begin{cases} 1, & \text{If resource } k \text{ transported at the beginning of time period } t \\ 0, & \text{Otherwise} \end{cases}, \text{ represents}$$

whether resource k is transported during time period t

3.1.5. Intermediate Variables

ST_j: Starting time of activity j

FT_j: Completion time of activity j

A_t: Activity set of ongoing jobs in time period t

S_{kst}: The remaining amount of resource k at demand point s at the end of the time period t

3.1.6. Assumptions

① The project contains j activities and two virtual activities, in which the two virtual activities represent the initial and final activities of the project, denoted as $j = 0$ and $j = J + 1$, respectively.

② Only when all the predecessor activities of the activity are completed can the activity begin.

③ Each activity can only execute one mode without interruption.

④ The supply capacity of the supply point and the inventory capacity of the demand point are limited and cannot be increased.

⑤ The loading and unloading costs and time of the transport vehicles were included in the corresponding transport costs and time.

⑥ Uncertain parameters, such as resource demand, project activity time, and unit transportation cost, are random variables.

⑦ Resources are consumed evenly in each time period.

3.2. Project Scheduling

Project scheduling occupies a dominant position in an engineering supply chain with the contractor as the core. In view of the project scheduling problem, under the condition of ensuring the quality of project, duration and cost are its three major objectives.

3.2.1. Schedule Objective

One of the most important goals in project scheduling is to minimize the project duration and complete the project as early as possible under all constraints. In this study, the completion time of the last activity $(J+1)$ can be used to describe the duration of the project; that is, the duration F_t can be expressed as Equation (1).

$$F_t = \sum_{t=EF_{(J+1)}}^{LF_{(J+1)}} \sum_{m=1}^{M_{(J+1)}} t x_{(J+1)mt} \tag{1}$$

3.2.2. Cost Objective

Cost is another important goal in project scheduling. Project costs are generally divided into direct and indirect portions. Among them, ordering, purchasing, and storage costs belong to direct costs; indirect costs belong to fixed costs in any time period and are related to project duration. In summary, the cost function F_c can be expressed by Equation (2):

$$F_c = \sum_{j=1}^{J} \sum_{m=1}^{M_j} \sum_{t=EF_j}^{LF_j} d_{jm} x_{jmt} \sum_{k=1}^{K} r_{jmk} Pc_k + \sum_{j=1}^{J} \sum_{m=1}^{M_j} \sum_{t=EF_j}^{LF_j} c_{jm} x_{jmt} + \sum_{k=1}^{K} \sum_{t=1}^{Ft} Oc_k z_{kt}$$
$$+ \frac{1}{2} \sum_{j=1}^{J} \sum_{m=1}^{M_j} \sum_{t=EF_j}^{LF_j} d_{jm} x_{jmt} \sum_{k=1}^{K} r_{jmk} Ic_k + \sum_{k=1}^{K} \sum_{t=1}^{Ft} Ic_k S_{kst} + c_0 Ft \tag{2}$$

3.2.3. Constraints

In full consideration of the actual situation of the engineering supply chain, the constraints are listed in this section. This will make the model more realistic.

$$\sum_{j \in R_s} \sum_{m=1}^{M_j} r_{jmk} \sum_{t'=t}^{t+d_{jm}-1} x_{jmt} \leq S_{ks(t-1)} + \sum_{i=1}^{I} v_{isk}(t), \forall k \in K, s \in S, t \in FT \tag{3}$$

$$\sum_{j \in A_t} \sum_{m=1}^{M_j} r_{jmk} \leq r_k, \forall k \in K, t \in FT \tag{4}$$

$$\sum_{m=1}^{M_j} \sum_{t=EF_j}^{LF_j} t x_{jmt} \leq \sum_{m=1}^{M_j} \sum_{t=EF_j}^{LF_j} (t - d_{jm}) x_{jmt}, \forall j \in P_j, j \in J \tag{5}$$

$$S_{kst} = S_{ks(t-1)} + \sum_{i=1}^{I} v_{isk}(t) - \sum_{j \in R_s} \sum_{m=1}^{M_j} r_{jmk} \sum_{t'=t}^{t+d_{jm}-1} x_{jmt'}, \forall k \in K, s \in S, t \in FT. \tag{6}$$

$$S_{ks0} = 0, S_{ksFt} = 0, \forall k \in K, s \in S \tag{7}$$

$$S_{kst} \geq 0, \sum_{k=1}^{K} S_{kst} \leq IC_s, \forall k \in K, s \in S, t \in FT \tag{8}$$

$$F_t \leq D, F_c \leq B \tag{9}$$

$$FT_j = \sum_{m=1}^{M_j} \sum_{t=EF_j}^{LF_j} t x_{jmt}, ST_j = FT_j - d_{jm}, \forall j \in J \tag{10}$$

$$\sum_{m=1}^{M_j} \sum_{t=EF_j}^{LF_j} x_{jmt} = 1, \forall j \in J \tag{11}$$

$$x_{jmt} \in \{0,1\}, \forall j \in J, t \in FT \tag{12}$$

$$z_{kt} = \begin{cases} 1, \sum_{i=1}^{I} \sum_{s=1}^{S} v_{isk}(t) > 0 \\ 0, \sum_{i=1}^{I} \sum_{s=1}^{S} v_{isk}(t) = 0 \end{cases} \tag{13}$$

Constraint condition Formula (3) represents the resource constraint. Equation (4) represents that the total consumption of resource k in each time period cannot exceed its maximum supply capacity. Equation (5) is the predecessor constraint. Equation (6) represents the remaining amount of available resources at the end of each time period, which can be regarded as a state transition variable. Equation (7) indicates that, to maximize the utilization of resources, the resource surplus should be zero at the beginning and end of the project. Equation (8) indicates that the resource surplus at the end of each period is greater than or equal to zero, and cannot exceed the inventory capacity. Equation (9) indicates the construction period and budget constraint. Equation (10) represents the start and end times of each activity. Equations (11)–(13) are logical constraints: each activity should be executed within the range of the earliest and latest completion times, and only one activity mode can be executed. Meanwhile, there are also characteristic constraints among the decision variables.

3.3. Resource Supply

After the project scheduling scheme is determined, the resource supplier seeks to minimize the total operational cost and transportation time by optimizing the transportation volume between the supply and demand points. The transportation model can be expressed as follows.

3.3.1. Operating Cost Target

The resource supplier transports the corresponding amount of resources to the demand point of the project. The total operating cost (i.e., Z_c) of the resource transport model is the transportation cost from the supply point to the demand point. Therefore, the total operational cost of this model can be expressed by Equation (14):

$$Z_c = \sum_{i=1}^{I} \sum_{s=1}^{S} \sum_{t=1}^{T} \sum_{k=1}^{K} c_{isk} v_{isk}(t) \tag{14}$$

3.3.2. Transport Time Target

Minimizing transportation time is an important goal. The transportation time on the transportation path (i, s) in the time period t can be expressed as $T_{is}(t) = \sum_{k=1}^{K} t_{isk} v_{isk}(t)$. Therefore, the total transportation time in this model can be expressed by Equation (15):

$$Z_t = \sum_{t=1}^{T} \max_{i,s} T_{is}(t) \tag{15}$$

3.3.3. Constraints

Equation (16) is the equation of state variable $B_{ik}(t)$, which represents the amount of resources k remaining at each resource supply point at the end of each time period t. Equation (17) demonstrates that the quantity of resources at each supply point is the maximum supply capacity of the supply point at the beginning, and that the quantity of resources is non-negative throughout the entire process. Equation (18) indicates that the quantity of resources transported to each demand point must satisfy the demand level of project scheduling in terms of the total quantity. Equation (19) indicates that the quantity of transported resources cannot exceed the maximum supply. Equation (20) represents the delivery-date constraint. Equation (21) is a logical constraint.

$$B_{ik}(t) = B_{ik}(t-1) - \sum_{s=1}^{S} v_{isk}(t), \forall i \in I \tag{16}$$

$$B_{ik}(0) = P_{ik}, B_{ik}(t) \geq 0, \forall i \in I \tag{17}$$

$$\sum_{i=1}^{I} v_{isk}(t) \geq \sum_{t=1}^{Ft} \sum_{j=1}^{J} \sum_{m=1}^{M_j} r_{jmk} d_{jm} x_{jmt}, \forall s \in S, \forall k \in K \tag{18}$$

$$\sum_{s=1}^{S} v_{isk}(t) \leq P_{ik}, \forall i \in I, s \in S \tag{19}$$

$$\max \sum_{k=1}^{K} t_{isk} v_{isk}(t) \leq T(t) \tag{20}$$

$$0 \leq v_{isk}(t) \leq R_k, \forall k \in K, t \in FT \tag{21}$$

3.4. Global Dynamic Game Optimization Model

After analyzing project scheduling and resource supply, the objective function and constraints are integrated into a dynamic game optimization model, which is more consistent with the coexistence of cooperation and conflict among supply chain members. This provides a theoretical basis for the sustainable operation of the engineering supply chain to improve technological innovation ability, cooperation, and management abilities among the upstream and downstream members.

When all constraints on project scheduling are set to A and resource supply constraints are set to B, then the overall dynamic game optimization model is as follows.

$$\min\{F_t, F_c\}$$
$$s.t. \begin{cases} A \\ \min\{Z_c, Z_t\} \\ s.t.\{B \end{cases} \tag{22}$$

4. Algorithm Design

The MRCPSP-MPSCP integrated system is an NP-hard problem. GA and PSO have been mentioned as the most practical methods to solve this kind of problem. For the problem with the 0–1 decision variable, Sakawa et al. [31] proved that a GA with Double Strings (GADS) shows superior convergence to the simple GA. Therefore, this study draws on several excellent algorithm ideas and proposes a hybrid GAPSO algorithm to solve the dynamic game optimization problem in the engineering supply chain. Specifically, GADS is used to solve the upper MRCPSP, and a Dynamically adjusted Inertial Weight PSO (DIWPSO) is used to solve the underlying MPSCP.

4.1. GADS

In this section, GADS is used to analyze and solve project scheduling. Its primary objective is to determine the execution priority of each activity and arrange the activities. Appropriate encoding methods and decoding rules were selected according to the characteristics of the problem, and the corresponding selection, crossover, mutation, and evolution termination conditions were designed.

4.1.1. Coding Design

To express the execution order of each activity and the characteristics of multiple models in the MRCPSP more reasonably, the algorithm uses the activity-linked list and the corresponding activity-mode-linked list as the code and composes the chromosome. To improve the efficiency of the algorithm, activity J was first stratified according to its priority. The level of each activity is determined as follows: the smaller the tier, the higher the priority of the activities within that tier. In the process of coding, the activities of the small level are always arranged before the activities of the large level, so that the chromosome can ensure the precedence constraint in the subsequent genetic operation and avoid the generation of infeasible solutions. As demonstrated in Table 1, there are nine activities on this chromosome, divided into four levels. The priority of the three activities in level one is higher than those of the other three levels, and the priority of the two activities in level two is higher than those of the three and four levels.

Table 1. Coding design.

Level	1			2			3		4
Activity J	1	3	2	5	4	6	8	7	9
Modes m_j	1	2	1	2	1	2	1	2	1

The priority of the project job is then encoded by numerical coding; that is, the length of the code is equal to the number of project activities, the position of the code represents the priority of activity J in this chromosome, and the number on this position represents the activity number. The higher the order of the activity J, the higher the priority. As indicated in Table 1, the priority of Activity 1 is $J_1 = 1$, which has the highest priority. Activity 9 has the lowest priority.

The job modes of an activity are encoded in a linked list of modes. m_j represents a set of modes of activity J.

4.1.2. Decoding Rules

Herein, a hybrid schedule generation scheme (HSGS) [32] was used as the decoding rule. The earliest start time of an activity can be determined when the predecessors of the activity have been completed and resource requirements have been met. HSGS is used to determine the completion time of each activity in turn and then calculate the total duration of the entire project.

Step 1. Let A_n be the set of activities that have been scheduled, and let U_n be the set of activities that have not been scheduled. When initialized, $A_n = \varnothing$ and $U_n = \{1, 2, \ldots, N\}$. First, the priority of each activity in U_n is sorted in descending order, and the activity with the highest priority is selected for the arrangement.

Step 2. Continue to select the highest priority activity from U_n and conduct a timing constraint judgment. If satisfied, proceed to the next step. If not, the next activity is selected for judgment until the activity that meets the conditions is determined.

Step 3. Conduct resource constraint judgment on the activity to determine whether it can be scheduled in parallel with scheduled activities. If so, proceed to the next step and arrange the activity into A_n; if not, go to Step 2.

Step 4. Update A_n and U_n, then repeat from Step 1 until all activities are scheduled, i.e., $A_n = \{1, 2, \ldots, N\}$ and $U_n = \varnothing$.

4.1.3. Fitness Function

Because there are two objective functions of duration and cost in the upper planning, the fitness function is constructed using the weighted aggregation method, to maintain the effectiveness of the multiple objectives. Let μ_1 and μ_2 represent the weights of the two objective functions; the fitness function can then be represented by Equation (23). After making the changes, the maximum fitness value is required.

$$Fitness(F) = \mu_1 \frac{F_t^{\max} - F_t}{F_t^{\max} - F_t^{\min}} + \mu_2 \frac{F_c^{\max} - F_c}{F_c^{\max} - F_c^{\min}} \tag{23}$$

4.1.4. Genetic Manipulation

Step 1. Set the parameters in the GADS: size L_1, maximum number of iterations T_1, crossover probability p_c, and mutation probability p_m.

Step 2. Initialize L_1 individuals as a group, set the initial iteration $\tau_1 = 0$, and use the coding program to generate the initial individuals $S_l(0)$.

Step 3. Through the elite roulette method to select individuals, according to the size of fitness, develop roulette with slots, and use the roulette to generate the next generation of individuals $(\tau_1 + 1)$. If the fitness function of the l chromosome in the population is $f(S_l)$, then the probability of chromosome S_l being selected is

$$P_l = \frac{f(S_l)}{\sum_{l=1}^{n} f(S_l)} \tag{24}$$

Step 4. Since the chromosome of the algorithm consists of an activity list and a mode list, it is necessary to cross these two lists in steps.

Step 4.1. The activity list is crossed using the alternating crossing method. First, the first gene from parent A is added to offspring A. Then, we select the first gene from parent B and judge whether it is duplicated with genes in offspring A. If it is duplicated or does not conform to the hierarchical order, it is discarded; if it is not repeated but conforms to the hierarchical order, it is added to offspring A.

Step 4.2. The second gene is selected from parent A to judge whether it is duplicated. Finally, the genes in the two parents are selected in turn to form offspring A. Similarly, the genes in parents B and A are selected to obtain child B.

Step 4.3. Then, the mode list was crossed by a single-point operation. Let i be the position of the gene, let N be the total length of the chromosome, and randomly select integer $n_1 < N$. If $1 \le i \le n_1$, then the mode of gene i of offspring A is equal to that of parent A; if $n_1 \le i \le N$, then the mode of child A is equal to that of parent B.

Step 4.4. Similarly, randomly select integer $n_2 < N$. If $1 \le i \le n_2$, then the mode of gene i of offspring B is equal to that of parent A; if $n_2 \le i \le N$, then the mode of child B is equal to that of parent B.

Assuming that $n_1 = 4, n_2 = 5$, a schematic diagram of the chromosome crossover operation is shown in Figure 2.

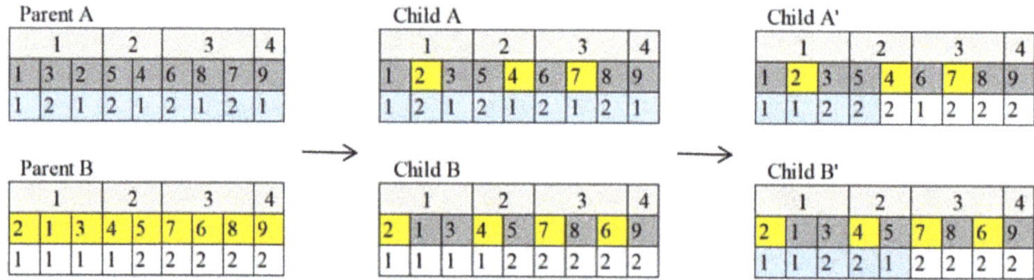

Figure 2. Schematic diagram of crossover operation.

Step 5. This step is concerned with mutation. For the variation of the activity list, on the premise of satisfying the hierarchy order, the mutation operation is carried out by the exchange mutation method, in which two mutation points are randomly selected from parents and genes are swapped at those two locations. However, activity modes do not change, as demonstrated in Figure 3. In Figure 3, activities 6 and 8 in level 3 are exchanged and mutated to obtain new individuals. For the variation in the mode list, an activity is randomly selected, and its activity mode is changed, as demonstrated in Figure 3.

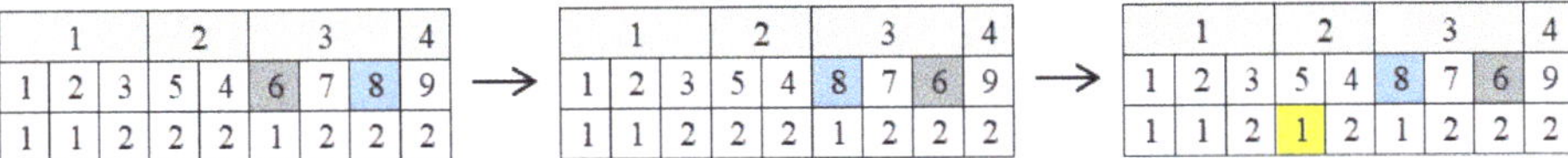

Figure 3. Variation operation diagram.

Step 6. Let the number of iterations be $\tau_1 = \tau_1 + 1$ and enter the next round of iterations until the maximum is reached.

4.2. DIWPSO

As a follower, the resource supplier must solve the problem of determining the resource allocation amount on each transportation path in each time period. Based on the characteristics of this problem, this section uses DIWPSO to solve the resource transportation policy.

4.2.1. Initial Code

In the existing research results, when solving the problem of resource transportation, the coding method mostly adopts the integer representation method, in which the customer (demand point) and the virtual distribution center are arranged together. In this study, the resource allocation quantity on the transportation path is adopted as the real number coding. Let $Y = v_{isk}(t)$ represent the position vector of each particle, initialize i particles as a population, and generate the ith particle with d-dimensional position vector Y_i; let its initial velocity $V_i = 0$, then the initial individual optimal is $P_i = Y_i^1$. The initial population is generated randomly so that it is distributed uniformly in the entire solution space as much as possible.

4.2.2. Fitness Function

The objective of resource supply is to minimize the running cost and transportation time, and the dimensions used are not the same. Therefore, the fitness function of the lower resource transportation model was constructed using the weighted aggregation method used in upper planning. Let β_1 and β_2 represent the weights of the two objective functions. The fitness function is shown in Equation (25), and the maximum fitness value is required.

$$Fitness(F_2) = \beta_1 \frac{z_c^{\max} - z_c}{z_c^{\max} - z_c^{\min}} + \beta_2 \frac{z_t^{\max} - z_t}{z_t^{\max} - z_t^{\min}} \tag{25}$$

4.2.3. Updating Policies

Step 1. Before the update operation, individuals are selected based on the elite strategy to increase the running speed of the algorithm. That is, the fitness of individuals generated in the population is first sorted from largest to smallest, and the top 50% of individuals are retained.

Step 2. DIWPSO is used for updating. Although the standard PSO has a fast convergence speed in the early stage, it is slow in later stages and easily converges locally. Therefore, this algorithm is improved from the perspective of the inertia weight. The inertia weight ω indicates the extent to which the original speed is retained; if ω is larger, the global search ability is stronger, and if ω is small, the local search ability is strong.

The update strategy is as follows: in the position vector, Equations (26) and (27) are used to update the particle velocity and position for the continuous factor $v_{isk}(t)$:

$$v_{id}(\tau + 1) = v_{id}(\tau)\omega + c_1 r_1(\tau)[p_{id}(\tau) - x_{id}(\tau)] + c_2 r_2(\tau)[g_d(\tau) - x_{id}(\tau)] \tag{26}$$

$$x_{id}(\tau+1) = x_{id}(\tau) + v_{id}(\tau+1) \tag{27}$$

$$\omega = \omega_{min} + (\omega_{max} - \omega_{min}) * e^{-\frac{\tau}{\tau_{max}}} + \sigma * betarnd(p,q) \tag{28}$$

where τ represents the current iteration number; τ_{max} represents the maximum number of iterations; ω_{max} represents the maximum inertia weight, which is set to 0.9; ω_{min} represents the minimum inertia weight, which is 0.1; σ is the inertia adjustment factor, which is 0.1; $p = 1, q = 3$; c_1 and c_2 are learning factors; r_1 and r_2 are uniform random numbers between [0,1]; $x_{id}(\tau)$ and $v_{id}(\tau)$ represent the position and velocity of the d dimension elements, respectively, of the i particle after the τ iteration; $p_{id}(\tau)$ represents the individual optimal position of the i particle in the d dimension; and $g_d(\tau)$ represents the global optimal position of all particles in the d dimension.

ω in the update strategy is an improved strategy for the dynamic adjustment of inertia weight [33], and the exponential function is used to control the change in inertia weight ω. With an increase in the number of iterations, $e^{-\tau/\tau_{max}}$ decreases nonlinearly; thus, ω can ensure the breadth of global search in the early stage and gradually decrease in the later stage to improve the ability of the local search and ensure its accuracy. Betarnd is a random number generator in MATLAB that can generate random numbers in line with the beta distribution. In addition, an inertia adjustment factor σ was added to control the deviation of the inertia weight, to make the adjustment more reasonable.

Step 3. Particle evaluation. To avoid generating infeasible particle positions and excessive velocities during the iteration, they must be within the corresponding limits.

$$v_{id} = \left\{ \begin{array}{l} v_{max}, v_{id} > v_{max} \\ v_{min}, v_{id} < v_{min} \end{array} \right. , x_{id} = \left\{ \begin{array}{l} x_{max}, x_{id} > x_{max} \\ x_{min}, x_{id} < x_{min} \end{array} \right. \tag{29}$$

Step 4. Particle adjustment. Since the fitness function is designed with the belief that larger is better, the individual $P_i(\tau)$ and the global $G(\tau)$ optimums are updated by calculating the fitness of the particles.

Step 4.1. For the individual optimum $P_i(\tau)$, if $Fitness[Y_i(\tau)] > Fitness[P_i(\tau-1)]$, update $P_i(\tau) = Y_i(\tau)$; otherwise, maintain the original value.

Step 4.2. For the global optimum $G(\tau)$, if $Fitness[P_i(\tau)] > Fitness[G(\tau-1)]$, update $G(\tau) = P_i(\tau)$; otherwise, maintain the original value, namely $G(\tau) = G(\tau-1)$.

Step 5. Premature particle determination. To judge the convergence degree of the particles, the population fitness variance [34] was introduced as the judgment mechanism of particle prematurity.

$$\delta^2 = \frac{1}{N}\sum_{i=1}^{N}\left(\frac{f_i - f_{avg}}{f}\right)^2 \tag{30}$$

$$f = \left\{ \begin{array}{l} \max|f_i - f_{avg}|, \max|f_i - f_{avg}| > 1 \\ 1, \text{ otherwise} \end{array} \right. \tag{31}$$

where δ^2 is the variance of population fitness; the larger δ^2 is, the better the population diversity, and vice versa. f_i is the fitness of the i particle; f_{avg} is the average fitness of the population, and f is the normalization factor, which limits the size of δ^2.

A population fitness judgment threshold δ_T^2 is selected for premature judgment: when $\delta^2 < \delta_T^2$, the particle enters premature convergence. δ_T^2 is generally much smaller than the fitness variance of the initial population; $\delta_T^2 = 0.001$ is taken here.

Step 6. The mutation operation exists to improve the ability of the algorithm to jump out of premature convergence, ensure the diversity of the population, and keep the algorithm from falling into local convergence in the later stage to stop searching for a better solution. The mutation mechanism of the differential evolution algorithm is used to mutate the identified premature particles.

$$V_i(\tau+1) = x_{r1}(\tau) + \eta[x_{r2}(\tau) - x_{r3}(\tau)] \tag{32}$$

$r1, r2, r3 \in (1, 2, \ldots, N)$ is a random number and $r1 \neq r2 \neq r3 \neq i$, and η is a scaling factor adjusted by adaptive strategy:

$$\eta = \eta_{\max} - \tau(\eta_{\max} - \eta_{\min}) / \tau_{\max} \tag{33}$$

where $\eta_{\max}$ and $\eta_{\min}$ are the upper and lower limits of the scaling factor, respectively.

4.3. Overall Process Framework of the Algorithm

The algorithm designed in this study is a two-layer GADS/DIWPSO hybrid algorithm. In the project scheduling problem, GADS is first used to initialize the feasible strategy and introduce it into lower-level planning. Then, DIWPSO is used to find the corresponding optimal solution of resource provisioning and the input to the upper planning is returned. Then, GADS is used to decode and generate the current optimal solution. This process is repeated until the upper optimal solution satisfies the stop condition. Through this dynamic interaction, the Stackelberg-Nash equilibrium strategy of the MRCPSP-MPMSP ensemble system is finally obtained.

The flow chart of this hybrid algorithm is shown in Figure 4, where the left part is the flow of GADS solving the upper-level project scheduling problem and the right part is the flow of DIWPSO solving the lower-level resource supply problem.

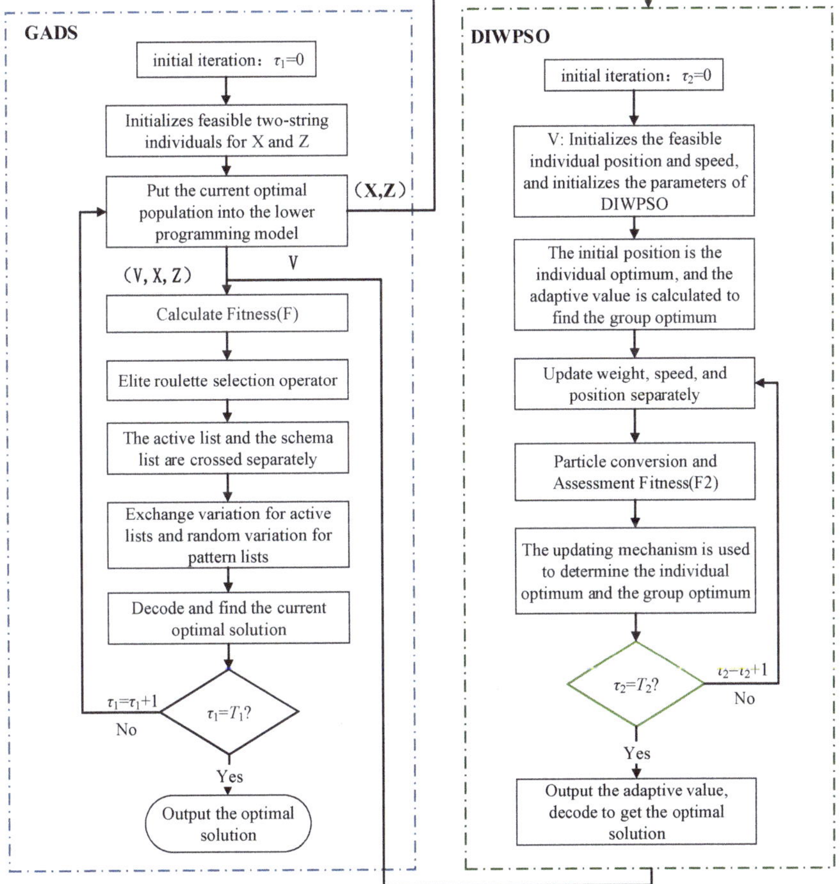

Figure 4. Flow chart of the hybrid algorithm.

5. Practical Application

The practical application and calculation test of a dam project verified the practicability and effectiveness of the proposed optimization method and provided decision-making guidance.

5.1. Project Description

In this study, a large hydropower project located in southeast China was considered as an application example. The project had a variety of hydraulic structures such as river dams, flood discharge structures, and hydraulic power generation systems. The river dam was a concrete double-curvature arch dam with a height of 610 m.

The concrete double-curvature arch dam construction project, which consists of 17 engineering activities, is the most important part. A flowchart is shown in Figure 5. Each activity has several optional modes, and each mode has a certain duration and resource demand. At the construction site, there are two large-scale resource demand points to allocate resources for each activity within the project, and the three resources required by the demand points are supplied by an external resource supplier with four resource supply points.

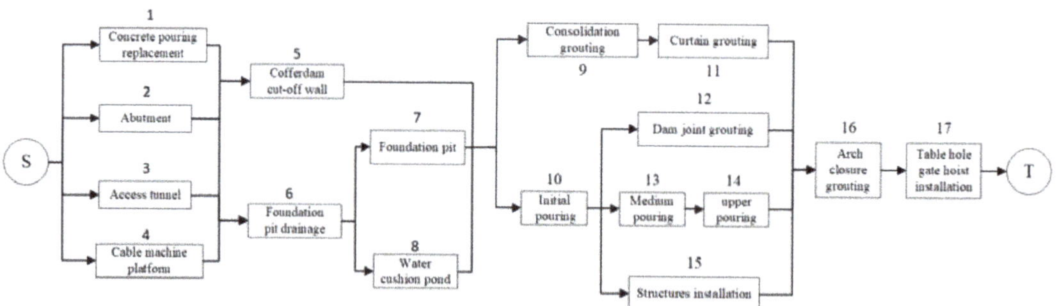

Figure 5. Construction flow chart of a concrete double-curvature arch dam.

5.2. Data Collection and Setting

5.2.1. Project Scheduling Data Processing

To collect relevant data for this practical application, we conducted interviews and surveys with relevant construction companies. The construction process of a concrete double-curvature arch dam can be divided into 17 activities, among which there are three types of common resources. Table 2 shows the activities in which each demand point is responsible for providing resources, and the other necessary data are shown in Table 3.

Table 2. Demand point-project activity mapping table.

Demand Points	Rs
1	1, 2, 3, 4, 6, 7, 8, 10
2	5, 9, 11, 12, 13, 14, 15, 16, 17

Table 3. Details on other parameters.

Resources	P_{ck}	I_{ck}	O_{ck}	r_k	R_k
$k = 1$	2.1	0.02	13.5	25	138
$k = 2$	3.6	0.03	21.6	18	105
$k = 3$	1.8	0.01	14.8	20	110

According to the preliminary data collected, the data of each activity in the project were processed in detail; specifically, uncertain variables were expressed in the form of random variables. The detailed processing data are shown in Table 4. In addition, the project planning period and available budget are $D = 52$ and $B = 8510$, respectively, the indirect cost of each time period is $c_0 = 5.8$, and the storage capacity of each period is $IC = 300$. The weights of the objective functions in the upper model were set to $\mu_1 = \mu_2 = 0.5$.

Table 4. Concrete double-curvature arch dam project activity details.

Activity	Mode	Resources r_{jmk}			Duration	Cost	Predecessors
j	m	$k = 1$	$k = 2$	$k = 3$	d_{jm}	c_{jm}	p_j
S	1	0	0	0	0	0	0
1	1	N(4.0,0.15^2)	N(4.1,0.20^2)	N(5.0,0.31^2)	N(3.1,0.15^2)	N(21.8,1.05^2)	S
	2	N(4.3,0.12^2)	N(4.6,0.20^2)	N(5.9,0.30^2)	N(2.8,0.15^2)	N(24.7,1.3^2)	S
2	1	N(12.8,0.40^2)	N(7.4,0.28^2)	N(6.7,0.21^2)	N(13.2,0.30^2)	N(84.6,2.0^2)	S
	2	N(13.6,0.45^2)	N(7.9,0.35^2)	N(7.1,0.20^2)	N(12.5,0.42^2)	N(87.8,1.8^2)	S
	3	N(14.8,0.30^2)	N(8.6,0.42^2)	N(7.6,0.13^2)	N(11.6,0.30^2)	N(91.5,1.7^2)	S
3	1	N(9.2,0.20^2)	N(8.2,0.28^2)	N(11.3,0.60^2)	N(5.8,0.32^2)	N(35.7,1.3^2)	S
	2	N(10.2,0.38^2)	N(9.1,0.35^2)	N(12.6,0.56^2)	N(5.2,0.20^2)	N(38.2,1.2^2)	S
4	1	N(7.3,0.10^2)	N(5.9,0.30^2)	N(7.3,0.23^2)	N(9.0,0.42^2)	N(29.5,1.5^2)	S
	2	N(8.0,0.15^2)	N(6.5,0.36^2)	N(8.1,0.35^2)	N(8.2,0.30^2)	N(32.3,1.0^2)	S
5	1	N(12.3,0.32^2)	N(7.8,0.45^2)	N(4.5,0.10^2)	N(9.3,0.25^2)	N(42.6,1.8^2)	1, 2, 3, 4
	2	N(13.1,0.32^2)	N(8.3,0.25^2)	N(4.8,0.20^2)	N(8.7,0.20^2)	N(46.5,1.7^2)	1, 2, 3, 4
6	1	N(3.7,0.08^2)	N(3.2,0.15^2)	N(8.7,0.20^2)	N(2.1,0.06^2)	N(15.7,1.08^2)	1, 2, 3, 4
7	1	N(7.0,0.35^2)	N(5.1,0.20^2)	N(10.7,0.40^2)	N(5.2,0.17^2)	N(38.0,1.0^2)	6
	2	N(7.5,0.16^2)	N(5.4,0.30^2)	N(11.6,0.40^2)	N(4.8,0.17^2)	N(39.2,1.3^2)	6
8	1	N(10.7,0.50^2)	N(8.6,0.32^2)	N(6.8,0.10^2)	N(4.0,0.07^2)	N(43.0,1.2^2)	6
	2	N(12.0,0.40^2)	N(9.8,0.42^2)	N(7.3,0.18^2)	N(3.5,0.10^2)	N(45.7,1.6^2)	6
9	1	N(6.8,0.20^2)	N(8.5,0.38^2)	N(9.1,0.28^2)	N(8.4,0.22^2)	N(62.5,1.7^2)	5, 7, 8
	2	N(7.2,0.15^2)	N(8.9,0.41^2)	N(9.7,0.30^2)	N(8.0,0.16^2)	N(65.0,1.7^2)	5, 7, 8
10	1	N(14.5,0.37^2)	N(8.4,0.20^2)	N(4.1,0.28^2)	N(4.3,0.11^2)	N(55.8,1.14^2)	5, 7, 8
	2	N(15.5,0.60^2)	N(9.0,0.18^2)	N(4.4,0.10^2)	N(4.0,0.06^2)	N(57.5,1.0^2)	5, 7, 8
11	1	N(6.0,0.18^2)	N(8.2,0.20^2)	N(9.3,0.28^2)	N(8.0,0.18^2)	N(48.3,1.4^2)	9
	2	N(6.4,0.16^2)	N(8.7,0.40^2)	N(10.0,0.30^2)	N(7.5,0.18^2)	N(50.8,1.6^2)	9
12	1	N(4.6,0.20^2)	N(3.4,0.15^2)	N(6.0,0.15^2)	N(15.5,0.26^2)	N(51.4,0.9^2)	10
	2	N(4.7,0.15^2)	N(3.5,0.10^2)	N(6.4,0.30^2)	N(15.0,0.37^2)	N(53.5,1.0^2)	10
	3	N(5.0,0.10^2)	N(3.7,0.12^2)	N(6.8,0.25^2)	N(14.2,0.35^2)	N(55.2,1.0^2)	10
13	1	N(10.1,0.20^2)	N(4.9,0.18^2)	N(3.5,0.10^2)	N(9.3,0.10^2)	N(72.4,1.2^2)	10
	2	N(10.8,0.5^2)	N(5.1,0.20^2)	N(3.9,0.10^2)	N(8.8,0.15^2)	N(74.8,1.8^2)	10
14	1	N(8.9,0.25^2)	N(4.9,0.10^2)	N(3.3,0.12^2)	N(3.0,0.05^2)	N(41.8,0.8^2)	13
	2	N(10.2,0.30^2)	N(6.0,0.30^2)	N(3.8,0.10^2)	N(2.6,0.06^2)	N(43.0,1.0^2)	1, 2, 3, 4
15	1	N(5.0,0.10^2)	N(2.9,0.05^2)	N(3.5,0.10^2)	N(2.8,0.05^2)	N(25.6,0.8^2)	10
	2	N(6.0,0.23^2)	N(3.5,0.20^2)	N(4.3,0.13^2)	N(2.5,0.06^2)	N(27.4,1.1^2)	10
16	1	N(9.2,0.25^2)	N(7.5,0.30^2)	N(8.7,0.32^2)	N(3.0,0.07^2)	N(36.2,1.2^2)	11, 12, 14, 15
	2	N(9.8,0.25^2)	N(8.0,0.30^2)	N(9.3,0.32^2)	N(2.8,0.07^2)	N(37.2,1.0^2)	11, 12, 14, 15
17	1	N(7.2,0.18^2)	N(5.3,0.10^2)	N(2.4,0.08^2)	N(4.2,0.08^2)	N(36.7,1.2^2)	16
	2	N(8.0,0.20^2)	N(5.9,0.20^2)	N(2.7,0.05^2)	N(3.8,0.05^2)	N(38.1,1.0^2)	16
T	1	0	0	0	0	0	17

5.2.2. Resource Supply Data Processing

All detailed engineering data on the resource supply were obtained from a hydropower project construction company in the watershed project. In a transportation network, the transportation of various resources is accompanied by the entire construction cycle. The entire transportation network can be divided into four supply and two demand points, and three shared resources can be transported from any supply to any demand point.

The maximum resource capacities of the four supply points were 723.4×10^4 m^3, 581.7×10^4 m^3, 528.3×10^4 m^3, and 790.2×10^4 m^3. The maximum resource capacity of the two demand points was 15×10^4 m^3. The project used dump trucks to transport three resources along different routes between different supply and demand points. The unit transport cost and time data for each resource are presented in Table 5.

Table 5. Unit transportation cost and time of resources.

Cost Parameters		Resource Types			Time Parameters		Resource Types		
		k_1	k_2	k_3			k_1	k_2	k_3
c_{isk}	c_{11k}	N(5.20,3.1)	N(6.00,4.2)	N(5.82,3.8)	t_{isk}	t_{11k}	N(0.34,0.21)	N(0.37,0.22)	N(0.32,0.18)
	c_{21k}	N(3.25,2.1)	N(3.66,2.2)	N(3.72,1.8)		t_{21k}	N(0.25,0.1)	N(0.26,0.15)	N(0.22,0.18)
	c_{31k}	N(4.23,1.7)	N(4.43,2.1)	N(4.59,2.4)		t_{31k}	N(0.23,0.12)	N(0.21,0.1)	N(0.19,0.13)
	c_{41k}	N(6.12,3.8)	N(6.44,4.2)	N(6.40,4.0)		t_{41k}	N(0.42,0.21)	N(0.44,0.32)	N(0.40,0.28)
	c_{12k}	N(5.57,2.8)	N(5.41,3.0)	N(5.77,4.2)		t_{12k}	N(0.27,0.11)	N(0.30,0.22)	N(0.24,0.12)
	c_{22k}	N(6.21,4.1)	N(6.33,4.2)	N(6.50,3.8)		t_{22k}	N(0.21,0.08)	N(0.23,0.12)	N(0.20,0.12)
	c_{32k}	N(5.60,3.1)	N(5.41,3.0)	N(5.77,4.2)		t_{32k}	N(0.28,0.37)	N(0.41,0.22)	N(0.37,0.24)
	c_{42k}	N(3.63,2.1)	N(3.84,2.2)	N(4.00,2.0)		t_{42k}	N(0.33,0.20)	N(0.34,0.22)	N(0.29,0.16)

5.3. Selection of Algorithm Parameters

These parameters are controllable factors that affect the convergence, effectiveness, and efficiency of the algorithm. To determine the most appropriate parameters, preliminary experiments and comparisons must be performed under different parameter settings. Herein, a fuzzy logic controller is used to automatically adjust the mutation rate of each generation, and the initial mutation rate is set as $p_m(0) = 0.1$. The inertia weight is adjusted with iteration according to equation (28), and previous studies [35] reveal that $\omega(1) = 0.9$ and $\omega(T) = 0.1$ are the most appropriate. The Taguchi method [36] was used to adjust the other parameters. Finally, the corresponding algorithm parameters were selected, as listed in Table 6.

Table 6. Hybrid algorithm parameter setting.

Parameters	GADS						DIWPSO						
	L_1	T_1	p_c	$p_m(0)$	L_2	T_2	c_1	c_2	σ	η_{max}	η_{min}	$\omega(1)$	$\omega(T_2)$
Values	100	300	0.7	0.1	100	200	2	2	0.1	0.6	0.2	0.9	0.1

5.4. Calculation Results

The designed hybrid algorithm was run in MATLAB(R2018b) on the collected data. After running the program 30 times, an optimal solution was obtained. The total project scheduling time and cost were 48.9 and 8326.54, respectively. The upper planning MRCPSP calculation results are listed in Table 7, showing the start-end time and mode selection of each activity; the corresponding Gantt chart is shown in Figure 6. The calculation result of the MPSCP of the lower planning is shown in Table 8, which defines the transportation volume of the three resources on each transportation route in each time period. The total transportation cost and time were 1144.38 and 13.73, respectively. The convergence iteration is 146 times, and the computation time is 956.3 s.

Table 7. MRCPSP calculation results.

Result	Project Activities																
	1	2	3	4	5	6	7	8	9	10	11	12	13	14	15	16	17
ST_j	0.0	0.0	8.1	0.0	13.2	13.2	18.8	15.3	24.1	24.1	32.1	28.1	30.9	39.7	28.1	42.3	45.1
LT_j	3.1	13.2	13.2	8.1	22.7	15.3	24.1	18.8	32.1	28.1	40.1	42.2	39.7	42.3	30.9	45.1	48.9
m	1	1	2	2	1	1	1	2	2	2	1	3	2	2	1	2	2

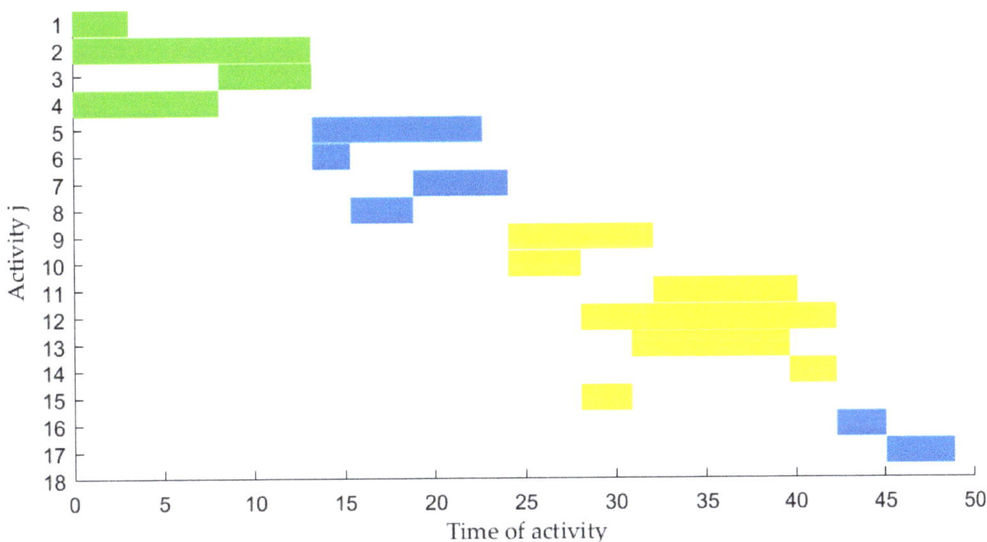

Figure 6. Gantt chart of MRCPSP.

Table 8. Resource transportation decision.

t		1	7	13	19	20	21	26	27	28	33	34	39	Others
v_{11}	k_1	38.04	36.82	14.76	7.39									0
	k_2	26.50	26.92	13.64		4.86								0
	k_3	29.00	30.20	30.60			13.63							0
v_{21}	k_1	30.21	29.24	28.01	29.16			30.49						0
	k_2	21.05	21.38	23.20		20.11			18.88					0
	k_3	23.03	23.98	24.30			22.23			4.76				0
v_{31}	k_1	27.44	26.56	25.44	26.49			15.76						0
	k_2	19.11	19.42	21.07		18.26								0
	k_3	20.92	21.78	22.07			20.19							0
v_{41}	k_1	41.04	39.72											0
	k_2	28.59	29.04											0
	k_3	31.29	32.58	2.74										0
v_{12}	k_1			20.51	29.33			38.39			36.73		37.99	0
	k_2			15.58		20.46			28.40		28.50		28.46	0
	k_3						14.37			30.60		27.65	28.00	0
v_{22}	k_1										29.17		30.18	0
	k_2								3.68		22.63		22.60	0
	k_3									19.55		21.96	22.24	0
v_{32}	k_1							11.94			26.50		27.41	0
	k_2								20.49		20.56		20.53	0
	k_3									22.07		19.95	20.16	0
v_{42}	k_1			38.05	39.62			41.42			39.63		40.99	0
	k_2			31.52		27.31			30.64		30.75		30.70	0
	k_3			30.28			30.20			33.02		29.84	30.16	0

5.5. Analysis and Discussion

5.5.1. Weight Analysis

Different weight settings (i.e., μ_1 and μ_2) represent different combinations of preferences for decision-makers. To further understand the influence of the weight setting in upper-level planning, a sensitivity analysis was carried out, and the corresponding results are presented in Table 9. Different weight settings led to different results in the upper and lower models, which indicates that the decisions of the two levels are greatly influenced by the upper weight settings and are closely related to each other.

Table 9. Weight sensitivity analysis.

Cases	Weight Values		Objective Function Values			
	μ_1	μ_2	F_t	F_c	Z_c	Z_t
case 1	0.7	0.3	47.35	8347.16	1150.97	13.51
case 2	0.6	0.4	48.04	8335.30	1147.75	13.58
case 3	0.5	0.5	48.86	8326.54	1144.38	13.73
case 4	0.4	0.6	49.40	8320.65	1141.21	13.85
case 5	0.3	0.7	50.36	8315.23	1138.83	13.97

5.5.2. Model Comparison

To verify the effectiveness of the model and the superiority of obtaining the optimal and satisfactory solution, the game model was compared with the single-layer model of the MRCPSP and MPSCP, which ignores the conflict.

To establish the corresponding single-layer model, project scheduling and resource supply were combined into a separate optimization problem. The objective function is the duration and cost of project scheduling, F_t and F_c, the decision variables are also (v, x), and the constraints include all the constraints in the upper planning. To calculate the comparative rationality of the results, the GADS proposed in the upper planning was also applied to the single-layer model and run in MATLAB(R2018b). Subsequently, the decision results are substituted into Z_c and Z_t to calculate the function value, and the objective function value of the single-layer model in the ideal state is obtained.

However, in practice, the lower-level planning MPSCP also has its own optimization objectives and constraints, and there are decision conflicts between the construction department and the resource supplier. Therefore, the ideal optimal solution obtained by the single-level planning model may not be a satisfactory solution for the MPSCP and will usually deviate. Therefore, the results obtained using the ideal single-layer model must be modified as follows:

In the first step, the decision result of the ideal single-layer model was used as the decision result of the upper MRCPSP. In the second step, considering the sequence of decisions, the decision results of the MRCPSP were substituted into the MPSCP to obtain the optimal transportation decision under this situation, namely, the modified solution. In the third step, the result of the transportation decision is substituted into the objective function of the MRCPSP to obtain the objective function value in this case.

In the dynamic game model, considering the hierarchical decision structure and the existence of decision conflicts, the above correction method is repeatedly used to obtain a satisfactory Stackelberg-Nash equilibrium solution. The corresponding calculation results are listed in Table 10, and Table 11 lists the comparison results of the algorithms.

Table 10. Selection of algorithm parameters.

Algorithms	Parameters												
	L_1	T_1	p_c	$p_m(0)$	L_2	T_2	c_1	c_2	σ	η_{max}	η_{min}	$\omega(1)$	$\omega(T_2)$
GA/PSO	100	300	0.75	0.15	100	200	1.8	2	×	×	×	0.9	0.1
GADS/PSO	100	300	0.7	0.1	100	200	1.8	2	×	×	×	0.9	0.1
GA/DIWPSO	100	300	0.75	0.15	100	200	2	2	0.1	0.6	0.2	0.9	0.1
GADS/DIWPSO	100	300	0.7	0.1	100	200	2	2	0.1	0.6	0.2	0.9	0.1

Figure 7 demonstrates the iterative process of the algorithm. The results of algorithm comparison reveal that: ① All four algorithms can obtain the optimal fitness in 200 iterations, and the hybrid GADS/DIWPSO algorithm has a higher fitness. ② The computation time and convergence speed of the four algorithms are acceptable, among which GADS/DIWPSO hybrid algorithm is faster than GADS/PSO but slightly slower than GA/DIWPSO and GA/PSO. ③ The GADS/DIWPSO hybrid algorithm has better standard deviation corresponding to fitness, convergence iteration times, and computation time than other algorithms, showing stable performance, which also reveals that the algorithm can effectively avoid infeasible solutions and reduce the probability of premature convergence. Therefore, the GADS/DIWPSO hybrid algorithm proposed in this study performs better than other algorithms in an acceptable computation time.

Table 11. Algorithm comparison results.

Algorithms		Fitness			Convergence Iteration Number			Computation Time		
		Best	Average	Standard Deviation	Best	Average	Standard Deviation	Best	Average	Standard Deviation
GA/PSO	F_1	0.957	0.940	0.0082	126	142	7.0	927.6	961.5	14.8
	F_2	0.905	0.890	0.0065						
GADS/PSO	F_1	0.968	0.961	0.0043	149	157	4.2	968.2	990.0	10.3
	F_2	0.926	0.916	0.0040						
GA/DIWPSO	F_1	0.963	0.954	0.0065	122	131	5.3	912.5	940.4	13.4
	F_2	0.935	0.929	0.0032						
GADS/DIWPSO	F_1	0.976	0.971	0.0035	146	152	3.8	956.3	975.6	9.5
	F_2	0.947	0.943	0.0020						

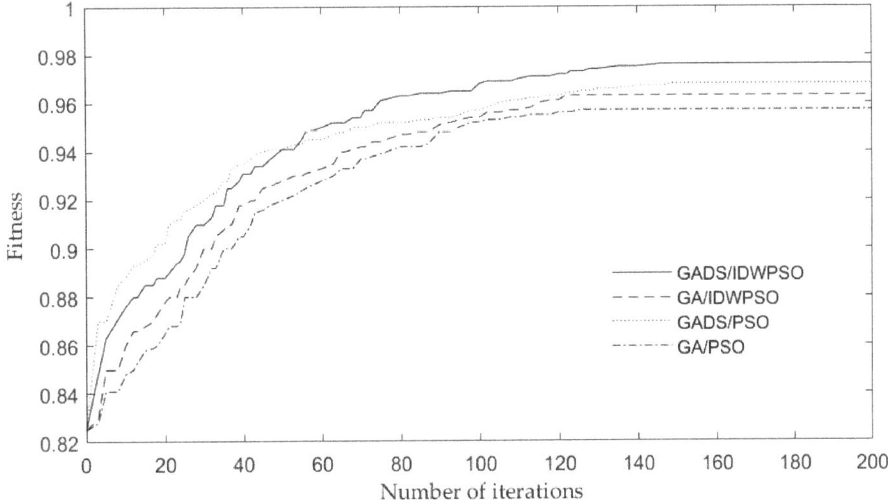

Figure 7. Algorithm iteration process.

5.6. Management Suggestions

Through the application of practical cases, some management suggestions are proposed for relevant departments from the perspective of the engineering supply chain:

① When making the project schedule, the decision maker of the engineering project shall ensure that the project schedule and resource supply are within a reasonable range so that the construction schedule based on materials, equipment, and labor force can meet the expected requirements. At the same time, it must be considered that too much or too little resource supply cannot ensure the schedule advancement, because the process sequence and intermittent time in the construction process of the project determine that the actual construction progress cannot violate the internal law of the project. Once the construction progress based on the process is exceeded, quality problems are likely to occur.

② The engineering supply chain generally involves multiple stakeholders such as owners, contractors, resource suppliers, and transportation agents. Different stakeholders are responsible for various professional tasks. These tasks are often interrelated, and if considered separately and while ignoring the conflicts of various stakeholders, they can lead to suboptimal solutions, which in turn can cause economic losses, construction delays, and other problems. Therefore, in the actual implementation of engineering projects, inherent conflicts and complex interactions must be identified and resolved.

③ In engineering practice, project managers must consider all kinds of resources, such as the labor force, materials, and equipment as a whole. The disharmony between any type of resource and other resources may cause resource redundancy or project stagnation at a certain link in an engineering project.

④ Modeling the decision-making process helps to understand the complexity and conflicts involved in the supply chain and then conducts quantitative analysis to determine a satisfactory equilibrium strategy. For example, the new Stackelberg dynamic game model proposed for the MRCPSP-MPSCP integrated system is more suitable than the corresponding single-layer model. In addition, the preference setting of the multi-objective function is important, and different preference combinations lead to different results.

6. Conclusions and Future Research

This study investigated the integration of multimode project scheduling and resource supply in an engineering supply chain. Resource constraint is not only a constraint condition of the engineering supply chain, but is often a separate optimization problem. Therefore, integrating resource supply into project scheduling is an MRCPSP-MPSCP integrated system with multi-agent decision-making characteristics and a hierarchical decision-making structure. Resolving conflicts in this integrated system helps ensure that the project runs successfully at an acceptable cost and is completed on time. On this basis, a Stackelberg dynamic game model was established, and a two-level multi-objective programming method was designed to further solve internal conflicts. Subsequently, a two-layer GADS/DIWPSO hybrid algorithm with an interactive evolution mechanism was proposed to solve the new Stackelberg model, and a satisfactory Stackelberg-Nash equilibrium solution was determined through a repeated dynamic interaction process. This provides theoretical significance for solving related problems of engineering supply chain.

In the context of the global impact of COVID-19, coordinated optimization and sustainable operation of the engineering supply chain play an important role in the recovery of the industrial economy. This study provides a theoretical basis and algorithm support for how engineering and construction departments and resource suppliers in the supply chain promote the optimization of overall benefits. For the engineering construction department, considering the limitation of resource supply, more thought is devoted to the project scheduling problem to ensure the overall operation of the project. For resource suppliers, considering the characteristics of master-slave decision-making, this study provides a reference for the formulation of a resource transportation strategy, and finally promotes mutual benefit on both sides to achieve better cooperation results.

After discussion and analysis, it can be discovered that in the engineering supply chain, the multi-period resource supply problem does have an impact on the project scheduling. Therefore, the dynamic game model for the MRCPSP-MPSCP integrated system is more realistic, and the proposed two-level multi-objective programming method and GADS/DIWPSO hybrid algorithm can solve the conflicts between stakeholders, and finally realize the Stackelberg-Nash equilibrium strategy. In conclusion, when solving similar problems, researchers should start from reality, fully consider the conflicts of interest among participants, and make reasonable assumptions. Only in this way can a better decision plan be generated.

However, there are still some limitations in this study: ① he scheduling problem of multiple projects is not considered; ② the mixed transportation of multi-type vehicles is not considered in terms of resource transportation; and ③ more participants can be considered in a large engineering supply chain, such as material manufacturers and transportation agents. These limitations will form the basis for future research.

Author Contributions: Conceptualization, C.F.; methodology, C.F. and S.H.; data analysis, C.F., S.H. and Y.M.; writing—original draft preparation, C.F. and S.H.; writing—review and editing, Z.L.; supervision, Z.L. All authors have read and agreed to the published version of the manuscript.

Funding: This research was supported by the National Natural Science Foundation of China (No. 71702167 and No. 72202056).

Institutional Review Board Statement: Not applicable.

Informed Consent Statement: Not applicable.

Data Availability Statement: All processed data used in the study have been shown in the article.

Acknowledgments: The authors gratefully acknowledge the funding and support provided by the National Natural Science Foundation of China (No. 71702167 and No. 72202056) and are grateful to the editors and the reviewers for their insightful comments and suggestions.

Conflicts of Interest: The authors declare no conflict of interest.

References

1. Li, C.-L.; Hall, N.G. Work Package Sizing and Project Performance. *Oper. Res.* **2019**, *67*, 123–142. [CrossRef]
2. Herroelen, W. Project Scheduling-Theory and Practice. *Prod. Oper. Manag.* **2005**, *14*, 413–432. [CrossRef]
3. Browning, T.R.; Yassine, A.A. Resource-constrained multi-project scheduling: Priority rule performance revisited. *Int. J. Prod. Econ.* **2010**, *126*, 212–228. [CrossRef]
4. Guoshan, L.; Min, W.; Zhuanxia, Z. A Bi-level Programming Problem Based on Time-window Delay for Resource-Constrained Project Scheduling. *Oper. Res. Manag. Sci.* **2021**, *30*, 6–12+27.
5. Kim, K.; Yun, Y.; Yoon, J.; Gen, M.; Yamazaki, G. Hybrid genetic algorithm with adaptive abilities for resource-constrained multiple project scheduling. *Comput. Ind.* **2005**, *56*, 143–160. [CrossRef]
6. Cheng, M.-Y.; Tran, D.-H. Opposition-based Multiple Objective Differential Evolution (OMODE) for optimizing work shift schedules. *Autom. Constr.* **2015**, *55*, 1–14. [CrossRef]
7. Demeulemeester, E.L.; Herroelen, W.S. An efficient optimal solution procedure for the preemptive resource-constrained project scheduling problem. *Eur. J. Oper. Res.* **1996**, *90*, 334–348. [CrossRef]
8. Vanhoucke, M.; Debels, D. The impact of various activity assumptions on the lead time and resource utilization of resource-constrained projects. *Comput. Ind. Eng.* **2008**, *54*, 140–154. [CrossRef]
9. Isah, M.A.; Kim, B.-S. Integrating Schedule Risk Analysis with Multi-Skilled Resource Scheduling to Improve Resource-Constrained Project Scheduling Problems. *Appl. Sci.* **2021**, *11*, 650. [CrossRef]
10. Schwarze, J. Activity networks: Project planning and control by network models. *Eur. J. Oper. Res.* **1979**, *3*, 167–168. [CrossRef]
11. Varma, V.A.; Uzsoy, R.; Pekny, J.; Blau, G. Lagrangian heuristics for scheduling new product development projects in the pharmaceutical industry. *J. Heuristics* **2007**, *13*, 403–433. [CrossRef]
12. Zhu, G.; Bard, J.F.; Yu, G. A Branch-and-Cut Procedure for the Multimode Resource-Constrained Project-Scheduling Problem. *INFORMS J. Comput.* **2006**, *18*, 377–390. [CrossRef]
13. Bellenguez-Morineau, O.; Neron, E. A Branch-and-Bound method for solving Multi-Skill Project Scheduling Problem. *RAIRO Rech. Opérationnelle* **2007**, *41*, 155–170. [CrossRef]
14. Sakawa, M.; Nishizaki, I. Interactive fuzzy programming for two-level nonconvex programming problems with fuzzy parameters through genetic algorithms. *Fuzzy Sets Syst.* **2002**, *127*, 185–197. [CrossRef]
15. Sarker, B.R.; Egbelu, P.J.; Liao, T.W.; Yu, J. Planning and design models for construction industry: A critical survey. *Autom. Constr.* **2012**, *22*, 123–134. [CrossRef]
16. Fang, X.; Zhe, X.; Jing, Y. Bi-objective optimization for the project scheduling problem with variable resource availability. *Syst. Eng. Theory Pract.* **2016**, *36*, 674–683.
17. Zhao, X.; Lü, X.; Mei, Q.; Feng, W. Project scheduling problem constrained by flexible resource with capability difference. *Ji Suan Ji Gong Cheng Yu Ying Yong* **2012**, *48*, 231–237. [CrossRef]
18. Schwindt, C.; Trautmann, N. Batch scheduling in process industries: An application of resource–constrained project scheduling. *OR Spektrum* **2000**, *22*, 501–524. [CrossRef]
19. Liu, S.-S.; Huang, H.-Y.; Risna Dyah Kumala, N. Two-Stage Optimization Model for Life Cycle Maintenance Scheduling of Bridge Infrastructure. *Appl. Sci.* **2020**, *10*, 8887. [CrossRef]
20. Asta, S.; Karapetyan, D.; Kheiri, A.; Özcan, E.; Parkes, A.J. Combining Monte-Carlo and hyper-heuristic methods for the multi-mode resource-constrained multi-project scheduling problem. *Inf. Sci.* **2016**, *373*, 476–498. [CrossRef]
21. Fang, X.; Hongbo, L.; Qingguo, B. Stochastic multi-mode resource constrained project scheduling. *Chin. J. Manag. Sci.* **2020**, 1–13. [CrossRef]
22. Peteghem, V.V.; Vanhoucke, M. A genetic algorithm for the preemptive and non-preemptive multi-mode resource-constrained project scheduling problem. *Eur. J. Oper. Res.* **2010**, *201*, 409–418. [CrossRef]
23. Nusen, P.; Boonyung, W.; Nusen, S.; Panuwatwanich, K.; Champrasert, P.; Kaewmoracharoen, M. Construction Planning and Scheduling of a Renovation Project Using BIM-Based Multi-Objective Genetic Algorithm. *Appl. Sci.* **2021**, *11*, 4716. [CrossRef]
24. Xie, L.; Chen, Y.; Chang, R. Scheduling Optimization of Prefabricated Construction Projects by Genetic Algorithm. *Appl. Sci.* **2021**, *11*, 5531. [CrossRef]

25. Kuo, R.J.; Huang, C.C. Application of particle swarm optimization algorithm for solving bi-level linear programming problem. *Comput. Math. Appl.* **2009**, *58*, 678–685. [CrossRef]
26. Kuo, R.J.; Lee, Y.H.; Zulvia, F.E.; Tien, F.C. Solving bi-level linear programming problem through hybrid of immune genetic algorithm and particle swarm optimization algorithm. *Appl. Math. Comput.* **2015**, *266*, 1013–1026. [CrossRef]
27. Amirtaheri, O.; Zandieh, M.; Dorri, B.; Motameni, A.R. A bi-level programming approach for production-distribution supply chain problem. *Comput. Ind. Eng.* **2017**, *110*, 527–537. [CrossRef]
28. Singh, V.; Ganapathy, L.; Pundir, A.K. An Improved Genetic Algorithm for Solving Multi Depot Vehicle Routing Problems. *Int. J. Inf. Syst. Supply Chain. Manag.* **2019**, *12*, 1–26. [CrossRef]
29. Kennedy, J.; Eberhart, R. Particle Swarm Optimization. In Proceedings of the IEEE International Conference on Neural Networks, Perth, Australia, 27 November–1 December 1995; Volume 1944, pp. 1942–1948.
30. Bidot, J.; Vidal, T.; Laborie, P.; Beck, J.C. A theoretic and practical framework for scheduling in a stochastic environment. *J. Sched.* **2008**, *12*, 315–344. [CrossRef]
31. Nishizaki, I.; Sakawa, M. *Cooperative and Noncooperative Multi-Level Programming*; Nishizaki, I., Sakawa, M., Eds.; Springer: New York, NY, USA, 2009; Volume 48.
32. Feng, C.; Ni, T.; Li, Z.; Cai, J.; Ma, Y. Conflict resolution towards an integrated project scheduling and material ordering system in a large-scale construction project. *Appl. Comput. Math.* **2019**, *18*, 202–217.
33. Tangqing, H.; Xuxiu, Z.; Xiaoyue, C. A hybrid particle swarm optimization with dynamic adjustment of inertial weight. *Electron. Opt. Control.* **2020**, *27*, 16–21.
34. Jianping, L. Hybrid particle swarm optimization algorithm based on chaos and differential evolution. *Comput. Simul.* **2012**, *29*, 208–212.
35. Feng, C.; Xu, J.; Yang, X.; Zeng, Z. Stackelberg-Nash Equilibrium for Integrated Gravelly Soil Excavation-Transportation-Distribution System in a Large-Scale Hydropower Construction Project. *J. Comput. Civ. Eng.* **2016**, *30*, 4016024. [CrossRef]
36. Taguchi, G.C.S.; Wu, Y. *Taguchi's Quality Engineering Handbook*; John Wiley & Sons, Inc.: Hoboken, NJ, USA, 2005.

Article

Rebar Fabrication Plan to Enhance Production Efficiency for Simultaneous Multiple Projects

Eunbin Hong [1], June-Seong Yi [1,*], JeongWook Son [1], MinYoung Hong [2] and YeEun Jang [1]

[1] Department of Architectural and Urban Engineering, Ewha Womans University, 52 Ewhayeodae-gil, Seodaemun-gu, Seoul 03760, Korea

[2] DL E&C, Donuimun, D Tower, 134 Tongil-ro, Jongno-gu, Seoul 03181, Korea

* Correspondence: jsyi@ewha.ac.kr

Abstract: A discrete-event simulation (DES) model was developed to enhance the reinforcing bar (rebar) fabrication efficiency for multiple simultaneous projects at different sites. The production volume and procedure of the actual rebar fabrication plant were compared to the simulation model to ensure its accuracy. By determining the loss rate and necessary processing time, the fabrication plan was then optimized. The rebar type and machine features, which influence the loss rate and time required for rebar fabrication, were configured as the parameters in a discrete-event simulation model. The model considers a situation in which a rebar fabrication plant simultaneously delivers rebars to multiple sites. In this manner, the model can quantify the loss rate and time required in the fabrication process. The determination of the loss rate according to the import ratio of raw steel, site combination, and length can help optimize the rebar fabrication plan and increase work efficiency. In the considered scenario, a two-site combination and import ratio of raw steel of 2:1 (8 m:10 m) was noted to corresponded to the maximum decrease in the loss rate and required time. By extending the proposed approach to the complete rebar process (processing–transportation–construction), the plant member production process can be optimized.

Keywords: rebar fabrication; discrete-event simulation; simultaneous delivery; rebar loss; production process optimization

Citation: Hong, E.; Yi, J.-S.; Son, J.; Hong, M.; Jang, Y. Rebar Fabrication Plan to Enhance Production Efficiency for Simultaneous Multiple Projects. *Appl. Sci.* **2022**, *12*, 9183. https://doi.org/10.3390/app12189183

Academic Editors: Albert P. C. Chan, Srinath Perera, Dilanthi Amaratunga, Makarand Hastak, Patrizia Lombardi, Sepani Senaratne, Xiaohua Jin and Anil Sawhney

Received: 10 August 2022
Accepted: 7 September 2022
Published: 13 September 2022

Publisher's Note: MDPI stays neutral with regard to jurisdictional claims in published maps and institutional affiliations.

1. Introduction

Reinforcing bars (rebars) represent a widely used component of reinforced concrete structures in construction projects that significantly influence the project cost and structural stability [1]. To reduce the loss rate, rebar fabrication techniques have been changed from the field to the plant. To avoid delivery delays, the processing time and loss rate in a rebar fabrication plant's production process must be estimated. However, the majority of research that is currently available on the fabrication of rebars has been concentrated on the necessary length and cutting processes of rebars, such as standardizing the length and shape of the rebar to reduce the loss rate, or planning the cutting of raw steel imported to the rebar fabrication plant [2–5]. Although the existing methods, such as an optimization algorithm with an NP-hard problem [6,7], can decrease the loss rates in rebar fabrication plants, the production plan cannot be optimized because such methods do not consider the actual scenario of rebar fabrication plants that perform simultaneous deliveries to multiple sites. For this reason, the goal of this study was to create a simulation model that takes into account the variables that affect work efficiency (such as the quantity of simultaneous cutting and bending operations, and the amount of time needed for cutting and bending) in scenarios involving the simultaneous delivery of rebars to various sites. Through risk processing, the rebar loss rate and processing time for the volume to be delivered were made clear. Even when the production volume from various sites increases, the proposed model can assist in creating an ideal production plan for situations involving numerous sites and varieties of raw steel.

2. Materials and Methods

2.1. Literature Review

The existing research on processing plant optimization may be broken down into two groups (Table 1): studies concentrating on timely production management and import systems of processed rebar for rebar building, and those on algorithms to decrease the loss rate.

Many researchers have performed case studies, simulation studies, and algorithm-based studies to optimize rebar fabrication plans or minimize the cost and material waste [1,2,4–6,8–10]. Other researchers attempted to enhance production plans by using simulation-based decision support systems based on rebar specifications [11]. Polat et al. (2007) used a simulation-model-based system to establish a decision support system that recommended lot sizes (large or small), scheduling strategies (optimistic, neutral, or pessimistic), and buffer sizes (large, medium, or small) considering project conditions. This system generated savings of 4.8% in just-in-case scenarios over just-in-time scenarios.

When a rebar fabrication plant must simultaneously process different types of rebars at multiple sites, a certain number of site combinations can be prioritized based on the experience and intuition of the planners. However, given the wide range of rebar types, the labor capacity and expertise may not be adequate. To identify the optimal plan to enhance productivity, various scenarios can be tested using simulation models.

2.2. Modeling with DES

The importance of the rebar type and site-specific combination affects how a rebar fabrication plant's production system evolves over time. To simulate a rebar production process that changes in terms of the productivity of intricate production systems, including numerous delivery combinations with a finite length, discrete-event simulation is an appropriate method. The DES model involves the operation of a system as a discrete sequence of events in time [12,13]. Each event occurs at a particular instant in time and marks a change of state in the system. DES has been used as an effective approach to better absorb complex interactions and uncertainties in construction operations [8,14]. Therefore, in this study, DES modeling was used to develop the best production schedule for rebar fabrication plants.

AnyLogic is a Java-based simulation program that may be used to assess different simulation-related research approaches [15]. Using the building blocks from AnyLogic's process modeling library, a production process simulation model of a rebar fabrication plant was created for this study. By testing different configurations related to the number of sites and raw steel ratio for each length, the developed model was used to reduce the loss rate. To further improve the production plan for the rebar fabrication plant, the processing time was also examined.

DESs were created based on the process depicted in Figure 1 in order to establish simulation models for rebar production plants and determine the optimization methods to lower the loss rates through various site-specific combinations and lengthwise cylindrical import rates. There are four stages to the modeling process. First, in the goal-setting stage, we identify the issues that arise in real-world settings, describe these issues, set goals, and create a study strategy. Second, the operating procedure and current state of the rebar fabrication plant are identified. Third, in order to simulate real-world systems, we gather multiple data points in operations and management during the model design process [16]. Fourth, we visit a real rebar fabrication plant to observe the rebar production process and gather data from each process to assess how well the current system is working. The goal of this study is to reduce the loss rate of the rebar; therefore, we construct a model by simplifying the systems that have an impact on that rate. We next test the model's viability to make sure it accurately captures the real system.

Table 1. Existing studies on rebar fabrication management and loss rate minimization.

Category	Author	Year	Contents
Management of product and rebar import	[13]	2007	Presented a simulation-based decision support system to assist contractors in selecting the most economical rebar management system before starting construction.
	[3]	2016	Described how the just-in-time (JIT) concept can help enhance the performance of transportation and material delivery activities in industrialized building system projects.
	[1]	2020	Proposed a framework based on takt time (to identify the optimal time for each process) and discrete-event simulation (DES) to integrate building information modeling (BIM) with JIT to realize realistic and optimal planning. Combined planning models with DES and BIM to simulate the dynamic environment to reduce uncertainties.
Loss rate minimization	[4]	2014	Developed novel problem-solving techniques, leading to effective cutting plans with a low trim loss and stock usage.
	[12]	2018	Proposed a novel approach for minimizing cutting waste from rebars by exploiting the slight flexibility in selecting the location of lap splices of rebars within reinforced concrete members, as specified by design codes.
	[9]	2020	Proposed a special-length-priority cutting waste minimization algorithm for rebars. A minimization method based on special and stock lengths was applied. The required rebar quantity was 6.04% lower than the actual quantity used.
	[8]	2021	Optimized the use of available market length rebars to minimize generated waste. Proposed a BIM-based automated framework integrated with mixed-integer linear programming (MILP). The trim waste could be rapidly and efficiently decreased using the BIM–MILP approach.
	[10]	2021	The cutting process was managed using optimization models based on three field variables: merging sequential demands, multiple stock lengths, and usable leftovers. The cutting waste in the case study decreased by more than 70%.

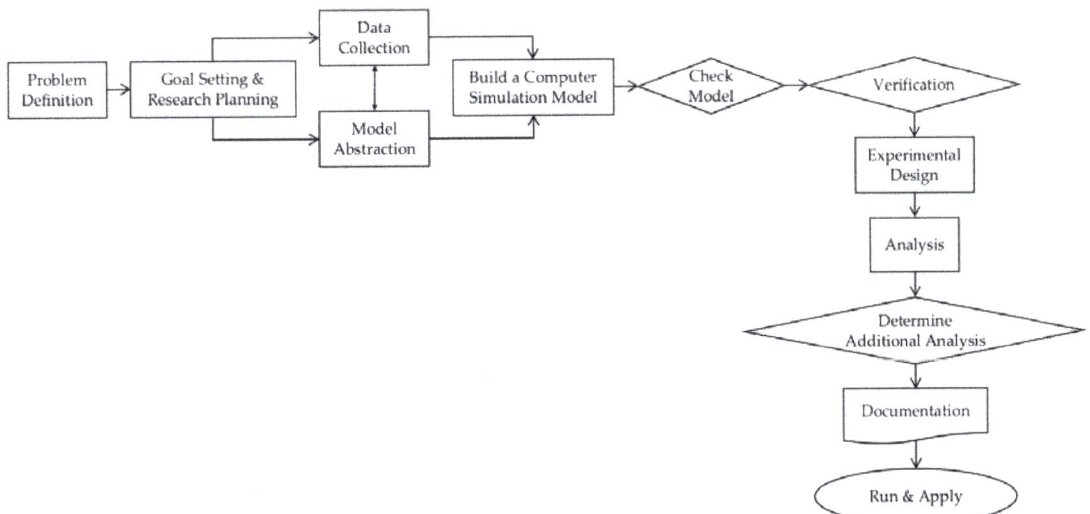

Figure 1. Simulation modeling process flow (Reprinted/adapted with permission from Ref. [17].

3. Problem Definition

The three steps that make up the rebar production process are shown in Figure 2: (1) importing, (2) cutting, and (3) bending. In the rebar fabrication plant, the required length and shape are calculated according to the strength and diameter of the rebar through design drawings. The received raw steel is processed according to the preceding process (Figure 2), the quality of the processed rebar is checked, and it is then shipped to individual sites. Although this process is straightforward, the loss rate is challenging to minimize because several variables associated with the (1) raw steel, (2) machine, (3) rebar shape, and (4) hoist must be considered. Additionally, we first set the variables that have a decisive effect on reducing the loss rate and necessary time. Firstly, the reason why time is considered the most important among these is that the model built is based on a discrete-event simulation. Secondly, considering the priority, time-related variables were set as major variables.

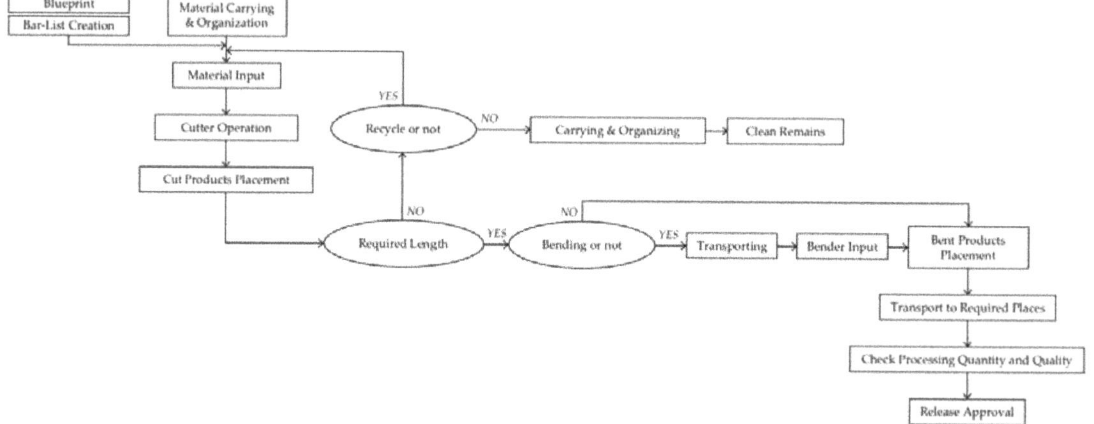

Figure 2. Processing system for rebar fabrication.

Imported raw steel is typically 8 m or 10 m long. With the increasing complexity of construction projects, the required lengths of the rebars at sites vary. The steel that is left over after fabrication is regarded as waste, which raises the project's loss rate or reduces the amount of raw materials accessible for other projects. As a result, plants frequently produce rebar in accordance with requests from several projects at once to reduce the loss rate.

As indicated in Table 2, rebar fabrication involves various types of machines for (1) transporting, (2) cutting, and (3) bending. The cutting machine typically cuts 18 or 24 rebars with a 10 mm diameter at once, or 13 or 21 rebars with a 13 mm diameter at once. There may be a slight delay because the bending machine normally bends five to nine rebars with a diameter of 10 mm, and three to seven rebars with a diameter of 13 mm at a time. Third, the rebar shapes to be fabricated vary across projects, which changes the machine capacities. For instance, the diameter of the raw steel affects how many rebars are produced simultaneously. Finally, depending on the transportation site and rebar shape, the hoist's maximum capacity fluctuates during the rebar transportation operation.

Therefore, by combining numerous locations to gather and process various types of rebars at once, a production plan must be designed that can decrease the loss rate that happens while processing raw steel with a limited length. This framework can also help ensure rebar quality. Notably, no specific standard exists for processing rebars to be delivered to multiple sites simultaneously. The processing of rebars typically relies on personnel experience. In such scenarios, the loss rate cannot be effectively minimized. In order to determine the loss rate, necessary processing time, and ideal production plans, a simulation model was created in this study based on the basic rebar fabrication process.

Table 2. Rebar fabrication machine.

Machine	Model	Features		
		Motor Capacity (kW)	Conveyor Speed (m/min)	# of Simultaneous Cuts (SD500)
Machine for automatically cutting rebars	HAAC-300B	11.50	5.5	D10: 27, D13: 31
	HAAC-300H	36.25	5.5	D10: 27, D13: 21
	TOYO-Japan TFC-M	9.65	4.3	D10: 18, D13: 13
Machine for automatic rebar bending	HAAB-25	9.75	-	D10: 9, D13: 7
Machine for bending rebars	HAAB-10-6	4.5	-	D10: 6, D13: 4
	HAAB-10-7	7.0	-	D10: 7, D13: 5
	TOYO-Japan TRB-10-5	5.25	-	D10: 5, D13: 3
	HAB-25	1.5	-	D10: 5, D13: 4

4. Modeling Simulation of the Rebar Production Plan

4.1. Model Overview

When designing a simulation model, it is very difficult to consider all the variables that explain the phenomenon; some variables that have a major influence on the phenomenon can be established and disestablished.

The model construction involves three steps. First, in order to reduce the loss rate and necessary time, it is necessary to identify the critical parameters: Following the importation of raw steel from the steel mill and the delivery to the plant, three factors must be considered: (1) site combination, (2) rebar type (the strength, diameter, length, and the form of rebar sent to the sites for actual building activity), and (3) machines (hoists, cutters, and benders appropriate for specific sites and rebars). Second, the raw steel is cut, and the leftover length is put to use again. The amount of rebars that each cutter can cut at once is shown by how the rebars are cut. The proposed approach was built to utilize the most cutting-edge resources possible. Reusing the rebars in this situation is crucial to lowering the loss rate. If the length of the rebar after it has been cut exceeds the length necessary for the type of rebar required at the site, it must be processed into a rebar with a reduced length. The loss rate in this process varies depending on the priority of each rebar type. The third step involves bending the cut rebar. Rebars of various shapes that must be bent are configured to pass through the bender.

By reflecting the abovementioned rebar machining process, an algorithm was established to enable the delivery of rebars to multiple sites simultaneously. To realistically reflect the rebar fabrication plant in the simulation model, the working schedule of an actual plant was used. The operating time of the machine was reflected in the model. In general, rebar fabrication plants operate night shifts during the week for order fulfillment. The plants operate until 6 p.m. on Saturdays and do not operate on Sundays. The simulation model was configured such that the machine was operated during working hours and not operated during breaks. The execution process of the DES model for the rebar production plan is shown in Figure 3.

Because the actual rebar production plant receives orders from and processes orders for up to four locations concurrently, four sites were specified as the maximum number of sites to which orders must be simultaneously supplied in the simulation model. It was also believed that 16 different types of rebars could be handled (eight types of SHD10 and eight types of SHD13, commonly used for apartment slabs and walls at each site, respectively).

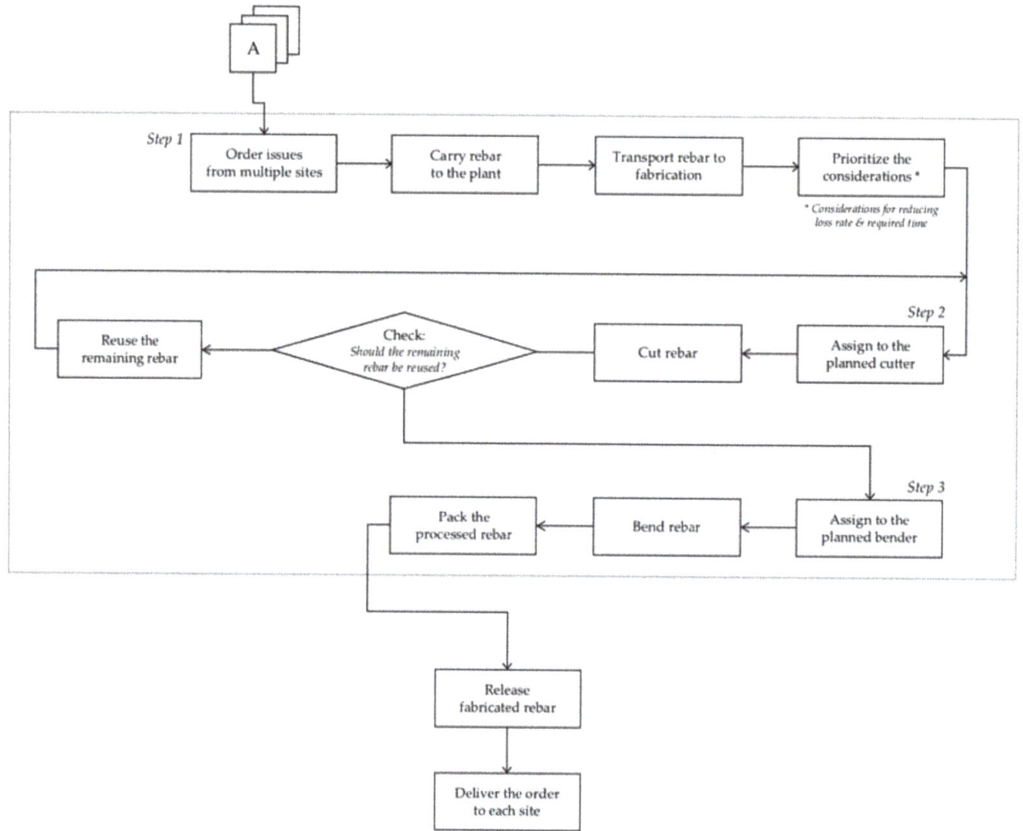

Figure 3. Process flow of the algorithm reflecting the rebar production process.

4.2. Data Collection

The procedure for the case selection was as follows: Korea's rebar production plant in Anseong, Gyeonggi-do, was chosen for model development, and a visit there was planned. With over 100 building sites in its delivery history, this plant can be considered to have a solid process. We then described the machine. The elements appropriate for the under-consideration situation were found among the different machine-specific characteristics of the rebar fabrication process described in the preceding section. Table 3 provides a summary of the characteristics of the hoists, cutters, and benders employed as resources in the simulation. The machine characteristics that have an impact on the processing times and loss rates were configured as parameters and applied to the simulation model. By adjusting the settings in accordance with the characteristics of each piece of machine, the loss rate and necessary time were discovered. The varieties of rebar required at various places were then identified. Table 4 provides a summary of the information introduced in the simulation of the rebar fabrication process based on the data of the real rebar length and number of cuts.

Table 3. Machine features.

Machine	Name	Utility	Capacity (ton)
Machine for rebar transport	Hoist	Raw steel assignment/Transportation	5
		Cutter stowage	3
		Bender stowage	3

Machine	Model	Single Cut Rate (s)	# of Simultaneous Cuts (SD500)
Machine for rebar cutting	HAAC-300B	5	D10: 27/D13: 21
	HAAC-300H	5	D10: 27/D13: 21
	TOYO-Japan TFC-M	5	D10: 18/D13: 13
Rebar bending machine	HAAB-25	10	D10: 9/D13: 7
	HAAB-10-6	10	D10: 6/D13: 4
	HAAB-10-7	10	D10: 7/D13: 5

Table 4. Rebar types for each site.

Site	Diameter	Shape	N	Length	Quantity	Weight
1	10	—	0	5.1	2538	7.249
	10	∟	1	4.9	1392	3.820
	13	—	0	4.3	2166	9.267
	13	—	0	3.27	369	1.288
2	10	⊏	2	1.06	1056	0.627
	10	□	4	1.33	456	0.340
	13	∟	1	1.45	192	0.277
	13	—	0	1.04	654	0.677
3	10	⊏	2	2.5	246	0.344
	10	□	4	2.02	258	0.292
	13	⊏	2	2.25	648	1.451
	13	⊏	2	2.12	345	0.728
4	10	⊏	2	0.93	1941	1.011
	10	—	0	0.76	636	0.271
	13	∟	1	0.84	678	0.567
	13	∟	1	0.5	360	0.179

4.3. A Simulation Model for Rebar Fabrication's Optimal Production Plans

Figure 4 shows the models for SHD10 and SHD13 and the corresponding verification processes. There are seven components to the simulation of rebar plant processing: (1) raw steel import, (2) transportation, (3) priority and rebar type setting, (4) cutting, (5) rebar reuse, (6) bending, and (7) machine schedule setting.

First, in the simulation model, a rebar with a particular diameter is put up as an agent to represent the raw steel imported into the rebar fabrication plant. Second, the rebar is transported and assigned to a cutter. It is believed that the hoist's maximum capacity can be gathered and transported all at once (for a hoist, the transport time typically varies with the transport distance). Third, different sites from which to collect various rebar types are taken into consideration before processing. Each agent is assigned a priority, which determines the order in which they move through the cutting process. By identifying various site-specific combinations in this step, the simulation can ultimately choose the optimum site-specific combination when prioritizing the agent. Fourth, the rebar is cut. Similar to the case of hoists, for which the capacity is set considering simultaneous transportation, the number of simultaneous cuts is set for the cutters.

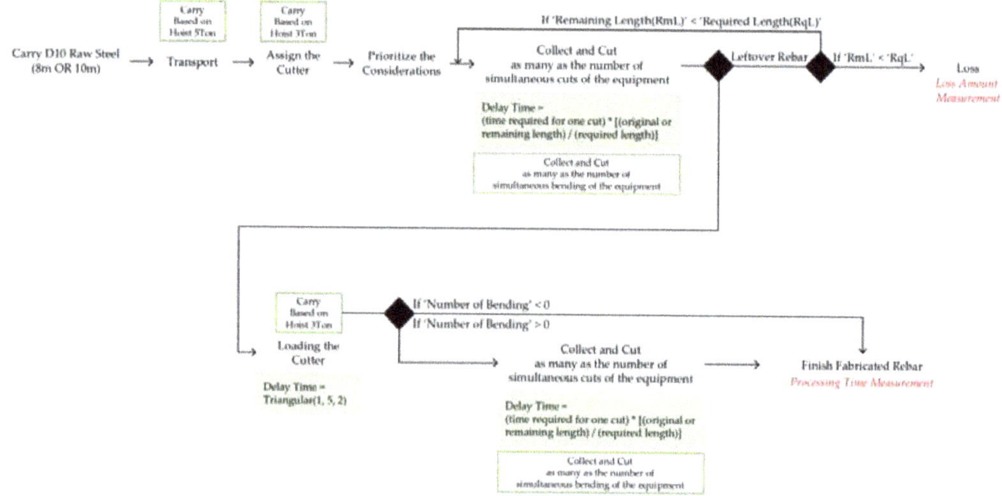

Figure 4. Simulation modeling that reflects processes within a rebar processing plant.

The cutting machine is employed, and the required time is input to represent the number of times the imported rebar can be cut (with the maximum amount) and the time needed to perform one cut, which adds time to the process. The length of the raw steel and the needed length are taken into consideration while defining the cut steel bars, which are then repeatedly moved to the cutting process. Fifth, we considered the reuse of the cut rebar. The following factors determine the loss rate: the cutting procedure is repeated to reuse the rebar if the original steel is cut and the residual rebar is longer than the type of rebar to be processed, and vice versa. Sixth, the rebars go through bending procedures. The long-term steel without the need for bending is designed to finish the rebar process, and the rebar bending process takes as much time as the number of bends to the specific rebar type. Seventh, the machine schedule employed in the process of fabricating rebar is taken into account. The machine operation in the plant under consideration begins at 8:00 a.m. and ends at 9:30 p.m., with three intervals in between. The suggested model takes into account the machinery turning on and off to match the actual state of a rebar production plant.

By simulating the combination of several locations and the ratio of raw steel by length, we were able to build an optimal production plan for the rebar fabrication plant by calculating the time required for rebar fabrication and the loss rate.

5. Results

5.1. Optimization

The situation of importing raw steel with a length ratio of 2:1 (8 m:10 m) corresponds to the largest reduction in the loss rate when taking into account both SHD10 and SHD13 scenarios. The loss accumulation pattern is comparable for length ratios of 1:1, 1:2, and 1:3, as shown in Table 5. Moreover, the loss accumulation patterns are similar for length ratios of 2:1 and 3:1.

To optimize the site combination, the loss rate was analyzed for different combinations of multiple sites. In the considered case, 16 rebar types were selected, and the maximum loss rate was observed when the length of the imported raw steel was only 8 m. A combination of the two sites in the rebar fabrication process corresponded to the optimal production plan.

Table 5. Analysis results of four site combinations.

Type of Rebar	8 m:10 m	Loss (ton)	Amount (ton)	Loss Rate (%)	Time (s)
SHD10	1:1	2.65	13.95	1.86	5512
	1:2	2.65	13.95	1.87	5508
	1:3	3.69	13.95	2.57	5841
	2:1	2.10	13.95	1.49	5492
	3:1	2.32	13.95	1.64	5621
SHD13	1:1	2.19	14.41	1.49	5943
	1:2	2.31	14.41	1.57	5888
	1:3	2.84	14.41	1.93	5467
	2:1	1.38	14.41	0.95	5521
	3:1	1.44	14.41	0.99	5564
Comprehensive result	1:1	4.84	28.39	1.68	5943
	1:2	4.96	28.39	1.72	5888
	1:3	6.54	28.39	2.25	5841
	2:1	3.49	28.39	1.21	5521
	3:1	3.77	28.39	1.31	5621

5.2. Model Verification

The data of the rebar fabrication plant were collected, and a model was built by reflecting the actual environment of the plant. The model was evaluated, and minor system errors were corrected. The model was additionally validated using information on the status of the rebars made at the actual rebar fabrication plant.

The simulation model was noted to exhibit a reasonable performance. According to Figures 5 and 6, the cutting process and flow of the processed rebars into the bending well were stopped as the machinery was shut off for one hour during lunch, 30 min during break time, and 30 min during dinner.

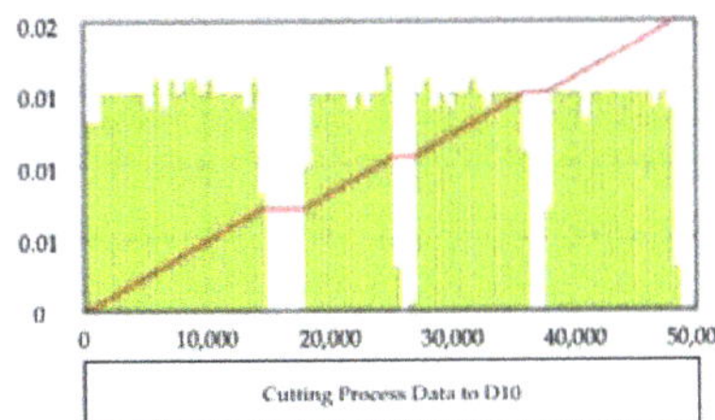

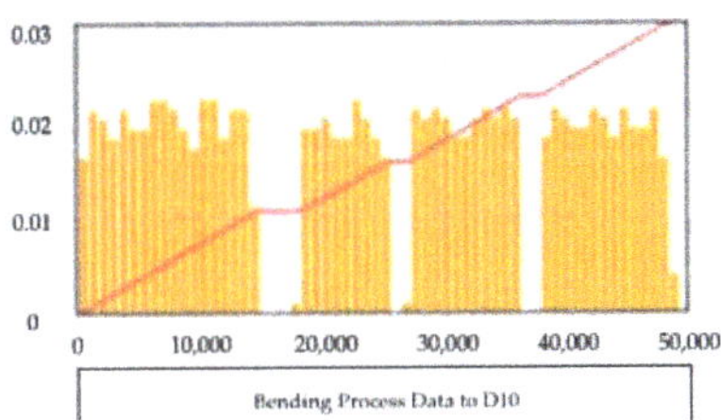

Figure 5. Model verification: hourly fabrication flow of D10 rebar.

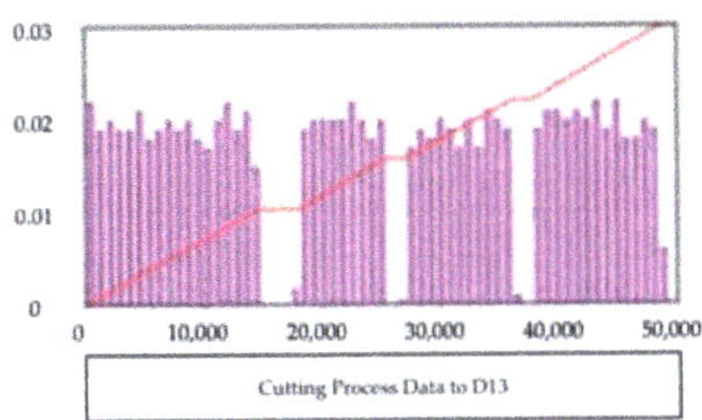

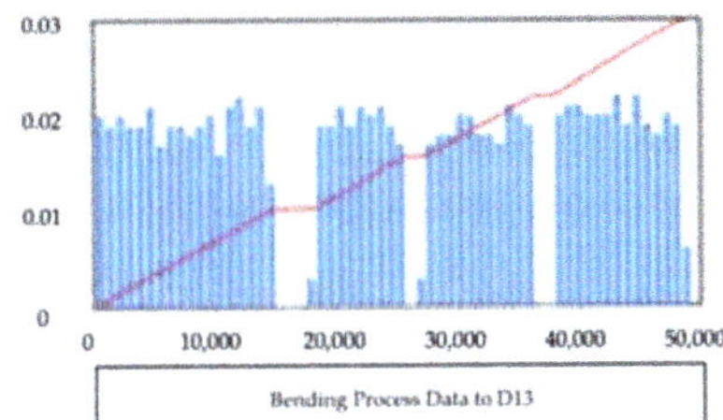

Figure 6. Model verification: hourly fabrication flow of D13 rebar.

5.3. Model Validation

The amount of processed rebar and the amount of loss from cutting and bending, as shown in Figures 7 and 8, rise over time. As a result, the individual phases in the processing of rebar can be reflected in the suggested simulation model.

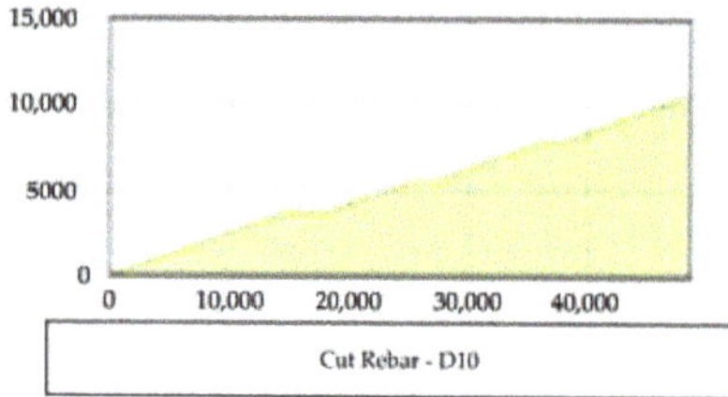
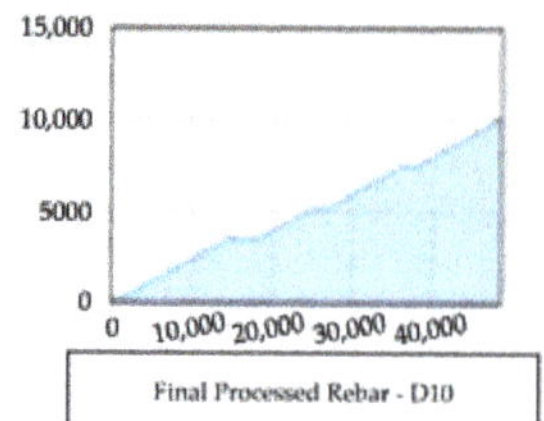
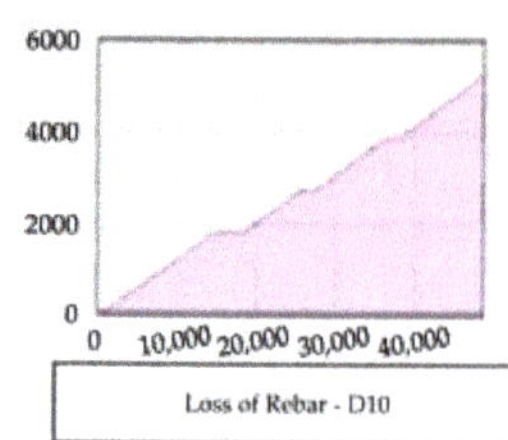

Figure 7. Model validation: D10 rebar time plot.

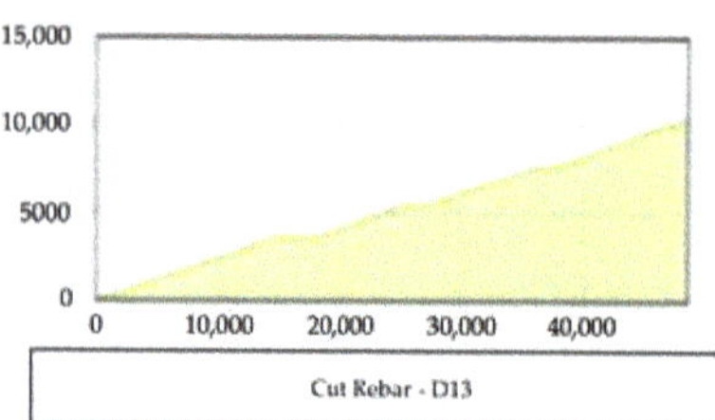
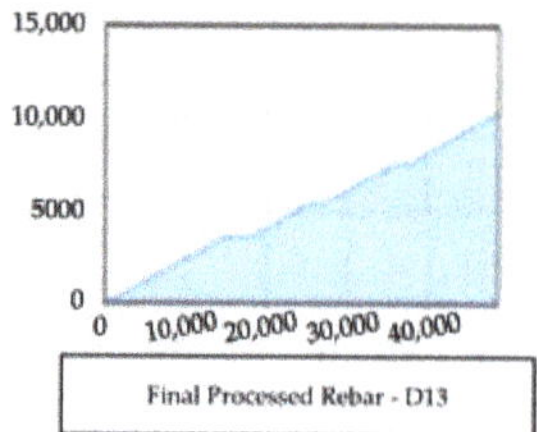
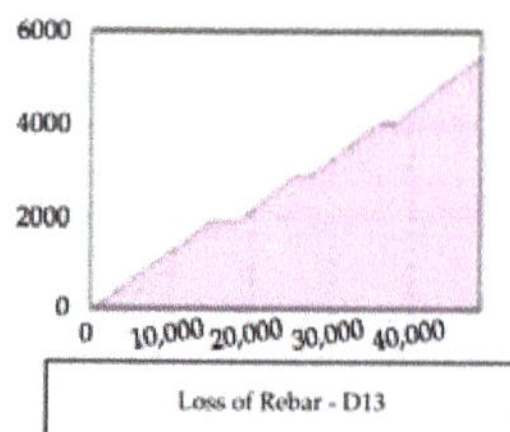

Figure 8. Model validation: D13 rebar time plot.

For the considered plant, the total number of working hours per day is 11.5 over weekdays, with 200 t of rebars processed per day according to the machine capacities. If the simulation model is run for 11.5 h, and 200 t of rebars are machined, the simulation model can be considered valid. The simulation results indicate that 202,856 kg, that is, approximately 203 t of rebar, is machined in 11.5 h. The results are shown as Table 6. Therefore, the optimal production plan and simulation model can be considered valid.

Table 6. Results of simulation validation.

Value	Diameter	Processed Rebar (ton)	Daily Output Overall (11.5 h) (ton)
Min.	D10	100.91	196.72
	D13	96.17	
Max.	D10	102.60	206.74
	D13	104.47	
Avg.			202.86

6. Conclusions

This paper proposes a DES model to identify the optimal combinations to decrease the loss rate and promote rebar fabrication to achieve simultaneous delivery to multiple projects. The optimal fabrication plan to minimize the loss rate was formulated based on the collected data. By setting up the rebar type and machine features that affect the loss rate and time needed for rebar fabrication as parameters, the simulation model was proven to be accurate.

Human aspects were not taken into account because the model's variables only included raw steel and machine characteristics. Consequently, the results obtained using the simulation model may be different from the loss rate of the actual rebar fabrication plant. Nevertheless, this study makes several valuable contributions. First, the suggested model may take into account the machine specifications for each piece of machinery that affects the processing speed and loss rate. By adjusting the parameter values to acquire the optimization results, the suggested model may, thus, be simply applied to any plant. Second, the manager's experience or intuition are often used to manage the loss rate. The suggested model can assist in creating an objective plan for the fabrication of rebar. Because it can

Appl. Sci. **2022**, *12*, 9183

reduce the percentage of material prices in construction projects, the decreased loss rate has a substantial economic impact. In future work, the proposed model can be extended to more diverse cases to optimize project costs in the construction industry.

Author Contributions: Conceptualization, E.H., J.-S.Y., M.H. and Y.J.; Formal analysis, E.H., J.-S.Y. and M.H.; Funding acquisition, J.-S.Y.; Investigation, E.H., M.H. and Y.J.; Methodology, J.-S.Y., J.S. and M.H.; Project administration, J.-S.Y. and J.S.; Resources, J.-S.Y.; Software, M.H.; Supervision, J.-S.Y. and J.S.; Validation, J.-S.Y., J.S. and M.H.; Visualization, E.H. and M.H.; Writing—original draft, E.H. and Y.J.; Writing—review & editing, J.-S.Y. and J.S. All authors have read and agreed to the published version of the manuscript.

Funding: This research was funded by the Korea Agency for Infrastructure Technology Advancement (KAIA) grant funded by the Ministry of Land, Infrastructure and Transport (Grant 22ORPS-B158109-03).

Institutional Review Board Statement: Not applicable.

Informed Consent Statement: Not applicable.

Conflicts of Interest: The authors declare no conflict of interest.

References

1. Abbasi, S.; Taghizade, K.; Noorzai, E. BIM-based combination of takt time and discrete event simulation for implementing just in time in construction scheduling under constraints. *J. Constr. Eng. Manag.* **2020**, *146*, 04020143. [CrossRef]
2. AnyLogic: Simulation Modeling Software Tools & Solutions for Business. Why Simulation Modeling? 2018. Available online: https://www.anylogic.com. (accessed on 2 November 2018).
3. Asri, M.A.N.M.; Nawi, M.N.M.; Nadarajan, S.; Osman, W.N.; Harun, A.N. Success factors of JIT integration with IBS construction projects—A literature review. *Int. J. Supply Chain. Manag.* **2016**, *5*, 71–76.
4. Benjaoran, V.; Bhokha, S. Three-step solutions for cutting stock problem of construction steel bars. *KSCE J. Civ. Eng.* **2014**, *18*, 1239–1247. [CrossRef]
5. Borxhchev, A.; Filippov, A. From System Dynamics and Discrete Event to Practical Agent Based Modeling: Reasons, Techniques, Tools. In Proceedings of the 22nd International Conference of the System Dynamics Society, Oxford, UK, 25–29 July 2014.
6. Ivanov, D. *Operations and Supply Chain Simulation with AnyLogic*; The AnyLogic Company: Hanau, Germany, 2017; p. 12.
7. Banks, J.; Carson, J.S., II; Nelson, B.L.; Nicol, D.M. *Discrete-Event System Simulation Fourth Edition*; Pearson Education India: Delhi, India, 2005.
8. Khondoker, M.T.H. Automated reinforcement trim waste optimization in RC frame structures using building information modeling and mixed-integer linear programming. *Autom. Constr.* **2021**, *124*, 103599. [CrossRef]
9. Lee, D.; Son, S.; Kim, D.; Kim, S. Special-length-priority algorithm to minimize reinforcing bar-cutting waste for sustainable construction. *Sustainability* **2020**, *12*, 5950. [CrossRef]
10. Melhem, N.N.; Maher, R.A.; Sundermeier, M. Waste-Based Management of Steel Reinforcement Cutting in Construction Projects. *J. Constr. Eng. Manag.* **2021**, *147*, 04021056. [CrossRef]
11. Mishra, S.P.; Parbat, D.K.; Modak, J.P. Field data-based mathematical simulation of manual rebar cutting. *J. Constr. Dev. Ctries.* **2014**, *19*, 111.
12. Nadoushani, Z.S.M.; Hammad, A.W.; Xiao, J.; Akbarnezhad, A. Minimizing cutting wastes of reinforcing steel bars through optimizing lap splicing within reinforced concrete elements. *Constr. Build. Mater.* **2018**, *185*, 600–608. [CrossRef]
13. Polat, G.; Arditi, D.; Mungen, U. Simulation-based decision support system for economical supply chain management of rebar. *J. Constr. Eng. Manag.* **2007**, *133*, 29–39. [CrossRef]
14. Robinson, S. *Simulation: The Practice of Model Development and Use*; Bloomsbury Publishing: London, UK, 2014.
15. Salem, O.; Shahin, A.; Khalifa, Y. Minimizing cutting wastes of reinforcement steel bars using genetic algorithms and integer programming models. *J. Constr. Eng. Manag.* **2007**, *133*, 982–992. [CrossRef]
16. Zekavat, P.R.; Moon, S.; Bernold, L.E. Holonic construction management: Unified framework for ICT-supported process control. *J. Manag. Eng.* **2015**, *31*, A4014008. [CrossRef]
17. Zheng, C.; Yi, C.; Lu, M. Integrated optimization of rebar detailing design and installation planning for waste reduction and productivity improvement. *Autom. Constr.* **2019**, *101*, 32–47. [CrossRef]

 *buildings*

Article

Decision-Making Framework for Construction Clients in Selecting Appropriate Procurement Route

Muhammed Bolomope [1,*], Abdul-Rasheed Amidu [2], Saheed Ajayi [3] and Arshad Javed [4]

[1] Department of Land Management and Systems, Lincoln University, Lincoln 7647, New Zealand
[2] Department of Property, The University of Auckland, Auckland 1010, New Zealand
[3] School of Built Environment, Leeds Beckett University, Leeds LS1 3HE, UK
[4] School of Economics and Finance, Massey University, Palmerston North 4442, New Zealand
* Correspondence: muhammed.bolomope@gmail.com

Abstract: Procurement decision-making is a crucial determinant of project success. Although several objective, stage-based models have been proposed to guide clients' procurement choices, little emphasis has been made on the subjective nature of construction clients. Recognizing the role of clients' experiences in justifying procurement routes, this study develops a decision-making framework that is capable of guiding construction clients in making informed procurement choices. Adopting a mixed-method approach, comprising semi-structured interviews and multi-objective optimization, relevant procurement options were appraised based on clients' specifications and project deliverables. The lived experiences of construction clients and the importance they attach to pre-defined selection rating criteria were subsequently evaluated, using a template that enables clients to prioritize procurement methods for different project types. The resultant framework offers a holistic, practical, and collaborative procurement selection process that promotes the efficient delivery of construction projects by reducing the cost overrun and delays associated with uninformed client decisions in construction procurement.

Keywords: construction; decision-making; construction clients; construction procurement

Citation: Bolomope, M.; Amidu, A.-R.; Ajayi, S.; Javed, A. Decision-Making Framework for Construction Clients in Selecting Appropriate Procurement Route. *Buildings* **2022**, *12*, 2192. https://doi.org/10.3390/buildings12122192

Academic Editors: Davide Settembre-Blundo and Jorge Pedro Lopes

Received: 26 October 2022
Accepted: 5 December 2022
Published: 12 December 2022

Publisher's Note: MDPI stays neutral with regard to jurisdictional claims in published maps and institutional affiliations.

1. Introduction

Decision-making in construction procurement is multi-faceted. It is influenced by different factors arising from clients' specifications, project peculiarities, and procurement options [1]. According to [2], the choices made by clients regarding the optimal procurement route remain a crucial, yet difficult hurdle in achieving overall project success. While there are active debates on construction project underperformance [3–5], previous studies have revealed that the basis of project inefficiency can be traced to the conception stage, when procurement decisions are made [6–8]. Consequently, scholars have emphasized the need for efficient procurement routes in the construction industry [1,4,9–11].

Aside from the objective procurement determinants of cost, quality, and time, Ref. [7] noted that client characteristics also influence procurement choices. The authors of [5,12] argued further that construction clients' (CCs') innovation in the procurement process is often driven by subjective feelings, knowledge, or past experiences on similar projects. Therefore, instead of evaluating procurement routes purely on objective models, the need for a subjective approach that assesses the distinct peculiarities of each project has been suggested [5]. According to [13], efforts should be made to develop decision-making tools that could match a range of project performance indicators to project peculiarities and client demands in order to achieve overall project success.

While many construction professionals believe overall project success to be a comprehensive assessment arising from the consensus of all key stakeholders [12], others believe that project success is much more complex [4,14] and that client's satisfaction with the final outcome is perhaps the most critical indicator of whether a project can be considered successful or not [15,16]. Although several scholars have attempted to simplify the construction procurement process [1,10,11,17], gap remains regarding the role and impact of CCs' subjective attributes in selecting suitable procurement routes for different project types. The main objective of this paper is, therefore, to explore the impact of CC's subjective experience in procurement selection and subsequently develop a framework that could guide CCs toward making informed procurement choices as they align their perceptions and experiences with the objective reality of the procurement selection process.

This study was executed in two stages. The first stage involved in-depth qualitative interviews which focused on understanding the impact of CCs' peculiar beliefs, experiences, and perceptions regarding the procurement selection process. Subsequently, the CCs were assigned the task of prioritizing different project parameters that inform their procurement decisions for different project types. The combined approach resulted in the development of a framework that could enhance the procurement selection process based on the robust consideration of client characteristics, goals, and project peculiarities. The remainder of the study is organized into six sections. The extant literature on procurement and construction clients was reviewed in Section 2, followed by the presentation of the adopted methodology in Section 3. The data collection and analysis were outlined in Section 4. Sections 5 and 6 outline the study findings and discussion, respectively, while the study culminates in Section 7.

2. Procurement and Construction Clients

2.1. Construction Clients

Unlike other industries, clients in the construction industry dictate the organizational and management pattern of project delivery [18,19]. Whereas there are standard practices and methods of delivering projects in industries like manufacturing and automobile [14], selecting procurement routes in the construction industry remains ambiguous [17,20]. According to [13], the procurement process in the construction industry is client-centered. CCs are individuals or organizations responsible for the provision, maintenance, and disposal of construction projects [21], and their actions or inactions influence the overall project outcome [7,15]. Further, CCs have different perspectives through which they assess construction procurement [22–24]. Ranging from the basic project requirements of cost, quality, and time, CCs' considerations have emerged to include factors such as project variation, risk perception, and end-user satisfaction [9,14].

According to [21], CCs are traditionally divided into two categories; public and private clients, but it has also been acknowledged that subdivisions of these two major categories exist based on clients' experiences and whether they are primary or secondary constructors [23]. The publications, 'Constructing the Team' [25], and 'Rethinking Construction' [26] have also evaluated the categorization of CCs to include expert, inexpert, etc. According to [12], the innovative role of CCs in the procurement process can be multi-dimensional. It could be an assertive role, where the client drives innovation; a cooperative role, where the construction team and the client jointly drive innovation; or a passive role, where the construction team drives innovation [2,14,19]. Appendix B-Figure A2 illustrates the various categorization of CCs.

As CCs consider project feasibility in the procurement process, they are also faced with some elements of project uncertainties [1]. Therefore, aside from focusing solely on project objectives and client attitude to risk, factors like clients' resources, peculiar project characteristics, ability to make changes, and ethical considerations have consistently influenced clients' innovation in the procurement process [23]. While these factors affect projects differently, based on their magnitude and complexity, the preferred procurement choice is often determined based on the factors that are of the most importance to the client [18]. Scholars have therefore, emphasized that there should be a harmonization of clients' objectives, the attributes of available procurement alternatives, and the expected project outcome [17].

According to [9], when clients' goals are clearly defined, selecting a procurement route should be a purely logical process. CCs should be able to choose project delivery options that suit their project expectations with ease [21]. However, the reality of clients' subjective attributes vis-a-vis the complexity of modern construction makes this unrealistic [14,23]. The demand for contemporary construction varies with clients' expertise and has continued to widen the gap between experienced and inexperienced clients [7]. Only clients that are up to date with the latest innovation and best practices are relevant in today's construction industry. This explains the difference in various clients' perspectives when rationalizing procurement options [5,18].

While the experienced client appreciates the significance of collaboration and sustainability in the overall project outcome [15], the inexperienced client may not emphasize these factors in selecting a project delivery route. However, irrespective of the categorization, CCs do not fully explore the procurement variants available to them [22]. Instead, they often rely on past occurrences, feedback from other stakeholders, and the impact of external factors like legal framework and public perception [13,15]. Unfortunately, this approach does not adequately appraise the peculiarity of particular projects, and it could lead to uninformed decisions by clients, which could ultimately result in poor quality of construction, cost overrun, or delay in project delivery [18]. In order to avert the consequences of inefficient decisions and their subsequent impact on the project outcome, it is essential that decision-making tools are made available to the clients, as is being suggested in this study.

2.2. Construction Procurement

CCs see procurement as a sequence of calculated risks that should be evaluated in order to emerge with a project that is safe, cost-effective, and fit for purpose [27]. This perspective has significantly influenced the evolution of construction procurement and has resulted in various forms of procurement options that have been mainly driven by project needs and clients' specifications [3,13]. However, rather than limiting the description of procurement as the process of acquiring with ease [22], the definition of procurement has remained dynamic and robust as an integral part of the project delivery process [6,17,28]. According to [29] (p. 107), procurement is "the acquisition of new buildings, or space within buildings, either by directly buying, renting, or leasing from the open market or by designing and building the facility to meet a specific need". The procurement process has also described by [10] as a clear approach to achieving clients' goals regarding project delivery over a given period, based on a mechanism that coordinates all stakeholders throughout the project lifecycle. As construction procurement continues to evolve, scholars are unanimous in the view that the selection of an appropriate procurement route is a major determinant of overall project outcome [1,6,9,16,30].

According to [5] (p. 20), an appropriate procurement selection technique involves a "set of rationalistic decisions within a closed environment, aiming to produce generic, prescriptive rules for clients and advisers to use to select the 'best' procurement route for their project." This process can be simply described as a framework within which construction is acquired or secured [31]. Although the outcome of previous construction project decisions could be extremely insightful in supporting CC's decision regarding similar projects [23], Ref. [32] argues that there are several procurement options available to the client. Within each procurement option, there are several variants, each of which may be possibly refined to accommodate client needs and project specifics [31]. Some of such procurement routes and their variants as they apply to clients in the construction industry are discussed below. Ref. [21] classified construction procurement options into four major groups as follows: (1) Separated or traditional procurement systems, (2) Design and build procurement systems, (3) Management-oriented procurement systems, and (4) Partnering/collaborative systems.

2.2.1. Separated or Traditional Procurement Systems

The evolution of project delivery routes has gone through different stages over the years [31,33]. Many projects constructed before the Second World War were procured through "traditional" means, which has remained in existence for more than 150 years [31,34]. Known as the oldest form of documented procurement option, Ref. [35] describes the traditional procurement route as having design as a separate function from construction. In adopting traditional procurement, the design is completed before the selection of a contractor to build the works [13], which is seen as the least-risk approach for the client as there is an inherent level of certainty about the project quality and construction duration if it is properly implemented [28]. Based on this, and assuming no variations are introduced, overall project costs can be determined with reasonable reliability before construction begins [34]. While lots of construction projects have been successfully delivered across the globe using traditional procurement [13], there are numerous reports of post-contract changes and delays, which often result in increased project costs and time overruns [17].

2.2.2. Design and Build Method

The end of the Second World War ushered in a season of consistent economic growth and human capital development [21]. To meet the societal demand at that time, there was a need for the timely delivery of public facilities, hence, the adoption of an integrated procurement strategy that combines the design and construction functions involved in project delivery [31]. This procurement option is termed "design and build" [35].

According to [13], design and build is a route wherein a single contractor assumes the risk and responsibility for the design and construction of projects, usually in return for a pre-determined price. It is generally regarded as a fast-track route because construction often commences before the comprehensive design is completed, with the contractor finalizing the design as the work progresses [35]. It can be deduced from the definition that this method reduces project time, assures cost certainty, and encourages integrated contractor contribution to the design and project planning [36]. A shift to design and build translates to the transfer of design responsibility from direct client control to organizations, whose core businesses are profit-focused [31]. The client in the design and build arrangement passes the legal obligation for both design and construction to an independent contractor [35]. This single contractor can be either an integrated firm with an in-house design and construction delivery team or a consortium of various design and construction firms, brought together for a particular bid [36]. While significant progress regarding the timely delivery of projects and cost certainty resulted from the evolution of the design and build procurement approach [13], there have been reported deficiencies in the quality of projects delivered through this route [18]. The need to advance the design and build procurement method led to other integrated forms of procurement.

2.2.3. Management-Oriented Method

The concept of management-oriented procurement was conceived to bridge the gap between traditional procurement and design and build [35]. This method evolved from the United Kingdom in the 1970s [21]. In adopting this procurement style, the CCs devolve the management of the design and construction of projects to an expert who acts as a management consultant on behalf of the client [31]. Management-oriented procurement enhances project quality and also accommodates design changes [36]. It ensures that the appointed managing consultants are responsible for the construction tasks without actually performing any of that work [13], at a cost to the client. This means that the consultants take over the construction process and ensure value for money on the project. The variants of management-based procurement include management contracting and construction management, with both sharing the main characteristic of appointing a managing party [31]. In construction management, the client appoints a construction manager (CM) to oversee the design and construction activities, using their expertise and experience to deliver the project for an agreed sum [36]. The role of the CM is mainly to ensure compliance to project specifications without any contractual link with the design team and contractors [22]. All contractual agreement remains between the client and the trade contractors [13]. However, for management contracting, the consultant bears part of the construction risk because they have an established contractual link with the package contractors [35]. A major benefit of management-oriented procurement is the participation of the expert consultant in the design and project planning [21,28]. Although uncertainty about project cost at the initial stage of the procurement is a major disadvantage of this option [22], early consultant involvement reduces the risk of project overrun while accommodating later design decisions as construction progresses. The use of more integrated procurement methods otherwise referred to as partnering, emerged in the early-90s [13].

2.2.4. Partnering/Collaborative Method

"Partnering in construction procurement is a structured management approach that enables teamwork, trust, long-term commitment, open culture, mutual objectives, customer focus, and innovation between contractual parties" [28] (p. 5). Apart from driving innovation through agreed mutual objectives, devising ways for conflict resolution, commitment to continuous improvement, measuring performance, and sharing gains [26], partnering suggests that efficient project outcome is better achieved through the collective effort of all stakeholders involved in project delivery [13,31]. Partnering, therefore, provides the premise required for the adoption of PPP (Public-Private Partnership) and PFI (Private Finance Initiative) procurement options.

According to [3], the fundamental rationale for PPP/PFI procurement is to establish a platform where the public and private sectors work together to realize optimum project outcomes while also managing project risks and disputes. Despite acknowledging that PPP/PFI procurement provides a wide range of benefits through innovative and collaborative practices amongst stakeholders in the delivery of public projects [13], scholars are of the view that the crucial considerations for successful PPP/PFI projects require appropriate risk assessment and allocation, transparency, adequate stakeholder engagement, strong legal framework, and availability of finance [3].

2.3. Factors Governing Procurement Route Selection

The selection of an appropriate construction procurement path is directly linked to project objectives [2,4,32]. While there are various procurement routes available for CCs to choose from, challenges arising from the dynamic construction environment, changing client objectives and expectations, increasing project complexity, lack of effective communication and disintegration within the construction industry have resulted in the constant debate on selecting appropriate procurement routes for construction projects [7,17]. For a suitable procurement choice to be made, it is essential to clearly understand the project objectives and relate their significance to the overall project's success [24,30]. Although comprehending CCs' rationale for undertaking construction projects may be complicated, the relevance of existing decision-making tools in selecting procurement routes also remains ambiguous [5,20]. To clearly address the complexity of construction procurement, it is essential to:

- Describe CCs and categorize them based on their relevance in the construction industry.
- Understand the rationale for clients' objectives and the resultant effect on project outcomes.
- Understand the dynamic nature of the construction procurement process.

Having previously identified various client types and procurement options, the characteristics, and expectations of specific projects are also expected to be clarified in order to differentiate the strengths and weaknesses of each procurement route. As shown in Table 1, studies that have previously investigated CCs and their attributes have established a set of commonly considered factors for construction procurement. Nevertheless, the selection of an appropriate procurement strategy has two components [13], viz:

1. Evaluating and establishing priorities for the project objectives and clients' attitudes to risk, and

2. Reviewing possible procurement options and selecting the most appropriate.

Table 1. Review of Factors Influencing Clients' Procurement Choices.

Source	Time	Cost Certainty	Quality	Risk	Complexity	Flexibility	Accountability	Competition	Dispute Resolution
[9]	✓	✓	✓	✓	✓	✓	✓	✓	-
[32]	✓	✓	✓	✓	✓	✓	-	✓	-
[8]	✓	✓	✓	✓	✓	✓	-	✓	✓
[37]	✓	✓	✓	-	✓	-	-	✓	-
[11]	✓	✓	✓	✓	✓	-	✓	-	-
[21]	✓	✓	✓	✓	-	-	✓	-	✓
[38]	✓	✓	✓	✓	✓	✓	✓	-	-
[39]	✓	✓	✓	-	-	✓	✓	✓	-
[12]	✓	✓	✓	✓	✓	-	✓	✓	✓
[27]	✓	✓	✓	✓	-	-	✓	-	✓
[40]	✓	✓	✓	✓	✓	✓	✓	✓	✓
[41]	✓	✓	✓	✓	-	-	-	-	-
[42]	✓	✓	✓	✓	✓	-	✓	-	-
[43]	✓	✓	✓	✓	✓	-	✓	✓	✓
[17]	✓	✓	✓	✓	✓	✓	✓	-	-
[44]	✓	✓	✓	✓	✓	✓	-	✓	✓

These two components are expected to be accessed holistically in line with best practices to develop a framework that could assist clients in making informed decisions.

While reflecting on, and corroborating the argument of [17] (p. 310) that "as far as known, apart from the work of [32], all other procurement decision-making charts were developed over a decade ago", Table 1 includes the recent work of [44] and [8] in providing an up-to-date review of factors influencing clients' procurement choices. Whereas all the studies suggest the significance of cost certainty, quality, and timely delivery of construction projects as major factors that influence CCs procurement choices, the reality of modern procurement also involves the consideration of factors such as risk, complexity, accountability, flexibility, and competition as shown in Table 1. According to [13], when the client type has been established, factors like the client's resources, project characteristics, ability to make changes, risk management, cost issues, timing, and quality assurance should be considered when evaluating the most appropriate procurement strategy. Although some of the factors may be in conflict and priorities need to be set, procurement route selection should consider the factors that are most important to the client [20,27,32].

The consideration for project factors in simplifying the procurement selection process can be traced to the National Economic Development Office report [42]. Subsequently, several studies including [27,37,43], have leveraged the NEDO report in proposing strategies that could be explored by CCs in rationalizing the construction procurement selection process. As shown in Table 1 scholars have established major factors that influence CCs' procurement choices and have subsequently proposed models to simplify the procurement selection task. For instance, Ref. [38] explored the effectiveness of a hierarchical process and multi-criteria screening in construction procurement evaluation, while [40], established the fuzzy function of different procurement selection criteria as a tool for improving procurement selection. Molenaar [41] also leveraged a multi-attribute analysis and regression model in predicting design and build procurement for public sector projects, and Ref. [39] clarified the objective relationship between financing, risk, and construction procurement in their study on private financing of construction projects and procurement systems. However, despite offering notable contributions that could ease construction procurement decision-making, the aforementioned studies are predominantly premised on the logical, systematic evaluation of project factors with limited consideration of the dynamic nature of CCs' motivation, experience, and subjective project requirements.

Nevertheless, few studies acknowledge the important role of client experience in procurement decision-making. According to a study on the participatory approach in the procurement selection of social infrastructure that was carried out by [32], it was established that efficient procurement decision-making require decision-makers to consistently reflect and evaluate project outcomes. The need to value client experience in analysing construction procurement options was also highlighted by [17]. However, while Ref. [32] focused on a particular client type, Ref. [17] did not clarify how CCs' experience could be integrated into the procurement selection process. Therefore, in advancing the current debate on construction procurement selection, this study proposes a holistic framework that recognizes and integrates different client types (as discussed in Section 2.1) with feasible procurement options (as discussed in Section 2.2), based on the project factors that have been established in the literature (as highlighted in Table 1) and listed as follows.

- Time Certainty
- Cost Certainty
- Project Quality
- Risk Evaluation
- Project Complexity
- Design Flexibility
- Accountability
- Competitive Bidding
- Dispute Resolution

To achieve the study objective, the highlighted factors above were considered alongside client expectations and project requirements through the briefing process described in the methodology section.

3. Methodology

A mixed-method approach involving qualitative interviews and multi-objective optimization (MOO) protocol was adopted in exploring the significance of CCs' perceptions in selecting procurement routes. This approach is particularly relevant to this study because it integrates the subjective influence of CCs' experience with the objective reality of the procurement selection process. The qualitative aspect of this study involved fourteen purposefully selected CCs across the public and private sectors in the United Kingdom (UK), with experiences spanning the various construction procurement phases highlighted previously. Their perception and assessment of different procurement routes were collated through in-depth semi-structured interviews. The UK is particularly suitable for this study because of its global influence in construction procurement innovation and its multiplicity of client types. The participants' sample size conforms with the suggestions of [45], with details of the interview respondents provided in Table 2. Further to the client categorization highlighted in Appendix B-Figure A2, the research participants were grouped into Public Experienced Primary Client (PEPC), Public Experienced Secondary Client (PESC), Private Experienced Primary Client (PrEPC), and Private Inexperienced Secondary Client (PrISC).

Following the qualitative aspect of the study, a decision-making chart illustrated in Appendix A-Figure A1 was used in collecting numeric data relating to the significance of various project factors to different client types through MOO. According to [46] (p. 82), multi-objective optimization requires the "definition of appropriate decision variables, objective functions and constraints, and finally, the selection of appropriate solution techniques." Unlike single-objective optimization, which sets out to identify the best amongst a series of alternatives, thereby recommending the superlative option, multi-objective optimization involves a more detailed comparison of various attributes of the available alternatives before choices are made [4,47]. For instance, instead of making a project decision based solely on cost consideration, MOO evaluates various dimensions of project expectations like cost reduction, timely delivery, quality assurance, best practice, safety considerations, etc., before substantiating a preferred procurement route. This further corroborates the opinion of [46] that MOO leads to various alternative solutions to a problem, with a compromise reached among the objectives considered.

Table 2. Background of research participants.

Respondents	Sector	Experience (Years)	Qualification	Current Position
PEPC-R1	Public (Housing)	33	BSc Civil Engr	Facilities Manager
PEPC-R2	Public (Energy)	19	BSc Building Tech	Project Manager
PEPC-R3	Public (Transport)	28	BTech Civil Engr	Project Director
PEPC-R4	Public (City Council)	14	MSc Civil Engr	Procurement Strategist
PESC-R5	Public (Health)	11	BSc Property	Asset Manager
PESC-R6	Public (Transport)	23	BSc Project Mgt	Project Manager
PESC-R7	Private (Retail Developer)	18	Diploma Project Mgt	Facilities Manager
PrEPC-R8	Private (Property Developer)	30	BSc Civil Engr	Construction Manager
PrEPC-R9	Private (University)	24	MSc Property	Project Manager
PrEPC-R10	Private (Housing Agency)	13	BSc Commerce	Portfolio Manager
PrEPC-R11	Private (Real Estate Investor)	17	BSc Property	Asset Manager
PrEPC-12	Private (Transportation)	11	MBA Management	Investment Manager
PrISC-R13	Private (Individual)	9	BSc Arch	Chief Executive
PrISC-R14	Private (Individual)	12	MSc Construction Mgt	General Manager

MOO has been widely used in facilitating objective decision-making in mathematics, business, science, and engineering [47–49]. It is also popular among scholars in the construction industry who have evaluated various aspects of decision-making [4,17]. In addressing the focus of this study, the MOO strategy relies on the client's prioritization of project deliverables with reference to the available procurement options. Being an objective decision-making strategy, MOO, therefore, complements the subjective opinion of CCs in rationalizing procurement routes.

4. Data Collection and Analysis

The identified research participants highlighted in Table 2 were engaged in a series of face-to-face discussions, and their experiences across various project types and procurement strategies were collated via recorded telephone interviews. The interviews were recorded to ensure a comprehensive data collection process and the research participants were assured of their confidentiality and anonymity. Details regarding how CCs make procurement decisions, factors that influence their procurement decisions, and the reason for their procurement preferences were collected and subsequently analyzed thematically. According to [50], thematic analysis recognizes flexibility in the data collection and reporting process by identifying direct and indirect ideas emanating from the data. Following the stage-based process suggested by [51], the recorded interviews were transcribed, and initial codes were identified. As stated by [52], codes are keywords or phrases that form the basis of participants' opinions, which are emphasized because they reflect the participants' intentions. The identified codes were evaluated and merged into initial themes reflecting participants' unique perspectives toward construction procurement through a process described by [51] as mapping.

As shown in Table 3, the initial set of themes was reviewed, paraphrased, and consolidated to create a set of robust, coherent, and established themes. The major themes that emerged from the analysis suggest that CCs' procurement decisions are influenced by their cognitive abilities, access to relevant information, and the dynamism of the built environment. These themes are subsequently discussed with reference to relevant quotes from the interview.

Further to the collation of qualitative data, CCs' ability to select appropriate procurement routes was assessed using MOO. In adopting MOO, researchers have proposed different stage-based approaches, which involve establishing a set of factors that influence the project outcome, evaluating these factors through aggregation, ranking, or weight-based techniques, and eventually choosing the most appropriate option among the available alternatives [48]. As previously demonstrated by [4], the MOO approach to this study was carried out in phases through the development of a decision-making chart. The decision-making chart was designed (as illustrated in Appendix A-Figure A1) as a working template for integrating and aligning client objectives and project peculiarities to appropriate procurement options. The decision-making chart comprises two sections, with the data collection process in section one entirely based on clients' input and requirements, in line with the project objectives. Section two involves the review of information provided by the CCs and the subsequent alignment of clients' preferences to create a pattern that suggests a suitable procurement route. Input from construction professionals in guiding CCs toward the most appropriate procurement route is considered at this stage through a briefing process. According to [53], the briefing process integrates the fragmented construction variables by evaluating client needs, project specifications, and professional inputs, toward the realization of an optimum project outcome.

Table 3. Factors that influence respondents' procurement preferences based on their experiences.

PEPC	PESC	PrEPC	PrISC	Code	Initial Theme	Established Theme
✓	✓	✓	-	Experience	Lessons from previous projects	
✓	✓	✓	✓	Perception	Personal conviction or belief	Construction client's cognition
✓	✓	✓	✓	Sentiment	Institutionalized preference	
✓	✓	✓	-	Bias	Process skipping	
✓	✓	✓	✓	Professionals	Availability of relevant skills	
✓	✓	✓	✓	Collaboration	Team influence	Access to relevant information
✓	✓	✓	✓	Legislation	Prevailing rules and regulations	
✓	✓	✓	✓	Demographics	Market or end-user projection	
✓	✓	✓	✓	Technology	Efficiency and adaptability	
✓	✓	✓	✓	Flexibility	Adjusting to market demand	Dynamic environment
✓	✓	✓	✓	Location	Project environment	
✓	✓	✓	✓	Disruptions	Uncertain events	

The decision-making chart drives the briefing process and provides a premise for actualizing the aim of this study, which is the development of a framework that is capable of guiding clients in making informed procurement choices. Based on the technique adopted from [17], the chart was used in evaluating clients' responses to questions pertaining to specific project factors identified in Table 1. Whereas the decision-making chart advances the previous works of [11,12] by collating data relating to basic project objectives, the project samples and factors highlighted in this study are illustrative, not exhaustive. These factors could be updated based on project complexity.

According to [21], the specifications of CCs could vary across project criteria and expectations. For instance, if the value for money spent (i.e., quality) is the crucial consideration for a particular project, CCs would rate the procurement criterion "quality" higher than the other criteria like timeliness and cost. Consequently, this study collated objective responses to structured questions from the research participants. The structured questions were asked across the various categorization of clients and different project types, as shown in the first and second rows of Appendix A-Figure A1, respectively. CCs' responses to the questions raised in the first section of the table were subsequently evaluated and coded. As suggested by [4], detailed and logical rules were set to analyse and code clients' influence on the various project objectives. This includes using numeric weighting techniques, as previously demonstrated by [9] and [49]. Weight was, therefore, assigned to each of the factors that influence the client's goals by using a numerical scale ranging from 0 to 100. Based on client's expectation and project peculiarity, CCs are expected to assign utility scores to each question in section 1 of Appendix A-Figure A1. In this study, utility scores are described as the values attached to the significance of project parameters by decision-makers. CCs' answers to questions on procurement factors were eventually coded, depending on how important the factors are to them. A response of 50 and below translates to "NO", while a response of 51 and above means "YES", as illustrated in Figure 1.

5. Findings

All the respondents except the PrISC acknowledged that they had procured several projects using various procurement strategies. Reflecting on their previous projects and their understanding of the procurement process, they emphasized that their experience is the major factor that influenced their procurement choices irrespective of the logical justification of the alternative procurement routes. The outcome of the qualitative analysis, which rationalizes the gap in this study, suggests that the relevance of CCs' experiences in making procurement decisions for different project types is influenced by their cognition, access to relevant information, and the dynamism of the built environment as clarified below.

5.1. Construction Client's Cognition

According to the respondents, individual and organizational perception, sentiment, or bias towards a specific procurement choice is responsible for most of their previous procurement decisions. Having operated in the construction industry for a long time, the participants argue that rather than concentrating on procurement choices, they focus on contractors' competence and adherence to due process. While some CCs argue that the consideration of some procurement routes often leads to a waste of valuable time as they are not relevant within their project scope, others believe that there are several routes toward achieving the same outcome. CCs, therefore, believe that less emphasis should be made on justifying a procurement preference over another as the ultimate assessment of project success is highly subjective. Referencing the construction of retail centers across the UK, R7 noted that:

"Design and build should not be considered a procurement option, in my opinion."

According to him, project complexity and stakeholder expectation make it unrealistic to entrust the credibility of such projects to an entity. Based on his experience, he suggested management contracting as a more appropriate procurement option that encourages collaborative inputs in the delivery of quality retail centers that are useful for their intended purpose.

R4 also stated thus:

"Everything comes down to value for money. Whichever procurement option that offers that is appropriate."

He noted that CCs are interested in the project outcome and how they can achieve the best result with the available resources. They, therefore, skip processes in evaluating procurement preferences based on their past experiences, available resources, and project deliverables. As a result, CCs often deviate from the logical, stage-based procurement selection process and rely on their cognition in selecting procurement routes.

5.2. Access to Relevant Information

The participants also revealed that the availability and access to relevant information affect their ability to make procurement decisions. They noted that the multiplicity of data available in today's built environment makes it difficult for CCs to be objective in making procurement decisions. Arising from different project stakeholders (e.g., contractors, local council, end-users, etc.), respondents argued that evaluating the variety of data in a timely and efficient way is unrealistic in an ideal situation. Respondents, therefore, stated that they engage in mental shortcuts, while accessing information for procurement purposes. Relying on their experience, they noted that negotiations and consistent stakeholder engagement influence their procurement preference. According to R6:

"It is not practicable to consider all the known factors that influence project success. Over the years, we have learnt to focus on the crucial factors that emerge from our consistent deliberations with stakeholders when making procurement decisions."

To the respondents, relevant information can only be timely and not absolute. It is, therefore, not feasible for them to follow a logical process when making procurement decisions. Their ability to reflect on previous experiences and anticipate possible challenges makes it easier for them to focus on pertinent information regarding project deliverables as they distinguish feasible procurement routes for different project types. According to R1:

> *"Accessing the right information is key. Various factors inform procurement choices, but a typical client will focus more on end-user satisfaction and project flexibility. Both of which cannot be measured objectively."*

5.3. Dynamic Environment

The respondents noted that the management of construction projects is a very dynamic and unpredictable practice that requires a value chain of activities across various sectors. As a result, it is often not realistic to make conclusive procurement decisions from the outset. Rather, clients' decision-making is premised on emerging project demands, as informed by current reality and inputs from other stakeholders. According to R14:

> *"Deciding on a procurement route is not a rigid process; it emerges with current reality."*

With technology, climate change, demographics and legislations constantly disrupting the procurement process, clients noted that the peculiarity of their immediate environment, their ability to adapt to possible changes, and the extent of competition in the delivery of similar projects are critical in the selection of procurement routes. In other to ensure that effective procurement choices are made amidst uncertainties, some respondents stated that they encourage collaborative practices with other stakeholders (e.g., construction professionals, contractors, suppliers, etc.) in the form of a special purpose vehicle (SPV), which often result in the adoption of a specific procurement route across various projects. Their experience in delivering previous projects, therefore, impacts their procurement preference, with minimal emphasis on logical assumptions. While narrating his experience, R9 opined that:

> *" … our organization works with a dedicated team of professionals with a track record of successful project delivery. Despite the variation in project complexity, our preference for management contracting is borne out of the success we have recorded in previous projects."*

Complementing the outcome of the qualitative study, respondents were assigned the task of selecting suitable procurement routes for different projects by following the logical, stage-based procedure highlighted in Appendix A-Figure A1. For clarification, using a PEPC as an example, responses from the client concerning the procurement of a prison facility are illustrated in the second column of the decision-making chart. The PEPC assigned values of 65, 50, and 88 to indicate their consideration of basic project factors of time, cost, and quality, respectively. The PEPC also acknowledges the complexity involved in prison construction and is willing to pay for the inherent risk, necessary expertise, and the possibility of project variation. The value of 90 provided by the client at the tendering stage suggests that the PEPC is solely responsible for the choices made, and his reluctance to explore competitive bidding was demonstrated in his preference for 40 as the benchmark for competition. Consideration for dispute resolution was subsequently deemed important by the PEPC, with a value of 55 assigned accordingly. The allotted values by the PEPC were converted to "YES" or "NO" and subsequently used to establish a procurement route.

This decision-making chart compliments the subjective nature of CCs and, therefore, forms the basis for a robust decision-making framework for CCs, which is the focus of this study. As shown in Figure 1 below, sample responses of different clients to project objectives were linked to relevant procurement options. For instance, the example illustrated above aligns with a management-oriented procurement option.

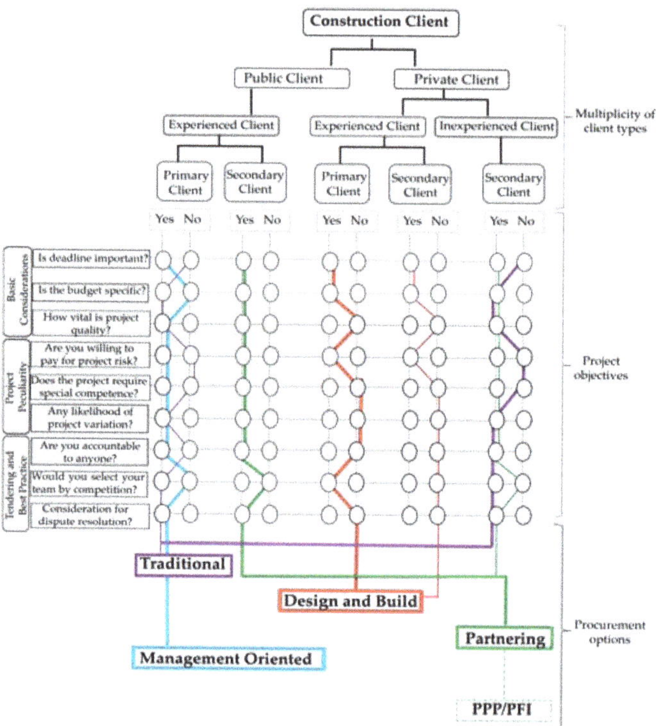

Figure 1. Decision-Making Framework for Construction Clients.

6. Discussion

A significant number of CCs rely on their experiences in selecting procurement methods. This is because the multi-dimensional relationship between clients' expectations, project specifications, project deliverables, and procurement routes often complicate the procurement selection process. Although the use of firm, process-based procurement practices have worked in other industries [14], the possible variation in project scope makes it unrealistic to adopt a rigid approach in the construction sector [27]. Scholars have, therefore, consistently emphasized the need to simplify the construction procurement process by aligning clients' subjective attributes to project deliverables and complexities [7,32], thus, providing a mechanism for efficient decisions to be made.

Whereas existing studies have attempted to model construction procurement by focusing primarily on client classification [12,18,42], project type [4], and methodological contributions [48], this study offers a more robust and practical framework that is universally applicable to all client categories and project types, irrespective of the complexity involved. The adopted data collection process also advances the hypothetical techniques previously explored by [17] by making use of qualitative interviews that complement the practicality of the research outcome.

As shown in Figure 1, the output of this study, which is the decision-making framework, is comprised of three distinct sections:

- Multiplicity of client types
- Project objectives
- Procurement options

Following the client classification by [21], the subjective nature of CCs, resulting from their cognitive ability, access to relevant information, and the dynamic nature of the built environment was explored in the prioritization of various project deliverables. The objective responses of CCs to structured questions regarding the project factors highlighted in Table 1 and outlined in the decision-making chart were also collated and codded accordingly, as part of the briefing process. The information provided on the decision-making chart was then aligned across the project objectives to arrive at suitable procurement routes for specified projects. Depending on the CCs' response to the basic project objectives of cost, quality, and time, "traditional" or "design and build" procurement options can be recommended for simple projects. However, for specialized projects that require unique expertise, collaborative practices, variation, etc., a more in-depth consideration of project aim and professional advice is essential.

Although clients' responses to the project brief suggest a procurement option, the decision-making chart also acknowledges the significance of professional advice in exploring the variants of the main procurement options. According to [33], expert advice in exploring the optimality of procurement options is vital for overall project success. This is particularly true for inexperienced clients, undertaking a complex construction project for the first time. The decision-making chart, therefore, accommodates informed professional expertise and advice on complex procurement issues relating to contracts, tendering, collaboration, and dispute management. For instance, to encourage collaboration and drive value for money, partnering could be recommended for large government projects. Experts' inputs in guiding CCs also serve as a medium for encouraging best practices across various procurement options. Advancing the view of [17] in their study of modern selection criteria for procurement methods in construction, this paper has been able to leverage CCs' subjective viewpoint in developing a decision-making framework that offers feasible procurement routes for different project types. The study outcome is also useful in comparing procurement preferences across various categories of CCs.

7. Conclusions

Scholars agree that the process of selecting construction procurement routes is not straightforward. Rather, it varies with project complexity, client type, and access to requisite information that drives project objectives. Understanding that the ability of CCs to make appropriate investment decisions is a critical factor that determines project success, the decision-making framework developed in this paper is capable of guiding CCs toward making informed decisions. This paper explored the literature to review the various categorizations of CCs, factors influencing the selection of procurement options, and some of the procurement choices available to CCs. It attests to existing arguments that there is a possibility of having more than one procurement route that will match specific client requirements.

While scholars have attempted to simplify the decision-making process for CCs, the practicality of the existing techniques have been challenged due to the dynamic nature of client objectives and the complexity of modern construction projects. Unlike previous scholarly contributions that are premised solely on clients' objectivity, this study acknowledges that clients' objectives are not static, and the reality of today's environment requires a flexible approach to CCs' decision-making. Therefore, adopting a mixed-method approach involving qualitative interviews and the MOO technique, this study leveraged CCs' experiences in selecting procurement routes through the value they attach to different project factors. Following the multiplicity of data gathered and evaluated in this study, this paper has achieved its aim of developing a framework (as shown in Figure 1) that provides different alternative routes to CCs, based on their experience and responses to the decision-making chart, used in rationalizing the significance of various procurement factors to specific project types.

The proposed framework applies to different real-life projects irrespective of project complexity and client categorization. It offers CCs a practical opportunity to be involved in the procurement process through a more in-depth approach that drives increased project success through effective collaboration and sustainable practices. Thus, enhancing and deepening their understanding of different procurement routes and their consequent contribution to project outcome. The framework also contributes to the ongoing debate on simplifying the construction procurement process by offering a platform for construction professionals and academics to drive innovation and best practices as a way of ensuring value for money in construction procurement.

Although this study offers a practical framework that is capable of guiding clients in selecting appropriate procurement options for different project types, the significance of the framework was not explored beyond the procurement context. The study scope is also limited to CCs in the UK and the decision-making framework itself will benefit from wider evaluation and validation across various case scenarios and different stages of project lifecycle. The effect of social, environmental, technological, and economic disruptions on the client's objectives and procurement path were also not covered in this research. Future studies should, therefore, consider the applicability of this framework across specific projects as a measure of the viability of project outcomes when compared to purely objective models. Researchers are also encouraged to investigate the effect of disruptions on CCs' procurement choices.

Author Contributions: Conceptualization, M.B.; methodology, M.B.; software, M.B.; validation, M.B., S.A. and A.-R.A.; formal analysis, M.B. and A.J.; data curation, M.B., A.-R.A., S.A. and A.J.; writing—original draft preparation, M.B.; writing—review and editing, M.B., A.-R.A., S.A. and A.J.; supervision, A.-R.A.; project administration, M.B. and S.A. All authors have read and agreed to the published version of the manuscript.

Funding: This research received no external funding.

Conflicts of Interest: The authors declare no conflict of interest.

Appendix A

Decision-making Chart

DECISION-MAKING CHART FOR BRIEFING CONSTRUCTION CLIENT ON THE APPROPRIATENESS OF AVAILABLE PROCUREMENT OPTIONS

INTERVIEW GUIDE

CLIENT TYPE	PEPC	PESC	PrEPC	PrISC
PROJECT SAMPLE	PRISON	LOCAL HOSPITAL	TEMPORARY HOUSING	OFFICE BLOCK
SECTION 1				
STAGE A: BASIC OBJECTIVES				
Is deadline important?	65	80	70	45
Is there a specific budget?	50	75	75	87
Do you require quality design and construction?	88	70	50	80
STAGE B: PROJECT PECULIARITY				
Are you willing to pay for the inherent project risk?	85	60	55	45
Does the project require any special competence?	88	60	44	50
Is there a likelihood of project variation?	82	55	40	55
STAGE C: TENDERING AND BEST PRACTICE				
Are you accountable to anyone other than yourself?	90	90	48	90
Would you choose your team by competition?	40	50	91	60
Consideration for dispute resolution?	55	78	50	75
SECTION 2				
OBSERVATION	Client's response to the brief is evaluated and linked to a procurement option			
ADVICE	Issues relating to project complexity, contract and dispute resolution are explored			
REVIEW	Best practices and modern procurement options are evaluated			
RECOMMENDATION	Most suitable procurement option is recommended			

Note: The overall score for each question is 100 and a response of 50 and below translates to "NO" while a response above 50 means "YES"

Key: PEPC (Public Experienced Primary Client)
PESC (Public Experienced Secondary Client)
PrEPC (Private Experienced Primary Client)
PrISC (Private Inexperienced Secondary Client)

Figure A1. Decision-Making Chart.

Appendix B

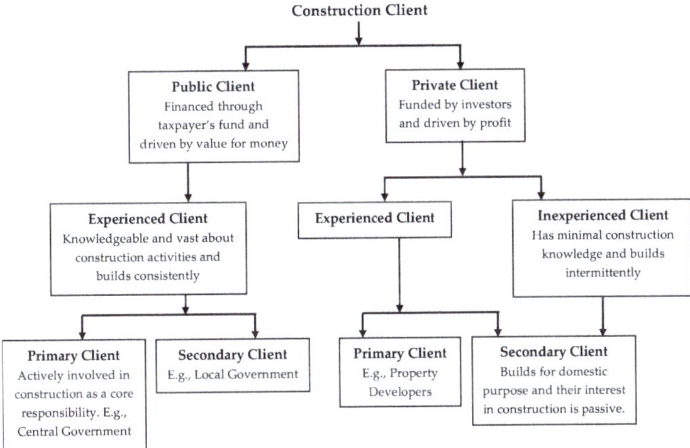

Figure A2. Client Categorization Adapted from [21].

References

1. Rajeh, M.A.; Tookey, J.; Rotimi, J. Procurement selection model: Development of a conceptual model based on transaction costs. *Australas. J. Constr. Econ. Build.-Conf. Ser.* **2014**, *2*, 56–63. [CrossRef]
2. Cheung, S.O.; Lam, T.I.; Leung, M.Y.; Wan, Y.W. An analytical hierarchy process-based procurement selection method. *Constr. Manag. Econ.* **2001**, *19*, 427–437. [CrossRef]
3. Bolomope, M.T.; Awuah, K.G.B.; Amidu, A.R.; Filippova, O. The challenges of access to local finance for PPP infrastructure project delivery in Nigeria. *J. Financ. Manag. Prop. Constr.* **2020**, *1*, 63–86. [CrossRef]
4. Li, T.H.; Thomas Ng, S.; Skitmore, M. Modeling multi-stakeholder multi-objective decisions during public participation in major infrastructure and construction projects: A decision rule approach. *J. Constr. Eng. Manag.* **2016**, *142*, 04015087. [CrossRef]
5. Tookey, J.E.; Murray, M.; Hardcastle, C.; Langford, D. Construction procurement routes: Re-defining the contours of construction procurement. *Eng. Constr. Archit. Manag.* **2001**, *8*, 20–30.
6. Asiedu, R.O.; Manu, P.; Mahamadu, A.M.; Booth, C.A.; Olomolaiye, P.; Agyekum, K.; Abadi, M. Critical skills for infrastructure procurement: Insights from developing country contexts. *J. Eng. Des. Technol.* **2021**. [CrossRef]
7. Manley, K.; Chen, L. The impact of client characteristics on the time and cost performance of collaborative infrastructure projects. *Eng. Constr. Archit. Manag.* **2016**, *23*, 511–532. [CrossRef]
8. Tang, Z.W.; Ng, S.T.; Skitmore, M. Influence of procurement systems to the success of sustainable buildings. *J. Clean. Prod.* **2019**, *218*, 1007–1030. [CrossRef]
9. Cheung, S.O.; Lam, T.I.; Wan, Y.W.; Lam, K.C. Improving objectivity in procurement selection. *J. Manag. Eng.* **2001**, *17*, 132–139. [CrossRef]
10. Naoum, S.; Egbu, C. Critical review of procurement method research in construction journals. *Procedia Econ. Financ.* **2015**, *21*, 6–13. [CrossRef]
11. Skitmore, M.; Marsden, D. Which procurement system? Towards a universal procurement selection technique. *Constr. Manag. Econ.* **1988**, *6*, 71–89.
12. Gibb, A.G.; Isack, F. Client drivers for construction projects: Implications for standardization. *Eng. Constr. Archit. Manag.* **2001**, *8*, 46–58. [CrossRef]
13. Morledge, R.; Smith, A.; Kashiwagi, D.T. *Building Procurement*, 1st ed.; Blackwell Pub: Malden, MA, USA, 2006.
14. Havenvid, M.I.; Hulthén, K.; Linné, Å.; Sundquist, V. Renewal in construction projects: Tracing effects of client requirements. *Constr. Manag. Econ.* **2016**, *34*, 790–807. [CrossRef]
15. Durdyev, S.; Ihtiyar, A.; Banaitis, A.; Thurnell, D. The construction client satisfaction model: A PLS-SEM approach. *J. Civ. Eng. Manag.* **2018**, *24*, 31–42. [CrossRef]
16. Love, P.E.D.; Skitmore, M.; Earl, G. Selecting a suitable procurement method for a building project. *Constr. Manag. Econ.* **1998**, *16*, 221–233. [CrossRef]
17. Naoum, S.G.; Egbu, C. Modern selection criteria for procurement methods in construction: A state-of-the-art literature review and a survey. *Int. J. Manag. Proj. Bus.* **2016**, *9*, 309–336. [CrossRef]
18. Masterman, J.W.E.; Masterman, J. *An Introduction to Building Procurement Systems*; Routledge: London, UK, 2013.
19. Windapo, A.; Adediran, A.; Rotimi, J.O.B.; Umeokafor, N. Construction project performance: The role of client knowledge and procurement systems. *J. Eng. Des. Technol.* **2021**, *20*, 1349–1366. [CrossRef]

20. Perera, G.P.P.S.; Tennakoon, T.M.M.P.; Kulatunga, U.; Jayasena, H.S.; Wijewickrama, M.K.C.S. Selecting suitable procurement system for steel building construction. *Built Environ. Proj. Asset Manag.* **2020**, *11*, 611–626. [CrossRef]
21. Masterman, J.W.E. *An Introduction to Building Procurement Systems*, 2nd ed.; Spon Press: London, UK, 2002.
22. Ashworth, A.; Hogg, K.; Willis, C.J. *Willis's Practice and Procedure for the Quantity Surveyor*, 12th ed.; Blackwell: Oxford, UK, 2007.
23. Briscoe, G.H.; Dainty, A.R.; Millett, S.J.; Neale, R.H. Client-led strategies for construction supply chain improvement. *Constr. Manag. Econ.* **2004**, *22*, 193–201. [CrossRef]
24. Othman, A.A.E.; Youssef, L.Y.W. A framework for implementing integrated project delivery in architecture design firms in Egypt. *J. Eng. Des. Technol.* **2020**, *19*, 721–757. [CrossRef]
25. Latham, S.M. *Constructing the Team: Final Report*; H.M.S.O-1: London, UK, 1994.
26. Egan. *Rethinking Construction: The Report of the Construction Task Force*; DTI (URN 98/1095), Construction Task Force; Department of Trade and Industry (DTI): London, UK, 1998.
27. Kumaraswamy, M.M.; Dissanayaka, S.M. Developing a decision support system for building project procurement. *Build. Environ.* **2001**, *36*, 337–349. [CrossRef]
28. Ruparathna, R.; Hewage, K. Review of contemporary construction procurement practices. *J. Manag. Eng.* **2015**, *31*, 04014038. [CrossRef]
29. Mohsini, R.; Davidson, C.H. Building procurement—Key to improved performance: Owner's procurement decisions have very real effect on the performance of the design team as it carries out various stages of building design and construction process. *Build. Res. Inf.* **1991**, *19*, 106–113. [CrossRef]
30. Luu, D.T.; Ng, S.T.; Chen, S.E. Formulating procurement selection criteria through case-based reasoning approach. *J. Comput. Civ. Eng.* **2005**, *19*, 269–276. [CrossRef]
31. Rahmani, F.; Maqsood, T.; Khalfan, M. An overview of construction procurement methods in Australia. *Eng. Constr. Archit. Manag.* **2017**, *24*, 593–609. [CrossRef]
32. Love, P.E.D.; Irani, Z.; Sharif, A. Participatory action research approach to public sector procurement selection. *J. Constr. Eng. Manag.* **2012**, *138*, 311–322. [CrossRef]
33. Yap, J.B.H.; Lim, B.L.; Skitmore, M.; Gray, J. Criticality of project knowledge and experience in the delivery of construction projects. *J. Eng. Des. Technol.* **2021**, *20*, 800–822. [CrossRef]
34. Greenhalgh, B.; Squires, G. *Introduction to Building Procurement*; Routledge: London, UK, 2011.
35. NEC. *NEC3 Procurement and Contract Strategies*; Thomas Telford: London, UK, 2009.
36. Morris, P.; Pinto, J.K. (Eds.) *The Wiley Guide to Project Technology, Supply Chain, and Procurement Management (Vol. 7)*; John Wiley and Sons: Hoboken, NJ, USA, 2010.
37. Franks, J. *Building Procurement Systems*; Chartered Institute of Building: Ascot, UK, 1990.
38. Alhazmi, T.; McCaffer, R. Project procurement system selection model. *J. Constr. Eng. Manag.* **2000**, *126*, 176–184. [CrossRef]
39. Chege, L.W.; Rwelamila, P.D. Private financing of construction projects and procurement systems: An integrated approach. In Proceedings of the CIB World Building Congress, Wellington, New Zealand, 4–6 April 2001.
40. Ng, T.; Luu, D.; Chen, S. Decision criteria and their subjectivity in construction procurement selection. *Constr. Econ. Build.* **2002**, *2*, 70–80. [CrossRef]
41. Molenaar, K.R. Selecting appropriate projects for design-build procurement. In *Profitable Partnering in Construction Procurement*; Routledge: London, UK, 2003; pp. 219–225.
42. National Economic Development Office N.E.D.O. *Thinking about Building*; H.M.S.O: London, UK, 1985.
43. Rowlinson, S. Selection criteria. In *Procurement Systems: A Guide to Best Practice in Construction*; Routledge: London, UK, 1999; pp. 276–299.
44. Walker, A. *Project Management in Construction*; John Wiley & Sons: Hoboken, NJ, USA, 2015.
45. Creswell, J.W.; Poth, C.N. *Qualitative Inquiry and Research Design: Choosing among Five Approaches*; Sage Publications: Thousand Oaks, CA, USA, 2017.
46. Asadi, E.; Da Silva, M.G.; Antunes, C.H.; Dias, L. Multi-objective optimization for building retrofit strategies: A model and an application. *Energy Build.* **2012**, *44*, 81–87. [CrossRef]
47. Muramatsu, M.; Kato, T. Selection guide of multi-objective optimization for ergonomic design. *J. Eng. Des. Technol.* **2019**, *17*, 2–24. [CrossRef]
48. Sariyildiz, I.S.; Bittermann, M.S.; Ciftcioglu, O. Multi-objective optimization in the construction industry. In Proceedings of the AEC 2008: 5th International Conference on Innovation in Architecture, Engineering and Construction, Antalya, Turkey, 23–25 June 2008.
49. Salama, T.; Moselhi, O. Multi-objective optimization for repetitive scheduling under uncertainty. *Eng. Constr. Archit. Manag.* **2019**, *26*, 1294–1320. [CrossRef]
50. Noble, H.; Smith, J. Qualitative data analysis: A practical example. *Evid.-Based Nurs.* **2014**, *17*, 2–3. [CrossRef] [PubMed]
51. Braun, V.; Clarke, V. Using thematic analysis in psychology. *Qual. Res. Psychol.* **2006**, *3*, 77–101. [CrossRef]
52. Feldman, G.; Shah, H.; Chapman, C.; Pärn, E.A.; Edwards, D.J. A systematic approach for enterprise systems upgrade decision-making: Outlining the decision processes. *J. Eng. Des. Technol.* **2017**, *15*, 778–802. [CrossRef]
53. Kamara, J.M.; Anumba, C.J.; Evbuomwan, N.F. Assessing the suitability of current briefing practices in construction within a concurrent engineering framework. *Int. J. Proj. Manag.* **2001**, *19*, 337–351. [CrossRef]

 *buildings*

Article

Development of a Classification Framework for Construction Personnel's Safety Behavior Based on Machine Learning

Shiyi Yin [1], Yaoping Wu [1], Yuzhong Shen [2,*] and Steve Rowlinson [3,4]

[1] College of Information, Mechanical and Electrical Engineering, Shanghai Normal University, Shanghai 201418, China
[2] College of Civil Engineering, Shanghai Normal University, Shanghai 201418, China
[3] Department of Real Estate and Construction, University of Hong Kong, Hong Kong, China
[4] School of Society and Design, Bond University, Robina, QLD 4226, Australia
* Correspondence: johnshen@shnu.edu.cn

Abstract: Different sets of drivers underlie different safety behaviors, and uncovering such complex patterns helps formulate targeted measures to cultivate safety behaviors. Machine learning can explore such complex patterns among safety behavioral data. This paper aims to develop a classification framework for construction personnel's safety behaviors with machine learning algorithms, including logistics regression (LR), support vector machine (SVM), random forest (RF), and categorical boosting (CatBoost). The classification framework has three steps, i.e., data collection and preprocessing, modeling and algorithm implementation, and optimal model acquisition. For illustrative purposes, five common safety behaviors of a random sample of Hong Kong-based construction personnel are used to validate the classification framework. To achieve high classification performance, this paper employed a combinative strategy, consisting of feature selection, synthetic minority oversampling technique (SMOTE), one-hot encoding, standard scaler and classifiers to classify safety behaviors, and multi-objective slime mould algorithm (MOSMA) to optimize parameters in the classifiers. Results suggest that the combinative strategy of CatBoost–MOSMA achieves the highest classification performance with the maximum average scores, including area under the curve of receiver characteristic operator (AUC) ranging from 0.84 to 0.92, accuracy ranging from 0.80 to 0.86, and F1-score ranging from 0.79 to 0.86. From the optimal model, a unique set of important features was identified for each safety behavior, and ten out of the 46 input indicators were found important for all five safety behaviors. Based on the findings, this study advocates using the machine learning strategy of CatBoost–MOSMA in future construction safety behavior research and makes concrete and targeted suggestions to cultivate different construction safety behaviors.

Keywords: classification; safety behavior; construction personnel; machine learning; MOSMA

Citation: Yin, S.; Wu, Y.; Shen, Y.; Rowlinson, S. Development of a Classification Framework for Construction Personnel's Safety Behavior Based on Machine Learning. *Buildings* **2023**, *13*, 43. https://doi.org/10.3390/buildings13010043

Academic Editor: Ahmed Senouci

Received: 24 November 2022
Revised: 19 December 2022
Accepted: 21 December 2022
Published: 24 December 2022

1. Introduction

Unsafe behaviors are the primary direct cause of construction accidents. Different types of accidents can be attributed to different sets of unsafe behaviors [1]. For example, to avoid falls from height the management should take care of unprotected holes/borders and correct workers' inappropriate use of personal protective equipment (PPE). Safety behavior is traditionally categorized as either safety compliance or safety participation. The former is an in-role task-related behavior, while the latter involves extra-role behaviors, which are voluntary and initiated by employees [2]. Griffin and Curcuruto further identify two categories of safety participation behavior: affiliative and proactive [3]. Helping and stewardship behaviors, civic virtue, and caring for safety are typical of affiliative safety participation behavior, whereas proactive safety participation behavior includes safety voice behaviors and initiating safety-related changes. Affiliative safety participation behavior is related to minor incidents, such as property damage and microinjuries, while proactive safety participation behavior is positively associated with near-miss reporting. Therefore, it can be hypothesized that different sets of drivers are accountable for different (un)safety behaviors. This paper attempts to validate this hypothesis with a machine-learning-enabled classification framework.

Besides the theoretical significance, this paper also has both a practical and a methodological significance as well. On the practical front, if different patterns of drivers for different safety behaviors are ascertained, targeted interventions can be proposed accordingly. Specifically, this paper selects five typical safety behaviors, i.e., the use of all necessary safety equipment to do the job (hereafter coded as SB1); following safety procedures in doing the job (hereafter coded as SB2); promoting safety programs willingly (hereafter coded as SB3); put in extra effort to improve workplace safety (hereafter coded as SB4); and help colleagues out when they are under risky conditions (hereafter coded as SB5). On the methodological front, as a subset of artificial intelligence, machine learning enables a system to learn from example data or past experience without explicit programming. Like traditional statistical modelling, it is also intended to seek solutions from data. Unlike traditional methods that are based on assumptions and ignore the nonlinear relationship among independent variables, machine learning methods are more flexible, have fundamental and simple assumptions, and take into consideration the complex relationship among independent variables. Machine learning has seen an increasing use by safety researchers in recent years. Construction workers' risk perceptions have a direct impact on their safety behavior. The traditional measurement of risk perceptions primarily relies on a post hoc survey-based assessment, which has limitations such as lack of objectivity and continuous monitoring ability. Given this, Lee et al. developed an automatic system to measure workers' risk perception using physiological signals obtained by wristband-type wearable biosensors in combination with a supervised learning algorithm [4]. Overexertion-induced work-related musculoskeletal disorders (WMSDs) are a primary cause of the nonfatal injuries for construction workers. To reduce overexertion, appropriate levels of physical loads need to be identified. In this regard, Yang et al. propose to employ a bidirectional long short-term memory algorithm to classify physical load levels, and investigate the feasibility of such an approach with a laboratory experiment [5]. In view of machine learning's advantage in predictive accuracy, Goh et al. use six supervised learning algorithms (i.e., support vector machine, random forest, K-nearest neighbor, naïve Bayes, artificial neural network, and decision tree) to assess the relative importance of different cognitive factors derived from the theory of reasoned action in affecting safety behavior [6].

Given the theoretical, practical, and methodological significance, a machine-learning-enabled safety behavior classification framework should be developed in order to improve construction safety performance in an efficient and effective way. In particular, this paper has two objectives, namely: (a) To identify drivers of different safety behaviors; (b) To propose new machine learning methods in predicting safety behaviors. The former intends to make targeted interventions for different safety behaviors based on the findings and the latter to explore new algorithms which are more suitable for analyzing safety-related behavioral data.

This paper is organized as follows. First, a safety behavior factor analysis and classification system is developed based on the literature review. Second, the sample, measures, machine learning models, and classification outputs are described. Third, results are presented, with an emphasis on model performance and factor importance analysis. Finally, both the contribution and limitations of the findings are discussed along with future research directions.

2. Safety Behavior Factor Analysis and Classification System

Safety behavior is an emergent property of a more complex system. Choi and Lee find that construction workers' safety behavior is a function of their socio-cognitive process and their interaction with the environment [7]. Based on bibliometric and content analyses of 101 empirical studies, Xia et al. propose a safety behavior antecedent analysis and classification system [8], which organizes the antecedents of safety behavior into five levels: (a) Self; (b) Work; (c) Home; (d) Work–home interface; (e) Industry/society. In addition, they put forward a resource flow model to explain how safety behavior emerges from such a complex system. Using Xia et al. 's framework [8], this study organizes influencing factors of construction safety behavior at four levels, i.e., client, project, group, and individual, and hence, develops a safety behavior factor analysis and classification system as well. The next section deliberates on the impact of these factors on safety behavior before presenting the system.

2.1. Client Level Factors

Among stakeholders in the construction supply chain, clients have the economic power to encourage other stakeholders to implement safety measures. Therefore, clients play a pivotal role in improving safety performance across construction projects. Specifically, client type and the extent of client involvement in safety management have implications for safety performance [9,10].

2.1.1. Client Type

Construction project clients can be categorized as either public or private according to their source of funding. Ma observes that safety records for the projects with public sector clients are better than those projects with private sector clients in Hong Kong [11], and believes that the reason is that most safety initiatives are mandatorily executed in public works' contracts, whereas they are voluntarily adopted in the private sector. In Nigeria, Umeokafor also notes that public clients' safety commitment and attitudes are better than their counterparts [12]. So, it is hypothesized that there are more safety behaviors in public projects than in private projects.

2.1.2. Client Involvement

Clients' direct involvement in safety management contributes to safety performance. In Australia, given the important contribution that clients can make to the safety performance of the construction projects, Lingard et al. develop a model client framework [13]. The framework establishes clients' safety roles throughout the life-cycle of the project. Using safety climate as a leading indicator of safety performance of small- and medium-sized construction projects, Votano and Sunindijo found that six of the clients' safety roles depicted by Lingard et al. are related to safety performance, and they are participation in the safety program, review and analysis of safety data, appointment of safety team, selection of safe contractors, safety specifications in tenders, and regular checks on plant/equipment [13,14]. Hence, this research postulates that client safety involvement is positively associated with workers' safety behaviors.

2.2. Project Level Factors

Safety management system at the project level has implications for workers' safety behavior. In order to curb unsafe acts, Shin et al. suggest that project management should offer a safety incentive as early as possible and facilitate effective communication about accidents in as much detail as possible [15]. Fang et al. propose a leadership–culture–behavior (LCB) approach, which maintains that leadership creates a safety culture, and hence, promotes safety behavior [16]. The LCB approach has been implemented in railway and residential projects in mainland China and Hong Kong, and has seen success. Among others, this paper focuses on the following project level factors: stage of project, contract sum, goal congruency, participative decision-making, professional development, organizational support, standardized safety rules and procedures, and safety climate.

2.2.1. Project Information

At least two project characteristics, namely, stage of project and contract sum, have bearing on construction project employees' safety behavior. Based on the percentage of construction works that has been completed, a project can be categorized into three stages, namely, start-up, advanced, and near close-out. At the start-up stage, the construction work has been completed by less than 30%. At the advanced stage, the construction work has been completed by 30–70%. At the near close-out stage, the construction work that has been completed is more than 70%. Employees usually exhibit more safety behaviors at the start-up and near close-out stages than at the advanced stage. This is because at the start-up stage, employees are new to the site, and act scrupulously. As time passes and production pressure increases, employees are more likely to take shortcuts and more unsafe behaviors ensue. When the project is being completed, as employees are more familiar with the site and some of their unsafe behaviors have been rectified, their safety behavior increase. Awolusi and Marks develop a safety activity analysis framework and tool, and validate the framework and tool using a case study project that is in the construction stage [17]. Over an eight-month period of the case study project, the occurrence rate of safety behavior experiences a U-shaped curve, initially decreasing from 45.7% to 37.0% and then increasing to 62.8%.

Contract sum is also related to employees' safety performance. Generally, in jurisdictions where mandatory safety incentive scheme is applied, projects with large a contract sum usually set aside more money on safety measures, and therefore, more safety behaviors result. Take Hong Kong as an example, due to the introduction of safety initiatives, such as the Pay for Safety Scheme (PFSS), the Safety Management System (SMS),the Independent Safety Auditing Scheme (ISAS), and the Site Supervision Plan System (SSPS), the construction industry has seen a dramatic decrease in accidents [18]. Hence, this paper hypothesizes that a large contract sum contributes to more safety behaviors.

2.2.2. Goal Congruency

Goal congruency has an impact on organizational behavior. Goal congruency is a scenario where employees at different levels of an organization share the same goal. When employees' personal goals are consistent with organizational goals, they feel more positive about the organization and expend more personal efforts to achieve those goals. Ukraine-based IT professionals De Clercq et al. found that goal congruence between employees and their supervisor negatively affects employees' organizational deviance, and the indirect effect of goal congruence on organizational deviance through work engagement is moderated by employees' emotional intelligence [19]. With 171 employees under the leadership of 24 supervisors, Bouckenooghe et al. found that supervisors' ethical leadership has a positive effect on followers' in-role job performance through the sequential mediation of goal congruence and psychological capital [20]. Hence, when project personnel, both the management and workers, take safety as the first priority, their safety behavior ensues.

2.2.3. Participative Decision-Making

Participative decision-making is positively associated with safety behavior. Participative decision-making refers to the extent to which employers allow or encourage employees to take part in organizational decision-making. Through participation in decision-making, employees bring different perspectives and frames of references to safety discussions and activities, and hence, can reduce all members' ignorance to hazards and signals of danger [21]. As employees are aware that their suggestions have been incorporated in safety decisions, they are more likely to take ownership of those decisions and act on them more proactively. As a leadership behavior, participative decision-making is associated with safety participation [22]. In the medical industry, Lee et al. found that empowering leaders who empower employees to participate in decision-making enhance employees' safety compliance [23].

2.2.4. Professional Development

Employees are the most valuable resource in construction projects. Despite the time and resource pressures preventing project managers from investing in employees' professional development, it pays off. Design for safety has been advocated for quite a long time, and designers need to receive safety training as part of their professional development. Toole elaborates on the opportunities and barriers in increasing designers' role in construction safety [24]. In another scenario, if a semi-skilled bar bender is sponsored to receive more professional training, s/he may bring more best safety practices to the crew and promote more safety behaviors.

2.2.5. Organizational Support

Organizational support is critical in creating a safety climate and, hence, safety behavior. Organizational support refers to employees' global beliefs about the extent to which their organization satisfies their needs and cherishes their contributions. It can be general or specific. Mearns and Reader found that general perceived organizational support has an impact on the UK's offshore workers' safety performance [25]. With Ghanaian industrial workers, Gyekye and Salminen found that general perceived organizational support is positively associated with compliance with safety procedures [26]. Guo et al. discovered that perceived supervisory and coworker support for safety reduces the negative impact of job insecurity on Chinese high-railway drivers' safety performance [27]. Tucker et al. found that urban bus drivers' perceived organizational support for safety exerts influence on their safety voice behavior through the mediation of their perceived coworkers' support for safety [28], highlighting the role played by coworkers.

2.2.6. Standardized Safety Rules and Procedures

Standardization in construction projects is difficult to achieve. Other high-risk industries, such as aviation and nuclear, usually have well-defined work procedures. Since the construction process is characterized by high variety and loose coupling, most of the construction work, to a significant extent, depends on employees' discretion and experience. Standardized safety rules and procedures make those rules and procedures easy to follow, and hence, contribute to an increase in safety behavior. However, the secondary effect of too much standardization should be restrained [29].

2.2.7. Safety Climate

Safety climate is a perceptual, collective, and multidimensional phenomenon, referring to individuals' shared perceptions of how safety is valued in the workplace [3]. The impact of a safety climate on safety behavior has been well-documented. Safety climates can exert a direct influence on safety behavior, and also can impact safety behavior through mediators, such as the psychological contract [30], safety knowledge and motivation [31], etc.

2.3. Group Level Factors

Construction workers usually move from project to project and may work with different main contractors, but they often work in a workgroups for a relatively long period. Therefore, compared with supervisors from the main contractor, workgroup supervisors usually have a bigger influence on construction workers [32]. This paper focuses on four phenomena at the workgroup level, i.e., supervisors' transformational leadership and contingent reward behavior (one aspect of their transactional leadership), leader–member exchange, and team–member exchange.

2.3.1. Transformational Leadership

Leadership refers to a process of motivating others to act toward shared goals. It involves setting goals, devising achievement methods, persuading others to accept these goals and achievement methods, and solving problems decisively and quickly. James M. Burns proposes two leadership styles: transactional and transformational. The transactional leader identifies the needs of employees and the organization, and then informs employees what to do to meet these needs. Beyond these needs, transformational leaders arouse and satisfy higher needs within each individual. A transactional–transformational leadership paradigm is broad enough to capture the leadership construct.

Transformational leadership is positively associated with safety behavior. Shen et al. propose and validate a sequential mediation model to explain the impact of supervisory transformational leadership on construction personnel's safety behavior [10]. Hoffmeister et al. found that different facets of transformational leadership have a different impact on different sample's safety behavior [33]. In particular, idealized influence has an impact on safety compliance behavior in both the apprentice and journeyman samples, but it has an impact on safety participation behavior only in the apprentice sample.

2.3.2. Contingent Reward

Contingent reward is a facet of transactional leadership, and refers to the leader clarifying which employee behaviors are desired, what the rewards for such behaviors will be, and rewarding the followers depending on task fulfilment and outcome. Behaviorism maintains that behavior is a function of its consequences. Leaders engage in contingent reward with regard to safety when they help employees appreciate safety-related goals, keep them focused on meeting these goals, and reward them for engaging in safety behaviors required by those goals [33]. Therefore, contingent reward should be associated with increased employee safety behaviors.

2.3.3. Leader–Member Exchange

Leader–follower relationships are an essential part of leadership effectiveness, and leader–member exchange refers to the follower's perceptions of the quality of the exchange between leader and followers [34]. Leader–member exchange is positively associated with safety behavior [35–37].

2.3.4. Team–Member Exchange

Similar to leader–member exchange, team–member exchange refers to an individual's perception of the quality of the exchange relationships within the team. It is positively associated with safety behavior [38,39].

2.4. Individual Level Factors

Safety behavior is complex, and an individual may work safely in some occasions and unsafely in others [40]. Hence, some individual differences may contribute to an individual's safety behavior. This study focuses on construction personnel's personal demographics, habit, affiliation, and safety motivation.

2.4.1. Personal Demographics

Personal demographics, including age, gender, marital status, educational level, number of dependents to support, and industrial experience, may have an influence on safety behavior [41]. Meng and Chan found that female poorly educated workers exhibit less safety citizenship behavior [42]. The level of safety citizenship behavior has seen an initial downtrend followed by an uptrend as industrial experience increases.

2.4.2. Habit

Alcohol and tobacco use are more prevalent in blue collar workers than in white collar workers. There is a strong association between unsafe behavior (e.g., infrequently using sunscreen) and smoking and risky drinking [43].

2.4.3. Affiliation

At least two affiliation-related factors, namely, affiliation type and hierarchical position in affiliation, are related to construction personnel's safety behavior. Personnel affiliated with clients exhibit more safety behaviors than those with contractors and consultants.

Personnel in managerial positions exhibit more safety behaviors than supervisory staff, who in turn exhibit more safety behavior than workers.

2.4.4. Safety Motivation

Safety motivation refers to an individual's readiness to expend effort to engage in safety behaviors and the valence associated with these behaviors. It directs, energizes and sustains safety behavior [3]. Griffin and Curcuruto view safety motivation as an outcome of safety climate and a determinant of safety behavior based on theories and empirical evidence [3].

Based on the arguments made earlier, the safety behavior factor analysis and classification system is proposed and shown in Figure 1.

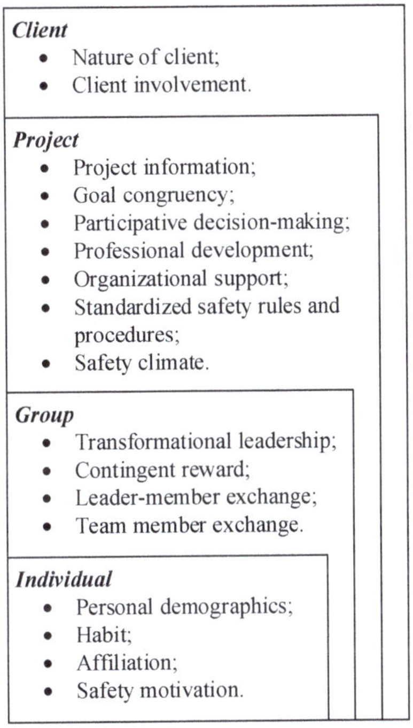

Client
- Nature of client;
- Client involvement.

Project
- Project information;
- Goal congruency;
- Participative decision-making;
- Professional development;
- Organizational support;
- Standardized safety rules and procedures;
- Safety climate.

Group
- Transformational leadership;
- Contingent reward;
- Leader-member exchange;
- Team member exchange.

Individual
- Personal demographics;
- Habit;
- Affiliation;
- Safety motivation.

Figure 1. Safety behavior factor analysis and classification system.

3. Materials and Methods

This study proposes a safety behavior classification framework that combines statistical analysis methods and machine learning algorithms. As shown in Figure 2, the framework has three steps, i.e., data collection and preprocessing, modeling and algorithm implementation, and optimal model acquisition. The data is processed automatically by the proposed combinative strategies. The proposed methods are described in detail as follows.

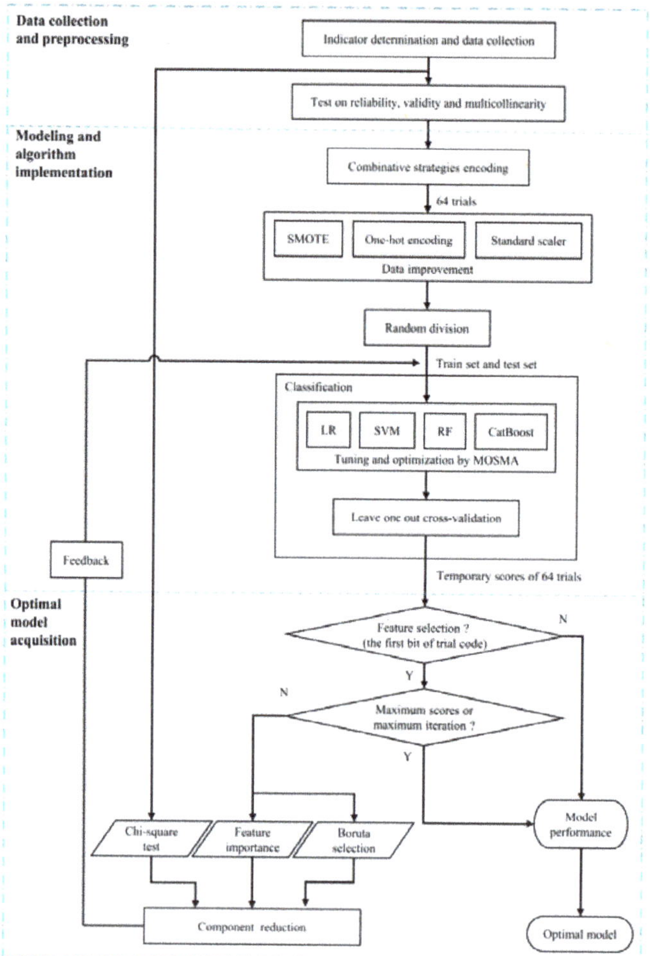

Figure 2. The safety behavior classification framework. First, users need to determine variables and indicators, and complete necessary preprocessing after data collection. Second, each trial has a unique code, and 64 models in total are trained and tuned automatically by specific methods mentioned in their codes. Last, the performance of the 64 models is output, and the model with maximum scores stands out as the optimal model. Meanwhile, users can also observe the results of feature selection to guide the analysis of the important factors of one risk behavior or the average important factors of certain risk behaviors.

3.1. Data Collection and Preprocessing

3.1.1. Data Collection

A questionnaire is used to collect data. The questionnaire has two parts. The first part is input variables, which have been shown in Figure 1. The second part is output variables, i.e., the five common safety behaviors. The sources of those indicators measuring these variables and the measures to ensure the questionnaire is self-contained and self-sufficient are recorded in Shen's work [41]. The sources of the indicators for each construct are also recorded in Shen's work [41]. The details of those input and output indicators are shown in Table A1 (Appendix A) and Table A2 (Appendix A), respectively.

The target population is Hong Kong construction personnel who are generally in three categories, i.e., contractor, consultant, and client. The contractor category includes management staff and direct laborers from main contractors and subcontractors. The consultant category covers engineers, architects, and quantity surveyors. The client category comprises both the public and private sectors. The target population size is unknown. The research team sets the confidence level at 90%, the margin of error at $\pm 5\%$, and the population proportion at 50%. Using Cochran's formula, the required sample size should be no less than 273. In order to secure sufficient, valid, and representative responses, the research team constructs a sampling frame consisting of construction personnel from local construction trade associations, professional bodies, governmental agencies, and property developers. Then, the research team sends hard-copy questionnaires to a random sample of 2996 construction personnel from the sampling frame. After two rounds of administration, the research team secures 292 valid responses. Non-response bias is not an issue [10].

3.1.2. Test on Reliability, Validity and Multicollinearity

For the purpose of highly reasonable and effective model training, data pre-processing is crucial in machine learning. As each record is collected by questionnaires, data need to pass both reliability and Bartlett's tests. In particular, the reliability of input second-level indicators (Cronbach's Alpha) is 0.82 above the threshold value of 0.7 [44]. Bartlett's test of those input second-level indicators is 0.81, indicating that feature selection can be done.

Additionally, one common issue in machine learning is that the large regression coefficients cannot be estimated precisely when the features are multicollinear. In accordance to Hair et al., variance inflation factor (VIF) is calculated to determine whether there is multicollinearity among independent variables [44]. In general, when the VIF values are lower than the common cutoff threshold of 10, multicollinearity is not a significant issue. The results of the multicollinearity test for all input second-level indicators are shown in Table 1, and it can be concluded that there is no multicollinearity among them.

Table 1. VIF values of input second-level indicators.

Variable *	VIF	Variable	VIF	Variable	VIF	Variable	VIF	Variable	VIF
NatClit	1.25	IndExpr	3.53	OS2	1.75	LMX2	2.21	CR1	2.37
StgProj	1.14	SmoHab	1.60	CI1	1.90	LMX3	2.47	CR2	2.47
ConSum	1.23	DriHab	1.34	CI2	2.00	LMX4	1.57	SC1	1.83
AffRes	1.41	GC1	1.61	CI3	1.90	TMX1	1.93	SC2	1.77
RespHier	1.73	GC2	1.83	CI4	2.07	TMX2	1.99	SC3	1.58
Gender	1.44	GC3	1.88	SSRP1	1.80	TMX3	1.92	SC4	1.64
Age	3.17	PDM1	2.00	SSRP2	1.80	TMX4	2.06	SM1	2.52
MarSts	1.53	PDM2	1.52	SSRP3	1.86	TL1	2.62	SM2	2.98
DeptRsp	1.32	PD	1.56	SSRP4	1.78	TL2	2.73	SM3	3.48
EduRsp	1.89	OS1	1.87	LMX1	2.21	TL3	1.75	SM4	2.81

* These codes refer to input second-level indicators, which are shown in Table A1 in Appendix A. For example, the code of 'GC1' refers to the first second-level indicator measuring the variable of goal congruency.

3.2. Modeling and Algorithm Implementation

3.2.1. Combinative Strategy Encoding and Data Improvement

In order to reach an optimal model, a combinative strategy, which contains five subprocesses, is proposed. The five subprocesses are feature selection, synthetic minority over-sampling technique (SMOTE), one-hot encoding, standard scaler, and classifiers. Feature selection is a process used to reduce the number of input variables in developing a classification model. This study simply divides the behaviors into Yes (high risk) and No (low risk), and this approach may result in an imbalanced distribution of each behavior. SMOTE is a proper method to address the imbalanced distribution issue [45]. The dataset contained nominal-categorical and ordinal-categorical features. One-hot encoding is used to create new binary features for each element in a categorical [46]. Moreover, all features are scaled at different intervals in the obtained dataset. By means of standard scaler, all features are converted, leading to a distribution with a mean value of 0 and a standard deviation of 1. Standard scaler helps limit the sample differences [46]. As a supervised learning concept, classification is a process of categorizing a set of data points into classes. In machine learning, a classifier is basically an algorithm that categorizes data into classes. This study used four classifiers, i.e., logistic regression (LR), support vector machines (SVM), random forest (RF), and categorical boosting (CatBoost).

This study tried 64 models, which are coded by the rules shown in Figure 3. The value of the first four bits is represented by the binary numbers 1 and 0, with 1 indicating *used* and 0 *unused*. The first part refers to feature selection, the second to SMOTE, the third to one-hot encoding, the fourth to standard scaler, and the last part is the first letter of the classifier's name. For example, a model code of "0101R" means that the model uses SMOTE, standard scaler and RF, and does not use feature selection and one-hot encoding.

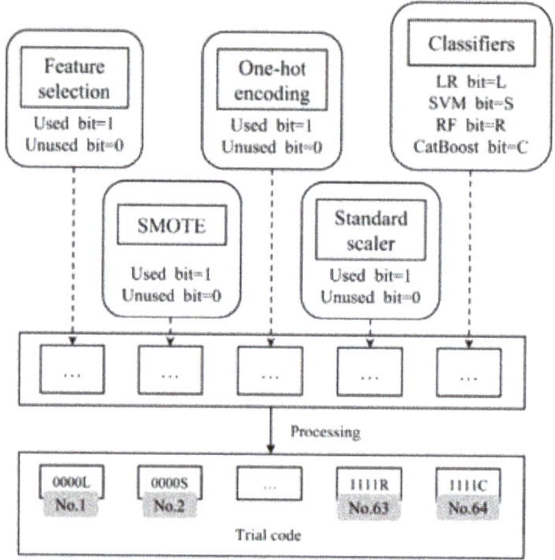

Figure 3. The process of encoding models.

3.2.2. Classification by Four Classifiers of Machine Learning

In terms of classification, there are many classic machine-learning algorithms, such as LR, SVM, etc. Recently, emerging algorithms are increasingly used, such as RF and CatBoost. In order to select a more suitable classifier, this study uses four classifiers, i.e., LR, SVM, RF and CatBoost.

Based on the natural logarithm, LR follows a logistic S-curve. Classification is determined by the probability of an outcome. SVM includes a set of related supervised learning methods to make prediction and regression. The statistical learning theory and structural risk minimization underlie the learning algorithms of SVM. According to Antwi-Afari et al. [47], SVM shows comparable or even better results than other machine-learning methods. RF is an ensemble of decision trees. It employs a bagging method to achieve classification. Each node is split using the best predictor from a subset of predictors chosen randomly at that node. As it is more robust in terms of generalizability than the decision trees, RF plays an important role in machine learning, such as the works of Niu et al. and Poh et al. [45,48]. Recently, decision trees have been extended to the family of gradient boosting algorithms, such as eXtreme Gradient Boosting (XGBoost), Light Gradient Boosting Machine (LightGBM), and Categorical Boosting (CatBoost). In particular, CatBoost is a framework based on oblivious trees. It has few parameters, supports categorical variables, and deals with categorical features in an efficient and reasonable manner. Furthermore, it modifies gradient computation to avoid a prediction shift in order to improve model accuracy. The results of a three-algorithm comparison show that CatBoost achieves the best results [49] despite the small differences among them.

3.2.3. Model Tuning and Hyperparameter Optimization by MOSMA and LOO

In some cases, over-fitting the data is an issue during the machine-learning process, resulting in poor generalizability. One of the most acceptable resolutions is to tune models and optimize parameters. This study uses an algorithm named slime mould algorithm (SMA) to tune the classifiers automatically. SMA is inspired by the behavior of slime mould, and has been applied in graph theory and path networks [50,51]. Since five behaviors are modeled in this study, a multi-objective SMA (MOSMA) is used to search the maximum average scores for these five behaviors. According to Houssein et al. [52], the MOSMA consumes significantly less training time than traditional optimization algorithms such as grid-search. Moreover, leave one out (LOO) cross-validation is fitting for those cases with a small sample size. For n samples, the number of training samples is n-1, while only one sample is left out for validation. This train-validation process is repeated for n times, and fully utilizes the dataset of the training dataset. Since there is no random sampling, bias is eliminated by LOO cross-validation [45]. Therefore, it is reasonable to combine LOO and MOSMA to find optimal settings in order to maximize the generalizability of the model.

3.2.4. Three Methods for Feature Selection

This study employs a combination of three traditional feature selection methods, i.e., feature importance (FI), Chi-square test (CT) and Boruta selection (BS).

When the variables in the dataset have varying degrees of influence on the five (un)safety behaviors, focusing on the most important features is critical for gaining a better understanding of them, respectively. To some extent, FI represents the diverse effects of various features. However, it does not entirely capture the association between the features and the safety behaviors, nor does it determine whether the feature has a positive or negative impact. In this regard, CT and odds ratio (OR) can make up for this deficiency, as they can not only calculate the correlation between features and safety behaviors, but also can reveal the nature of the impact (i.e., positive or negative). BS is a novel feature-selection algorithm for finding all relevant variables [53]. According to Poh et al. [45], BS has a critical advantage over ordinary feature-selection techniques in that it may pick the input variable in a robust and unbiased manner by using bagging schemes and including statistical confidence tests into its selection process.

Features are preserved in each iteration if more than half of the votes are in favor of passing. On the contrary, they are returned to the prediction part of modeling until the maximum score is achieved. For instance, Table 2 explains how to make selection decisions regarding three input indicators, i.e., NatClit, DeptRsp, and TMX1.

Table 2. Feature-selection method.

Variables	Methods	SB1	SB2	SB3	SB4	SB5	Votes	Result
NatClit	FI	√ *	√	√	√	√	9	Retain
	BS			√	√			
	CT		√		√			
DeptRsp	FI	√	√	√	√	√	10	Retain
	BS	√	√	√	√			
	CT					√		
TMX1	FI	√					2	Cut
	BS							
	CT					√		

* The variable obtains one vote if it is shown as an important feature for one behavior.

3.3. Optimal Model Acquisition

There are many indicators to evaluate the final training model's performance. For simplicity and efficiency, this study employs common indicators, including area under the curve of receiver characteristic operator (AUC), accuracy, precision, recall, and F1-score [48]. Accuracy, precision, recall, and F1-score are partial performance indicators, whereas AUC is a comprehensive indicator. They are defined by the following functions, which are based on the confusion matrix.

$$AUC = \frac{\sum I\left(P_{\text{positive}}, P_{\text{negative}}\right)}{M \times N} \tag{1}$$

where $I\left(P_{\text{positive}}, P_{\text{negative}}\right) = \begin{cases} 1, & P_{\text{positive}} > P_{\text{negative}} \\ 0.5, & P_{\text{positive}} = P_{\text{negative}} \\ 0, & P_{\text{positive}} < P_{\text{negative}} \end{cases}$, M and N are the numbers of positive and negative samples in the dataset, respectively.

$$Accuracy = \frac{TP + TN}{TP + TN + FP + FN} \tag{2}$$

$$Precision = \frac{TP}{TP + FP} \tag{3}$$

$$Recall = \frac{TP}{TP + FN} \tag{4}$$

$$F1 - score = \frac{2 \times Precision \times Recall}{Precision + Recall} \tag{5}$$

4. Results

4.1. Necessity of Tuning Models and Optimizing Parameters by MOSMA

Since the sample was randomly divided into training and test sets, it is necessary to limit the error of the model by tuning models and optimizing parameters. This study used the MOSMA method, which is rarely employed in the construction safety domain. Using the average of the outcomes of 10 random divisions as the final performance score, this study compared the performance of the MOSMA and the traditional grid-search method. Figure 4 shows the average AUC scores of the four classifiers for the five behaviors, and Figure 5 shows the average accuracy and F1-scores. From these two figures, it can be concluded that CatBoost–MOSMA has the maximum classification performance, and hence, is used for feature importance analysis later on.

	LR	SVM	RF	CatBoost
SB1	0.61	0.71	0.75	0.78
SB2	0.7	0.76	0.88	0.85
SB3	0.73	0.81	0.88	0.86
SB4	0.69	0.77	0.92	0.93
SB5	0.78	0.8	0.89	0.9
AVG	0.702	0.77	0.864	0.864

(a) Without MOSMA

	LR	SVM	RF	CatBoost
SB1	0.73	0.75	0.87	0.84
SB2	0.74	0.76	0.88	0.89
SB3	0.72	0.79	0.86	0.88
SB4	0.78	0.8	0.92	0.92
SB5	0.77	0.83	0.88	0.9
AVG	0.748	0.786	0.882	0.886

(b) With MOSMA

Figure 4. AUC of classifiers with(out) MOSMA.

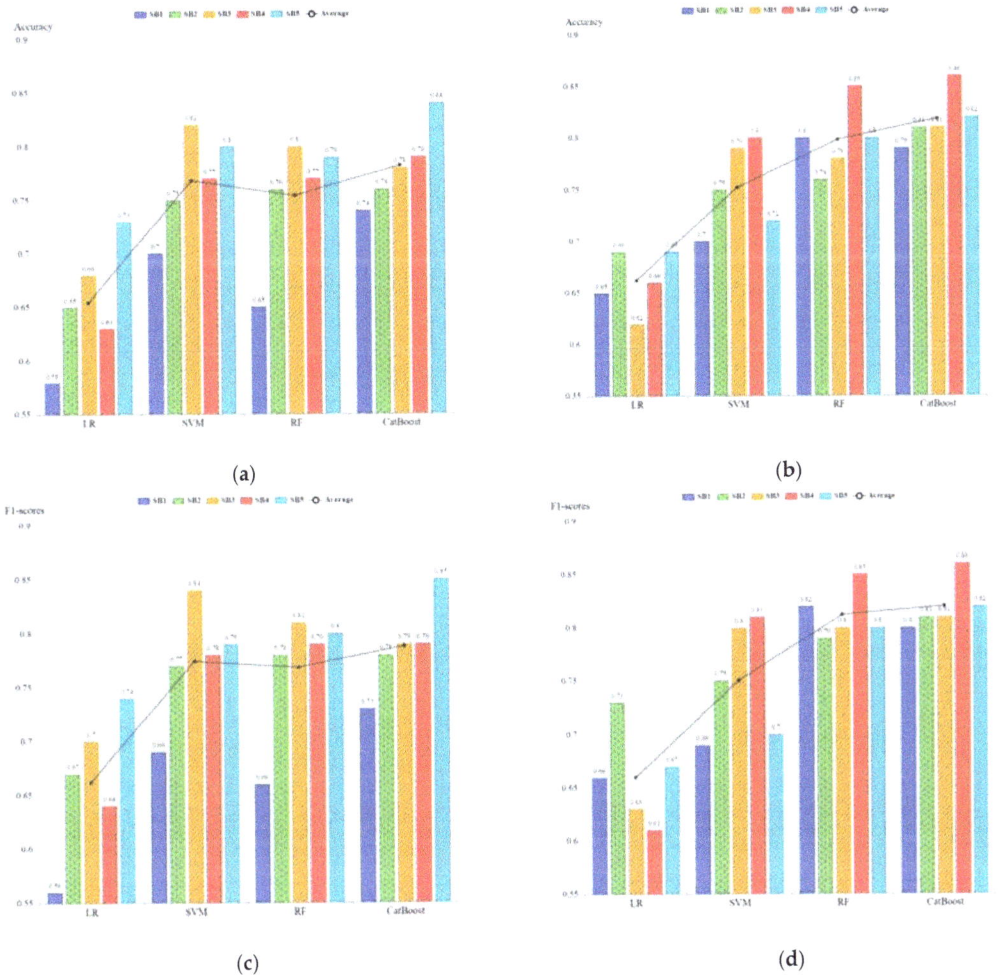

Figure 5. Accuracy and F1-scores of classifiers with(out) MOSMA. (**a**) Accuracy without MOSMA. (**b**) Accuracy with MOSMA. (**c**) F1-scores without MOSMA. (**d**) F1-scores with MOSMA.

4.2. Performance of Different Models

As mentioned above, this study has tried 64 models. Figure 6 depicts their performance in terms of AUC, accuracy, and F1-score. As can be seen from Figure 6, the models coded as "1111C" (No. 64) and "1010C" (No. 40) have satisfactory performance. In the former model, four methods (i.e., feature selection, SMOTE, one-hot encoding, and standard scaler) and the classifier of CatBoost are employed. In the latter model, two methods (i.e., feature selection and one-hot encoding) and CatBoost are used. The former model yields the maximum AUC of 0.9175, accuracy of 0.8075, and F1-score of 0.6497. Although the F1-score of 0.6497 is not the highest, it ranks the upper-middle among all models. The latter model yields the AUC of 0.8970, accuracy of 0.8583, and the maximum F1-score of 0.7725. Since No. 64 model garners the maximum AUC, which is a comprehensive performance indicator, the following sections report results from it.

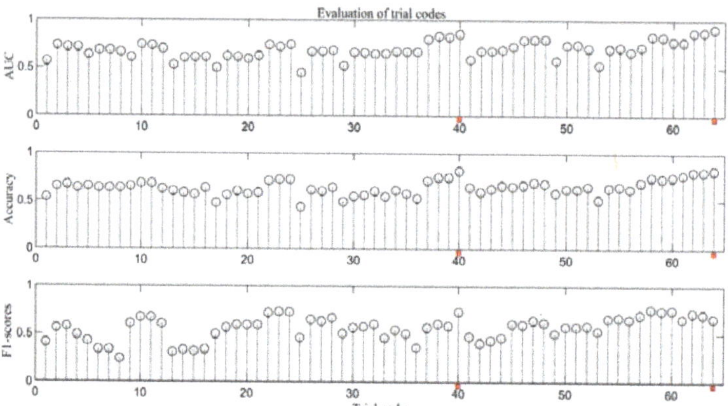

Figure 6. Performance of 64 models.

4.3. Feature Selection

After feature selection, different numbers of input indicators are supposed to account for different safety behaviors. As shown in Figure 7, SB1 needs to consider the fewest input indicators (i.e., 25), while SB5 needs to consider the most input indicators (i.e., 35). Despite that, there are ten input indicators that account for all of the five safety behaviors in common. The ten input indicators are affiliation (coded as AffRes), contract value (coded as ConSum), clients setting safety goals (coded as CI2), very clear safety rules, policies, and procedures (coded as SSRP2), safety rules not allowed to be violated (coded as SSRP3), colleagues understanding my job needs (coded as TMX4), project managers seeking safety suggestions (coded as SC2), timely accident reporting (coded as SC4), safety ownership (coded as SM2), and risk reduction at workplace (coded as SM4).

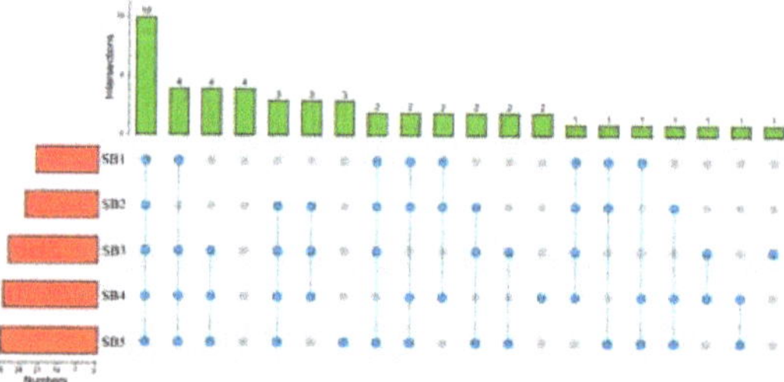

Figure 7. Upset plot for variables after feature selection.

4.3.1. Feature Importance

The importance of all input indicators for all the five safety behaviors is shown in Table 3. The top three important indicators for the five safety behaviors are highlighted. For example, regarding SB5, the top three important indicators are contract value (coded as ConSum), project managers seeking safety suggestions (coded as SC2), and affiliation (coded as AffRes). This indicates that construction personnel on projects with larger contract value, construction personnel on projects where project managers seek more safety suggestions, and those personnel from the client are more likely to use all necessary safety equipment on site.

Table 3. Feature importance of the five safety behaviors.

	SB1	SB2	SB3	SB4	SB5
NatClit	0.06	0.05	0.03	**0.09 (3rd)**	0.08
CI1	0.00	0.04	0.00	0.03	0.03
CI2	**0.10 (2nd)**	**0.10 (3rd)**	0.01	0.03	0.01
CI3	0.05	0.02	0.04	0.07	0.04
CI4	0.00	0.00	0.01	0.00	0.02
ConSum	0.04	0.02	**0.13 (2nd)**	**0.09 (3rd)**	**0.17 (1st)**
GC3	0.00	0.00	0.01	0.00	0.00
PDM1	0.00	0.00	0.00	0.00	0.00
PDM2	0.02	0.01	0.02	0.00	0.00
PD	0.02	0.00	0.01	**0.10 (2nd)**	0.02
OS1	0.00	0.00	0.01	0.00	0.00
OS2	0.01	0.01	0.02	0.00	0.00
SSRP1	0.01	0.03	0.01	0.00	0.01
SSRP2	**0.08 (3rd)**	0.01	0.01	0.02	0.05
SSRP3	0.02	0.01	0.00	0.00	0.01
SSRP4	0.05	0.03	0.00	0.02	0.00
SC1	0.00	0.00	0.03	0.01	0.00
SC2	0.02	0.00	0.01	0.00	**0.16 (2nd)**
SC3	0.00	0.00	0.01	0.00	0.00
SC4	0.02	0.02	**0.21 (1st)**	0.02	0.00
TL1	0.04	0.01	0.00	0.00	0.00
TL2	0.00	0.00	0.00	0.00	0.00
TL3	0.05	0.04	0.01	0.06	0.00
LMX1	0.00	0.00	0.00	0.00	0.00
LMX2	0.00	0.01	0.00	0.00	0.00
LMX3	0.00	0.01	0.01	0.01	0.00
LMX4	0.00	0.03	0.01	0.03	0.01
TMX2	0.00	0.01	0.02	0.00	0.01
TMX3	0.00	0.02	0.00	0.00	0.01
TMX4	0.03	0.01	0.04	0.04	0.04
Age	0.05	**0.10 (3rd)**	0.05	0.08	0.06
DeptRsp	0.02	0.04	0.05	0.05	0.04
AffRes	**0.14 (1st)**	**0.13 (2nd)**	**0.12 (3rd)**	**0.18 (1st)**	**0.13 (3rd)**
SM1	0.02	0.03	0.09	0.01	0.00
SM2	**0.08 (3rd)**	0.00	0.00	0.01	0.04
SM3	0.02	0.04	0.01	0.01	0.00
SM4	0.05	**0.16 (1st)**	0.03	0.04	0.04
SUM	1.00	1.00	1.00	1.00	1.00

4.3.2. Correlation and OR Values

As mentioned earlier, FI reflects the relative importance of different input indicators for each safety behavior but it does not show whether they exert positive influence or negative influence. In order to make up for this deficiency, correlation analysis based on CTs with OR values is carried out. Table 4 shows the results of correlation analysis for SB1 (i.e., use all necessary safety equipment to do the job). If the p-value is significant and the OR is above 1.0 along with the confidence interval, then with feature SB1 is more likely to take place. If the p-value is significant and the OR is below 1.0 along with the confidence interval, then feature SB1 is less likely to happen. From Table 5, it can be concluded that the drivers of SB1 are GC1, SSRP3, CI3, LMX1, TMX4, SC2, and SM2, among others. OR values between the five safety behaviors and all of the input indicators are shown in Figure 8. At least two points deserve mentioning. First, different sets of drivers are behind different safety behaviors. For example, ConSum has more impacts on SB3 and SB4 than on SB1. Second, some indicators can be omitted in establishing the classification framework, such as StgProj, Gender, Age, EduRsp, and DriHab, because they have no bearing on any of the five safety behaviors.

Table 4. Chi-square test and OR values.

Features	Chi-Square Test		OR	95% CI	
	χ^2	p		Lower Limit	Upper Limit
Age	0.81	0.368	0		
GC1	4.66	0.031	1.68	1.05	2.71
SSRP3	26.10	0.000	5.39	2.70	10.78
CI3	20.07	0.000	3.04	1.86	4.99
LMX1	7.34	0.007	2.65	1.28	5.45
TMX1	1.71	0.191	0		
TMX4	8.60	0.003	2.12	1.28	3.53
SC2	21.25	0.000	4.10	2.19	7.68
SM2	32.56	0.000	4.19	2.53	6.94

Table 5. Comparison with previous studies.

Reference	Method of Tuning and Optimization	Label	Classifier	Cross-Validation	Accuracy	F1-Score
Poh et al. [45]	Fixed parameters	Trichotomy	RF	LOO	0.78	/
	Fixed parameters	Trichotomy	LR	LOO	0.59	/
	Fixed parameters	Trichotomy	SVM	LOO	0.44	/
	Fixed parameters	Trichotomy	DT *	LOO	0.71	/
	Fixed parameters	Trichotomy	KNN *	LOO	0.73	/
Niu et al. [48]	Grid search	Binary	GBDT *	10 folds	0.80	0.61
	Grid search	Binary	RF	10 folds	0.77	0.67
Lee et al. [4]	BPSO *	Binary	GSVM *	10 folds	0.81	0.81
	BPSO	Binary	KNN	10 folds	0.79	/
	BPSO	Binary	DT	10 folds	0.71	/
Koc and Gurgun [46]	Trial error	Quartering	XGBoost	/	/	0.61
Proposed	MOSMA	Binary	CatBoost	LOO	0.86	0.86
	MOSMA	Binary	RF	LOO	0.85	0.85
	MOSMA	Binary	SVM	LOO	0.80	0.81
	MOSMA	Binary	LR	LOO	0.69	0.73

* BPSO, binary particle swarm optimization; DT decision tree; KNN, k-nearest neighbor; GBDT, gradient boosting decision tree; GSVM, Gaussian support vector machine; Bi-LSTM, bidirectional long short-term memory.

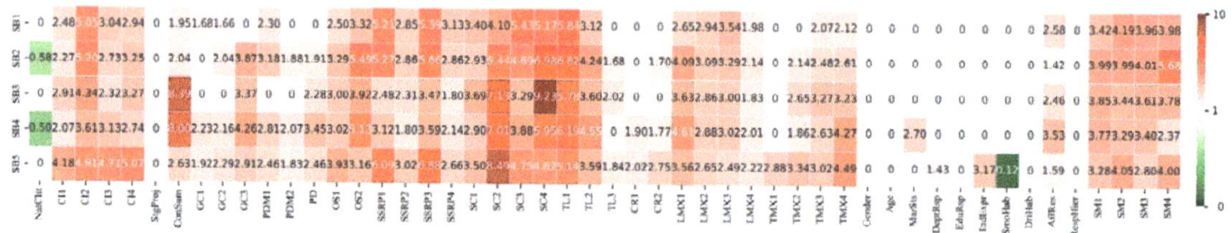

Figure 8. OR values between the five safety behaviors and all of the input indicators.

5. Discussion

5.1. Findings

This study has achieved the two objectives mentioned earlier, and has theoretical, practical, and methodological implications.

First, in theory, safety behavior as an emergent property of a complex socio-technical system has different drivers. Using machine learning, this study supports the proposition. In particular, this study found that in order to encourage personnel to use all necessary safety equipment on the job (i.e., SB1), clients should set examples for contractors and consultants, safety motivation should be enhanced, and clients, private clients in particular, are encouraged to be involve in safety management as early as possible. In projects with a large contract sum, older personnel with more dependents to support is more likely to follow safety procedures on the job (i.e., SB2). In projects with a large contract sum, construction personnel are more likely to promote safety programs willingly (i.e., SB3) with clients actively engaging in safety management. In public projects with a large contract sum, personnel is encouraged to pursue professional development, and hence, more likely to put in extra effort to improve workplace safety (i.e., SB4). In projects with a sound safety climate and more client involvement, personnel is more likely to help colleagues who are in risky conditions (i.e., SB5). Based on the findings, practicable and targeted measures are proposed to promote the five safety behaviors, respectively.

Second, machine learning has advantage over traditional statistical methods in addressing more complex interrelations among independent variables [6]. To garner this advantage, this study first evaluates the performance of four common machine-learning methods. Although these four methods achieve the comparatively satisfactory performance, this study develops a combinative method, CatBoost–MOSMA, to train and test the data again. This is because MOSMA has achieved superior performance in hyperparameter tuning, and this study attempts to introduce it into the safety research domain. Through 64 trials, the combinative method has achieved the maximum classification performance, and therefore, is used to establish factor importance. Furthermore, as noted by Poh et al. [45], the imbalanced distribution of the classes is usually an issue in previous research. This combinative method adopts the SMOTE technique to address this issue and obtains more robust results. This is shown in Table 5, which compares the classification performance between the proposed combinative method and other classification methods. Compared with other methods of tuning and optimization, MOSMA achieves a higher accuracy score when using the same classifiers. When the performance of classifiers is not significantly different, MOSMA achieves a higher F1-score. Hence, it can be concluded that the proposed combinative strategy of MOSMA-CatBoost is effective and efficient in classifying binary construction safety behavioral data.

5.2. Limitations and Future Research Directions

Although the study has achieved its objectives, it has limitations. First, the sample size can be further enlarged. Although a new machine-learning strategy is developed specifically to tackle the small sample size issue and some seminal studies have used a smaller sample set, it is highly recommended that future researches collect more data. Second, the study uses a sample from Hong Kong, and whether the findings can be extrapolated to other countries/regions needs more research efforts. Third, the factors affecting safety behaviors mentioned in the study are not exhaustive, and their interrelationship is not clearcut. Hence, more in-depth research needs to be undertaken in this regard. Fourth, similar to the third one, this study attempts to propose a generic classification framework, and different construction sites are encouraged to tailor the framework to cater for their own needs. Fifth, this study employs a combination of three feature-selection methods, including FI, CT and BS. Only those input indicators that obtain over half votes were retained. In other word, this approach may omit some input indicators that are strongly correlated with some safety behavior. For instance, the input indicator SmoHab is strongly negatively correlated with SB5, but does not correlate with other safety behaviors. Therefore, it has been deleted. It can be seen in the experiment results that this method generally benefits all of the safety behaviors as a whole since the classification performance improves after deleting those input indicators that were only correlated with certain safety behaviors.

Despite these limitations, the classification framework is highly recommended for future research efforts, given its satisfactory performance.

5.3. Practical Use of the Research

The proposed methods can be used in safety management practice on construction sites, as shown in Figure 9. A survey is conducted with a representative sample of construction personnel on the site, and the data are stored into a safety behavioral database. After training, safety staff is charged with modeling and algorithm implementation and deriving model results, which suggest different safety behavioral orientation associated with different feature patterns. Using a combination of their experience and this data-driven clue, safety staff shall be able to predict a newcomer's safety behavioral orientation, and then propose and implement targeted interventions. When the prediction performance turns out to be unsatisfactory, a new round of survey begins, and more data are stored in the database. Complemented with their gut feeling, this data-driven decision support system is supposed to help deter unsafe behaviors on construction sites in an efficient and effective way.

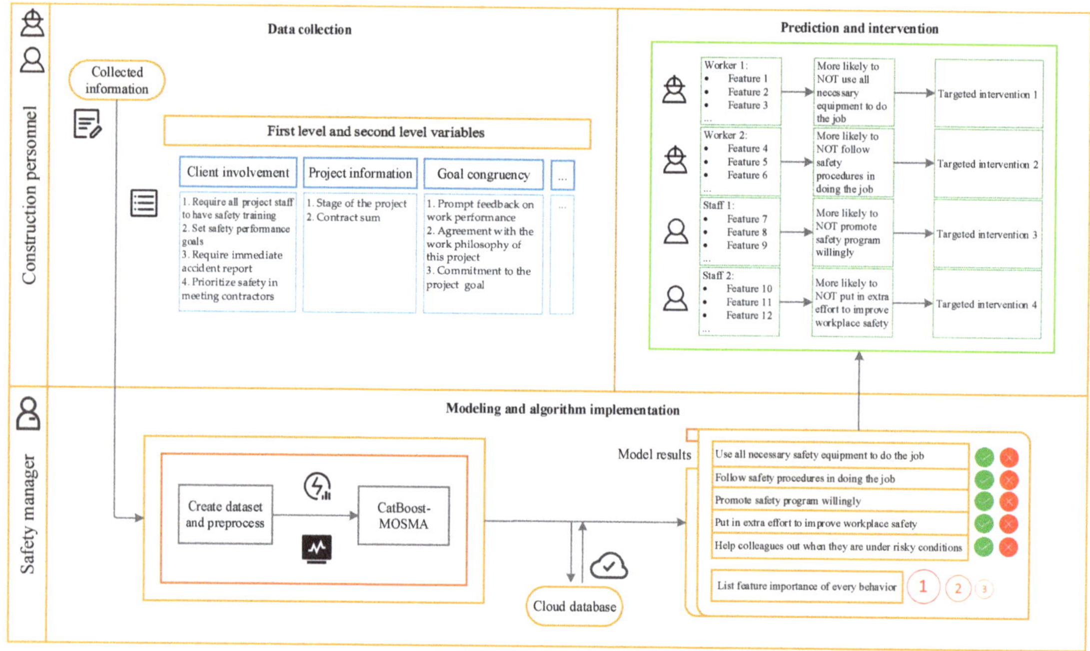

Figure 9. Practical use of the research.

6. Conclusions

Different sets of drivers underlie different safety behaviors, and uncovering such complex patterns, help formulate targeted measures to cultivate safety behaviors. Machine learning can explore such complex patterns among safety behavioral data. Given the theoretical, methodological and practical significance, this paper attempts to develop a classification framework for construction personnel's safety behaviors with machine-learning algorithms, including LR, SVM, RF, and CatBoost. The classification framework has three steps, i.e., data collection and preprocessing, modeling and algorithm implementation, and optimal model acquisition. For illustrative purposes, five common safety behaviors of a random and representative sample of Hong Kong-based construction personnel are used to validate the classification framework. To achieve a high classification performance, this paper employs a combinative strategy of CatBoost–MOSMA. Results support this combinative strategy in dealing with construction safety behavioral data. From the derived optimal model, a unique set of important features can be identified for each safety behavior, and ten out of the 46 input indicators are found important for all the five safety behaviors. Based on the findings, safety staff is supposed to make concrete and targeted interventions to individual construction personnel on site, and improve safety performance in a more efficient and effective way.

Author Contributions: Conceptualization, S.Y. and Y.S.; methodology, S.Y. and Y.W.; validation, Y.S. and Y.W.; formal analysis, Y.S.; investigation, S.Y., Y.W., Y.S. and S.R.; writing—original draft preparation, S.Y., Y.W. and Y.S.; writing—review and editing, Y.S., S.Y. and S.R.; funding acquisition, Y.S. All authors have read and agreed to the published version of the manuscript.

Funding: This study is supported by a grant from the National Natural Science Foundation of China (Project No.: 71701130).

Institutional Review Board Statement: The study was conducted in accordance with the Declaration of Helsinki, and approved by the Institutional Review Board (or Ethics Committee) of the University of Hong Kong (protocol code EA011011 and 6 October 2011).

Informed Consent Statement: Informed consent was obtained from all subjects involved in the study.

Data Availability Statement: Not applicable.

Conflicts of Interest: The authors declare no conflict of interest.

Appendix A

Table A1. Input indicators.

First-Level Dimensions	Second-Level Indicators	Label	Value	Frequency	Percent (%)	Code
Nature of client	Client type	Public	1	205	70.2	NatClit
		Private	2	87	29.8	
Client involvement	Require all project staff to have safety training	Yes	1	132	45.2	CI1
		No	0	160	54.8	
	Set safety performance goals	Yes	1	96	32.9	CI2
		No	0	196	67.1	
	Require immediate accident report	Yes	1	152	52.1	CI3
		No	0	140	47.9	
	Prioritize safety in meeting contractors	Yes	1	130	44.5	CI4
		No	0	162	55.5	
Project information	Stage of the project	Start-up (less than 30%)	1	77	26.4	StgProj
		Advanced (30–70%)	2	117	40.1	
		Near close-out (greater than 70%)	3	98	33.6	
	Contract sum	$\leq$99 millions	1	67	22.9	ConSum
		100–499 millions	2	98	33.6	
		500–999 millions	3	40	13.7	
		$\geq$1000 millions	4	87	29.8	
Goal congruency	Prompt feedback on work performance	Yes	1	146	50	GC1
		No	0	146	50	
	Agreement with the work philosophy of this project	Yes	1	134	45.9	GC2
		No	0	158	54.1	
	Commitment to the project's goal	Yes	1	34	11.6	GC3
		No	0	258	88.4	

Table A1. *Cont.*

First-Level Dimensions	Second-Level Indicators	Label	Value	Frequency	Percent (%)	Code
Participative decision-making	Satisfaction with the decision-making process	Yes	1	26	8.9	PDM1
		No	0	266	91.1	
	Have opportunity to participate in decision making	Yes	1	132	45.2	PDM2
		No	0	160	54.8	
Professional development	Encouraged to seek further professional development	Yes	1	118	40.4	PD
		No	0	174	59.6	
Organizational support	Support from colleagues	Yes	1	43	14.7	OS1
		No	0	249	85.3	
	Support from the leadership	Yes	1	49	16.8	OS2
		No	0	243	83.2	
Standardized safety rules and procedures	Performance standards are very clear.	Yes	1	37	12.7	SSRP1
		No	0	255	87.3	
	Rules, policies, and procedures are very clear.	Yes	1	144	49.3	SSRP2
		No	0	148	50.7	
	Rules cannot be violated.	Yes	1	47	16.1	SSRP3
		No	0	245	83.9	
	Rules are enforced strictly.	Yes	1	130	44.5	SSRP4
		No	0	162	55.5	
Safety climate	Accidents and incidents are always reported.	Yes	1	106	36.3	SC1
		No	0	186	63.7	
	The project manager encourages staff to make suggestions to improve safety.	Yes	1	54	18.5	SC2
		No	0	238	81.5	
	The project manager genuinely cares about the staff's safety.	Yes	1	53	18.2	SC3
		No	0	239	81.8	
	All the project staff are fully committed to safety.	Yes	1	46	15.8	SC4
		No	0	246	84.2	
Transformational leadership	My supervisor suggests new ways.	Yes	1	29	9.9	TL1
		No	0	263	90.1	
	My supervisor suggests different angles.	Yes	1	33	11.3	TL2
		No	0	259	88.7	
	My supervisor teaches and coaches.	Yes	1	113	38.7	TL3
		No	0	179	61.3	
Contingent reward	My supervisor rewards my achievement.	Yes	1	115	39.4	CR1
		No	0	177	60.6	
	My supervisor recognizes my achievement.	Yes	1	145	49.7	CR2
		No	0	147	50.3	

Table A1. *Cont.*

First-Level Dimensions	Second-Level Indicators	Label	Value	Frequency	Percent (%)	Code
Leader–member exchange	Supervisor understands my job problems and needs.	Yes	1	35	12.0	LMX1
		No	0	257	88.0	
	Supervisor recognizes my potential.	Yes	1	44	15.1	LMX2
		No	0	248	84.9	
	Supervisor helps me out with all his might.	Yes	1	43	14.7	LMX3
		No	0	249	85.3	
	My working relationship with supervisor is very good.	Yes	1	129	44.2	LMX4
		No	0	163	55.8	
Team–member exchange	My colleagues are willing to help me with my assignment.	Yes	1	97	33.2	TMX1
		No	0	195	66.8	
	My colleagues recognize my potential.	Yes	1	129	44.2	TMX2
		No	0	163	55.8	
	My colleagues let me know if I interfere with their work.	Yes	1	115	39.4	TMX3
		No	0	177	60.6	
	My colleagues understand my job problems and needs.	Yes	1	89	30.5	TMX4
		No	0	203	69.5	
Demographic information	Gender	Male	1	269	92.1	Gender
		Female	2	23	7.9	
	Age	<20	1	0	0	Age
		20–30	2	20	6.8	
		31–40	3	51	17.5	
		41–50	4	99	33.9	
		>50	5	122	41.8	
	Marital status	Married	1	246	84.2	MarSts
		Single	2	46	15.8	
	Number of dependents	0	1	21	7.2	DeptRsp
		1–2	2	132	45.2	
		3–4	3	123	42.1	
		5–6	4	12	4.1	
		>6	5	4	1.4	
	Educational level	Below primary	1	1	0.3	EduRsp
		Primary	2	5	1.7	
		Secondary	3	22	7.5	
		Certificate/diploma	4	17	5.8	
		College or higher	5	247	84.6	
	Industrial experience	<3	1	10	3.4	IndExpr
		3–10	2	29	9.9	
		11–15	3	36	12.3	
		16–20	4	37	12.7	
		>20	5	180	61.6	

Table A1. *Cont.*

First-Level Dimensions	Second-Level Indicators	Label	Value	Frequency	Percent (%)	Code
Habit	Smoking habit	Smoke even at work	1	9	3.1	SmoHab
		Smoke, but not at work	2	24	8.2	
		Do not smoke	3	259	88.7	
	Drinking habit	Drink even at work	1	0	0	DriHab
		Drink, but not at work	2	104	35.6	
		Do not drink	3	188	64.4	
Affiliation	Type of affiliation	Contractor	1	119	40.8	AffRes
		Consultant	2	89	30.5	
		Client	3	84	28.8	
	Hierarchical position	Worker	1	18	6.2	RespHier
		Supervisory staff	2	115	39.4	
		Management	3	159	54.5	
Safety motivation	Workplace health and safety is important.	Yes	1	147	50.3	SM1
		No	0	145	49.7	
	It is beneficial to me to maintain or improve my personal safety.	Yes	1	144	49.3	SM2
		No	0	148	50.7	
	Maintaining safety at all times is important.	Yes	1	170	58.2	SM3
		No	0	122	41.8	
	To reduce the risk of workplace accidents and incidents is very important.	Yes	1	173	59.2	SM4
		No	0	119	40.8	

Table A2. Output indicators.

Safety behavior	Use all necessary safety equipment to do the job	Yes	1	114	39.0	SB1
		No	0	178	61.0	
	Follow safety procedures in doing the job	Yes	1	105	36.0	SB2
		No	0	187	64.0	
	Promote safety program willingly	Yes	1	76	26.0	SB3
		No	0	216	74.0	
	Put in extra effort to improve workplace safety	Yes	1	66	22.6	SB4
		No	0	226	77.4	
	Help colleagues out when they are under risky conditions.	Yes	1	90	30.8	SB5
		No	0	202	69.2	

References

1. Guo, S.; He, J.; Li, J.; Tang, B. Exploring the Impact of Unsafe Behaviors on Building Construction Accidents Using a Bayesian Network. *Int. J. Environ. Res. Public Health* **2020**, *17*, 221. [CrossRef] [PubMed]
2. Mullen, J.; Kelloway, E.K.; Teed, M. Employer Safety Obligations, Transformational Leadership and Their Interactive Effects on Employee Safety Performance. *Saf. Sci.* **2017**, *91*, 405–412. [CrossRef]

3. Griffin, M.A.; Curcuruto, M.M. Safety Climate in Organizations. *Annu. Rev. Organ. Psychol. Organ. Behav.* **2016**, *3*, 191–212. [CrossRef]
4. Lee, B.G.; Choi, B.; Jebelli, H.; Lee, S. Assessment of Construction Workers' Perceived Risk Using Physiological Data from Wearable Sensors: A Machine Learning Approach. *J. Build. Eng.* **2021**, *42*, 102824. [CrossRef]
5. Yang, K.; Ahn, C.R.; Kim, H. Deep Learning-Based Classification of Work-Related Physical Load Levels in Construction. *Adv. Eng. Inform.* **2020**, *45*, 101104. [CrossRef]
6. Goh, Y.M.; Ubeynarayana, C.U.; Wong, K.L.X.; Guo, B.H.W. Factors Influencing Unsafe Behaviors: A Supervised Learning Approach. *Accid. Anal. Prev.* **2018**, *118*, 77–85. [CrossRef]
7. Choi, B.; Lee, S. An Empirically Based Agent-Based Model of the Sociocognitive Process of Construction Workers' Safety Behavior. *J. Constr. Eng. Manag.* **2018**, *144*, 04017102. [CrossRef]
8. Xia, N.; Xie, Q.; Griffin, M.A.; Ye, G.; Yuan, J. Antecedents of Safety Behavior in Construction: A Literature Review and an Integrated Conceptual Framework. *Accid. Anal. Prev.* **2020**, *148*, 105834. [CrossRef]
9. Onubi, H.O.; Yusof, N.; Hassan, A.S. The Moderating Effect of Client Types on the Relationship between Green Construction Practices and Health and Safety Performance. *Int. J. Sustain. Dev. World Ecol.* **2020**, *27*, 732–748. [CrossRef]
10. Shen, Y.Z.; Ju, C.J.; Koh, T.Y.; Rowlinson, S.; Bridge, A.J. The Impact of Transformational Leadership on Safety Climate and Individual Safety Behavior on Construction Sites. *Int. J. Environ. Res. Public Health* **2017**, *14*, 45. [CrossRef]
11. Ma, Y.H. *Client's Contributions to Project Safety Performance: A Comparison between Public and Private Construction Projects*; The University of Hong Kong: Hong Kong, 2006.
12. Umeokafor, N. An Investigation into Public and Private Clients' Attitudes, Commitment and Impact on Construction Health and Safety in Nigeria. *Eng. Constr. Archit. Manag.* **2018**, *25*, 798–815. [CrossRef]
13. Lingard, H.; Blismas, N.; Cooke, T.; Cooper, H. The Model Client Framework: Resources to Help Australian Government Agencies to Promote Safe Construction. *Int. J. Manag. Proj. Bus.* **2009**, *2*, 131–140. [CrossRef]
14. Votano, S.; Sunindijo, R.Y. Client Safety Roles in Small and Medium Construction Projects in Australia. *J. Constr. Eng. Manag.* **2014**, *140*, 04014045. [CrossRef]
15. Shin, M.; Lee, H.S.; Park, M.; Moon, M.; Han, S. A System Dynamics Approach for Modeling Construction Workers' Safety Attitudes and Behaviors. *Accid. Anal. Prev.* **2014**, *68*, 95–105. [CrossRef] [PubMed]
16. Fang, D.; Huang, Y.; Guo, H.; Lim, H.W. LCB Approach for Construction Safety. *Saf. Sci.* **2020**, *128*, 104761. [CrossRef]
17. Awolusi, I.G.; Marks, E.D. Safety Activity Analysis Framework to Evaluate Safety Performance in Construction. *J. Constr. Eng. Manag.* **2017**, *143*, 05016022. [CrossRef]
18. Choi, T.N.Y.; Chan, D.W.M.; Chan, A.P.C. Perceived Benefits of Applying Pay for Safety Scheme (PFSS) in Construction-A Factor Analysis Approach. *Saf. Sci.* **2011**, *49*, 813–823. [CrossRef]
19. De Clercq, D.; Bouckenooghe, D.; Raja, U.; Matsyborska, G. Unpacking the Goal Congruence–Organizational Deviance Relationship: The Roles of Work Engagement and Emotional Intelligence. *J. Bus. Ethics* **2014**, *124*, 695–711. [CrossRef]
20. Bouckenooghe, D.; Zafar, A.; Raja, U. How Ethical Leadership Shapes Employees' Job Performance: The Mediating Roles of Goal Congruence and Psychological Capital. *J. Bus. Ethics* **2015**, *129*, 251–264. [CrossRef]
21. Nævestad, T. Safety Cultural Preconditions for Organizational Learning in High-risk Organizations. *J. Contingencies Crisis Manag.* **2008**, *16*, 154–163. [CrossRef]
22. Wu, C.; Wang, F.; Zou, P.X.W.; Fang, D. How Safety Leadership Works among Owners, Contractors and Subcontractors in Construction Projects. *Int. J. Proj. Manag.* **2016**, *34*, 789–805. [CrossRef]
23. Lee, Y.-H.; Lu, T.-E.; Yang, C.C.; Chang, G. A Multilevel Approach on Empowering Leadership and Safety Behavior in the Medical Industry: The Mediating Effects of Knowledge Sharing and Safety Climate. *Saf. Sci.* **2019**, *117*, 1–9. [CrossRef]
24. Toole, T.M. Increasing Engineers' Role in Construction Safety: Opportunities and Barriers. *J. Prof. Issues Eng. Educ. Pract.* **2005**, *131*, 199–207. [CrossRef]
25. Mearns, K.J.; Reader, T. Organizational Support and Safety Outcomes: An Un-Investigated Relationship? *Saf. Sci.* **2008**, *46*, 388–397. [CrossRef]
26. Gyekye, S.A.; Salminen, S. Workplace Safety Perceptions and Perceived Organizational Support: Do Supportive Perceptions Influence Safety Perceptions? *Int. J. Occup. Saf. Ergon.* **2007**, *13*, 189–200. [CrossRef] [PubMed]
27. Guo, M.; Liu, S.; Chu, F.; Ye, L.; Zhang, Q. Supervisory and Coworker Support for Safety: Buffers between Job Insecurity and Safety Performance of High Speed Railway Drivers in China. *Saf. Sci.* **2019**, *117*, 290–298. [CrossRef]
28. Tucker, S.; Chmiel, N.; Turner, N.; Hershcovis, M.S.; Stride, C.B. Perceived Organizational Support for Safety and Employee Safety Voice: The Mediating Role of Coworker Support for Safety. *J. Occup. Health Psychol.* **2008**, *13*, 319–330. [CrossRef] [PubMed]
29. Dekker, S.W. The Bureaucratization of Safety. *Saf. Sci.* **2014**, *70*, 348–357. [CrossRef]
30. Newaz, M.T.; Davis, P.; Jefferies, M.; Pillay, M. The Psychological Contract: A Missing Link between Safety Climate and Safety Behaviour on Construction Sites. *Saf. Sci.* **2019**, *112*, 9–17. [CrossRef]
31. Neal, A.; Griffin, M.A.; Hart, P.M. The Impact of Organizational Climate on Safety Climate and Individual Behavior. *Saf. Sci.* **2000**, *34*, 99–109. [CrossRef]
32. Oswald, D.; Lingard, H.; Zhang, R.P. How Transactional and Transformational Safety Leadership Behaviours Are Demonstrated within the Construction Industry. *Constr. Manag. Econ.* **2022**, *40*, 374–390. [CrossRef]

33. Hoffmeister, K.; Gibbons, A.M.; Johnson, S.K.; Cigularov, K.P.; Chen, P.Y.; Rosecrance, J.C. The Differential Effects of Transformational Leadership Facets on Employee Safety. *Saf. Sci.* **2014**, *62*, 68–78. [CrossRef]

34. Gottfredson, R.K.; Wright, S.L.; Heaphy, E.D. A Critique of the Leader-Member Exchange Construct: Back to Square One. *Leadersh. Q.* **2020**, *31*, 101385. [CrossRef]

35. He, C.Q.; Jia, G.S.; McCabe, B.; Sun, J.D. Relationship between Leader-Member Exchange and Construction Worker Safety Behavior: The Mediating Role of Communication Competence. *Int. J. Occup. Saf. Ergon.* **2021**, *27*, 371–383. [CrossRef]

36. He, C.Q.; McCabe, B.; Jia, G.S. Effect of Leader-Member Exchange on Construction Worker Safety Behavior: Safety Climate and Psychological Capital as the Mediators. *Saf. Sci.* **2021**, *142*, 105401. [CrossRef]

37. Li, N.W.; Bao, S.W.; Naseem, S.; Sarfraz, M.; Mohsin, M. Extending the Association Between Leader-Member Exchange Differentiation and Safety Performance: A Moderated Mediation Model. *Psychol. Res. Behav. Manag.* **2021**, *14*, 1603–1613. [CrossRef] [PubMed]

38. Chen, S.-Y.; Lu, C.-S.; Ye, K.-D.; Shang, K.-C.; Guo, J.-L.; Pan, J.-M. Enablers of Safety Citizenship Behaviors of Seafarers: Leader-Member Exchange, Team-Member Exchange, and Safety Climate. *Marit. Policy Manag.* **2021**, 1–16. [CrossRef]

39. Shen, Y.; Tuuli, M.M.; Xia, B.; Koh, T.Y.; Rowlinson, S. Toward a Model for Forming Psychological Safety Climate in Construction Project Management. *Int. J. Proj. Manag.* **2015**, *33*, 223–235. [CrossRef]

40. Beus, J.M.; Taylor, W.D. Working Safely at Some Times and Unsafely at Others: A Typology and within-Person Process Model of Safety-Related Work Behaviors. *J. Occup. Health Psychol.* **2018**, *23*, 402–416. [CrossRef]

41. Shen, Y. An Investigation of Safety Climate on Hong Kong Construction Sites. Ph.D. Thesis, The University of Hong Kong, Hong Kong, 2013.

42. Meng, X.; Chan, A.H. Demographic Influences on Safety Consciousness and Safety Citizenship Behavior of Construction Workers. *Saf. Sci.* **2020**, *129*, 104835. [CrossRef]

43. Strickland, J.R.; Wagan, S.; Dale, A.M.; Evanoff, B.A. Prevalence and Perception of Risky Health Behaviors among Construction Workers. *J. Occup. Environ. Med.* **2017**, *59*, 673–678. [CrossRef] [PubMed]

44. Hair, J.F.; Black, W.C.; Babin, B.J.; Anderson, R.E. *Multivariate Data Analysis*, 8th ed.; Cengage Learning, EMEA: Hampshire, UK, 2019.

45. Poh, C.Q.X.; Ubeynarayana, C.U.; Goh, Y.M. Safety Leading Indicators for Construction Sites: A Machine Learning Approach. *Autom. Constr.* **2018**, *93*, 375–386. [CrossRef]

46. Koc, K.; Gurgun, A.P. Scenario-Based Automated Data Preprocessing to Predict Severity of Construction Accidents. *Autom. Constr.* **2022**, *140*, 104351. [CrossRef]

47. Antwi-Afari, M.F.; Li, H.; Yu, Y.; Kong, L. Wearable Insole Pressure System for Automated Detection and Classification of Awkward Working Postures in Construction Workers. *Autom. Constr.* **2018**, *96*, 433–441. [CrossRef]

48. Niu, Y.; Li, Z.; Fan, Y. Analysis of Truck Drivers' Unsafe Driving Behaviors Using Four Machine Learning Methods. *Int. J. Ind. Ergon.* **2021**, *86*, 103192. [CrossRef]

49. Bentéjac, C.; Csörgő, A.; Martínez-Muñoz, G. A Comparative Analysis of Gradient Boosting Algorithms. *Artif. Intell. Rev.* **2021**, *54*, 1937–1967. [CrossRef]

50. Becker, M. On the Efficiency of Nature-Inspired Algorithms for Generation of Fault-Tolerant Graphs. In Proceedings of the 2015 IEEE International Conference on Systems, Man, and Cybernetics, Hong Kong, 9–12 October 2015.

51. Li, S.; Chen, H.; Wang, M.; Heidari, A.A.; Mirjalili, S. Slime Mould Algorithm: A New Method for Stochastic Optimization. *Future Gener. Comput. Syst.* **2020**, *111*, 300–323. [CrossRef]

52. Houssein, E.H.; Mahdy, M.A.; Shebl, D.; Manzoor, A.; Sarkar, R.; Mohamed, W.M. An Efficient Slime Mould Algorithm for Solving Multi-Objective Optimization Problems. *Expert Syst. Appl.* **2022**, *187*, 115870. [CrossRef]

53. Kursa, M.B.; Rudnicki, W.R. Feature Selection with the Boruta Package. *J. Stat. Softw.* **2010**, *36*, 1–13. [CrossRef]

Article

Automated Schedule and Cost Control Using 3D Sensing Technologies

Ahmed R. ElQasaby *, Fahad K. Alqahtani * and Mohammed Alheyf

Department of Civil Engineering, College of Engineering, King Saud University, P.O. Box 800, Riyadh 11421, Saudi Arabia
* Correspondence: ahmedrashed43@yahoo.com (A.R.E.); bfahad@ksu.edu.sa (F.K.A.)

Abstract: Nowadays, many construction projects in KSA still struggle with cost overruns and delay in activities. Therefore, automatic monitoring approaches are needed in the construction progress monitoring domain (CPM) to address these concerns. Thus, this paper proposed a system integrating a BIM-planned model with site laser scans, as laser scanners showed massive potential in the CPM domain. The algorithms of the proposed system recognized 3D objects based on the intersection between models, alignment accuracy, and Lalonde features. The proposed system combined 3D object recognition technology with 5D information data into a 5D progress tracking system using earned value (EV) principles. The reason behind that is a lack of research regarding conducting a 5D assessment integrated BIM with 3D sensing technology in the CPM domain. The proposed system was verified using field data from a superstructure construction project where the object recognition indicators showed a 98% recall and 99% precision in recognizing 3D objects. The proposed system also used a color-coding system to address the condition of each element based on its recognition and scheduling state and address any occlusions while calculating the recognized objects. The results also revealed an automatically updated status of the project's progress in terms of schedule(4D) and cost(5D). The automated results were also validated with a manual calculation, where a slight variation (1.35%) was observed between those calculations. This system demonstrates a degree of accurate progress tracking, automatically exceeding manual performance with less computational time.

Keywords: automated progress tracking; 5D BIM; laser scanning; integration; EV principles

Citation: ElQasaby, A.R.; Alqahtani, F.K.; Alheyf, M. Automated Schedule and Cost Control Using 3D Sensing Technologies. *Appl. Sci.* **2023**, *13*, 783. https://doi.org/10.3390/app13020783

Academic Editor: Dimitris Mourtzis

Received: 29 November 2022
Revised: 25 December 2022
Accepted: 29 December 2022
Published: 5 January 2023

1. Introduction

The success of projects is evaluated through project completion within constraints of time, scope, cost, and quality. According to the KSA vision 2030 report, in 2017 alone, approximately 60% of construction projects were 20% behind schedule. In addition, more than 35% of the project time was spent collecting and analyzing data [1]. Further, approximately 15% of the construction cost was for rework activities. Therefore, time and cost were wasted in collecting and analyzing data, making as-built plans, monitoring the project, and fixing errors [1]. Therefore, researchers turned their attention to automated inspection to increase the response time of delayed activity rather than manual inspection [2,3]. Another example is that researchers were inclined to use automation methods to track and monitor the construction progress for better visualization after 2007 [4].

In this context, the construction progress monitoring domain (CPM) has developed massively in the last two decades. The exponential increase in computational capacities has allowed the architectural, engineering, and construction (AEC) industry to develop and implement automated methods in the CPM field. Lately, the development of the CPM field has depended on two primary methods: building information modeling (BIM) and 3D sensing technologies. BIM is focused on accurately establishing "as planned" 3D models. An as-planned model can also generate a spatial representation of project components.

Then, integrating a 3D model with the project's information could produce an accurate 3D BIM-based model [5,6]. The four-dimensional model was also recognized as the schedule model (4D). The scheduling model has been designated to establish the activities' sequence over time. Cost information on the project's activities is another dimension of BIM known as 5D BIM. Activities' completion and cost over time have been simulated in a virtual environment. There have been limitations to the current BIM-based cost model, such as the cash flow analysis [7]. Some researchers also identified the sixth dimension as the facility phase. However, other studies referred to the sixth dimension as sustainability and its implementation in smart cities [8].

2. Previous Studies

3D sensing technologies are another crucial aspect that improved immensely track and monitor the progress of construction components. These technologies included radio frequency identification (RFID), an ultra-wideband system (UWB), a global positioning system (GPS), image processing methods, and laser scanners (LS) [9]. Previously, researchers managed to assemble "as-built" models [10], where a developed as-built model was created to restore, record, and improve historic buildings. Another study investigated integrating BIM and remote sensing instruments where a BIM-based model with a laser scanner was integrated for quality control in real-time to reduce schedule and cost overrun [11].

Among the 3D sensing technologies, A laser scanner, also known as Light Detection and Ranging (LiDAR), is one of the AEC industry's most recognized technology. Laser scanning aims to map 3D objects into point cloud datasets [12]. Similarly, researchers monitored and controlled the infrastructure components using as-built data using a laser scanner [13]. Laser-based methods have also been used in recognizing construction applications such as workspace modeling, asset management, and worker tracking [14]. Another laser scanner application tracks buildings' temporary or secondary components [15]. Although there are other 3D sensing technologies, laser scanners (LS) are one of the best-fitted technologies to track and monitor the 3D status of projects accurately [16–19]. In addition, researchers used automated methods as they have lesser limitations and could save much work and time in assessing the progress of construction projects [16].

3D spatial technologies were used to monitor and control the progress in the CPM domain. Some researchers applied the RFID system to form an as-built model, while others used UWB systems [20,21]. The image processing technology was also used in the CPM field using digital images or UAVs of construction activities [22,23]. Point cloud data sets were similarly used to evaluate the progress in construction buildings through laser scanning technology [19,24]. However, researchers used more than one sensing technology for more robustness and better results; for example, a study conducted by [25] used more than one 3D sensing technology (UWB system and laser scanner). Another study used a combination of RFID and laser scanning technology [26]. Other studies have used a fusion of image processing methods and laser scanners [24,27–29].

Furthermore, some latest review articles discussed different insights to recognize knowledge gaps and recommend future directions in the CPM field. For example, the methodology in [4] applied scientometric analysis to point out a broad picture of CPM. Another example is a systematic literature survey conducted to automate indoor progress monitoring [30]. The 3D model reconstruction and geometry quality inspection were also discussed comprehensively using the point cloud datasets [31]. Meta-analysis was estimated to review the quality of studies of object recognition performance indicators in the CPM field between 2007 and 2021 [32]. Previous studies recommended the usage of 5D assessment for future research to address the gap in the BIM integrated with 3D sensing technologies in CPM applications.

Therefore, the contribution of this paper is to propose an automated construction progress tracking system for schedule and cost control. The reason behind that is the lack of research on conducting 5D assessment in the CPM domain using EV principles [4,19,30–32]. The proposed system automatically implements a 5D assessment: progress feedback regarding schedule and cost per scan. The 5D assessment enables reviewing the progress and states the project condition through EV principles (schedule performance index, cost performance index). The proposed system outcomes were also compared to the manual system to validate the accuracy of the proposed system.

3. Methodology

This paper illustrates an automated progress tracking system in construction progress monitoring to assess the updated information in schedule and cost. A BIM-planned model was established from 2D shop drawings. Then, the as-built model was established by collecting scans using laser scanning technology, processing them, and registering them to a common coordinate system. The as-built model would also be evaluated and assessed to determine the quality of point cloud sets based on three KNN searching algorithms: a fixed number of nearest neighbors, a fixed neighborhood radius, and an adaptive neighborhood radius. Once the integration between those models was automatically established, two algorithms were designed and developed to recognize the as-built objects. Another two algorithms were developed to review the progress in terms of schedule and cost (4D, 5D) using EV principles. The flowchart of the proposed approach is shown in Figure 1. Further explanation of the methodological steps is provided in the following sections.

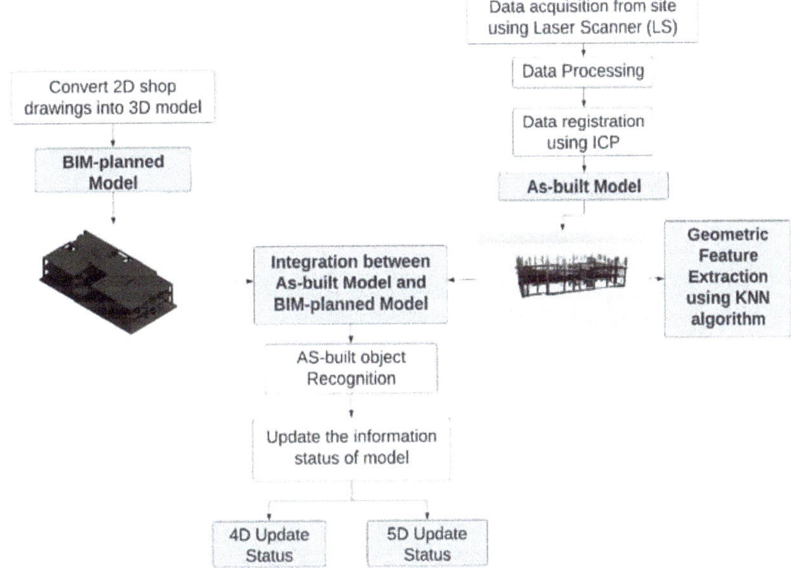

Figure 1. Steps of Methodology.

3.1. Tools

In order to apply the methodology mentioned above, a set-up of a BIM-planned model and an as-built model was crucial to be established. Revit interface was used to establish the BIM-planned model by converting 2D shop drawings to 3D models. Then, the planned schedule and cost models were manually established based on material, equipment, and labor costs for each project milestone. Material costs include supplies or materials purchased for the project, such as concrete, walls, and rebars. The transportation and storage cost is also included in the cost of materials.

The data acquisition was conducted on-site using a laser scanner Faro Focus^{3D} because laser scanners are mainly accurate and efficient [33,34] (See Section 3.3). Then, datasets were processed, the scattered points were transformed into a range image, and laser scans were registered using project reference points from one of the local coordinate systems of multiple scans to a common coordinate system [35,36]. Iteratively closest point (ICP) was then used in registration where correspondences between points of a scan called the source and points of another called the target was established to minimize the spatial distance between points in each pair. The reason behind using ICP was to achieve satisfactory registration results [37]. The outcome of the previous procedures was to generate the as-built model.

Once the as-built model was established, the next step was to indicate the strength or the weakness of the spatial distribution of the datasets by feeding the as-built model with K-nearest neighbors search algorithms (fixed number of nearest neighbors (Method I), fixed neighborhood radius (Method II), and adaptive neighborhood radius (Method III)) [38,39]. One million points were used as a reasonable sample because using the original point clouds is not computationally feasible. Firstly, the KNN algorithm based on a fixed number of neighboring points was set as [500, 5000] with an interval of 50. Secondly, the KNN algorithm based on a fixed neighborhood radius was explored using a neighborhood threshold between 5 cm–50 cm with a step length of 5 cm. The neighborhood threshold considered the further analysis of geometric features of columns, beams, and slabs when setting the threshold radius. Finally, the KNN method based on adaptive radius was set between 1–30 cm with an interval of 1cm by calculating the information entropy of the neighboring point cloud set [40]. The chosen lower band was set based on the point cloud noise, density, sensor specification, and computational constraints. However, the chosen upper band was set based on the most significant object in the scene (facades for LS-data sets) [41]. The geometric features shown in Table 1 were obtained to illustrate the spatial distribution of the datasets based on the KNN searching algorithms [41–43].

Table 1. Definitions of geometric features.

Geometric Feature	Equation	Definition
Linear Index L_λ	$\frac{\lambda_1 - \lambda_2}{\lambda_1}$	represents the linear features of the neighboring point cloud clusters
Planar Index P_λ	$\frac{\lambda_2 - \lambda_3}{\lambda_1}$	represents the planar features of the neighboring point cloud clusters
Scatter Index S_λ	$\frac{\lambda_3}{\lambda_1}$	represents the scattering features of the neighboring point cloud clusters

3.2. Methods

3.2.1. Three-Dimensional Object Recognition

As soon as the point cloud assessment was completed, the as-built model was incorporated into the Revit interface. Therefore, a transformation matrix was established where a planned model was fixed. The point cloud model was then transformed to match the reference model automatically. It was stated that the point cloud was clumsy enough to be recognized. Thus, the point cloud set was transformed into a geometry-based model, as mentioned thoroughly in Algorithm 1.

After the geometry-based model was established, the proposed approach was introduced to initiate this recognition system representing the correspondence between the BIM-planned model and the as-built model. The proposed approach depended on three main aspects. Firstly, the alignment accuracy between the two models was vital to object recognition. Secondly, the recognition approach was based on three distinctive features called the Lalonde features [44]. It was also used for the linerarness, surfaceness, and scatterness of a 3D point cloud set [45,46]. Finally, at least 95% of an element would intersect with the geometry model to be considered a recognized element, as declared thoroughly in Algorithm 2. In other words, Algorithm 2 searches through the BIM-planned model to find the closest geometry to each BIM-placed object. If the BIM-planned object is found, the actual component is classified based on the object type in the BIM-planned model.

Algorithm 1: Transformation of point cloud model into geometry-based model

	Input: Point cloud P, Point cloud model where $P \in P_M$,
	*Output: Object **O**, Geometry-based model G_M, Structural elements **E**,*
1	*Get Pointcloud Instance From **P** file*
2	*If pointCloudInstance $\neq$ null*
3	*Then*
4	*P = pointCloudInstance.GetPoint()*
5	*For Each **P** in P_M*
6	*O = CreateSphereSolid(**P**)*
7	*ObjectsList = append(O)*
8	*End G_M = DirectShape.CreateElement (**ObjectList**)*
9	*End If*
10	*End*

Algorithm 2: Comparison between the geometry-based model and BIM-planned elements

	Input: Geometry-based model G_M, BIM-planned model BIM_M
	*Output: Structural elements **E**, Linkstructural elements L_E*
1	*Get E from **P** file*
2	*Get L_E from **P** file*
3	*For Each E in BIM_M*
4	*If E does not intersect with G_M*
5	*Set Elementcolor **RED***
6	*RedList = append (E)*
7	*End If*
8	*End*
9	*For Each L_E in BIM_M*
10	*If E does not intersect with L_E **and** L_E intersects with G_M*
11	*Set Elementcolor **Green***
12	*GreenList = append (L_E)*
13	*End If*
14	*End*
15	*If E does not intersect with L_E **and** L_E does not intersect with G_M*
16	*Set Elementcolor **Yellow***
17	*YellowList = append (LE)*
18	*End If*
19	*End*
20	*If YellowList >> GreenList **Then***
21	*Set Elementcolor **Blue***
22	*Bluelist = append (L_E)*
23	*End If*
24	*End*

Then, a color-coding system was established to demonstrate the condition of each element based on its recognition and scheduling state, as illustrated in Table 2. Each color would represent the recognition and scheduling state and determine whether it would be included in the calculation for automated schedule and cost.

Table 2. Color Coding of elements according to Algorithm 2.

	Recognized	Not Recognized
Constructed	the color of the material	Red
Not yet Constructed	Green	Yellow
Not yet fully constructed	Blue	Brown

3.2.2. Automated Schedule and Cost Control

To update the project's status in terms of schedule and cost (4D, 5D), the authors developed two algorithms based on the results of the object recognition system. On one hand, Algorithm 3 calculated the 4D updated status based on the BCWS and BCWP estimated from the BIM-planned model and the geometry-based model, respectively, where budgeted unit cost was inserted into the algorithm. The element's color would also determine whether its cost would be included. Then, the schedule performance index (SPI) was calculated automatically to review the schedule status of a project.

Algorithm 3: Calculate the automated schedule progress

Input: Structural elements E, Linkstructural elements L_E, Budget unit cost B_{Cost} Concrete Volume V_c
Output: Geometry Model Cost G_M Cost, BIM-planned model total cost BIM_M TC, SPI

1   **For Each** Category **in** E
2 | **Get** V_c **For** category
3 | **Calculate** Category cost **From** B_{Cost} and V_c
4 | BIM_M TC = Category Cost
5   **End**
6   **For Each** L **in** Redelement
7 | **Calculate** Red TC **From** B_{Cost} and $V_{c\,Red}$
8   **End**
9   **For Each** L_E **in** Greenelement
10 | **Calculate** Green TC **From** B_{Cost} and $V_{c\,Green}$
11 **End**
12 **For Each** L_E **in** Yellowelement
13 | **Calculate** Yellow TC **From** B_{Cost} and $V_{c\,Yellow}$
14 **End**
15 G_M Cost = BIM_M TC − Red TC + Green TC + Yellow TC
16 SPI = G_M Cost/BIM_M TC
17 **If** SPI > 1
18 | **Then Print** *"Ahead of schedule."*
19 **Else If** SPI < 1
20 | **Print** *"Behind schedule."*
21 **Else Print** *"Within schedule."*
22 **End If**
23 **End**

On the other hand, Algorithm 4 calculated the 5D updated status based on the BCWP and ACWP estimated from the geometry-based model and the revised BIM-planned model, respectively, where the actual unit cost was inserted into the algorithm. The element's color would determine whether its cost would be included. Then, the cost performance index (CPI) was calculated automatically to review the cost status of the project.

Algorithm 4: Calculate the automated Cost progress

Input: Structural elements E, Actual unit cost A_{Cost}, Geometry Model Cost G_M Cost
Output: Actual Total cost Actual TC, CPI

1   **For Each** Category **in** E
2 | **Get** V_C **For** Category
3 | **Calculate** Category cost **From** A_{Cost} and V_C
4 | **Actual** TC = Category Cost
5   CPI = G_M Cost/Actual TC
6   **If** CPI > 1
7 | **Then Print** *"Under Budget"*

Algorithm 4: *Cont.*
8   *Else If CPI < 1*
9   │ *Then Print "Over Budget"*
10 *Else Print "On Budget"*
11 *End If*
12 *End*

3.3. Case Study

The data comprises a set of four field laser scans obtained from an investment building in The Rawda Administration Center, mainly consisting of reinforced concrete frame structure and Hardy slabs. The project location is [24.795813, 46.839646] beside Shaikh Isa Bin Salman Al Khalifah Rd, Al Maizilah, Riyadh. The site image of the case study is shown in Figure 2.

Figure 2. Site Image of the case study.

The construction site was scanned using Faro Focus^{3D} [47] between 25 December 2020 and 20 January 2021. The weather on the days of the survey was hot; however, with a clear sky and low wind.

4. Results and Discussions

4.1. Point Cloud Characteristics

Regarding the evaluation of registration quality, the point cloud characteristics should be thoroughly discussed. Table 3 summarizes the dataset's characteristics in all scans. The findings showed a relatively high average result of RMSE. The first reason behind that is Faro Focus 3D usually has a higher ranging error [+2 mm at 10 m and 25 m each at 90% and 10% reflectivity excluding the noise], according to the data from the manual of Faro Focus^{3D} [47]. The second reason for a higher RMSE is that fewer tie points were not scanned when the scans were conducted initially. As a result, manual point matching was used, leading to a relatively higher registration error, as previously confirmed by [33]. However, the RMSE results in [19] showed a lower RMSE/Scan of 1.68 mm than the registration results conducted in this paper due to the usage of signalized targets on presurveyed site control points.

Table 3. Datasets characteristics in the case study.

Scans	Scan Date	Stations/Scan	Point Cloud (Millions)	Standard Deviation σ (mm)	RMSE (mm)	Min Overlap (%)	Inclinometer Mismatch (°)
Scan 1	25 December 2020	11	14.7	3.07	4.52	41.26	0.012
Scan 2	6 January 2021	13	11.72	3.74	6.30	40.1	0.021
Scan 3	14 January 2021	11	9.82	3.73	6	31.7	0.0735
Scan 4	20 January 2021	11	11.22	3.71	5.5	35.47	0.021

The noise of the point cloud revealed an average of 3.6 mm, within the threshold of the range noise. The results also showed a minimum overlap of more than 30% for the four scans. The inclinometer mismatch error of all scans also indicated lower results, implying a good-quality scan registration within the sensor specification.

4.2. Point Cloud Assessment

KNN search algorithms were used to evaluate the quality of the point cloud datasets by extracting specific geometric features (linear index L_λ, planar index P_λ, and scattered index S_λ). Table 4 illustrates the geometric features obtained from the KNN search algorithms (see Section 3.1). The sample set was calculated respectively based on the eigenvalues. The results of a method I revealed that the salience features of the sample points were (linear, planar = 42.9%). However, the salience feature changed to (planar = 50%) and (planar = 74%) in methods II and III, respectively. The feature extraction values in Method III were more accurate than in Method I and Method II due to the use of entropy information that led to less unpredictability of data (more distribution).

Table 4. Geometric Features Extraction according to KNN methods.

Methods	L_λ	P_λ	S_λ
Method I (K = 5000)	0.429	0.429	0.143
Method II (r = 50 cm)	0.333	0.50	0.168
Method III	0.20	0.740	0.060

Therefore, the results showed that the majority of the sample sets were classified as linear and planar with a small index of scattereness, which was reflected in the robust distribution of the datasets. However, previous studies pointed out that the sample set on forests was divergent, where the salience feature changed based on the type of objects. Some objects, such as stems, exhibited a linear index or planar index, while others, such as leaves, exhibited a scatterness index [39,43]. Similarly, in this paper, the results of the geometric features indicated the structure of the sample set where most structural elements were classified as linear or planar.

4.3. Three-Dimensional Object Recognition

The proposed approach's object recognition results were demonstrated using recall and precision rates for two incorporated scans. The recall and precision results achieved exceptionally satisfactory performance, 98%, and 99%, respectively, on average between scans. The minor errors result from objects with only a few points acquired in the scans or temporary objects with a few points wrongly recognized (false negative and false positive rates).

Figure 3 shows the object recognition results obtained from the scan on 25 December 2020, where the foundations were not recognized because the data acquisition date was after the backfilling activities, making the foundations invisible for the laser to recognize. As a result, the foundations were colored red. However, the second-floor columns showed visible progress in the schedule as the work performed exceeded the schedule; hence they are colored green. Some of these columns failed to be recognized because of occlusions; hence they are colored yellow.

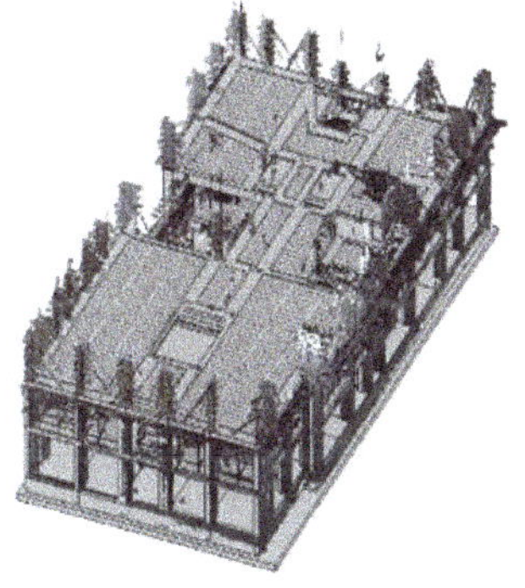

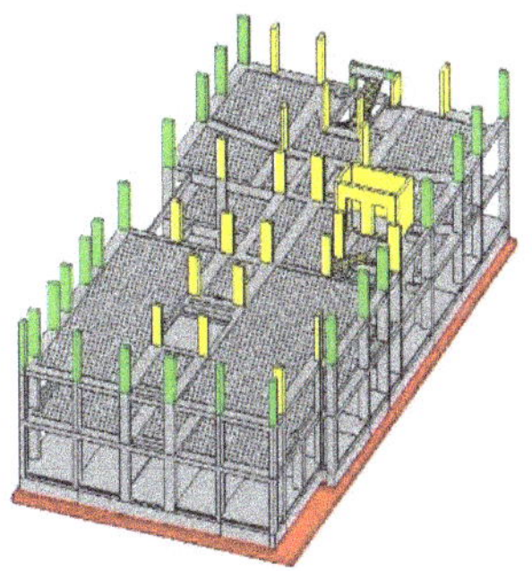

(**a**) Models before generating the code   (**b**) Models before generating the code

Figure 3. The object recognition results between models on 25 December 2020.

Similarly, from Figure 4, it was observed that the object recognition results obtained from the scan on 20 January 2021, where the foundations were not recognized, as mentioned in the previous paragraph. Nevertheless, the second-floor slab showed modest progress in the schedule; hence they are colored green. While the second-floor slab is not yet fully constructed; therefore, they are colored blue.

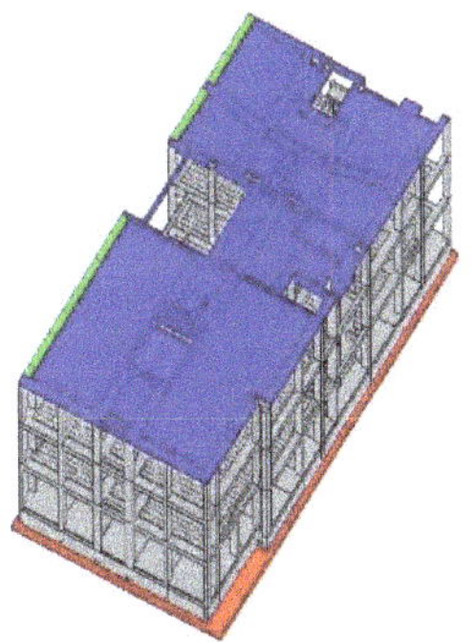

(**a**) Models before generating the code   (**b**) Models before generating the code

Figure 4. The object recognition results between models on 20 January 2021.

Three-dimensional object recognition is built mainly on the similarities between the attributes and properties in as-planned and as-actual models. Therefore, researchers used various approaches to recognize the 3D point clouds. Therefore, the case study findings and previous studies regarding object recognition results [17,19,37] are compared. The case study findings show a higher precision and recall than those presented by [17,37]. However, it is in agreement with the findings reported by [19].

4.4. Schedule and Cost Control

The proposed system generates a user interface where the calculation of the progress tracking for schedule and cost can be measured. The user interface estimates the progress of schedule and cost based on the principles of EV using the budgeted cost of work scheduled (BCWS), Budgeted cost of work performed (BCWP), and Actual cost of work performed (ACWP). Figure 5 shows the progress tracking results for the scan acquired on 25 December 2020 using the schedule and cost of concrete work in the case study. Figure 5 also shows the total cost from unrecognized elements in BCWS, BCWP, and ACWP.

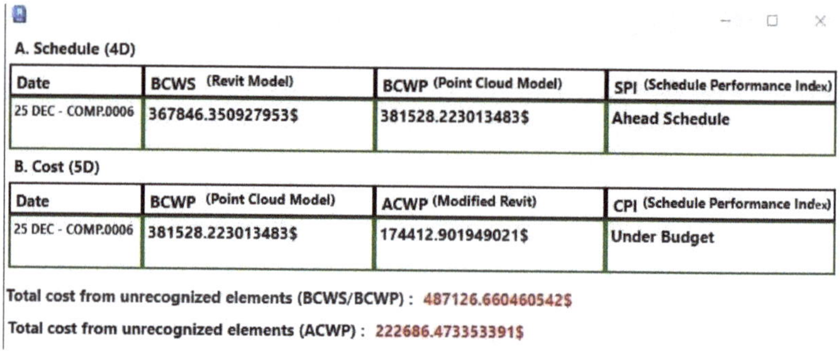

A. Schedule (4D)

Date	BCWS (Revit Model)	BCWP (Point Cloud Model)	SPI (Schedule Performance Index)
25 DEC - COMP.0006	367846.350927953$	381528.223013483$	Ahead Schedule

B. Cost (5D)

Date	BCWP (Point Cloud Model)	ACWP (Modified Revit)	CPI (Schedule Performance Index)
25 DEC - COMP.0006	381528.223013483$	174412.901949021$	Under Budget

Total cost from unrecognized elements (BCWS/BCWP) : 487126.660460542$

Total cost from unrecognized elements (ACWP) : 222686.473353391$

Figure 5. The proposed system's user interface Calculation of SPI and CPI on 25 December 2020.

However, as previously illustrated, the foundations were considered the only unrecognized elements in this case study. They must be included in the schedule and cost estimation because construction activities depend on the foundations' completion. In this case study, due to the 100% completion of the unrecognized elements (the foundations), BCWS is equivalent to BCWP as presented below Cost (5D) in Figure 5.

Table 5 demonstrates the results of the 4D progress for the case study, including the costs of unrecognized elements. Using the calculated SPI, the schedule performance of the whole project at chosen scans was determined. The results showed a fast-track project where the two scans were ahead of schedule, as the SPI was larger than one in both scans. Fewer studies were conducted to update the schedule automatically [15,27,37]. These studies mainly depended on the construction schedule to show the progress. Meanwhile, this paper demonstrates the automation of an updated schedule based on the visible recognized elements and their budget unit cost to calculate the schedule performance index.

Table 5. Result of the earned value (SPI) to determine the project's 4D progress (Including cost from unrecognized elements).

Scan Date	BCWS ($)	BCWP ($)	SPI	4D Performance
25 December 2020	854973	868655	1.016	Ahead
20 January 2021	888250	894171	1.007	Ahead

In addition, Table 6 demonstrates the results of the 5D progress for the case study, including the costs of unrecognized elements. The cost performance of the whole project was obtained using the calculated SPI. The results showed a saving project where the two scans were under budget, as the CPI was larger than one. To the author's best knowledge, this paper is the first to address the 5D assessment in the CPM domain integrated with BIM because previous studies indicated the lack in this field, as presented in [30–32].

Table 6. Result of the earned value (CPI) to determine the project's 5D progress (Including cost from unrecognized elements).

Scan Date	BCWP ($)	ACWP ($)	CPI	5D Performance
25 December 2020	868655	397099	2.188	Under Budget
20 January 2021	894171	408763	2.188	Under Budget

4.5. Comparison between Manual and Automated Calculations

To validate the accuracy of the automated user interface, the authors compare manual and automated techniques in progress calculations. Figure 6a,b show the comparison results between the manual and automated calculations on 25 December 2020 and 20 January 2021. Firstly, on 25 December 2020, the calculations are approximately similar in BCWS, but they differ in both BCWP and ACWP calculations due to the included cost of formwork and steel fixed rebar of stairs on the second floor in manual calculations. The same issue happened on 20 January 2021, where the manual result of BCWP and ACWP was slightly higher than the automated result because the manual results included the cost of the formwork of the third-floor slab.

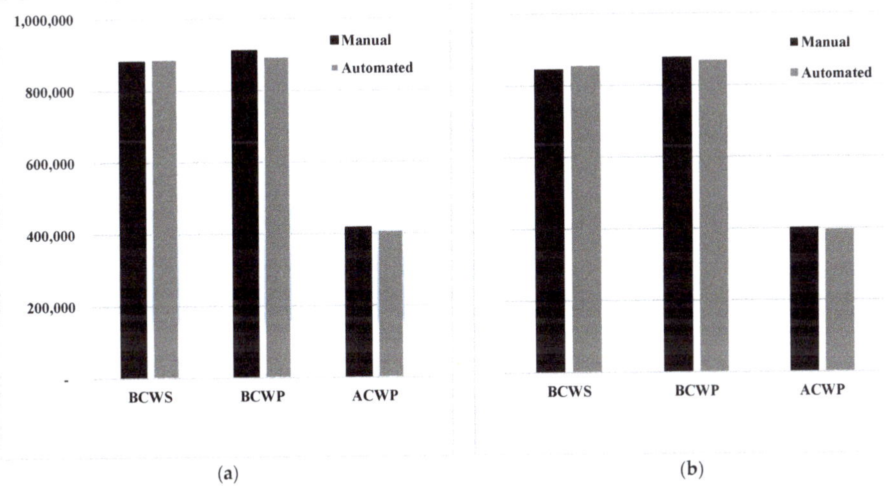

Figure 6. The results between manual and automated calculations on two selected dates: (**a**) 25 December 2020, (**b**) 20 January 2021.

The proposed system does not calculate the objects that are not fully constructed (colored in blue) but takes the occlusion objects that failed to be recognized (colored in yellow) into account in its estimation formula. The automated results differ only by 1.35% from the manual calculation. This variation comprises only the objects that are not fully constructed and the wrongly recognized objects.

Further, Table 7 compares manual and automated calculations regarding SPI and CPI, where the results showed a slight difference in SPI results between manual and automated calculations, even though it does not affect the project's status. At the same time, the CPI results showed approximate results between manual and automated calculations. The study findings above prove the validity of the proposed system. Hence, this system can be adapted to construction projects, enhancing the monitoring and controlling process as well as increasing the efficiency of schedule and cost updates with less time. This system can be developed outside the Autodesk platform, expanding the knowledge beyond one platform. Additionally, this system can be expanded to incorporate more domain knowledge (sustainability 6D, facility management 7D). Further, this system can be conducted in different types of projects, which could highlight other factors for improvement.

Table 7. The results of manual and automated calculation in terms of SPI and SPI.

Scan Date	Manual		Automated	
	SPI	**CPI**	**SPI**	**CPI**
25 December 2020	1.037	2.183	1.016	2.188
20 January 2021	1.034	2.183	1.007	2.188

5. Conclusions

This paper presented an automated progress-tracking system that integrates 5D information data with laser scanning using the data collected from a superstructure construction project. The proposed system algorithms were based on the intersection percentage between models, alignment accuracy, and Lalonde features. The proposed system also automatically estimates the construction progress and updates the project status in schedule and cost with less computational time. The main findings of the case study revealed that the object recognition indicators (recall and precision) achieved a remarkably decent performance of 98% and 99%, respectively. The proposed system also uses a color-coding system to address the different conditions of elements. Additionally, it also considers occlusions when calculating the recognized progress.

The proposed system also shows that the automated calculations of updated schedules and costs can improve progress estimation results compared to manual calculations, where there is a slight variation of only (1.35%) between manual and automated calculations. The reason is that the current system's estimation formula does not consider the cost of not fully constructed objects until they are completed. Thus, as future work, the current system should be evaluated in other construction buildings to declare a guideline and improvement. The authors acknowledge that the current approach has some limitations (i.e., the system is only available via the Autodesk Revit platform, and the laser scanner needs experienced labor). However, there is sufficient improvement using this approach to monitor the progress along with Earned Value principles.

Author Contributions: Conceptualization, A.R.E., F.K.A. and M.A.; methodology, A.R.E.; software, A.R.E.; validation, A.R.E., F.K.A. and M.A.; formal analysis, A.R.E., investigation, A.R.E., resources, A.R.E., F.K.A. and M.A.; data curation, A.R.E.; writing—original draft preparation, A.R.E.; writing—review and editing, A.R.E., F.K.A. and M.A.; visualization, A.R.E.; supervision, F.K.A. and M.A.; project administration, F.K.A. and M.A.; funding acquisition, F.K.A. All authors have read and agreed to the published version of the manuscript.

Funding: This research was funded by the Researchers Supporting Project number (RSP2023R264), King Saud University, Riyadh, Saudi Arabia.

Institutional Review Board Statement: Not applicable.

Informed Consent Statement: Not applicable.

Data Availability Statement: The data presented in this study are available on request from the corresponding author. The data are not publicly available due to privacy or ethical restrictions.

Acknowledgments: The authors extend their appreciation to the Researchers Supporting Project number (RSP2023R264), King Saud University, Riyadh, Saudi Arabia for funding this work.

Conflicts of Interest: The authors declare no conflict of interest.

Abbreviations

List of abbreviations and acronyms used in the paper.

AEC	Architectural, Engineering, and Construction
BIM	Building Information Modeling
LS	Laser scanner
LiDAR	Light Detection and Ranging
GPS	Global Positioning System
RFID	Radio Frequency Identification
UWB	Ultra-wideband
CPM	Construction Progress Monitoring
KNN	K-nearest neighborhood
UAV	Unmanned Aerial Vehicle
ICP	Iteratively Closest Point
PCA	Principal Component Analysis
RMSE	Root Mean Square Error
EV	Earned Values
SPI	Schedule Performance Index
CPI	Cost Performance Index
ACWP	Actual Cost Work Performed
BCWP	Budgeted cost of work Performed
BCWS	budgeted cost of work Scheduled

References

1. KSA Vision 2030. Available online: https://www.vision2030.gov.sa/ar/ (accessed on 28 December 2022).
2. Yi, W.; Chan, A.P. Critical review of labor productivity research in construction journals. *J. Manag. Eng.* **2014**, *30*, 214–225. [CrossRef]
3. Zavadskas, E.K.; Vilutienė, T.; Turskis, Z.; Šaparauskas, J. Multi-criteria analysis of Projects' performance in construction. *Arch. Civ. Mech. Eng.* **2014**, *14*, 114–121. [CrossRef]
4. Patel, T.; Guo, B.H.; Zou, Y. A scientometric review of construction progress monitoring studies. *Eng. Constr. Archit. Manag.* **2021**, *29*, 3237–3266. [CrossRef]
5. Azhar, S. Building information modeling (BIM): Trends, benefits, risks, and challenges for the AEC industry. *Leadersh. Manag. Eng.* **2011**, *11*, 241–252. [CrossRef]
6. Kensek, K.M. *Building Information Modeling*, 1st ed.; Routledge: London, UK, 2014.
7. Lu, Q.; Won, J.; Cheng, J.C. A financial decision-making framework for construction projects based on 5D Building Information Modeling (BIM). *Int. J. Proj. Manag.* **2016**, *34*, 3–21. [CrossRef]
8. Alqahtani, F.K.; El Qasaby, A.R.; Abotaleb, I.S. Urban Development and Sustainable Utilization: Challenges and Solutions. *Sustainability* **2021**, *13*, 7902. [CrossRef]
9. Alizadehsalehi, S.; Yitmen, I. A concept for automated construction progress monitoring: Technologies adoption for benchmarking project performance control. *Arab. J. Sci. Eng.* **2019**, *44*, 4993–5008. [CrossRef]
10. Baik, A.H.A.; Yaagoubi, R.; Boehm, J. Integration of Jeddah historical BIM and 3D GIS for documentation and restoration of historical monument. *Int. Soc. Photogramm. Remote Sens. (ISPRS)* **2015**, *XL-5/W7*, 29–34.
11. Wang, J.; Sun, W.; Shou, W.; Wang, X.; Wu, C.; Chong, H.Y.; Liu, Y.; Sun, C. Integrating BIM and LiDAR for real-time construction quality control. *J. Intell. Robot. Syst.* **2015**, *79*, 417–432. [CrossRef]
12. Zhang, C.; Arditi, D. Automated progress control using laser scanning technology. *Autom. Constr.* **2013**, *36*, 108–116. [CrossRef]
13. Son, H.; Bosché, F.; Kim, C. As-built data acquisition and its use in production monitoring and automated layout of civil infrastructure: A survey. *Adv. Eng. Inform.* **2015**, *29*, 172–183. [CrossRef]
14. Chen, J.; Cho, Y.K. Point-to-point comparison method for automated scan-vs-bim deviation detection. In Proceedings of the 2018 17th International Conference on Computing in Civil and Building Engineering, Tampere, Finland, 5–7 June 2018; pp. 5–7.
15. Turkan, Y.; Bosché, F.; Haas, C.T.; Haas, R. Tracking secondary and temporary concrete construction objects using 3D imaging technologies. In *Computing in Civil Engineering, Proceedings of the ASCE International Workshop on Computing in Civil Engineering, Los Angeles, CA, USA, 23–25 June 2013*; American Society of Civil Engineers: Reston, VA, USA, 2013.
16. Cheok, G.S.; Stone, W.C.; Lipman, R.R.; Witzgall, C. Ladars for construction assessment and update. *Autom. Constr.* **2000**, *9*, 463–477. [CrossRef]
17. Turkan, Y.; Bosche, F.; Haas, C.T.; Haas, R. Automated progress tracking using 4D schedule and 3D sensing technologies. *Autom. Constr.* **2012**, *22*, 414–421. [CrossRef]
18. Bosché, F.; Ahmed, M.; Turkan, Y.; Haas, C.T.; Haas, R. The value of integrating Scan-to-BIM and Scan-vs-BIM techniques for construction monitoring using laser scanning and BIM: The case of cylindrical MEP components. *Autom. Constr.* **2015**, *49*, 201–213. [CrossRef]

19. Maalek, R.; Lichti, D.D.; Ruwanpura, J.Y. Automatic recognition of common structural elements from point clouds for automated progress monitoring and dimensional quality control in reinforced concrete construction. *Remote Sens.* **2019**, *11*, 1102. [CrossRef]

20. Fang, Y.; Cho, Y.K.; Zhang, S.; Perez, E. Case study of BIM and cloud-enabled real-time RFID indoor localization for construction management applications. *J. Constr. Eng. Manag.* **2016**, *142*, 05016003. [CrossRef]

21. Rashid, K.M.; Louis, J.; Fiawoyife, K.K. Wireless electric appliance control for smart buildings using indoor location tracking and BIM-based virtual environments. *Autom. Constr.* **2019**, *101*, 48–58. [CrossRef]

22. Golparvar-Fard, M.; Peña-Mora, F.; Savarese, S. D4AR–a 4-dimensional augmented reality model for automating construction progress monitoring data collection, processing and communication. *J. Inf. Technol. Constr.* **2009**, *14*, 129–153.

23. Dimitrov, A.; Golparvar-Fard, M. Vision-based material recognition for automated monitoring of construction progress and generating building information modeling from unordered site image collections. *Adv. Eng. Inform.* **2014**, *28*, 37–49. [CrossRef]

24. Kim, H.E.; Kang, S.H.; Kim, K.; Lee, Y. Total variation-based noise reduction image processing algorithm for confocal laser scanning microscopy applied to activity assessment of early carious lesions. *Appl. Sci.* **2020**, *10*, 4090. [CrossRef]

25. Shahi, A.; Safa, M.; Haas, C.T.; West, J.S. Data fusion process management for automated construction progress estimation. *J. Comput. Civ. Eng.* **2015**, *29*, 04014098. [CrossRef]

26. Hajian, H.; Becerik-Gerber, B. A research outlook for real-time project information management by integrating advanced field data acquisition systems and building information modeling. In *Computing in Civil Engineering, Proceedings of the ASCE International Workshop on Computing in Civil Engineering, Austin, TX, USA, 24–27 June 2009*; Caldas, C.H., O'Brien, W.J., Eds.; American Society of Civil Engineers: Reston, VA, USA, 2009; pp. 83–94.

27. Han, K.K.; Golparvar-Fard, M. Appearance-based material classification for monitoring of operation-level construction progress using 4D BIM and site photologs. *Autom. Constr.* **2015**, *53*, 44–57. [CrossRef]

28. Teizer, J. Status quo and open challenges in vision-based sensing and tracking of temporary resources on infrastructure construction sites. *Adv. Eng. Inform.* **2015**, *29*, 225–238. [CrossRef]

29. Han, K.; Degol, J.; Golparvar-Fard, M. Geometry-and appearance-based reasoning of construction progress monitoring. *J. Constr. Eng. Manag.* **2018**, *144*, 04017110. [CrossRef]

30. Ekanayake, B.; Wong, J.K.W.; Fini, A.A.F.; Smith, P. Computer vision-based interior construction progress monitoring: A literature review and future research directions. *Autom. Constr.* **2021**, *127*, 103705. [CrossRef]

31. Wang, Q.; Kim, M.K. Applications of 3D point cloud data in the construction industry: A fifteen-year review from 2004 to 2018. *Adv. Eng. Inform.* **2019**, *39*, 306–319. [CrossRef]

32. ElQasaby, A.R.; Alqahtani, F.K.; Alheyf, M. State of the Art of BIM Integration with Sensing Technologies in Construction Progress Monitoring. *Sensors* **2022**, *22*, 3497. [CrossRef]

33. Bosche, F.; Haas, C.T.; Akinci, B. Automated recognition of 3D CAD objects in site laser scans for project 3D status visualization and performance control. *J. Comput. Civ. Eng.* **2009**, *23*, 311–318. [CrossRef]

34. Kopsida, M.; Brilakis, I.; Vela, P.A. A review of automated construction progress monitoring and inspection methods. In Proceedings of the 32nd CIB W78 Conference, Eindhoven, The Netherlands, 27–29 October 2015; Volume 2015, pp. 421–431.

35. Alheyf Aldosari, M.D. Mobile LiDAR for Monitoring MSE Walls with Smooth and Textured Precast Concrete Panels. Ph.D. Dissertation, Purdue University Graduate School, West Lafayette, IN, USA, 2020.

36. Bosché, F. Plane-based registration of construction laser scans with 3D/4D building models. *Adv. Eng. Inf.* **2012**, *26*, 90–102. [CrossRef]

37. Kim, C.; Son, H.; Kim, C. Automated construction progress measurement using a 4D building information model and 3D data. *Autom. Constr.* **2013**, *31*, 75–82. [CrossRef]

38. Liang, X.; Kankare, V.; Yu, X.; Hyyppä, J.; Holopainen, M. Automated stem curve measurement using terrestrial laser scanning. *IEEE Trans. Geosci. Remote Sens.* **2013**, *52*, 1739–1748. [CrossRef]

39. Xia, S.; Wang, C.; Pan, F.; Xi, X.; Zeng, H.; Liu, H. Detecting stems in dense and homogeneous forests using single-scan TLS. *Forests* **2015**, *6*, 3923–3945. [CrossRef]

40. Brodu, N.; Lague, D. 3D terrestrial lidar data classification of complex natural scenes using a multi-scale dimensionality criterion: Applications in geomorphology. *ISPRS J. Photogramm. Remote Sens.* **2012**, *68*, 121–134. [CrossRef]

41. Demantké, J.; Mallet, C.; David, N.; Vallet, B. Dimensionality based scale selection in 3D lidar point clouds. *Laserscanning* **2011**. [CrossRef]

42. Lin, Y.J.; Benziger, R.R.; Habib, A. Planar-based adaptive down-sampling of point clouds. *Photogramm. Eng. Remote Sens.* **2016**, *82*, 955–966. [CrossRef]

43. Lin, W.; Fan, W.; Liu, H.; Xu, Y.; Wu, J. Classification of handheld laser scanning tree point cloud based on different knn algorithms and random forest algorithms. *Forests* **2021**, *12*, 292. [CrossRef]

44. Lalonde, J.F.; Vandapel, N.; Huber, D.F.; Hebert, M. Natural terrain classification using three-dimensional ladar data for ground robot mobility. *J. Field Robot.* **2006**, *23*, 839–861. [CrossRef]

45. Himmelsbach, M.; Luettel, T.; Wuensche, H.J. Real-time object classification in 3D point clouds using point feature histograms. In Proceedings of the IEEE/RSJ International Conference on Intelligent Robots and Systems, St. Louis, MO, USA, 11–15 October 2009; pp. 994–1000.
46. Nguyen, K.; Jain, A.K.; Allen, R.L. Automated gland segmentation and classification for gleason grading of prostate tissue images. In Proceedings of the 20th International Conference on Pattern Recognition, Istanbul, Turkey, 23–26 August 2010; pp. 1497–1500.
47. FARO Technologies Inc. _Faro_Laser_Scanner_Focus3D_Manual Faro Focus 3D, Lake Mary. October 2011. Available online: https://downloads.faro.com/index.php/s/CY5BS9Jd2JEf8YY (accessed on 28 December 2022).

 *buildings*

Article

Developing a BIM Single Source of Truth Prototype Using Blockchain Technology

Amer A. Hijazi [1,*], Srinath Perera [1], Ali M. Alashwal [1] and Rodrigo N. Calheiros [2]

[1] Centre for Smart Modern Construction, School of Engineering, Design, and Built Environment, Western Sydney University, Kingswood, NSW 2747, Australia

[2] School of Computer, Data and Mathematical Sciences, Western Sydney University, Parramatta, NSW 2150, Australia

* Correspondence: a.hijazi@westernsydney.edu.au

Abstract: Blockchain technology has been proposed as a potential solution for coordinating information and trust to aid the development of a single source of the truth data model, going beyond peer-to-peer cash transactions. It is, therefore, argued that the construction supply chain (CSC) will resolve issues related to the lack of reliable platforms for construction and asset management operations once blockchain technology and Building Information Modelling (BIM) are integrated. Though there is no longer any debate about the importance of integrating blockchain technology with BIM, there is still a lack of academic literature on its proof of concept. This study aims to create a thorough proof of concept for integrating blockchain technology and BIM for supply chain data delivery. It demonstrated a step-by-step methodology starting from understanding the current business scenario and proposing logical system architecture, followed by selecting a blockchain platform, designing system architecture related to technologies, prototyping, and evaluating through a virtual business scenario. The software prototype presented in this paper helps establish the technological viability of a single source of the truth data model for integrating blockchain technology and BIM. The supply chain data delivery for handover was considered in this software prototype. However, the process used to create this software prototype can be replicated in future work on blockchain technology-based built environment applications or digital transformation in the built environment research.

Keywords: BIM; blockchain; construction supply chain; single source of truth; software prototype

Citation: Hijazi, A.A.; Perera, S.; Alashwal, A.M.; Calheiros, R.N. Developing a BIM Single Source of Truth Prototype Using Blockchain Technology. *Buildings* **2023**, *13*, 91. https://doi.org/10.3390/buildings13010091

Academic Editor: Muhammad Shafique

Received: 28 November 2022
Revised: 23 December 2022
Accepted: 26 December 2022
Published: 30 December 2022

1. Introduction

Construction is a project-based industry [1] where stakeholders temporarily come together to complete one-off projects [2–6]. Construction projects encounter performance-related challenges throughout asset management operations, final product quality, and stakeholder conflicts [7]. For instance, a subcontractor typically has no direct obligations to anyone other than the main contractor [2]. Therefore, the construction industry has become less trustworthy with more adversarial relationships, which is considered an obstacle to industry performance and innovation [1,8]. Adopting digitalized and smart solution tools will assist in resolving performance issues and contribute to the success of construction projects by increasing industry productivity [9]. However, Mason and Escott [10] stated that transitioning to the smart construction industry still has a path full of challenges due to the traditional procurement mindsets and lack of mature/trusted methodology to enable this. Perera, Ingirige [11] stated that the efficiency of the data workflow and the ability of stakeholders to have a transparent data exchange are critical factors in enabling information and communications technology (ICT) in the construction supply chain (CSC).

Building Information Modeling (BIM) is widely acknowledged as the primary driver of the industry's digital transformation [12,13]. There have been previous studies that used BIM in the CSC to accomplish integrated information delivery all the way through the project's life cycle [3,9,14,15]. Globally, BIM plays a crucial role in facilitating the digitalization of design and related workflow in the construction industry [14–17]. Through BIM coordination tools and collaboration processes, CSC stakeholders are better informed of one another's activities [18]. There are still several obstacles to the widespread use of BIM in supply chain operations, including the construction industry's inherent complexities, a lack of openness, adversarial relationships, fragmented data, and disputes among players [2–5,8,13,14,19,20]. Because *"Construction Operation Building Information Exchange"* (COBie) data is maintained centrally by a single actor, disagreements arise across CSC parties and prevent BIM from being a legally recognized delivery model [4,8,15,19]. Furthermore, BIM tools cannot generate digital proofs for various transactions [21]. In addition, traceability issues and the inability of BIM to contain all project compliance and product data are also among the limitations of BIM [14]. Deng, Ren [22] emphasized that any future framework for the CSC's BIM integration must include trust as a prerequisite for effective communication between project stakeholders. Consequently, any future CSC research should take reliable data exchange into account.

Blockchain technology has been heralded as revolutionary [4,23–25] and is set to upend many facets of enterprises that rely on coordinating information and trust to allow a trustworthy database architecture with multiple control entities [26,27]. Blockchain is a type of distributed ledger technology, which securely records information in cryptographically sealed blocks replicated across a peer-to-peer network [14,27,28]. Due to its tamper-resistant properties and suitability for data auditability and transparency, blockchain is being recognized by a growing number of sectors as a key innovation with a trusted data exchange platform [14,27,29]. However, according to *"Australia's Department of Industry, Science, Energy, and Resources National Blockchain Roadmap"*, the construction industry is lagging behind other industries in terms of the proportion of business activities involving blockchain technology [30]. For the design and development phases of a BIM model, existing academic studies have validated the authoring copyright, but not the ownership of CSC product data or the supply and manufacturer data node (*which includes production data, compliance data, reliability data, maintenance, and warranty*) [21,31,32]. Due to poor CSC data quality and reliability, the advancement of BIM implementation for operations may not produce the desired results for its higher levels of digitalization [8,12,19,33], as it is not possible to act on a digital asset that cannot be trusted [33].

This study aims to create a prototype proof of concept for integrating blockchain and BIM for construction supply chain data delivery. The software prototype presented in this paper helps establish the technological feasibility of a single source of truth model for integrating blockchain technology and BIM. The concept of a single source of truth model was borrowed from the ICT sector [34]. The *"single source of truth"* is defined as an *"authoritative source of its data that offers data services to other entities while ensuring that business entity decisions are based on the same datasets"* [34–36]. This paper demonstrates a step-by-step methodology starting from understanding the current business scenario and proposing logical system architecture, followed by selecting a blockchain technology platform, designing system architecture related to technologies, prototyping, and evaluating it through a virtual business scenario. The process stated by Qing and Yu-Liu [37] detailing the development of software prototypes was used in this research for developing a software prototype. The scope of this paper is limited to the delivery of CSC data in preparation for handover and operation. Because it does not influence the supplier and manufacturer data node for the handover stage, Building Information Modeling (BIM) for the design stage, which is where the process is centered on BIM 3D model authoring, is excluded.

The paper is structured as follows. First, the authors conduct an in-depth analysis of the current situation of BIM and blockchain integration to investigate how blockchain technology might be used to solve the issues faced by the CSC (Section 2). Then, the authors introduce the research method, design, and tools for the BIM single source of truth prototype model development using blockchain technology (Section 3). Next, the paper presents step-by-step software prototype development and how it was comprehensively evaluated to ensure the accurate execution of the smart contract using a virtual business scenario that included external validation (Section 4). Following that, the discussion is presented, including practical implications, managerial insights, limitations of this study, and future research directions (Section 5). Finally, the authors present the conclusions and contributions of this study (Section 6).

2. Literature Review

Though there is no longer any debate about the importance of integrating blockchain technology with BIM, there is still a lack of academic literature on its proof of concept [38–41]. By demonstrating the reliability of the construction supply chain data across all supply chain actors, the integration of blockchain technology and BIM has the potential to alter the definition of BIM supply chain data delivery for facility management from information-centered 3D modeling based on coordination tools to a trusted data exchange model based on a reliable workflow [3,36,42]. However, most academic literature has used hypothetical cases to support this integration, with a significant gap in software prototype approaches [43]. Future works are recommended to demonstrate a step-by-step methodology of integrating BIM and blockchain technology to help various practical scenarios, given that BIM is the best route to implement any emerging technology in the construction industry [28,44,45]. Perera, Hijazi [28] presented a step-by-step methodology for implementing a blockchain in a built environment, which helps to provide the foundation for developing technological feasibility of proofs of concept relevant to land registry transactions; however, the proposed model does not deal with supply chain data. The proposed model [28] could result in introducing blockchain to solve transparency challenges in some built environment applications. Still, data source affects adoption, and this is where BIM needs to be in action with blockchain to create reliable supply chain data [27,35]. Regarding this, Li and Kassem [46] stated that the construction industry is not yet sufficiently digitized to fully benefit from blockchain technology. Due to this, there is still a lack of maturity in the use of blockchain technology with BIM for supply chain data delivery [14,47–49].

Current academic literature focuses primarily on how this integration may occur, presenting blockchain as a new technological tool only for aiding transparent transactions for BIM 3D modeling files in the form of *"project-centric"* 3D modeling files [32,46,50]. Celik, Petri [51] proposed a blockchain technology-based BIM model; it integrates by saving the IFC (Industry Foundation Classes) file hash code (a BIM file) and its action using smart contracts. However, this integration approach drastically restricts the utilization of blockchain technology's potential for value transfer in the form of a digital ecosystem of connected databases with multiple control entities; it also isolates BIM delivery in siloed electronic files, such as Revit files [8]. Revit is one of part of the BIM software that includes 3D modeling graphical and non-graphical information to enable the project delivery through coordination by avoiding gaps and overlap in team members' work utilizing an electronic file-based model [52]. The integration needs to ensure an ecosystem of linked databases within the blockchain technology (a decentralized database) to be connected, not isolated, to BIM (a centralized database) [36,50]. This results in the advancement of BIM delivery toward machine-readable datasets *"enabling an ecosystem of connected databases based on consistently organised datasets"*, as viable solutions for the automation of operations and facilities management [12,53]. A recent study by Hijazi, Perera [36] provided the data model for integrating BIM with blockchain technology to help construct organized and trustworthy datasets for BIM supply chain data delivery; nevertheless, the study does not illustrate in-depth validation of the suggested model and its technological feasibility. In the industrial scenario, *"BIMCHAIN"* was developed by a French startup that presents a solution for integrating blockchain technology capabilities with BIM [21,35]. It intends to generate digital evidence of different BIM transactions. However, the blockchain technology in this scenario only keeps a hash of the digital Revit file's alteration record, not the actual data update, which means a single party (the model authoring stakeholder) is responsible for reconciling diverse BIM transactions. Thus, this solution might validate the 3D BIM model authoring file copyright for the design delivery [21], or ensure a confidentiality-minded framework between the project members in case of a sensitive BIM design collaboration model [54], but not the ownership for the supply chain product data where there are multiple control entities in the 3D BIM model during the handover phase [8]. Copyright is only one type of intellectual property (IP) protection; the contract could ensure it as it comes under the responsibility definition, such as the case of the Construction Industry Council (CIC) BIM protocol, which detailed how the BIM model and objects should be created, and where BIM delivery is still struggling to be reliable by ensuring the ownership of the supply chain product data [36,55].

A blockchain platform is one of the most secure database platforms because it is a decentralized and distributed database, and its data is immutable [3,46,56]. Therefore, the blocks (i.e., transactions) within a blockchain platform are copied across numerous computers, ensuring that the data contents of each block cannot be modified. Moreover, the algorithm can verify and validate the block's proof-of-work by itself [23]. In contrast, Coyne and Onabolu [57] stated that the blockchain struggles to overcome the privacy issue. There are different blockchain platforms available. *"Ethereum"* is the first public blockchain platform to allow smart contracts for general consumption, and it is now utilized mostly by the financial industry [29,50,54,58–60]; however, it is not appropriate for many types of businesses, such as the construction industry, where data privacy is crucial [28,50]. Thus, the blockchain should guarantee the veracity and accessibility of information while protecting its confidentiality [3,45,61]. Privacy and identity management methods have and will continue to have a substantial influence on business blockchain development [62]. Thus, the *"FIBREE"* blockchain industry report [63] strongly recommended "Hyperledger Fabric" as a blockchain platform with private permissions to meet the privacy requirements for a broad range of industry use cases, including data transactions in the construction industry [60,64]. Multiple software development kits based on modular and pluggable components [50,64] are provided by Hyperledger Fabric to accommodate varied applications and ease participant buy-in [60]. Section 3 explains why Hyperledger Fabric blockchain technology was selected as the best platform for constructing a system prototype that integrates BIM and blockchain technology.

A smart contract can be programmed to process a self-executing contract by translating the rules from the terms of the agreement into lines of code using a *"Generalized Adaptive Framework"* (GAF) [65] for a neutral data standard, such as the *"International Foundation Class"* (IFC). This allows for the automation of code verification procedures for routing data that must be stored in the blockchain network. The suggested GAF concept for automated processes entails the construction of a computable representation of predefined laws, as well as means for transferring data between the framework's various components (blockchain network) and BIM data [66]. A further approach might be a smart contract that secures the enforceability of transparency by executing the CSC data, thus making the CSC data immutable and accurate. The prototype proposed in this article implemented the second alternative by directly linking CSC data from external entities to a blockchain network and implementing transactions on top of a blockchain ledger using a smart contract solution. The section that follows describes, in detail, the methodologies used to illustrate the technical viability of a single source of truth approach for integrating BIM with blockchain technology.

3. Research Method, Design, and Tools

The development process of the software prototype follows the procedure stated by Qing and Yu-Liu [37] for the development of software prototypes. Several other researchers, such as Perera, Hijazi [28], Xue and Lu [67], and Ahmadisheykhsarmast and Sonmez [68], working on blockchain at the application layer, have also used similar developmental steps. Thus, we designed the development process of the BIM single source of truth prototype using blockchain by performing the following steps: understand the current business scenario and propose logical system architecture, select the blockchain platform, design the physical system architecture related to technologies, and develop the prototype and evaluate it, as illustrated in Figure 1 and described below.

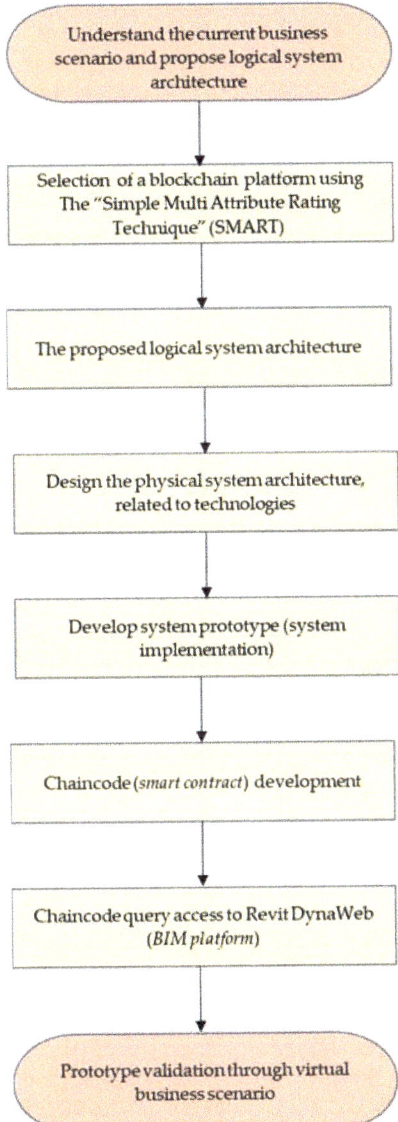

Figure 1. The development process of the BIM single source of truth prototype using blockchain technology.

To understand the current business scenario, the data flow diagram (DFD) was used to structurally identify the existing process of the BIM construction supply chain data delivery by describing data flows of a system at various detail levels and propositioning logic models that express data transformation in a system [36,37,69,70]. The information delivery for the current business scenario is set up to work in line with the *"Common Data Environment (CDE)"* workflow, which was outlined in *"ISO-19650 Part 1, Section 12"*, and

"BIM maturity Level 2" deliverables adhere to the guidelines of the *"PAS 1192-2-2013"*. The current business scenario is described in Section 4.1 and the proposed system overview for the logical system architecture, independent of technology is elaborated in Section 4.2.

The *"Simple Multi-Attribute Rating Technique"* (SMART), which is a common tool for assisting in correct decision-making to solve a problem and find the best solution [71], was used to rank and identify the most suited blockchain platform among the identified platforms. Permissioned networks were chosen as the best sort of blockchain for CSC data transactions because stakeholders should be identified and held responsible for their conduct. For CSC activities, the blockchain database for a project contains sensitive information that organizations intend to keep private, such as commercial information [72]. As a result, the list of candidates with blockchain platforms included *Corda R3* [73], *Elements* [74], *Hyperledger Fabric* [75], *IBM Blockchain* [76], and *NEM* [77]. The Hyperledger Fabric blockchain platform was chosen as the best fit for the defined needs since it obtained the greatest value. In its design and implementation, the Hyperledger Fabric architecture provides great levels of flexibility and secrecy, making it applicable in a wide range of environment applications [25,50,78,79]. It aids in attaining privacy since it needs permission to read and write through the permission model, which is characterized by the capacity to modify the state of the ledger community [79].

By separating transaction processing into three steps, the Hyperledger Fabric design provides auditability. Phase one: utilizing distributed logic processing via Chaincode services to create a smart contract, which is the business logic code of a transaction on the Hyperledger Fabric platform; phase two: utilizing transaction ordering via consensus services to create blocks of transactions and facilitate network trust; and phase three: transaction validation via membership services [50,80,81]. The outputs of the selection of a blockchain platform, *the Hyperledger Fabric*, directly contributed to the system architecture (Section 4.3). This step includes designing the physical system architecture, identifying the technology that will be used, and determining where the described system processes used Hyperledger Fabric terminology. According to the process flow stated by Qing and Yu-Liu [37], the next step in developing system architecture is validation by developing a system prototype including the *"process sequence"*, which is a process planning of a successful transaction within the system prototype for integrating BIM, blockchain, and Chaincode (smart contract) development, as explained in Section 4.4.

Finally, the software prototype was comprehensively evaluated to ensure the accurate execution of the smart contract using a virtual business scenario that included external validation. The attributes of the virtual business scenario were set to conduct the test run. Cladding attributes, as a sample of the CSC data delivery, were considered the CSC element to execute the smart contract. The cladding has become an incandescent topic of attention in several countries as an example of a CSC object that is not considered a "structural" part of the building [82], but it could cause a threat to the safety of the residences during the operation phase [83], such as what happened at the Grenfell Tower (London), which led to an unprecedented loss of life [84,85], and the fire incidents at the Lacrosse Apartment Building (Melbourne) and the Torch Tower (Dubai), which led to unavoidable multi-million-dollar bills for property owners [83]. Even when several ad-hoc actions have been implemented to combat non-compliant cladding products, the Australasian Procurement and Construction Council mentioned that more than 50% of cladding products might still be non-compliant, with the majority of stakeholders completely unaware of the financial burden it could present [83,86]. Keeping the preceding discussion in mind, cladding was deemed an excellent example for the virtual business scenario to help introduce a BIM single source of truth prototype using blockchain. The external entity's (supplier) role was performed by the researchers, whereas the main contractor's role was performed by a participant organisation. The protocol for the virtual business scenario, selection criteria, data collection, data analysis, and the test case and its results are explained in detail in Section 4.5.

4. Develop, Validate, and Test the Proof of Concept

4.1. The Current Business Scenario

In the main process of the current BIM supply chain data flow, the BIM model is developed during a project's construction phase in response to requirements set out in the "Employer's Information Requirements" (EIR) to work in line with the Common Data Environment workflow, which is outlined in "ISO-19650 Part 1, Section 12", and "BIM maturity Level 2" deliverables adhere to the guidelines of the "PAS 1192-2-2013 specification for information management for the capital/delivery phase of construction projects using building information modelling". In this process, suppliers are required to prepare an IFC file for the product data (CSC element) to be sent to a subcontractor or consultant (the author of the IFC BIM model). The "BIM execution plan" (BEP) then details how the 3D BIM model is to be delivered to the client through the main contractor in order to fulfill the client's asset information demands (AIR). In the construction phase, the "responsible" party is the main contractor, who is in charge of coordinating the transfer of the federated 3D BIM model used to create the "asset information model" (AIM). This process is decomposed into three main subprocesses. The first subprocess (Process 1.0) is to acquire CSC data. Each task team is required to send the IFC model data to the main contractor through the CDE. This subprocess will create considerable ambiguity about the ownership and authoring liability of the IFC data between the suppliers, subcontractors, and the main contractor for the CSC elements and/or model. The second subprocess (Process 2.0) is to review the CSC data provided by the main contractor and share it with other appropriate task teams or delivery teams or with the appointing party. In this subprocess, the main contractor is responsible for managing the construction phase's workflow and for transferring the federated BIM model to generate the AIM. As they are not responsible for the model components, there is a benefit in preserving an immutable record of where the CSC model elements were obtained. The federated BIM model must reflect the facility as built and be enhanced with CSC data (as defined in the EIR) before being "published" as the third subprocess (Process 3.0). In this subprocess, the AIM is derived through a combination of federated BIM model deliverables and COBie datasheets, all of which are underpinned by Uniclass 2015. As mentioned in Section 2, BIM is yet to be considered a reliable delivery model, as the COBie data are "centrally stored" by a single actor [3,4,10,15]. In this scenario, the main contractor has three subprocesses for acquiring the CSCs that were developed by their originator or task team (consultants, subcontractors, or suppliers). Process 1.1 will be decomposed into three child diagrams (subprocesses), Process 1.2 will be decomposed into six child diagrams, and Process 1.3 will be decomposed into six child diagrams, as illustrated in Figure 2. The decomposition involves the top-down development of a DFD to reduce the complexity of the system [37]. This will help to provide a clear picture of the existing information delivery system (existing work packages) and create a functional process network for developing the proposed model. It will also demonstrate the relationship between the CSC stakeholders and how the function of reconciling different data is still usually undertaken by one or a limited number of parties. This is a key point in understanding how the proposed solution needs to be adapted to the existing work packages. However, analyzing information delivery for more than one CSC element to aid in the understanding of the existing information delivery system would be redundant, as all elements have the same work package. A single CSC element can be used to mimic the current data flow.

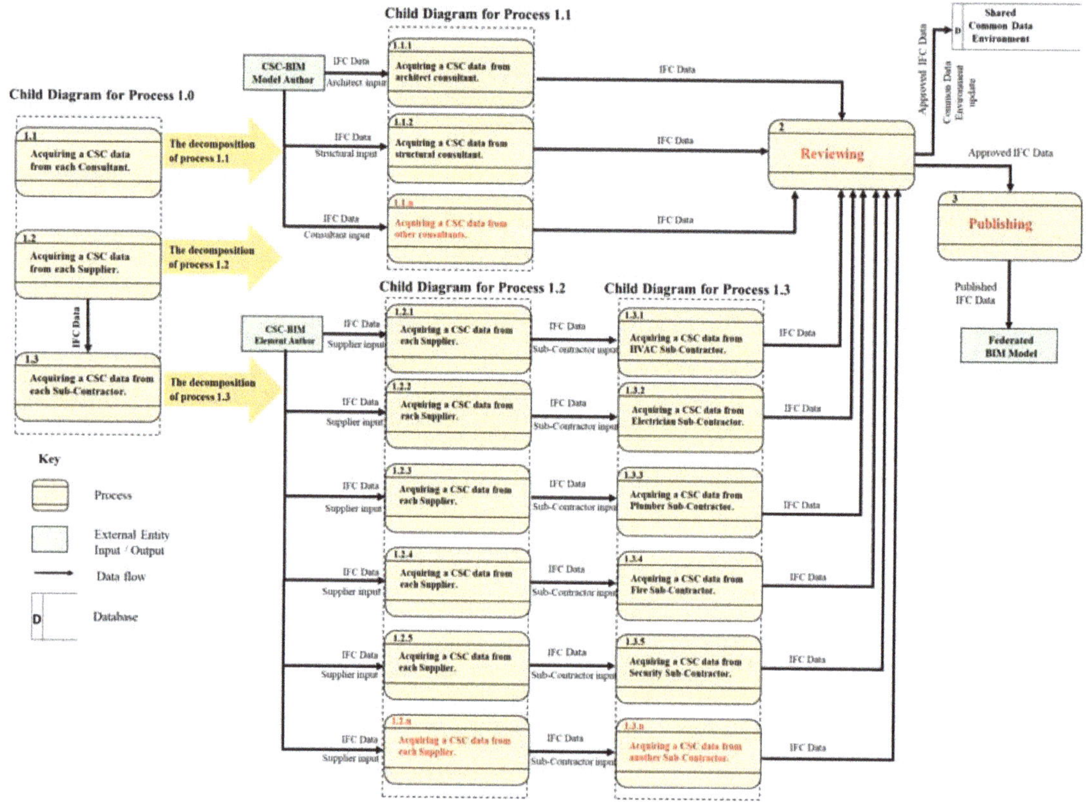

Figure 2. The data flow diagram (DFD) for BIM construction supply chain data delivery.

4.2. The Proposed Logical System Architecture

In the proposed logical system architecture, independent of technology, as illustrated in Figure 3, the CSC external entities, including consultants, sub-contractors, and suppliers, deploy CSC data directly to a blockchain using a smart contract. As discussed in Section 2, the proposed prototype in this study would directly connect supply chain data from *"external entities"* to a blockchain by deploying transactions on top of the ledger utilizing a smart contract method.

In order to centrally link the supply chain data delivery that is operating in blockchain in a BIM platform, blockchain and BIM are connected through a *"REST Application Programming Interface"* (API) that allows access to the CSC data transaction. This allows for the blockchain to centrally link the supply chain data delivery that is operating in blockchain. This subprocess makes it possible to quickly retrieve the history of the data delivery throughout a supply chain for a project. Subsequently, the findings of the selection of a blockchain platform were utilized to translate the independent system architecture (logical system architecture) into the software components of Hyperledger Fabric (*physical system architecture related to technologies*), which will be explained in the following sub-section.

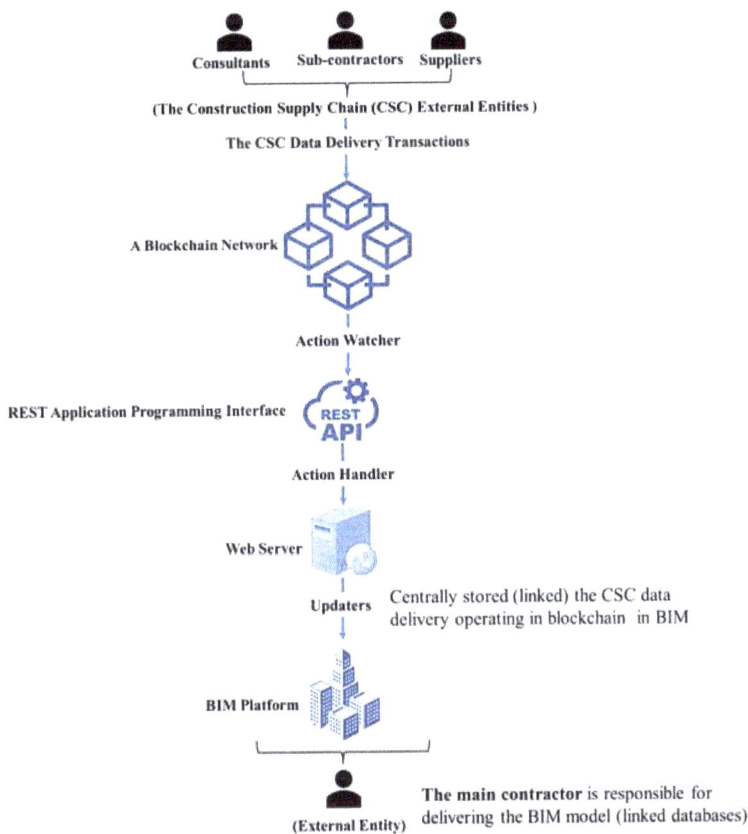

Figure 3. The proposed logical system architecture is independent of technology for the BIM single source of truth prototype using blockchain technology.

4.3. The Proposed Physical System Architecture

The physical system architecture related to technologies has the same components as the logical system architecture: the blockchain network, the REST Application Programming Interface (API), the web server, and the BIM platform. However, the components are demonstrated with a particular technology with sufficient detail for system implementation in the physical system architecture.

In the *"physical system architecture"*, the client or system user *"external entity"* refers to an application that is separate from the blockchain and that connects to the ledger in order to perform supply chain data transactions [75]. All authorized prototype users will have immediate access to the blockchain through the *"Node.js console application"* or via graphical user interfaces in the client application, the *"Node.js web application"*. For this technologically-related physical system design, the Node.js console application is made available for authorized users so that they may interact directly with the blockchain via the usage of the console [87]. The ordering service and the peers who each have a copy of the ledger and Chaincode are connected to one another over the blockchain network. The BIM platform *"Revit DynaWeb"* accesses the blockchain system using the REST API, supported by a web server on the local machine, providing easy access to the project's CSC data history recorded in the blockchain. It is crucial in today's web-enabled world to provide machine-readable forms of text content, commonly in "Extensible Markup Language" (XML) [88]. The physical system architecture is illustrated in Figure 4. In a blockchain system, the user (client application) initiates communication with the blockchain. The Chaincode (smart contract) is executed, which comprises the application logic for the CSC attributes' smart contract, in order to obtain or update ledger data (depending on the data model requirements).

(Step 1) The command to invoke the chaincode is known as a transaction proposal (the CSC attributes), which is sent by the client application (system user- external entity) to the endorsing peers on the blockchain network via Node.js console application.

(Step 2) The endorsing peers verify the transaction proposal and execute it by invoking the chaincode.

(Step 3) The endorsing peers create a proposal response including the transaction results and peer's signature and send it to the client application.

(Step 4) If the transaction proposal was only a query, the process ends by displaying the result to the user.

(Step 5) If a ledger update is required, the client application packages the transaction proposal and endorsed responses into a transaction and broadcasts it to the ordering service.

(Step 6) The ordering service receives transactions from the entire network, orders the transactions, and creates a block of transactions.

(Step 7) The ordering service transmits the block to the leading peer,

(Step 8) which then distributes the block to all the other peers.

(Step 9) The peers validate the transactions within the block and tag the transactions as valid or invalid. If the transaction is valid, then the world state is updated, whereas all valid and invalid transactions are added to the blockchain, which ensures auditability.

(Step 10) The peers emit an event to notify the client that the transaction is valid or invalid, and that it has been added to the blockchain, and the result is displayed to the user .

(Step 11) API Server Node runs the API Server code. The web server allows access to the data stored in the blockchain. This ensures that the proposed system architecture solution is vendor-agnostic.

Finally, *(Step 12)* the server code accesses Revit DynaWeb to enable the BIM user (external entity- sink) to access the data (CSC attributes). DynaWeb is a *dynamo* package providing support for interaction with the innerweb in general and with REST APIs in particular.

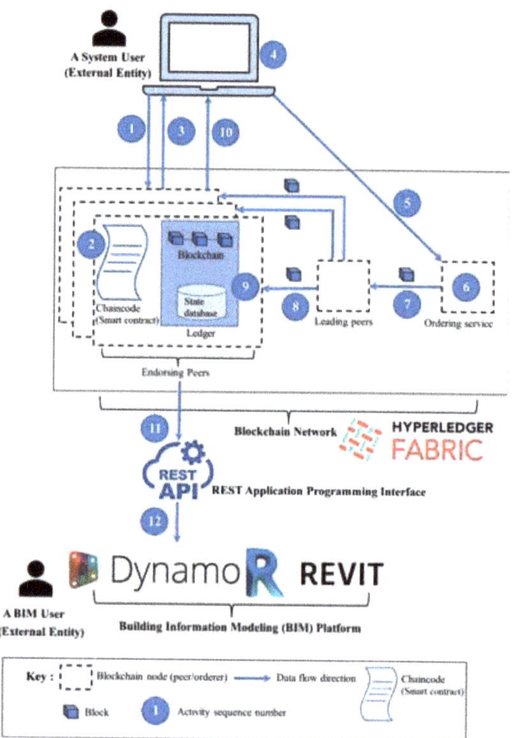

Figure 4. Physical system architecture for integrating BIM with the Hyperledger Fabric platform (*a blockchain technology platform*).

According to the process flow stated by Qing and Yu-Liu [37], the subsequent step to the development of system architecture is validation by developing a system prototype (system implementation) which will be explained in the following section.

4.4. System Prototype (System Implementation)

The system prototype is used to translate the validated data model into a technological solution utilizing the physical system architecture developed in the previous section. Figure 5 shows the steps involved in a transaction within the proposed system prototype, from the time a data owner (suppliers, sub-contractors, or consultants) invokes (writes) the on-chain attributes on the blockchain network to the time the on-chain attributes are centrally stored (linked) to the BIM platform by the main contractor. For simplicity, the interactions between the REST API and the web server were excluded from the diagram; these operations would occur between the API server node and the BIM platform. The prototype also handles unsuccessful searches and changes, albeit these procedures are not displayed in the figure for readability.

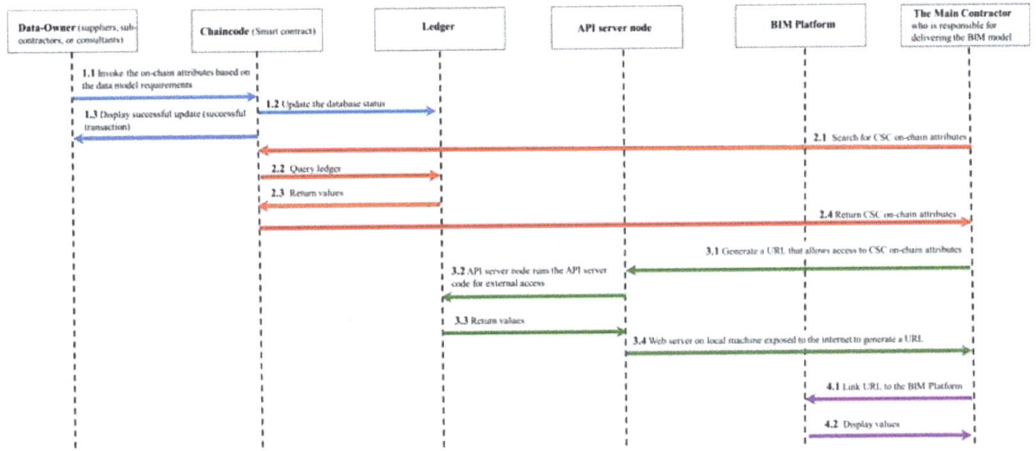

Figure 5. The process sequence for a successful transaction in the BIM–blockchain system prototype. Implementing the system prototype included installing "*Hyperledger Fabric v2.3.3* [89]", generating the network to implement and write a Chaincode (smart contract) using the "*JavaScript* [90]" programming language, and establishing an API server node to access "*Revit* [52] *DynaWeb* [91]". This section demonstrates and discusses the smart contract (Hyperledger Fabric Chaincode) development to hand over on-chain data delivery and is linked to the BIM platform, in this case "*Revit DynaWeb*", to provide ready access to the record of on-chain supply chain data delivery transactions.

4.4.1. Chaincode (Smart Contract) Development

In the Hyperledger Fabric network, the Chaincode can be written in one of the following programming languages: *Go, Java, TypeScript*, or *JavaScript* [92]. For this prototype, the *JavaScript* programming language was used to write the Chaincode (smart contract) to support the development of the system prototype for API applications. The Chaincode requires supply chain stakeholders to submit transactions using the invoke application to provide on-chain data. The query application assures the supply chain element's transaction history (amendment and revision) by using the Chaincode or smart contract. As illustrated in Figure 5, for the process sequence of a successful transaction, the query application return values provide ready access to the record of on-chain data delivery transactions. Given this, the primary methods used in the proposed Chaincode were defined below.

initLedgedr(ctx): The constraint *ctx* is a set of *ChaincodeStub* structures in JSON format. This function initializes on-chain data delivery based on the provided data model attributes (or the proposed business logic) in JSON format, and stringifies and stores all those data into *ctx* using <async> *putState* (key, value) [87]. Thus, to submit the on-chain CSC data transaction, *invokeProcess (ctx, myArgs[0] .... myArgs[20])* includes all the arguments (the data attributes of the business logic for the smart contract) that need to be entered by the CSC stakeholder passing through the Node.js console application. Suppose an argument (attribute) has been missed, or its value has been wrongly entered. In this case, the Node.js console application will display this message as a *console error* (*failed to submit transaction*). However, if it is run successfully, the Node.js console application will display this message as a *console log* (*transaction has been submitted*), as illustrated in Figure 6.

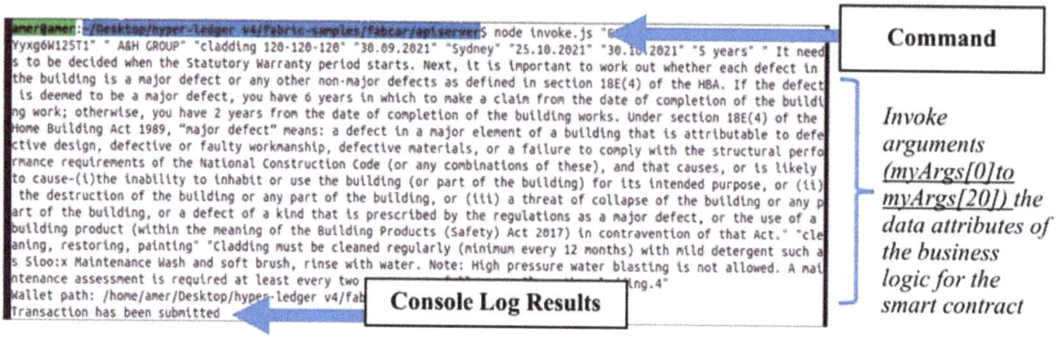

Figure 6. We invoked the application's arguments, as shown in the Git console.

queryID (ctx, myArgs): This method calls the return values for on-chain CSC data delivery attributes linked to a specific *invokeProcess* (submitted transaction) with the value of *myArgs* (a specific attribute) [75,87]. All values with that *queryValue* will be returned in stringified JSON format by calling *<async> putState*(key, value). To run this method, the *queryID* application needs to be passed through the Node.js console application. The return values are displayed on the Git console, as illustrated in Figure 7.

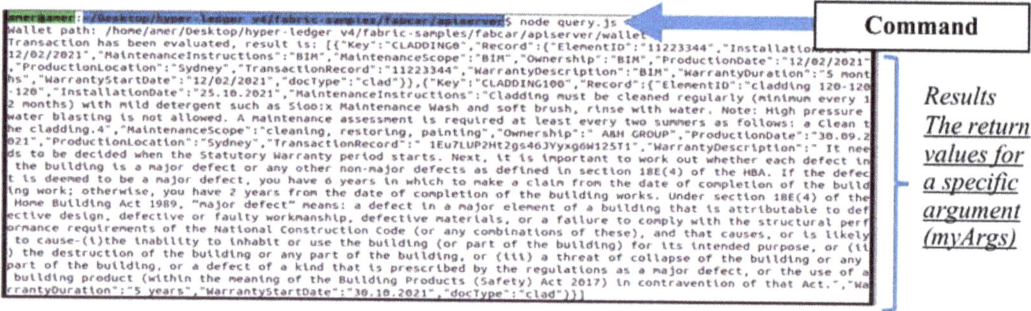

Figure 7. We successfully ran the queryID application and the return values for a specific argument, as shown in the Git console.

queryAll (ctx): In this method, all values (all on-chain CSC data delivery for a project) inside this blockchain will be returned after calling this method. To run this method, the *queryAll* application needs to pass through the Node.js console application. Once a block is created after running the *invokeProcess* application, it cannot be modified. However, a block can have more than one transaction, and tracking its history of transactions can be detailed after running the *queryAll* application.

The prototype API supports two methods. The first method, GET, returns values for CSC data delivery (return values of query application). The second method, POST, provides access to add a new record to the ledger. The supply chain element's transaction history may be shown using the GET method again (amendment and revision). To connect the API server node to Revit Dynamo (the next step, Section 4.4.2.), a web server solution was implemented. To enable access from the Internet to the machine, a reverse proxy, Ngrok, was used.

Then, the URL link generated by the reverse proxy was used to access the server. The operation ran successfully and the result was the HTTP request–response from other machines. At this level, the system prototype was dealing only with the blockchain platform and exposing a web server on the local machine to the internet for its return values of query applications. Given the physical system architecture, as explained in Section 4.4.2., the following sub-section demonstrates how the reverse proxy was linked to Revit DynaWeb to access the transaction (return values of query application) by the BIM user (external entity sink).

4.4.2. Chaincode Query Access to Revit DynaWeb (BIM Platform)

This section illustrates the last step required to centrally integrate (store) on-chain data delivery with the BIM platform or tools. *"Revit DynaWeb"* is used as a BIM platform to demonstrate this integration. As explained previously, *DynaWeb* is a *"dynamo package providing support for interaction with the interwebz in general and with REST APIs in particular"* [91]. Thus, it assists in retrieving (GET) information from the web and sending (POST) information to the web. It also contains some handy JSON de/serialization nodes for using web data directly in Dynamo graphs as native types [91]. Using DynaWeb, which is included in the Dynamo package, simplifies routine tasks by providing a centralized interface for linking on-chain data transfer with the BIM model (Revit model). Therefore, this solution guarantees that the distributed database (from the Hyperledger Fabric Network) may be linked to several BIM models (Revit models) to store supply chain data delivery. At the same time, the traditional BIM CDE will go operating as a hub for a network of datasets sources. The *Dynamo* workplace environment was accessed from the *Manage* tab in the *Revit* visual programming panel and by clicking on *Dynamo*. Then, the *DynaWeb* package was successfully implemented in the *Dynamo* workplace environment. The *"DynaWeb"* package includes *"WebRequest: the web request that gets executed"*, *"WebClient: the context in which a request is executed"*, *"WebResponse: this contains the response from the server as well as additional metadata about the response and server itself"*, *"Execution: this provides nodes that simply execute requests, making it easier and clearer to use standard Hypertext Transfer Protocol (HTTP) verbs such as GET"*, and the *"Helpers: a few helper nodes, with a particular focus on deserialisation"*. In addition, the above *DynaWeb* package was extended to add new scripts (new nodes) to automate repetitive processes that check if the on-chain dataset has been linked centrally to the *Revit* model or not, as shown in Figure 8. To check if selecting a CSC element on the Revit model has linked the on-chain dataset (return values of query application), the watch function displays the message *"URL found and opened"* and, if it is not found, the watch function displays the message *"Valid URL not found"*, as the on-chain dataset is not linked to this item.

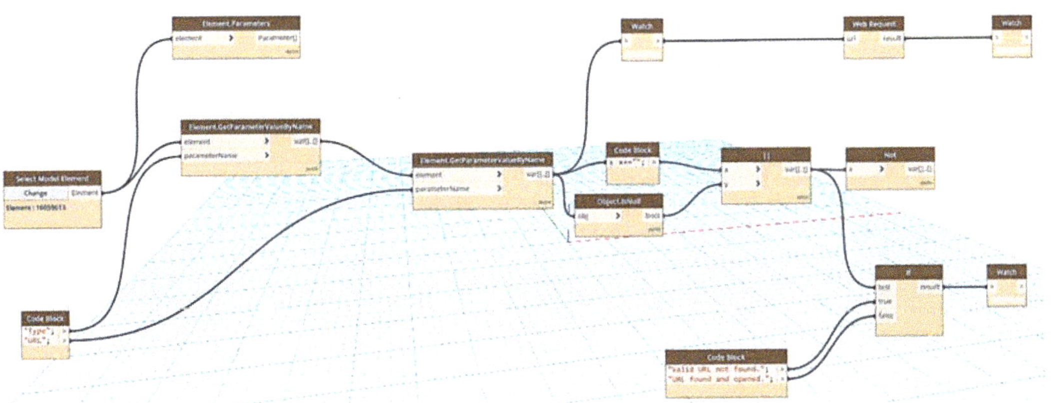

Figure 8. Implementation of the DynaWeb package in the Dynamo workplace environment.

Assimilating the above, the on-chain dataset based on the URL that was created from *Ngrok* (as explained in the previous Section 4.4.1) was copied in the *String* node, and the data resulted in the *Watch* box with JSON format. At the same time, as shown in Figure 9, the BIM user may access the on-chain dataset from inside *Revit* by selecting the element's attributes and then choosing the *URL property*.

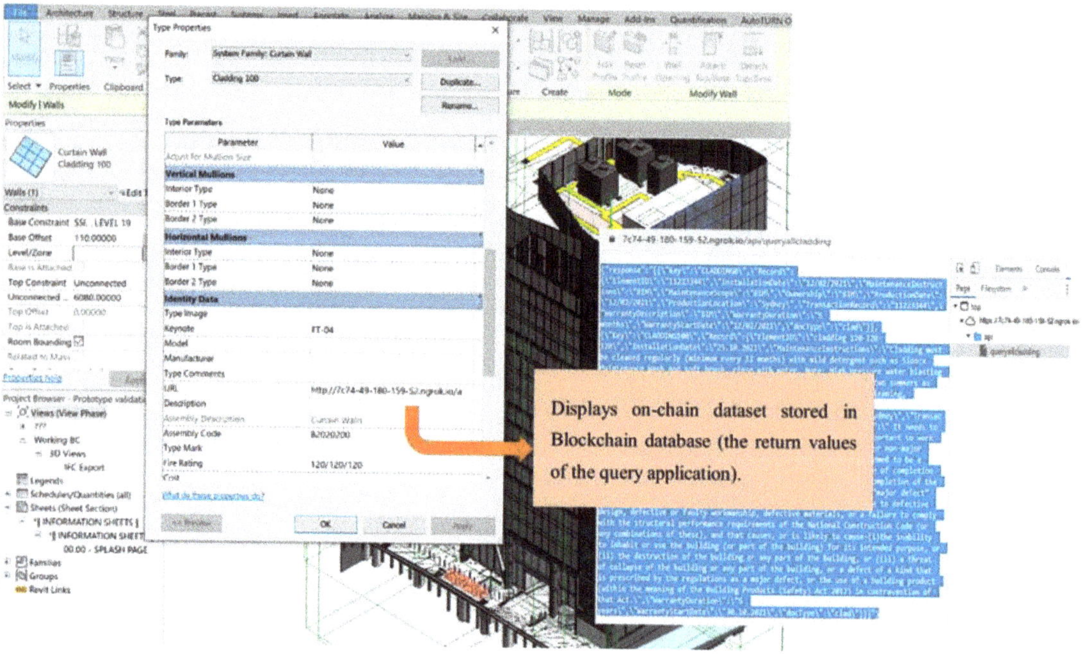

Figure 9. Successful linking of the on-chain dataset (the return values of the query application) to Revit (the BIM platform) and displaying (accessed) its transaction values.

Considering this, the system prototype successfully demonstrated the solution for or integrated blockchain and BIM using *Hyperledger Fabric v2.3* and *Revit DynaWeb v2021*. According to Qing and Yu-Liu [37], evaluating the prototype is the next step for which a virtual business scenario is utilized in this research, as explained in the following section.

4.5. Prototype Validation with a Virtual Business Scenario

The prototype was comprehensively evaluated to ensure accurate execution of the smart contract using a virtual business scenario that involves external validation. The attributes of the virtual business scenario were set to conduct the test run. A building project was considered as a project type, and cladding attributes, as discussed in Section 3, were considered the CSC element to execute the smart contract. The role of the external entity (supplier) was performed by the researcher, while a participant organization performed the role of the main contractor. It was ensured that the information delivery system used by the participant organization is configured to work in accordance with the workflows for the CDE that are outlined in *"ISO-19650 Part 1, Section 12"*, and the deliverables for the *"BIM maturity Level 2"* were based on the *"PAS 1192-2-2013"* standard. Table 1 illustrates the protocol for the virtual business scenario.

Table 1. Protocol for the virtual business scenario.

Attributes	Attribute Description
Project Type	Building
CSC Element Type	Cladding
External Entity (Supplier)	Researchers
Main Contractor	Participant Organisation
Developer	Researchers
BIM Platform	Revit v2021.
BIM API	Revit DynaWeb v2021.
Blockchain Platform	Hyperledger Fabric v2.3.

Data Collection: The prototype developed, as discussed in Section 4.4, was used to validate the integration of BIM and blockchain. Subsequently, the actors invoked and queried to test the system. The virtual business scenario concluded with a structured interview to complete the evaluation of the prototype.

Data Analysis: The test case scenario was used and the results of the test run were tabulated in the form of a checklist. This was performed to confirm that the prototype fulfills all of the system criteria and provides a software prototype for the integration of BIM and blockchain, as illustrated in Table 2.

Table 2. A virtual business scenario test case and its results.

Action ID	The Test Case Scenario Taken	Output	Test Result
#1	Developer (*Researcher*) brings the Hyperledger Network up by running/startFabric.sh javascript	The Hyperledger Network generated and started running	Pass
#2	Developer (*Researcher*) enrolls the admin and imports it into the wallet and registers the user	Successfully registered and enrolled admin user "appUser" and imported it into the wallet	Pass
#3	Developer (*Researcher*) sets up the API Server for the Hyperledger Fabric Network and runs the webserver that links to Revit DynaWeb	Successfully set up a channel from the public internet to a port on the local machine	Pass
#4	Supplier (*Researcher*) is required to enter the following attributes of the CLADDING element	Successfully ran the Invoke application and entered its arguments based on the proposed Entity–Relationship Diagram data model attributes	Pass

Table 2. Cont.

Action ID	The Test Case Scenario Taken	Output	Test Result
#5	Main Contractor (*Participant Organization*) queries to read the CSC element on-chain dataset submitted by the supplier by using any one of the attributes (myArgs[0] . . . . mArgs[20]) from the Hyperledger Fabric	Successfully ran the queryID application and the return values for a specific argument (myArgs)	**Pass**
#6	Main Contractor (*Participant Organization*) uses Dynamo to select the CSC element on the Revit model, the watch function displays "URL found and opened"	Successfully ran the DynaWeb package in the Dynamo and the watch function displayed this message "URL found and opened"	**Pass**
#7	Main Contractor (*Participant Organization*) reads the on-chain dataset from Dynamo by using the watch function	Successfully displayed and read the on-chain dataset (return values of query application)	**Pass**
#8	Main Contractor (*Participant Organization*) clicks on the CSC element on the Revit model, goes to properties, and clicks on the URL in the URL field to read the on-chain dataset	Successfully linked the on-chain dataset (return values of query application) to Revit (BIM platform) and displayed (accessed) its transaction values	**Pass**
#9	Main Contractor (*Participant Organzsation*) uses the Dynamo change button to select a different CSC element on the Revit model, the watch function displays "Valid URL not found"	The watch function successfully displayed the message "Valid URL not found" as the on-chain dataset (the return values of the query application) is not linked to this item	**Pass**

There was consensus from the Participant Organization that the software prototype ensures the reliability of supply chain data delivery by enforcing the supply chain stakeholders to hand over on-chain data delivery. Further, the *Participant Organization* acknowledged that the software prototype for the integration of BIM and blockchain paves the way for the progression of the 3D BIM model toward the *digital engineering framework* [93] by enabling "machine-readable data" in the form of a reliably structured dataset. The *Participant Organization* acknowledged that there is a benefit in centrally integrating the on-chain supply chain data delivery utilizing the URL (XML format) because it assures that the software prototype delivers a vendor-agnostic solution that is compatible with several BIM vendor's software packages. The test run concluded with the recommendations to adopt the software prototype solution; the delivery partner should submit all project deliverables as structured datasets; thus, it should be contractually bound and represented as part of the Professional Services Agreement (PSA). The BIM execution plan needs to include the implementation costs and time. The software prototype is a new technological solution and the hardware requirements of blockchain are very expensive.

5. Discussion

Transparency, traceability, and a lack of supply chain data are some of the deficiencies of BIM use. Blockchain may alleviate the shortcomings of BIM by providing transparency and accountability and by ensuring ownership through smart contracts. However, its adoption is contingent on the data source, where BIM may play a significant role when paired with blockchain to provide verifiable data for supply chain operations [27,67]. By not restricting BIM to isolated electronic files, the prototype proposed in this study exploits blockchain's potential for value transfer toward a reliable ecosystem of interconnected databases. In addition, it facilitates machine-readable data in the form of uniformly formatted databases, opening the way for the progression of BIM toward digital engineering. The results of the test run validated the technical viability of the BIM single source of truth prototype using blockchain to ensure the delivery of trustworthy supply chain data in blockchain that is centrally tied to BIM. As it links and stores on-chain supply chain data delivery through a URL, the system prototype offers a vendor-agnostic solution that is suitable for a variety of BIM software applications. The URL-XML format ensures semantic consistency across all disciplines of a project. The participating organization noted that the system prototype ensures a constant and predictable dataset of on-chain supply chain data delivery by using a logical and familiar navigation and use interface. The deployment of the system prototype for the virtual company scenario proved that businesses are keen to commercialize this solution not only to ensure consistent digital output but also to maintain a competitive advantage. The scalability of the suggested method was not, however, explored in this study. The development of the BIM single source of truth prototype utilizing blockchain technology will pave the way for the BIM handover model to enable the definition of the digital twin, as it has received significant attention as an emerging technology and is now regarded as a crucial component of Industry 4.0 [94]. The trust, which is the main quality of this prototype, will improve the BIM data quality and ensure its reliability for operation and facilities management, which was also introduced by the Centre for Digital Built Britain as one of the Gemini principles pillars placed on the digital twins for being a single source of information for operation and facilities management [33]. However, as the digital twin model supports the future vision of smart cities, the remaining question is, what about the sustainability position of this prototype? This paper does not study the sustainability impact of blockchain; however, this prototype is considered an industrial application (Blockchain 3.0) that operates without miners through the help of nodes, unlike the cryptocurrencies, such as Bitcoin (Blockchain 1.0) and smart contracts and financial applications, such as Ethereum (Blockchain 2.0) [95].

Despite the significance of this paper's results, it is important to acknowledge a few limitations. The system prototype does not account for variations that may result from different procurement procedures, in which the role of each stakeholder in the logical system may be modified. In order for the integration of BIM and blockchain to be really effective, however, individuals and organizations must rethink their traditional procurement practices and boost their investment in innovation. Future research could examine the relationship between the various procurement systems (Public-Private Partnership, Alliancing/Integrated Project Delivery, Design Bid Build, Partnering, Traditional, Management or Early Contractor Engagement, etc.), and the proposed method for the integration of BIM and blockchain. Therefore, the system prototype should emphasize the use of the "Generalized Adaptive Framework" (GAF) to automate the code verification processes for routing the data that must be deposited in the blockchain system. This may need the development of a computable representation of the rules and techniques for the interchange of data between the different framework components and the BIM data.

6. Conclusions

This article advanced the state of knowledge primarily by the contribution of a proof of concept prototype for integrating blockchain and BIM for supply chain data delivery. In doing so, this paper demonstrated a step-by-step methodology starting from understanding the current business scenario and proposing logical system architecture, selecting a blockchain platform, designing system architecture related to technologies, and prototyping development and evaluation. The deployment of the system prototype for the virtual business scenario revealed that organizations are keen on transforming this solution to the commercialization stage not only to guarantee a trustworthy digital delivery but also to preserve a competitive advantage. There is a value in centrally linking the on-chain supply chain data delivery using the URL (XML format) as it ensures that the software prototype offers a vendor-agnostic solution that is interoperable with various BIM vendor's software packages. The system prototype proved Hyperledger Fabric's suitability for integration and provided a solution compatible with many BIM software vendors. The Hyperledger Fabric architecture enables high levels of flexibility and privacy in its design and implementation, making it suited for a variety of built environment applications. It helps achieve privacy by requiring permission to read and write using a permission paradigm that is characterized by the capacity to modify the state of the ledger community.

This study's recommended strategy for the integration of BIM and blockchain took into account construction supply chain data delivery for handover and operation throughout the building phase. Future research might leverage these results to build methodologies for other project phases, such as design (pre-construction) or facilities management (post-construction). Future development of the system prototype might concentrate on using the GAF to automate code verification techniques for routing data that must be saved in the blockchain system. This might need the creation of a computable representation of stated rules and data exchange protocols between the various framework components and BIM data. Legality is one of the most daunting blockchain-related concerns, according to industry insiders. Future construction research should concentrate on collaborating closely with legal specialists to investigate the legal implications of blockchain technology, the use of smart contracts in addition to conventional contracts, and the necessary revisions to contract language.

Author Contributions: Conceptualization, A.A.H., S.P., R.N.C. and A.M.A.; methodology, A.A.H., S.P., R.N.C. and A.M.A.; validation, A.A.H.; formal analysis, A.A.H.; writing—review and editing, A.A.H., S.P., R.N.C. and A.M.A.; supervision, S.P., R.N.C. and A.M.A. All authors have read and agreed to the published version of the manuscript.

Funding: This research was funded by the Centre for Smart Modern Construction and the School of Engineering, Design, and Built Environment, Western Sydney University, Australia.

Data Availability Statement: Interview transcripts and interpreted statements supporting this study's findings and the smart contract codes are available from the corresponding author upon reasonable request.

Acknowledgments: The authors acknowledge the contributors of the Centre for Smart Modern Construction, the School of Engineering, Design, and Built Environment, Western Sydney University, Australia, and the case study organizations that participated in the web-based maturity assessment.

Conflicts of Interest: The authors declare no conflict of interest.

Abbreviations

AIM	Asset Information Model
API	Application Programming Interface
BEP	BIM Execution Plan
BIM	Building Information Modelling
BIMXP	BIM Execution Plan
CDE	Common Data Environment
CIC	Construction Industry Council
COBie	Construction Operation Building Information Exchange
CSC	Construction Supply Chain
DFD	Data Flow Diagram
EIR	Employer's Information Requirements
ERD	Entity–Relationship Diagram
GAF	Generalized Adaptive Framework
HTTP	Hypertext Transfer Protocol
ICT	Information and Communications Technology
IFC	Industry Foundation Class
IP	Intellectual Property
KECS	Knowledge Elicitation Case Study
PAS	Publicly Available Specifications
PDBB	Project Data Building Blocks

SCM	Supply Chain Management
SMART	Simple Multi Attribute Rating Technique
SSoT	Single Source of Truth
URL	Uniform Resource Locator
WMS	Web Map Service

References

1. Pryke, S. *Construction Supply Chain Management: Concepts and Case Studies*; Wiley-Blackwell: Chichester, UK, 2009.
2. Amade, B.; Peter, E.O.A.; Amaeshi, U.F.; Okorocha, K.A.; Ogbonna, A.C. Delineating Supply Chain Management (SCM) Features in Construction Project Delivery: The Nigerian Case. *Int. J. Constr. Supply Chain Manag.* **2017**, *7*, 1–19. [CrossRef]
3. Kinnaird, C.; Geipel, M. *Blockchain Technology: How the Inventions Behind Bitcoin are Enabling a Network of Trust for the Built Environment*; ARUP: London, UK, 2017.
4. Lamb, K. *Blockchain and Smart Contracts: What the AEC Sector Needs to Know*; CDBB Research Bridgehead Report; The Centre for Digital Built Britain, The University of Cambridge: Cabridge, UK, 2018.
5. Mathews, M.; Robles, D.; Bowe, B. BIM+ Blockchain: A solution to the Trust Problem in Collaboration? CITA BIM Gathering, 11. 2017, p. 2017. Available online: https://arrow.tudublin.ie/cgi/viewcontent.cgi?article=1032&context=bescharcon (accessed on 11 November 2022).
6. Ashworth, A.; Perera, S. *Cost Studies of Buildings*; Routledge: London, UK, 2015.
7. Gould, N.; King, C.; Britton, P. *Mediating Construction Disputes: An Evaluation of Existing Practice*; Centre for Construction Law, Kings College: London, UK, 2010.
8. Hijazi, A.A.; Perera, S.; Calheiros, R.N.; Alashwal, A. Rationale for the Integration of BIM and Blockchain for the Construction Supply Chain Data Delivery: A Systematic Literature Review and Validation through Focus Group. *J. Constr. Eng. Manag.* **2021**, *147*, 03121005. [CrossRef]
9. Getuli, V.; Ventura, S.M.; Capone, P.; Ciribini, A.L.C. A BIM-based Construction Supply Chain Framework for Monitoring Progress and Coordination of Site Activities. *Procedia Eng.* **2016**, *164*, 542–549. [CrossRef]
10. Mason, J.; Escott, H. Smart contracts in construction: Views and perceptions of stakeholders. In Proceedings of the FIG Conference, Istanbul, Turkey, 6–11 May 2018.
11. Perera, S.; Ingirige, B.; Ruikar, K.; Obonyo, E. *Advances in Construction ICT and e-Business*; Taylor & Francis: London, UK, 2017.
12. *This Australian Standard® AS 7739 Digital Engineering for Rail—Part 1: Concepts and Principles*; Rail Industry Safety and Standards Board (RISSB) [Digital Engineering for Rail]: Brisbane, Australia, 2022.
13. Rajabi, M.S.; Radzi, A.R.; Rezaeiashtiani, M.; Famili, A.; Rashidi, M.E.; Rahman, R.A. Key assessment criteria for organizational BIM capabilities: A cross-regional study. *Buildings* **2022**, *12*, 1013. [CrossRef]
14. Li, J.; Greenwood, D.; Kassem, M. Blockchain in the built environment and construction industry: A systematic review, conceptual models and practical use cases. *Autom. Constr.* **2019**, *102*, 288–307. [CrossRef]
15. Mason, J. The BIM Fork—Are smart contracts in construction more likely to prosper with or without BIM? *J. Leg. Aff. Disput. Resolut. Eng. Constr.* **2019**, *11*, 02519002. [CrossRef]
16. Mahamadu, A.-M.; Mahdjoubi, L.; Booth, C.A. Critical BIM qualification criteria for construction pre-qualification and selection. *Archit. Eng. Des. Manag.* **2017**, *13*, 326–343. [CrossRef]
17. Mahdjoubi, L.; Brebbia, C.A.; Laing, R. *Building Information Modelling (BIM) in Design, Construction and Operations*; WIT Transactions on the Built Environment; Ashurst Lodge: Ashurst, UK, 2015.
18. Deng, Y.; Gan, V.J.L.; Das, M.; Cheng, J.C.P.; Anumba, C. Integrating 4D BIM and GIS for Construction Supply Chain Management. *J. Constr. Eng. Manag.* **2019**, *145*, 4019016. [CrossRef]
19. Nguyen, B.; Buscher, V.; Cavendish, W.; Gerber, D.; Leung, S.; Krzyzaniak, A.; Robinson, R.; Burgess, J.; Proctor, M.; O'Grady, K.; et al. *Blockchain and the Built Environment*; Arup Group Limited: London, UK, 2019.
20. Penzes, B. Blockchain technology in the construction industry. In *Digital Transformation for High Productivity*; Institution of Civil Engineers: London, UK, 2018.
21. Pradeep, A.S.E.; Amor, R.; You, T.W. Blockchain Improving Trust in BIM Data Exchange: A Case Study on BIMCHAIN. In *Construction Research Congress 2020: Computer Applications*; American Society of Civil Engineers: Reston, VA, USA, 2020.
22. Deng, Z.; Ren, Y.; Liu, Y.; Yin, X.; Shen, Z.; Kim, H.-J. Blockchain-based trusted electronic records preservation in cloud storage. *Comput. Mater. Contin.* **2019**, *58*, 135–151. [CrossRef]
23. Turk, Ž.; Klinc, R. Potentials of Blockchain Technology for Construction Management. *Procedia Eng.* **2017**, *196*, 638–645. [CrossRef]
24. Nanayakkara, S.; Perera, S.; Bandara, H.D.; Weerasuriya, G.T.; Ayoub, J. Blockchain technology and its potential for the construction industry. In Proceedings of the 43rd Australasian Universities Building Education Association (AUBEA) Conference: Built to Thrive: Creating Buildings and Cities that Support Individual Well-being and Community Prosperity, Noosa, QLD, Australia, 6–8 November 2019; pp. 662–672.
25. Perera, S.; Nanayakkara, S.; Rodrigo, M.; Senaratne, S.; Weinand, R. Blockchain technology: Is it hype or real in the construction industry? *J. Ind. Inf. Integr.* **2020**, *17*, 100125. [CrossRef]
26. Davies, E.; Kirby, N.; Bond, J.; Grogan, T.; Moore, A.; Roche, N.; Rose, A.; Tasca, P.; Vadgama, N. *Towards a Distributed Ledger of Residential Title Deeds in the UK*; University College London: London, UK, 2020.

27. Vadgama, N. *Distributed Ledger Technology in the Supply Chain*; UCL Centre for Blockchain Technologies: London, UK, 2019.

28. Perera, S.; Hijazi, A.A.; Weerasuriya, G.T.; Nanayakkara, S.; Rodrigo, M.N.N. Blockchain-Based Trusted Property Transactions in the Built Environment: Development of an Incubation-Ready Prototype. *Buildings* **2021**, *11*, 560. [CrossRef]

29. Hileman, G.; Rauchs, M. *Global Blockchain Benchmarking Study*; Cambridge Centre for Alternative Finance, University of Cambridge: Cambridge, UK, 2017; p. 122.

30. Government of Australia. *National Blockchain Roadmap*; Australian Government: Canbera, Australia, 2020.

31. Li, J.; Kassem, M.; Watson, R. A Blockchain and Smart Contract-Based Framework to Increase Traceability of Built Assets. In Proceedings of the 37th CIB W78 Information Technology for Construction Conference (CIB W78); Available online: https://www.researchgate.net/publication/340547769_A_Blockchain_and_Smart_Contract-Based_Framework_to_Increase_Traceability_of_Built_Assets_EasyChair_Preprint (accessed on 11 November 2022).

32. Dounas, T.; Lombardi, D.; Jabi, W. Framework for decentralised architectural design BIM and Blockchain integration. *Int. J. Archit. Comput.* **2020**, *19*, 1478077120963376. [CrossRef]

33. Bolton, A.; Enzer, M.; Schooling, J. *The Gemini Principles*; Centre for Digital Built Britain and Digital Framework Task Group: Cambridge, UK, 2019.

34. Pang, C.; Szafron, D. Single Source of Truth (SSOT) for Service Oriented Architecture (SOA). In *International Conference on Service-Oriented Computing*; Springer: Berlin, Germany, 2014; pp. 575–589.

35. Hijazi, A.A.; Perera, S.; Alashwal, A.; Calheiros, R.N. Enabling a single source of truth through BIM and blockchain integration. In Proceedings of the 2019 International Conference on Innovation, Technology, Enterprise and Entrepreneurship (ICITEE 2019), Bahrain, Kingdom of Bahrain, 24–25 November 2019; pp. 385–393.

36. Hijazi, A.A.; Perera, S.; Calheiros, R.N.; Alashwal, A. A data model for integrating BIM and blockchain to enable a single source of truth for the construction supply chain data delivery. *Eng. Constr. Archit. Manag. online ahead of print.* 2022.

37. Qing, L.; Yu-Liu, C. *Modeling and Analysis of Enterprise and Information Systems*; Beijing Higher Education Press: Beijing, China, 2009.

38. Li, X.; Lu, W.; Xue, F.; Wu, L.; Zhao, R.; Lou, J.; Xu, J. Blockchain-Enabled IoT-BIM Platform for Supply Chain Management in Modular Construction. *J. Constr. Eng. Manag.* **2022**, *148*, 04021195. [CrossRef]

39. Hamledari, H.; Fischer, M. Role of blockchain-enabled smart contracts in automating construction progress payments. *J. Leg. Aff. Disput. Resolut. Eng. Constr.* **2021**, *13*, 04520038. [CrossRef]

40. Adamska, B.A.; Blahak, D.; Abanda, F.H. Blockchain in Construction Practice. In *Collaboration and Integration in Construction, Engineering, Management and Technology*; Springer: Berlin, Germany, 2021; pp. 339–343.

41. Liu, Z.; Chi, Z.; Osmani, M.; Demian, P. Blockchain and building information management (BIM) for sustainable building development within the context of smart cities. *Sustainability* **2021**, *13*, 2090. [CrossRef]

42. Manzione, L. Proposition for a Collaborative Design Process Management Conceptual Structure Using BIM. Ph.D. Thesis, Universidade de Sao Paulo, Polytechnic School, Sao Paulo, Brazil, 2013.

43. Nawari, N.O.; Ravindran, S. Blockchain technology and BIM process: Review and potential applications. *J. Inf. Technol. Constr.* **2019**, *24*, 209–238.

44. Wu, L.; Lu, W.; Xue, F.; Li, X.; Zhao, R.; Tang, M. Linking permissioned blockchain to Internet of Things (IoT)-BIM platform for off-site production management in modular construction. *Comput. Ind.* **2022**, *135*, 103573. [CrossRef]

45. Hijazi, A.A.; Perera, S.; Alashwal, A.; Calheiros, R.N. Blockchain Adoption in Construction Supply Chain: A Review of Studies Across Multiple Sectors. In Proceedings of the International Council for Research and Innovation in Building and Construction (CIB) World Building Congress 2019 June—Constructing Smart Citie, Hong Kong, China, 17–21 June 2019.

46. Li, J.; Kassem, M. Applications of distributed ledger technology (DLT) and Blockchain-enabled smart contracts in construction. *Autom. Constr.* **2021**, *132*, 103955. [CrossRef]

47. Amaludin, A.E.; Bin Taharin, M.R. Prospect of blockchain technology for construction project management in Malaysia. *ASM Sci. J.* **2018**, *11*, 199–205.

48. Bukunova, O.V.; Bukunov, A.S. *Tools of Data Transmission at Building Information Modeling*; IEEE: Piscataway, NJ, USA, 2019; pp. 1–6.

49. Parn, E.A.; Edwards, D. Cyber threats confronting the digital built environment: Common data environment vulnerabilities and block chain deterrence. *Eng. Constr. Archit. Manag.* **2019**, *26*, 245–266. [CrossRef]

50. Nawari, N.O. Blockchain Technologies: Hyperledger Fabric in BIM Work Processes. In Proceedings of the International Conference on Computing in Civil and Building Engineering, Sao Paulo, Brazil, 18–20 August 2020; pp. 813–823.

51. Celik, Y.; Petri, I.; Barati, M. Blockchain supported BIM data provenance for construction projects. *Comput. Ind.* **2023**, *144*, 103768. [CrossRef]

52. Revit: BIM Software for Designers, Builders and Doers. Available online: https://www.autodesk.com.au/products/revit/overview?term=1-YEAR&tab=subscription (accessed on 19 December 2022).

53. Alonso, R.; Borras, M.; Koppelaar, R.H.; Lodigiani, A.; Loscos, E.; Yöntem, E. SPHERE: BIM digital twin platform. *Proceedings* **2019**, *20*, 9.

54. Tao, X.; Liu, Y.; Wong, P.K.-Y.; Chen, K.; Das, M.; Cheng, J.C. Confidentiality-minded framework for blockchain-based BIM design collaboration. *Autom. Constr.* **2022**, *136*, 104172. [CrossRef]

55. Mason, J. Intelligent contracts and the construction industry. *J. Leg. Aff. Disput. Resolut. Eng. Constr.* **2017**, *9*, 04517012. [CrossRef]

56. Belle, I. The architecture, engineering and construction industry and blockchain technology. In Proceedings of the DADA 2017 International Conference on Digital Architecture, Nanjing, China, September 2017; Available online: https://www.researchgate.net/publication/322468019_The_architecture_engineering_and_construction_industry_and_blockchain_technology (accessed on 19 December 2022).

57. Coyne, R.; Onabolu, T. Blockchain for architects: Challenges from the sharing economy. *Archit. Res. Q.* **2017**, *21*, 369–374. [CrossRef]

58. Brown, K. Bitcoin and Ethereum: Empirical Evidence on Node Distribution. 2018. Available online: https://www.semanticscholar.org/paper/Bitcoin-and-Ethereum%3A-Empirical-Evidence-on-Node-Brown/1f7c0ec3d972f24ea36cd19ec7cccc35b345e724 (accessed on 19 December 2022).

59. Nasarre-Aznar, S. Collaborative housing and blockchain. *Administration* **2018**, *66*, 59–82. [CrossRef]

60. Ranjith, V.; Sathyajith, R.; Naveen Kumar, U.; Pushpalatha, M. A Performance Comparison of Different Security Hashing in a Blockchain Based System that Enables a More Reliable and SWIFT Registration of Land. *Int. J. Adv. Sci. Technol.* **2020**, *29*, 1–6.

61. Bowmaster, J.; Rankin, J.; Perera, S. *E-Business in the Architecture, Engineering and Construction (AEC) Industry in Canada: An Atlantic Canada Study;* Western Sydney University: Sydney, Australia, 2016.

62. Oxford Saïd Business School. What stakeholders are involved in the blockchain strategy system? In *Oxford Blockchain Strategy;* Oxford Saïd Business School: Oxford, UK, 2018; Volume 2021.

63. FIBREE. *Industry Report Blockchain Real Estate;* FIBREE: Berlin, Germany, 2021.

64. Nanayakkara, S.; Rodrigo, M.; Perera, S.; Weerasuriya, G.T.; Hijazi, A.A. A methodology for selection of a Blockchain platform to develop an enterprise system. *J. Ind. Inf. Integr.* **2021**, *23*, 100215. [CrossRef]

65. Nawari, N.O. A Generalized Adaptive Framework (GAF) for Automating Code Compliance Checking. *Buildings* **2019**, *9*, 86. [CrossRef]

66. Dimyadi, J. *Automated Compliance Audit Processes for Building Information Models with an Application to Performance-Based Fire Engineering Design Methods;* ResearchSpace@: Auckland, New Zeland, 2016.

67. Xue, F.; Lu, W. A semantic differential transaction approach to minimizing information redundancy for BIM and blockchain integration. *Autom. Constr.* **2020**, *118*, 103270. [CrossRef]

68. Ahmadisheykhsarmast, S.; Sonmez, R. A smart contract system for security of payment of construction contracts. *Autom. Constr.* **2020**, *120*, 103401. [CrossRef]

69. Chong, H.Y.; Balamuralithara, B.; Chong, S.C. Construction contract administration in Malaysia using DFD: A conceptual model. *Ind. Manag. Data Syst.* **2011**, *11*, 1449–1464. [CrossRef]

70. Rodrigo, M.; Perera, S.; Senaratne, S.; Jin, X. Systematic development of a data model for the blockchain-based embodied carbon (BEC) Estimator for construction. *Eng. Constr. Archit. Manag.* **2021**. Available online: https://researchdirect.westernsydney.edu.au/islandora/object/uws:60599/ (accessed on 19 December 2022). [CrossRef]

71. Kasie, F.M. Combining simple multiple attribute rating technique and analytical hierarchy process for designing multi-criteria performance measurement framework. *Glob. J. Res. Eng.* **2013**, *13*, 15–30.

72. Wang, Y.; Han, J.H.; Beynon-Davies, P. Understanding blockchain technology for future supply chains: A systematic literature review and research agenda. *Supply Chain Manag. Int. J.* **2019**, *24*, 62–84. [CrossRef]

73. R3 Technical Documentation. Available online: https://docs.corda.net/ (accessed on 19 December 2022).

74. Elements Project. Available online: https://github.com/ElementsProject (accessed on 19 December 2022).

75. Hyperledger-Fabric for the Enterprise. Available online: https://hyperledger-fabric.readthedocs.io/en/release-2.2/ (accessed on 19 December 2022).

76. Getting Started with IBM Blockchain Platform. Available online: https://cloud.ibm.com/docs/blockchain (accessed on 19 December 2022).

77. NEM Blockchain Platform. Available online: https://nemplatform.com/developers/ (accessed on 19 December 2022).

78. Cachin, C. Architecture of the hyperledger blockchain fabric. In *Workshop on Distributed Cryptocurrencies and Consensus Ledgers;* Headquarters Location: Wuhan, China, 2016; Volume 310, pp. 1–4.

79. Sajana, P.; Sindhu, M.; Sethumadhavan, M. On blockchain applications: Hyperledger fabric and ethereum. *Int. J. Pure Appl. Math.* **2018**, *118*, 2965–2970.

80. Androulaki, E.; Barger, A.; Bortnikov, V.; Cachin, C.; Christidis, K.; De Caro, A.; Enyeart, D.; Ferris, C.; Laventman, G.; Manevich, Y. Hyperledger fabric: A distributed operating system for permissioned blockchains. In Proceedings of the Thirteenth EuroSys Conference, Porto, Portugal, 23–26 April 2018; pp. 1–15.

81. Cocco, S.; Singh, G. Top 6 Technical Advantages of Hyperledger Fabric for Blockchain Networks. Available online: https://developer.ibm.com/technologies/blockchain/articles/top-technical-advantages-of-hyperledger-fabric-for-blockchain-networks/ (accessed on 28 July 2020).

82. Berry, S.; Marker, T. Residential energy efficiency standards in Australia: Where to next? *Energy Effic.* **2015**, *8*, 963–974. [CrossRef]

83. Hudson, F.S.; Sutrisna, M.; Chawynski, G. A certification framework for managing the risks of non-compliance and non-conformance building products: A Western Australian perspective. *Int. J. Build. Pathol. Adapt.* **2020**, *39*. [CrossRef]

84. Gorse, C.; Sturges, J. Not what anyone wanted: Observations on regulations, standards, quality and experience in the wake of Grenfell. *Constr. Res. Innov.* **2017**, *8*, 72–75. [CrossRef]

85. Hills, R. Cladding audits: The problem of combustible cladding and the wider problem of NCBPs and non-compliant building work. *J. Build. Surv. Apprais. Valuat.* **2018**, *6*, 312–321.
86. Government of Western Australia. *Industry Regulation and Safety "Self-Certification"*; Government of Western Australia: Perth, Australia, 2018.
87. Ye, X. Combining BIM with Smart Contract and Blockchain to Support Digital Project Delivery and Acceptance Processes. Master Thesis, Ruhr-Universität Bochum, Computational Engineering, Bochum, Germany, 2019.
88. Ghannad, P.; Lee, Y.-C.; Dimyadi, J.; Solihin, W. Automated BIM data validation integrating open-standard schema with visual programming language. *Adv. Eng. Inform.* **2019**, *40*, 14–28. [CrossRef]
89. Hyperledger Fabric v2.3.3. Available online: https://hyperledger-fabric.readthedocs.io/en/release-2.3/whatsnew.html#what-s-new-in-hyperledger-fabric-v2-3 (accessed on 19 December 2022).
90. JavaScript Libraries. Available online: https://www.javascript.com/ (accessed on 19 December 2022).
91. DynaWeb with REST APIs in Particular. Available online: https://radumg.github.io/DynaWeb/ (accessed on 19 December 2022).
92. Hyperledger. Writing Your First Application. Available online: https://hyperledger-fabric.readthedocs.io/en/release-2.2/write_first_app.html (accessed on 19 December 2022).
93. NSW. *Digital Engineering Framework Interim Approach*; Transport for NSW: Sydney, Australia, 2018.
94. Naderi, H.; Shojaei, A. Civil Infrastructure Digital Twins: Multi-Level Knowledge Map, Research Gaps, and Future Directions. *IEEE Access* **2022**, *10*, 122022–122037. [CrossRef]
95. Hoy, M.B. An introduction to the Blockchain and its implications for libraries and medicine. *Med. Ref. Serv. Q.* **2017**, *36*, 273–279. [CrossRef]

Article

Research on BIM Application Two-Dimensional Maturity Model

Changqing Sun [1,2]**, Hong Chen** [3,4,*]**, Ruyin Long** [3,4,*] **and Ruihua Liao** [5]

[1]   School of Economics and Management, China University of Mining and Technology, Xuzhou 221116, China
[2]   Zhongye Changtian International Engineering Co., Ltd., Changsha 410125, China
[3]   School of Business, Jiangnan University, Wuxi 214122, China
[4]   Institute of National Security and Green Development, Jiangnan University, Wuxi 214122, China
[5]   School of Information Technology, Hunan First Normal University, Changsha 410205, China
*   Correspondence: 8202105496@jiangnan.edu.cn (H.C.); 8202105495@jiangnan.edu.cn (R.L.);
    Tel.: +86-133-7221-0769 (H.C.); +86-139-5229-8058 (R.L.)

Abstract: The existing project management maturity models and BIM maturity models have obvious deficiencies in evaluating the management level of engineering projects with BIM applications. This study aimed to use accepted assessment indexes to design an innovative BIM application maturity model suitable for different projects with BIM applications. This study proposes the concept of BIM Application Two-Dimensional Maturity (BATM), which simultaneously emphasizes the project business management (PBM) and project BIM application (PBA) maturities. The BATM model assesses the PBM and PBA maturities based on eight performance domains and 37 desired outcomes of PMBOK 7th edition. The application case shows that the use of the BATM model is simple and its effect is obvious. This study is the first to assess the BIM application maturity from the two dimensions of PBM and PBA, and provides new insights into the project BIM application maturity assessment. The application case sets an example for other companies to assess and improve their BATM.

Keywords: project management; project BIM application; two-dimensional maturity; maturity model; PMBOK

Citation: Sun, C.; Chen, H.; Long, R.; Liao, R. Research on BIM Application Two-Dimensional Maturity Model. *Buildings* **2022**, *12*, 1960. https://doi.org/10.3390/buildings12111960

Academic Editor: Amos Darko

Received: 20 October 2022
Accepted: 8 November 2022
Published: 11 November 2022

Publisher's Note: MDPI stays neutral with regard to jurisdictional claims in published maps and institutional affiliations.

1. Introduction

In recent years, adopting building information modeling (BIM) has become increasingly popular in the design, construction, operations, and maintenance phases of the building's life cycle [1,2]. BIM is a digital representation of physical and functional characteristics of a facility and a shared knowledge resource for information about a facility, forming a reliable basis for decisions during its life cycle [3]. In the engineering industry, owners, designers, builders, and managers have already reported the benefits of adopting the BIM methodology, which has led to its increasing acceptance at a global level [4].

Maturity models, which originated from total quality management [5] and are widely used in various industries [6], are primarily based upon the capability maturity model (CMM) of the Software Engineering Institute. Maturity models allow individuals and organizations to self-assess the maturity of various aspects of their processes against benchmarks [7], and enable organizations to accelerate the enhancement in their capabilities in fields such as business process management [8], software research and development [9], digital government [10], knowledge management [11], and project management [12]. Maturity models assume predictable patterns in every evolutionary phase of organization development [13]. These distinctive phases, with each later phase being superior to a previous phase, provide a roadmap for organizational improvement. The continuous progress of an organization on the evolutionary path implies gradual improvements in the organizational capabilities. The maturity levels represent a staged path for the performance and process improvement efforts of organizations [14].

BIM maturity can be defined as the level of "quality, repeatability and degree of excellence" in relation to performing a BIM-related task or delivering a BIM service or output [15]. Different BIM maturity models have been created to measure BIM maturity in the architecture, engineering, and construction industries [16]. Some models focus on assessing BIM against projects, while others target evaluating organizations [17].

However, the existing BIM maturity models have two main inadequacies. Firstly, most maturity models tend to evaluate the BIM application maturity at specific project phases under specified conditions, such as the design phase [18,19], construction phase [20,21], and facility management [22,23]. Few models can be applied to all phases from design and construction to operations and maintenance. Secondly, each model puts forward its own differing assessment indexes and has its own definition of maturity levels, and there is no commonly accepted model. As a result, it is difficult for users to choose a suitable model for their BIM maturity assessment [24].

In the era of digital twins, for the engineering industry, BIM represents the virtual world, whereas engineering construction represents the real world, and the two are like twins. Therefore, there is a need to study the BIM maturity and the project management maturity at the same time; however, there is no literature in this regard at present.

To this end, this study proposes an innovative BIM maturity model called the BIM Application Two-Dimensional Maturity (BATM) model, which combines the functions of the project management maturity model (PMMM) and BIM maturity model, simultaneously emphasizes project business management (PBM) and project BIM application (PBA) from the two dimensions of the real world (PBM) and virtual world (PBA), and achieves the effect of $1 + 1 > 2$. The application of the BATM model helps in enhancing the maturity level of the project management and BIM application, improving the efficiency of organizational production management, and promoting organizational advancement along a maturity ladder.

The remainder of this paper is organized as follows. The next section provides the background of project management maturity models and BIM maturity models such that the BATM framework can be better understood. The Methods section introduces the BATM concept, definition, model structure, and the related questionnaire, as well as an example in which the BATM model is applied. Subsequently, the important functions and innovations of the study are discussed. Finally, the Conclusions section presents the theoretical and practical implications of the study.

2. Research Background

2.1. Project Management Maturity Models

The maturity in managing projects implies the established, proven, and innovative practices and procedures that lead to success in planning and completing projects [25]. Companies in various industries are pursuing improvements in project management maturity. A PMMM can enable an organization to seek perfect project management by implementing gradual maturity improvement processes within the organization [26]. PMMMs are regarded as the useful tools for evaluating an organization's current project management capability [27]. Project management capability is the competence required to ensure an organization remains competitive when conducting projects [28]. Capability frameworks are the basis for maturity models that address how capabilities can be developed along an anticipated, desired, or logical path [29].

The successful application of the CMM in the software industry inspired the development of the maturity model for project management. A PMMM is a complete framework and a comprehensive tool for evaluating the maturity level of project management. Since its creation in the 1990s, it has been used to systematically improve the maturity level of project management. As higher project management maturity levels represent the ability of organizations to obtain better results from their projects, the stakeholders of organizations are willing to assess their current project management maturity status for future development and improve to the next phase if desired [30].

After more than 30 years of development, many PMMMs have become available. The Microframe project management maturity model is one of the earliest PMMMs to be applied practically [31]. The Berkeley project management process maturity model, known as PM2 and presented by Ibbs and Kwak, determines and positions an organization's relative project management level based on those of other organizations [32]. A five-scale PMMM known as K-PMMM [33], which was established by Kerzner, analyzes the efficiency of project management organization, drawing attention to the importance of strategic project management to improve know-how in the marketplace. PMS-PMMM, which was released by Project Management Solutions in 2001, combines the five maturity levels proposed by the Software Engineering Institute and the project management knowledge areas proposed by the Project Management Institute (PMI) to form a comprehensive, easy-to-accept project management maturity improvement model [34]. The organizational project management maturity model (OPM3) introduced by the PMI not only provides a systematic assessment and improvement method for the enterprise from a single project to entrepreneur portfolio projects, but also introduces and solidifies the best practice in every business process [27]. The portfolio, programme, and project management maturity model (P3M3), which was developed by the UK's Office of Government Commerce, comprises three independent sub-models (portfolio, programme, and project) and considers all seven processes as equally important. In the P3M3, the lowest maturity of the seven processes is the maturity of the organization [6]. In 2016, the International Project Management Association (IPMA) developed a methodology called "IPMA Delta" in order to certify the ability of an organization to use project management techniques. The assessment results of IPMA Delta show, in detail, the room for improvement, also giving recommendations for the future areas that need to be refined. MMM, focusing on a strategy of continuous improvement and following the four steps of the PDCA cycle to put this approach into practice, was developed by Langston and Ghanbaripour [35].

The only feature on which almost all models seem to converge is the determination of five maturity levels, even if they are not perfectly equal, either in the contents or in the denominations [36]. These levels and the corresponding main models can be summarized as follows:

- Level 1: initial or basic with awareness (OPM3, IPMA Delta, P3M3, CMMI, PM2, PMS-PMMM, MMM).
- Level 2: structured, managed, or repeatable (OPM3, P3M3, CMMI, PMS-PMMM, MMM).
- Level 3: defined, standardized, or institutionalized (OPM3, IPMA Delta, P3M3, CMMI, PMS-PMMM, MMM).
- Level 4: fully managed at the corporate level (OPM3, IPMA Delta, P3M3, CMMI, PM2, PMS-PMMM, MMM).
- Level 5: optimized with continuous learning and improvement (OPM3, IPMA Delta, P3M3, CMMI, PM2, PMS-PMMM, MMM).

In recent years, PMMM research has been expanded, and many scholars have investigated project risk management models [37,38]. Silvius and Schipper developed a sustainable PMMM as a practical tool for the assessment and development of the integration of sustainability in projects [39]. Seelhofer and Graf extended the concept of organizational project management maturity to the national context and developed a systematic framework of national project management maturity and the national PMMM [40].

Since most PMMMs are based on a guide to the project management body of knowledge (PMBOK) of PMI [41], by adopting PMMMs, organizations can systematically plan and improve their project management capabilities and benchmark their performance in accordance with the industry standards [31]. The assessment of maturity through PMMMs enables further improvement directions to be identified [26].

Despite their similarities, PMMMs differ from each other in terms of their assessment methodology. Hence, selecting an appropriate PMMM is a crucial managerial decision, and the organizational environment and project characteristics must be considered well to ensure the suitability of the selected model [30].

2.2. BIM Maturity Model

Over the past decade, a large number of BIM maturity models have been developed to measure the performance of BIM application. BIM maturity models are mainly divided into two categories [17,42]: one is the project BIM maturity model focusing on project application performance, and the other is the organizational BIM maturity model focusing on enterprise implementation capability. The famous project BIM maturity models include NBIMS CMM, iBIM, and VDC Scorecard, and the famous organizational BIM maturity models include BIM PM, BIM MM, BIM Quick Scan, and BIM AP. In addition, there are individual models that can be applied to the BIM maturity assessment of both organizations and projects.

The U.S. National BIM Standard (NBIMS) was published in 2007 and provided information as a guide for the adoption, implementation, and application of BIM to enable core principles. The Capability Maturity Model (CMM) of NBIMS is a matrix with 11 interest areas on the x-axis and 10 maturity levels on the y-axis [43], and is a useful tool for the strategic management in the BIM implementation of an organization [44]. The interactive capability maturity model (ICMM) is a further enhancement of CMM, developed to meet the growing need for an accurate and up-to-date model [45]. Bew and Richards developed the iBIM maturity model in 2008. Its assessment indexes focus on technology, standards, guidelines, classification, delivery, etc. Its maturity is divided into four levels. Level 0 is characterized by paper-based medium delivery methods. Level 1 represents structural elements by 2D or 3D digital objects. Model-based collaboration occurs in Level 2 between different parties, and network-based integration occurs in Level 3 [46]. VDC Scorecard was designed to measure the performance of the projects of virtual design firms with four major areas, 10 divisions, and 74 measures. Its distinct feature is the establishment of confidence levels to measure the degree of objective compliance [47]. VDC Scorecard has both quantitative and qualitative assessment methods with multiple choice and open-ended questions. It assesses performances of BIM projects against the industry benchmark and has five maturity levels.

The BIM Proficiency Matrix (BIM PM) was developed by Indiana University Architect's Office to score the performance of BIM services of designers and contractors in Indiana University projects [48]. BIM PM is composed of 32 measures of eight areas and five maturity levels [17]. It has also been criticized for its heavy focus on the technical aspects of BIM implementation rather than process and protocol. The BIM maturity matrix (BIM MM) is multi-dimensional and can be represented by a tri-axial knowledge model comprising BIM Fields, BIM Stages, and BIM Lenses [49]. The model proposes five BIM maturity levels: initial, defined, managed, integrated, and optimized [50]. BIM MM assessment can be provided by the online BIM Excellence platform. The question number of assessment varies according to the assessment granularity level, and a maturity score is compiled related to 12 positions grouped into five areas [51]. BIM Quick Scan developed by TNO (The Netherlands Organization for applied scientific research) is a benchmarking tool for organizational performance with a reasonably extensive scope covering 44 measures in four main areas, including: organization and management, mentality and culture, information structure and information flow, and tools and applications [43]. It can combine quantitative and qualitative assessments of the "hard" and "soft" aspects of BIM, and distinguish the strengths and weaknesses of BIM application for an organization. Organizational BIM Assessment Profile (BIM AP) was created by Pennsylvania State University Computer Integrated Construction (CIC) Research Program in 2012. Its maturity is measured by 20 planning elements with six themes: Strategy, BIM Uses, Process, Information, Infrastructure and Personnel, Companies. Their maturity levels range from 0 (Non-Existent) to 5 (Optimizing) [22,51].

The multifunctional BIM maturity model (MPMM) focuses on BIM maturity at different scales from individual projects to an organization's full projects portfolio, covering measurements across three domains: technology, process, and protocol. Detailed, operable rubrics enable the assessment of each subdomain of each domain, and the assessment result points to four maturity levels (0–3) [52].

From the above introduction, it can be seen that these existing studies mainly focus on the technology capability maturity of BIM from different perspectives and conditions, and ignore the digital twin relationship between BIM and projects. In addition, a large number of BIM maturity models have different assessment indexes and different level definitions. As a result, users can be confused and do not know how to choose assessment models.

2.3. Research Gaps

This study aims to fill several main literature gaps, as follows:

- Each of the existing BIM maturity models was developed to achieve a specific BIM assessment purpose [17]; moreover, many models can only be applied to an individual project phase from scheme design to facility maintenance, which leads to the situation of different assessment indexes and different level definitions for different models, increases the difficulty for users to choose an appropriate model [42], and then affects the popularization and use of these models. Determining how to design a BIM maturity model with a generally acceptable assessment index and maturity level system, and make the model applicable to all projects with BIM application, are problems worth studying.

- In the PMMM literature, the assessment indexes of most studies were based on PM-BOK [31,41], but PMBOK was not used in the study of the PBA maturity. Because the objects of BIM services are projects and BIM implementation processes are also similar to project management processes, PMBOK should be of guiding and reference value for PBA maturity research. In addition, PMBOK 7th edition pays close attention to eight performance domains and 37 desired outcomes, which is more closely combined with the BIM application and creates favorable conditions for PBA maturity assessment based on PMBOK. However, to date, there is no research about the PBM maturity model or the PBA maturity model based on the PMBOK 7th edition.

- The PBM and PBA maturities are two important aspects of modern project management, and reflect the digital twin relationship from the two dimensions of the virtual and real worlds. In modern project management, the PMMMs can no longer ignore the existence of BIM, and the BIM maturity model also needs to consider the contribution of BIM in project management. However, there is no research that simultaneously assesses their maturities, or that identities problems and highlights directions for the improvement in PBM and PBA by a maturity assessment.

3. Method

3.1. Concept of BATM

As previously stated, the existing project management maturity models and BIM maturity models have obvious deficiencies in evaluating the management level of engineering projects with BIM applications. Against the background in which the idea of digital twins has become popular and 37 desired outcomes of the PMBOK 7th edition have become an acceptable global standard, in order to enable an organization to accurately understand the capabilities of its project management and BIM applications and then take effective measures for improvement, the PBM and PBA maturities can be evaluated simultaneously based on each desired outcome.

It is in this context that the BATM model is proposed. The BATM is expressed by a two-dimensional value such as (x, y), in which x reflects the PBM maturity level, and y reflects the PBA maturity level. The BATM model consists of the following items: maturity level definition, assessment indexes, questionnaire, maturity calculation and problem identification methods, improvement advice, etc.

The purpose of putting forward the concept of BATM is to use the PBM and PBA maturities to reflect the level of enterprises' project management and project BIM applications; and, through the BATM model, determine enterprises' strengths and weakness in the PBM and PBA aspects, identify the improvement directions, and promote the advancement of enterprises' PBM and PBA maturities.

It is noteworthy that the two-dimensional maturity in this study differs significantly from the multi-dimensional maturity in other studies. The two dimensions investigated in this study were the PBM of the real world and the PBA of the virtual world. For example, considering "Effective management of procurements" which is one desired outcome of the project work performance domain in PMBOK 7th edition, the BATM value of (3, 1) means that the PBM maturity level of procurement management in the real world is 3 and the PBA maturity level in the virtual world is 1; it also means that there are obvious deficiencies in its online capabilities for supporting offline project management. However, the multiple dimensions in other studies refer to several aspects of pure project management. For example, Hu presented a three-dimensional PMMM constituted by best practice maturity, process maturity, and organization system maturity [53].

3.2. Classification of Maturity Level

The BATM assessment is inseparable from the level definition of the PBM and PBA maturities. The classification and definition of PBM and PBA maturity level should be clear.

The classification of the PBM maturity levels refers to that of the PMS-PMMM model, and its maturity levels, from 1 to 5, are the initial, managed, defined, quantitatively managed, and optimizing levels, respectively.

The classification of the PBA maturity levels refers to the information technology governance maturity model under the COBIT 4.1 framework. In COBIT 4.1, the maturity levels from 0 to 5 are the non-existent, initial, repeatable but intuitive, defined process, managed and measurable, and optimized levels, respectively [54], which match well with the maturity levels of PMS-PMMM.

The specific feature definitions of the PBM and PBA maturity level are shown in Table 1.

Table 1. Definition of maturity level.

Level	Definition of PBM Maturity Level	Definition of PBA Maturity Level
Level 0		• The organization does not apply BIM in engineering projects and does not possess PBA awareness.
Level 1	• Project management is temporary, and even chaotic occasionally. • Organization rarely provides a stable environment for implementing projects. • The success of a project primarily depends on the efforts of the individual, rather than the standardized management processes of the organization, and organization has begun to realize these problems in project management.	• The organization has realized the importance of PBA. Some applications of BIM tools and software exist in some projects, but the effect of PBA is fragmented. • No defined and standardized processes exist.
Level 2	• Organization has established basic processes to track projects, and encourage and support other projects to use these standardized processes. • The management primarily depends on personal knowledge or general tools, and the actual effect varies significantly by project.	• The organization has purchased systematic PBA tools or constructed a PBA software platform, and has established the corresponding PBA processes, on which the project members can perform their work. • The organization has no mandatory requirement for PBA in project management, and the effect of PBA depends entirely on the individual's ability and responsibility.
Level 3	• Project management processes have been institutionalized and standardized, as well as extended to all projects. • The management depends on industry standards, and the organization can master the summary information and detailed information of each project.	• The project management processes have been integrated into the PBA processes, the tools and platform of PBA are reliable and verified, and the processes and requirements of PBA have been standardized and documented. • The organization requires employees to follow the PBA processes, however, the management of PBA deviation is insufficient.
Level 4	• Project management processes are combined with the organizational processes. • The management depends on organizational standards, the project implementation is under control, and project management decisions are made using project data. • The organization has standardized analysis methods to evaluate project performance.	• The organization can monitor and measure the implementation and deviation of the PBA, and detailed results of the PBA performance can be acquired and analyzed statistically. • PBA is under good control and constant improvement, and various intelligent equipment and software of project management are combined with PBA, thereby contributing to PBM.
Level 5	• The organization has established and performed the processes for evaluating the efficiency and effectiveness of project implementation. • The processes for improving project performance are implemented, and continuous improvement is the focus of project management.	• Based on quantitative feedback and continuous improvement, PBA processes have been refined to the level of "best practices". • PBA is indispensable for improving the efficiency and effect of PBM in an organization.

3.3. Definition of Model Structure

PMBOK is an excellent reference for analyzing project management capabilities, in which an abundance of "best practice" information is outlined in the document [34]. Because the knowledge content of each PMBOK performance domain is abundant, each performance domain is categorized into several key desired outcomes [55]. In total, there are eight performance domains and 37 desired outcomes in the PMBOK 7th edition.

In the BATM model, the eight performance domains are used as the first level assessment indexes, and the 37 desired outcomes are used as the second level assessment indexes; this constitutes the assessment index system of BATM model. The second level assessment indexes are used to measure the PBM and PBA maturities from the two dimensions of the virtual and real worlds, and the maturities of the first level assessment indexes are summarized from the second level indexes. For example, under the delivery performance domain, the maturity level is measured using five desired outcomes. They include: (1) projects contribute to business objectives and advancement of strategy; (2) projects realize the outcomes they were initiated to deliver; (3) project benefits are realized in the time frame in which they were planned; (4) the project team has a clear understanding of requirements; (5) stakeholders accept and are satisfied with project deliverables. The specific assessment indexes are shown in Table 2.

Table 2. Assessment index system of BATM.

Performance Domains	Code	Desired Outcomes
Stakeholder performance domain	D_{11}	A productive working relationship with stakeholders throughout the project.
	D_{12}	Stakeholder agreement with project objectives.
	D_{13}	Stakeholders who are project beneficiaries are supportive and satisfied while stakeholders who may oppose the project or its deliverables do not negatively impact project outcomes.
Team performance domain	D_{21}	Shared ownership.
	D_{22}	A high-performing team.
	D_{23}	Applicable leadership and other interpersonal skills demonstrated by all team members.
Development approach and life cycle performance domain	D_{31}	Development approaches that are consistent with project deliverables.
	D_{32}	A project life cycle consisting of phases that connect the delivery of business and stakeholder value from the beginning to the end of the project.
	D_{33}	A project life cycle consisting of phases that facilitate the delivery cadence and development approach required to produce the project deliverables.
Planning performance domain	D_{41}	The project progresses in an organized, coordinated, and deliberate manner.
	D_{42}	There is a holistic approach to delivering the project outcomes.
	D_{43}	Evolving information is elaborated to produce the deliverables and outcomes for which the project was undertaken.
	D_{44}	Time spent planning is appropriate for the situation.
	D_{45}	Planning information is sufficient to manage stakeholder expectations.
	D_{46}	There is a process for the adaptation of plans throughout the project based on emerging and changing needs or conditions.
Project work performance domain	D_{51}	Efficient and effective project performance.
	D_{52}	Project processes are appropriate for the project and the environment.
	D_{53}	Appropriate communication with stakeholders.
	D_{54}	Efficient management of physical resources.
	D_{55}	Effective management of procurements.
	D_{56}	Improved team capability due to continuous learning and process improvement.
Delivery performance domain	D_{61}	Projects contribute to business objectives and advancement of strategy.
	D_{62}	Projects realize the outcomes they were initiated to deliver.
	D_{63}	Project benefits are realized in the time frame in which they were planned.
	D_{64}	The project team has a clear understanding of requirements.
	D_{65}	Stakeholders accept and are satisfied with project deliverables.
Measurement performance domain	D_{71}	A reliable understanding of the status of the project.
	D_{72}	Actionable data to facilitate decision making.
	D_{73}	Timely and appropriate actions to keep project performance on track.
	D_{74}	Achieving targets and generating business value by making informed and timely decisions based on reliable forecasts and assessments.
Uncertainty performance domain	D_{81}	An awareness of the environment in which projects occur, including, but not limited to, the technical, social, political, market, and economic environments.
	D_{82}	Proactively exploring and responding to uncertainty.
	D_{83}	An awareness of the interdependence of multiple variables on the project.
	D_{84}	The capacity to anticipate threats and opportunities and understand the consequences of issues.
	D_{85}	Project delivery with little or no negative impact from unforeseen events or conditions.
	D_{86}	Opportunities are realized to improve project performance and outcomes.
	D_{87}	Cost and schedule reserves are utilized effectively to maintain alignment with project objectives.

The BATM structure can be classified into three layers: project layer, performance domain layer, and desired outcome layer. Because the mode structure is based entirely on PMBOK 7th edition, we did not conduct an empirical analysis of it.

3.4. Questionnaire

The questionnaire consists of three parts. The first part introduces the purpose and requirements of the questionnaire, the second part explains the definition of PBM and PBA maturity levels, and the third part is the scoring table for experts to score. The left side of the scoring table lists the assessment indexes of the BATM model, including eight performance domains and their corresponding 37 desired outcomes. The right side is the selection area of PBM and PBA maturity levels, and the weights of desired outcomes can be set according to their performance domains.

In the questionnaire, a scale of 1 to 5 for PBM maturities and a scale of 0 to 5 for PBA maturities were adopted to measure the responses. Level 1 corresponds to score 1, level 2 corresponds to score 2, etc. Because the first, third, and fifth levels of the PBM and PBA maturities are the initial level, the defined level, and the optimized level, respectively, and their second and fourth levels are also similar, we can deem that their five maturity levels are relatively consistent. In addition, as there may be no PBA in some projects, the PBA maturity levels include the non-existent level. The format of the scoring table is shown in Table 3.

Table 3. Scoring table of BATM questionnaire.

Assessment Indexes		PBM Maturity						PBA Maturity						
		L1	L2	L3	L4	L5	Weight	L0	L1	L2	L3	L4	L5	Weight
Stakeholder performance domain	D_{11}	☐	☐	☐	☐	☐	___	☐	☐	☐	☐	☐	☐	___
	D_{12}	☐	☐	☐	☐	☐	___	☐	☐	☐	☐	☐	☐	___
	D_{13}	☐	☐	☐	☐	☐	___	☐	☐	☐	☐	☐	☐	___
Team performance domain	D_{21}	☐	☐	☐	☐	☐	___	☐	☐	☐	☐	☐	☐	___
	D_{22}	☐	☐	☐	☐	☐	___	☐	☐	☐	☐	☐	☐	___
	D_{23}	☐	☐	☐	☐	☐	___	☐	☐	☐	☐	☐	☐	___
Development approach and life cycle performance domain	D_{31}	☐	☐	☐	☐	☐	___	☐	☐	☐	☐	☐	☐	___
	D_{32}	☐	☐	☐	☐	☐	___	☐	☐	☐	☐	☐	☐	___
	D_{33}	☐	☐	☐	☐	☐	___	☐	☐	☐	☐	☐	☐	___
Planning performance domain	D_{41}	☐	☐	☐	☐	☐	___	☐	☐	☐	☐	☐	☐	___
	D_{42}	☐	☐	☐	☐	☐	___	☐	☐	☐	☐	☐	☐	___
	D_{43}	☐	☐	☐	☐	☐	___	☐	☐	☐	☐	☐	☐	___
	D_{44}	☐	☐	☐	☐	☐	___	☐	☐	☐	☐	☐	☐	___
	D_{45}	☐	☐	☐	☐	☐	___	☐	☐	☐	☐	☐	☐	___
	D_{46}	☐	☐	☐	☐	☐	___	☐	☐	☐	☐	☐	☐	___
Project work performance domain	D_{51}	☐	☐	☐	☐	☐	___	☐	☐	☐	☐	☐	☐	___
	D_{52}	☐	☐	☐	☐	☐	___	☐	☐	☐	☐	☐	☐	___
	D_{53}	☐	☐	☐	☐	☐	___	☐	☐	☐	☐	☐	☐	___
	D_{54}	☐	☐	☐	☐	☐	___	☐	☐	☐	☐	☐	☐	___
	D_{55}	☐	☐	☐	☐	☐	___	☐	☐	☐	☐	☐	☐	___
	D_{56}	☐	☐	☐	☐	☐	___	☐	☐	☐	☐	☐	☐	___
Delivery performance domain	D_{61}	☐	☐	☐	☐	☐	___	☐	☐	☐	☐	☐	☐	___
	D_{62}	☐	☐	☐	☐	☐	___	☐	☐	☐	☐	☐	☐	___
	D_{63}	☐	☐	☐	☐	☐	___	☐	☐	☐	☐	☐	☐	___
	D_{64}	☐	☐	☐	☐	☐	___	☐	☐	☐	☐	☐	☐	___
	D_{65}	☐	☐	☐	☐	☐	___	☐	☐	☐	☐	☐	☐	___
Measurement performance domain	D_{71}	☐	☐	☐	☐	☐	___	☐	☐	☐	☐	☐	☐	___
	D_{72}	☐	☐	☐	☐	☐	___	☐	☐	☐	☐	☐	☐	___
	D_{73}	☐	☐	☐	☐	☐	___	☐	☐	☐	☐	☐	☐	___
	D_{74}	☐	☐	☐	☐	☐	___	☐	☐	☐	☐	☐	☐	___
Uncertainty performance domain	D_{81}	☐	☐	☐	☐	☐	___	☐	☐	☐	☐	☐	☐	___
	D_{82}	☐	☐	☐	☐	☐	___	☐	☐	☐	☐	☐	☐	___
	D_{83}	☐	☐	☐	☐	☐	___	☐	☐	☐	☐	☐	☐	___
	D_{84}	☐	☐	☐	☐	☐	___	☐	☐	☐	☐	☐	☐	___
	D_{85}	☐	☐	☐	☐	☐	___	☐	☐	☐	☐	☐	☐	___
	D_{86}	☐	☐	☐	☐	☐	___	☐	☐	☐	☐	☐	☐	___
	D_{87}	☐	☐	☐	☐	☐	___	☐	☐	☐	☐	☐	☐	___

3.5. Calculation of BATM Level

After experts return the scoring table of the BATM questionnaire, the questionnaire organizers first identify whether experts' responses are qualified (in the two-dimensional maturity options of a management process of the scoring table, multiple selections and no selection are regarded as unqualified), then deal with the qualified data to obtain the three-layer BATM. The courses of evaluating the BATM involve the maturity determination of desired outcome layers based on the two dimensions of PBA and PBM, and that of performance domain layers and the project layer according to certain weight ratios. In the following expressions, w_{ikj} is the weight of expert k in the desired outcome j of performance domain i, i.e. the weight in the scoring table; w_{ik}, decided by the organizers, is the weight of expert k in performance domain i (for simplicity, the weights among experts can be considered not to change with performance domains); w_i, also decided by the organizers, is the weight of performance domain i; n is the number of desired outcomes in performance domain i; and l is the number of experts. When determining the expert weight w_{ik}, the questionnaire organizers need to consider the basic information, such as the experts' education background, their corresponding positions, and their working years in each position. However, in order to simplify the statistical workload, experts' scores can also be treated equally; that is, the default value 1 can be used as the experts' weights.

The BATM value of each desired outcome $(m^{ij}_{PBM},\ m^{ij}_{PMA})$ can be obtained by Formulas (1) and (2), which are equal to the weighted average of the desired outcome maturity values provided by the experts and the corresponding expert weights.

$$m^{ij}_{PBM} = \frac{\sum_{k=1}^{l}\left(m_{PBM_{ikj}}.w_{ik}\right)}{\sum_{k=1}^{l} w_{ik}} \tag{1}$$

$$m^{ij}_{PMA} = \frac{\sum_{k=1}^{l}\left(m_{PMA_{ikj}}.w_{ik}\right)}{\sum_{k=1}^{l} w_{ik}} \tag{2}$$

The BATM value of each performance domain $(m^{i}_{PBM},\ m^{i}_{PMA})$ can be obtained using Formulas (3) and (4). They are equal to the weighted average of the experts' performance domain maturity values and corresponding expert weights, in which the experts' performance domain maturity values equal the weighted average of the desired outcome maturity values and corresponding desired outcome weights provided by the experts in their scoring tables.

$$m^{i}_{PBM} = \frac{\sum_{k=1}^{l}\left(\dfrac{\sum_{j=1}^{n}\left(m_{PBM_{ikj}}.w_{ikj}\right)}{\sum_{j=1}^{n} w_{ikj}}.w_{ik}\right)}{\sum_{k=1}^{l} w_{ik}} \tag{3}$$

$$m^{i}_{PMA} = \frac{\sum_{k=1}^{l}\left(\dfrac{\sum_{j=1}^{n}\left(m_{PMA_{ikj}}.w_{ikj}\right)}{\sum_{j=1}^{n} w_{ikj}}.w_{ik}\right)}{\sum_{k=1}^{l} w_{ik}} \tag{4}$$

The BATM value of the project layer (m_{PBM}, m_{PMA}) is equal to the weighted average of all performance domain maturity values and the corresponding performance domain weight. They can be obtained by Formulas (5) and (6), in which m_{PBM} and m_{PBA} represent the project layer's PBM and PBA maturities, respectively.

$$m_{PBM} = \frac{\sum_{i=1}^{8}\left(m^{i}_{PBM}.w_{i}\right)}{\sum_{i=1}^{8} w_{i}} \tag{5}$$

$$m_{PMA} = \frac{\sum_{i=1}^{8}\left(m^{i}_{PMA}.w_{i}\right)}{\sum_{i=1}^{8} w_{i}} \tag{6}$$

4. Results

4.1. Survey Method

In 2021, The BATM assessment was applied to an engineering company engaged in general contracting. Data were acquired using the questionnaire survey method, and 50 participants were selected for the survey from project staff who had worked in the company for more than 5 years. About half of the participants had used BIM software or participated in BIM training, and the other half were project managers and other management personnel who knew something about BIM but had no experience of operating BIM software. Their information is listed in Table 4.

Table 4. Participant information table.

Gender		Educational Background			Post	Work Experience	
Male	Female	Undergraduate	Graduate	BIM	Project Management	<10	≥10
42	8	37	13	6	44	11	39

After understanding the purpose of the survey, the participants carefully determined the PBM and PBA maturities of 37 desired outcomes according to the definition of maturity levels and their understanding of the actual PBM and PBA maturities in the company. Ultimately, a total of 49 valid questionnaires were acquired (in an unqualified questionnaire, some options were not answered).

4.2. Statistical Analysis

Content validity analysis: Because the BATM questionnaire was developed based on the PMBOK 7th edition, which is an acceptable standard, the researchers organized a pilot study to evaluate the internal validity. In this pilot study, researchers conducted comprehensive interviews with seven project personnel who had participated in at least two projects adopting BIM. Based on the positive assessment of these project managers, the conclusion obtained from them was that the content of the questionnaire was closely related to the PBM and PBA maturities, its structure was simple and clear, and its operability was appropriate.

Reliability analysis: The reliability analysis was conducted using Cronbach's alpha coefficient. According to Kim and Feldt [56], when the internal consistency coefficient of the data reaches 0.70 or higher, the data can be considered to have sufficient reliability. In the study, the item scale was internally consistent because all of Cronbach's alpha coefficients exceeded the threshold value (0.70) (Table 5).

Table 5. Descriptive statistics, Cronbach's alpha, and correlation coefficients.

		Minimum	Maximum	Mean	Std. Deviation	Cronbach Alpha	Correlation Coefficient	t
Stakeholder	PBM	1.00	3.00	2.15	0.72	0.75	0.67	6.27
	PBA	0.00	2.00	1.29	0.70	0.78		
Team	PBM	1.33	3.33	2.71	0.51	0.93	0.53	4.40
	PBA	0.00	2.00	1.47	0.66	0.95		
Development	PBM	2.00	3.33	2.73	0.44	0.91	0.77	8.44
	PBA	0.33	2.00	1.63	0.48	0.76		
Planning	PBM	1.50	3.00	2.43	0.41	0.76	0.79	9.13
	PBA	0.50	2.00	1.50	0.46	0.73		
Project work	PBM	2.00	3.00	2.69	0.33	0.81	0.62	5.55
	PBA	0.00	2.17	1.73	0.47	0.90		
Delivery	PBM	1.00	3.00	2.47	0.43	0.86	0.51	4.14
	PBA	0.00	2.00	1.79	0.45	0.95		
Measurement	PBM	1.00	3.50	2.30	0.56	0.83	0.61	5.34
	PBA	0.00	2.25	1.52	0.52	0.84		
Uncertainty	PBM	1.00	3.43	2.45	0.65	0.94	0.84	10.9
	PBA	0.00	2.00	1.35	0.53	0.87		

Data analysis: To understand the data more effectively, we conducted a descriptive data analysis and a correlation analysis, and the results are listed in Table 5. The mean values reflect the maturity of each performance domain, in which the PBM maturities change from 2.15 to 2.71 and PBA maturities change from 1.29 to 1.79. Correlation analyses were conducted to verify the correlations between the PBM and PBA maturities. As can be seen from Table 5, the correlation coefficients in all performance domains are between 0.51 and 0.84; the *t*-test statistics are between 4.14 and 10.9, which are greater than $t_{a/2}$ ($\alpha = 0.05$; $t_{a/2} = 2.008$) and indicate clear correlations.

Table 6 shows the significant difference among different performance domain data and among different participant data, in which the results were obtained by the analysis of variance with two factors. The fact that all of the F-values were greater than the values of "F-crit," and all of the *p*-values were less than the significance level of 0.05, implies that statistically significant differences existed among the maturities of the eight performance domains and among the feedback of the 49 participants.

Table 6. Analysis of variance with two factors.

Dimension	Source of Difference	Sum of Squares	Degree of Freedom	Mean Square	F	*p*-Value	F-Crit
PBM	Performance domains	15.28	7	2.18	9.44	1.073×10^{9}	2.037
	Participants	38.53	48	0.80	3.47	1.513×10^{11}	1.396
PBA	Performance domains	6.50	7	0.93	4.11	2.409×10^{4}	2.037
	Participants	52.60	48	1.10	4.84	1.735×10^{18}	1.396

4.3. Problems and Measures

The BATMs of the eight performance domains are shown in Figure 1. The PBM and PBA lines represent the PBM and PBA maturities achieved in 2021, respectively. Based on the figure, we discovered the following problems: (1) the PBM maturities of two performance domains (i.e. stakeholder and measurement) are obviously lower than 2.5, and the rest are close to or more than 2.5; (2) the PBA maturities of the stakeholder and uncertainty domains are obviously lower than 1.5, and the rest are close to or more than 1.5; (3) the maturity gaps between PBM and PBA are large, in which the uncertainty and team domains are particularly prominent (their BATM values are (2.45, 1.35) and (2.71, 1.47), and the ratios of their gaps to the PBA maturities are 81.4% and 84.6%, respectively).

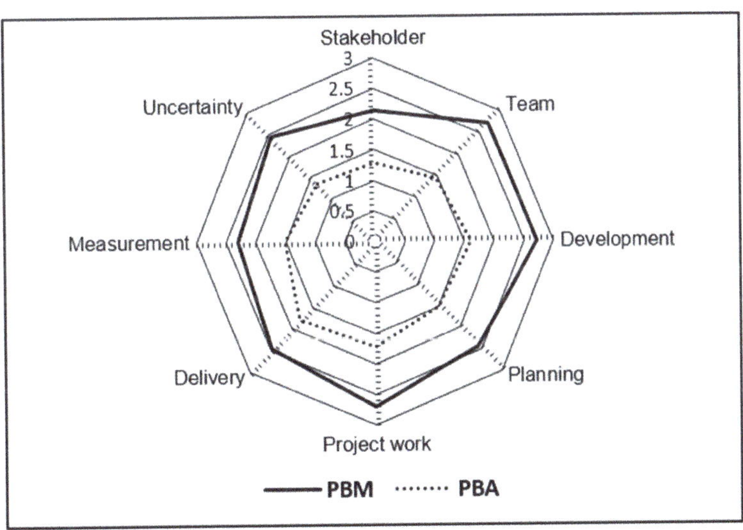

Figure 1. Assessment result of BATM model.

The obvious gaps between the PBM and PBA maturities shows that PBA not only does not guide and support PBM, but also lags far behind BPM, and there is a significant room for PBA to develop in the project, performance domain, and desired outcome layers. The organization should strengthen its support for PBA, ensure the PBA maturity catches up with the PBM maturity, and form a good interaction situation.

Through this assessment, the problems faced by the company were determined and the improvement directions were identified. In terms of the PBA maturity, the improvement priorities were the stakeholder and uncertainty domains. Regarding the PBM maturity, the improvement priorities were the stakeholder and measurement domains. In order to narrow the gap between PBA and PBM, the improvement priorities were the uncertainty and team domains.

The results of the BATM assessment aroused significant attention of the company leaders, and the company took a series of measures to improve the PBM and PBA maturities, referring to the improvement priorities provided by the BATM assessment.

5. Discussion

BIM technology has been developed globally for a period of more than 20 years. Because BIM applications exist in different phases of the project life cycle, including design, construction, operation, and maintenance, and BIM applications involve many stakeholders, such as the designers, equipment supplies, construction companies, consulting-related enterprises, project owners, and even relevant government departments, the whole process of BIM applications is complex. As a result of the complexity of BIM applications, they are not as good as expected under many circumstances [57], although BIM has been well applied in some countries and some projects. To promote BIM applications, various BIM application maturity assessment models based on different enterprise perspectives and different project phases have been developed [17,22,42–52].

However, all of the BIM application maturity models have several obvious deficiencies. First, the assessment indexes of all models are different, and a unified assessment index system has not been derived. Second, these models also have huge discrepancies in the definition of maturity levels, with the number of maturity levels ranging from three to ten. Third, these models are built under specific business perspectives and specified project conditions, and have application limitations under other project situations. These deficiencies make it difficult for various BIM application maturity models to be popularized and applied. BIM application maturity models do not have the good effect expected by the public, and create difficulties for users in choosing these models [24]. Hence, a holistic model enabling BIM maturity assessments is necessary [17].

Considering the digital twin relationship between BIM applications and engineering projects, and that the purpose of BIM applications is to achieve the project objectives and BIM application processes are deeply integrated with project implementation processes, in this study, the eight performance domains and 37 desired outcomes of PMBOK 7th edition were selected as the assessment indexes of BATM. The selection of such indexes not only ensures that the BATM indexes are consistent with the processes and objectives of BIM applications and project management, but also avoids the dilemma of designing different assessment indexes for different purposes; thus, projects and organizations can then promote the improvement in BIM application maturity under the unified standard. At the same time, the assessment indexes based on PMBOK also indicate that BATM is applicable to all types of projects, whether they are building projects or highway projects. The selection of BATM assessment indexes is an innovation of this study. In addition, it should be noted that the BIM-related software, hardware, personnel, standards, and other indexes are not the assessment indexes of the project BIM application maturity, but are those of the organizational BIM capability maturity.

In the engineering industry, the main purpose of BIM applications is to serve the engineering projects. It is clear that the high level of BIM applications plays an assisting and supporting role in the project implementation, and the high level of project management creates higher requirements for BIM applications. In the context of the continuous development of digital technology, BIM applications and project implementations are increasingly embodied in a digital twin relationship. Through the comparison of the project management maturity in the real world and the BIM application maturity in the virtual world, it is easier to identify the inadequacies and problems in project management and BIM applications. Under the guidance of this idea, this study proposes the BATM model for BIM maturity assessment from the two dimensions of the virtual and real worlds. The two-dimensional maturity ideology of the virtual and real worlds is another innovation of this study.

With regard to the maturity level, this paper refers to a large number of documents and selects the five-level scheme, which has a more intuitive definition of the levels and is more commonly used [6,27,34–36,50]. The five levels of PBA maturity are the initial, repeatable but intuitive, defined process, managed and measurable, and optimized levels (the non-existent level occurs without PBA), and the five levels of PBM maturity are the initial, managed, defined, quantitatively managed, and optimizing levels. The five level schemes of the two dimensions are basically the same.

The application case of the BATM model shows that the model is simple and its effect is obvious. Based on the statistical analysis of expert scores and the radar graph of the two-dimensional maturity values, the deficiencies existing in the PBM and PBA can be visually and accurately identified. Furthermore, the priority improvement direction of BIM application can be determined immediately, and organizations and projects can advance along the BATM ladder.

In the case of multiple BATM assessments, the dynamic changes in two-dimensional maturity can be clearly observed, which will be more effective for the improvement in maturity.

6. Conclusions

6.1. Theoretical Implications

In terms of theoretical contribution, most importantly, this study provides a scheme using generally accepted indexes to assess the BIM maturity for projects with BIM applications. This is the first study to present the BATM concept with the PBM and PBA maturities from the two dimensions of the virtual and real worlds, and the first to assess the PBM and PBA maturities based on the eight performance domains and 37 desired outcomes of PMBOK 7th edition. Furthermore, this study establishes a complete BATM assessment system by combining with an application case. Finally, the study further develops the maturity theory of project management and BIM applications.

6.2. Practical Implications

The practical implications are as follows. Firstly, because the assessment indexes of the BATM model are based on the PMBOK 7th edition, the BATM model can be applied to different project types and different project phases, which eliminates the user's difficulty in choosing a model from various PBA models with different assessment indexes, and increases the practical value of the BATM model. Secondly, the eight performance domains and 37 desired outcomes of PMBOK are generally accepted, so they are more convenient and easier to use than the indexes of other models. Thirdly, the comparison between the PBM and PBA maturities makes it easier to identify the deficiencies and problems in project management and BIM applications, and then to take effective measures to improve their maturities. Fourthly, the actual application case showed that the questionnaire can be used easily by the participants, and set an example for other companies to improve their PBM and PBA capability.

6.3. Limitations and Future Studies

This paper focuses on the project BIM application maturity. The limitation of the paper is that it does not research the organizational BIM capability maturity. Similar research of the organizational BIM capability maturity can be carried out in the future. In addition, this paper takes an engineering company engaged in general contracting business as an example; however, the effect of BATM application to design or construction enterprises needs to be further validated.

Author Contributions: Conceptualization, H.C.; Methodology, R.L. (Ruyin Long); Writing—original draft, C.S.; Writing—review and editing, R.L. (Ruihua Liao). All authors have read and agreed to the published version of the manuscript.

Funding: This research received no external funding.

Data Availability Statement: Some or all data, models, or code that support the findings of this study are available from the corresponding author upon reasonable request.

Conflicts of Interest: The authors declare no conflict of interest.

Abbreviations

Comparison Table of Important Abbreviations.

No.	Abbreviation	Meaning
1	BIM	Building information modeling
2	BATM	BIM application two-dimensional maturity
3	PBM	Project business management
4	PBA	Project BIM application
5	PMBOK	Project management body of knowledge
6	CMM	Capability maturity model
7	PMMM	Project management maturity model
8	PMS-PMMM	The PMMM leased by Project Management Solutions
9	PMI	Project Management Institute
10	IPMA	The International Project Management Association
11	NBIMS	U.S. National BIM Standard
12	OPM3	Organizational project management maturity model
13	P3M3	The portfolio, programme, and project management maturity model
14	CMMI	Capability maturity model integration
15	PM2	The Berkeley project management process maturity model
16	MMM	The management maturity model developed by Langston and Ghanbaripour
17	COBIT	Control objectives for information and related technology

References

1. Wang, T.; Feng, J. Linking BIM Definition, BIM Capability Maturity, and Integrated Project Delivery in the AECO Industry: The Influences of BIM Diffusion and Moral Hazard. *J. Urban Plan. Dev.* **2020**, *148*, 04022025. [CrossRef]
2. Alirezaei, S.; Taghaddos, H.; Ghorab, K.; Tak, A.N.; Alirezaei, S. BIM-augmented reality integrated approach to risk management. *Autom. Constr.* **2022**, *141*, 104458. [CrossRef]
3. Vitiello, U.; Ciotta, V.; Salzano, A.; Asprone, D.; Manfredi, G.; Cosenza, E. BIM-based approach for the cost-optimization of seismic retrofit strategies on existing buildings. *Autom. Constr.* **2019**, *98*, 90–101. [CrossRef]
4. Alcínia Zita Sampaio, Maturity of BIM Implementation in the Construction Industry: Governmental Policies. *Int. J. Eng. Trends Technol.* **2021**, *69*, 92–100. [CrossRef]
5. Cooke-Davies, T. The 'real' success factors on projects. *Int. J. Proj. Manag.* **2002**, *20*, 185–190. [CrossRef]
6. Young, M.; Young, R.; Romero Zapata, J. Project, programme and portfolio maturity: A case study of Australian Federal Government. *Int. J. Manag. Proj. Bus.* **2014**, *7*, 215–230. [CrossRef]
7. Demir, C.; Kocabaş, I. Project Management Maturity Model (PMMM) in educational organizations. *Procedia Soc. Behav. Sci.* **2010**, *9*, 1641–1645. [CrossRef]
8. Szelagowski, M.; Berniak-Woźny, J. The adaptation of business process management maturity models to the context of the knowledge economy. *Bus. Process Manag. J.* **2020**, *26*, 212–238. [CrossRef]
9. Rashid, N.; Khan, S.U.; Khan, H.U.; Ilyas, M. Green-Agile Maturity Model: An Evaluation Framework for Global Software Development Vendors. *IEEE Access* **2021**, *9*, 71868–71886. [CrossRef]

10. Rytova, E.; Verevka, T.; Gutman, S.; Kuznetsov, S. Assessing the Maturity Level of the Digital Government of Saint Petersburg. *Int. J. Technol.* **2020**, *11*, 1081–1090. [CrossRef]
11. Alghail, A.; Yao, L.; Abbas, M.; Baashar, Y. Assessment of knowledge process capabilities toward project management maturity: An empirical study. *J. Knowl. Manag.* **2022**, *26*, 1207–1234. [CrossRef]
12. Jia, G.; Ni, X.; Chen, Z.; Hong, B.; Chen, Y.; Yang, F.; Lin, C. Measuring the maturity of risk management in large-scale construction projects. *Autom. Constr.* **2013**, *34*, 56–66. [CrossRef]
13. Kazanjian, R.K.; Drazin, R. An empirical test of a stage of growth progression model. *Manag. Sci.* **1989**, *35*, 1489–1503. [CrossRef]
14. Hidayati, P.B.; Budiardjo, E.K.; Solichah, I. Global Software Development and Capability Maturity Model Integration: A Systematic Literature Review. In Proceedings of the 3rd International Conference on Informatics and Computing; ICIC: Roxbury, MA, USA, 2018. [CrossRef]
15. Succar, B.; Sher, W.; Williams, A. Measuring BIM performance: Five metrics. *Archit. Eng. Des. Manag.* **2012**, *8*, 120–142. [CrossRef]
16. Smits, W.; van Buiten, M.; Hartmann, T. Yield-to-BIM: Impacts of BIM maturity on project performance. *Build. Res. Inf.* **2017**, *45*, 336–346. [CrossRef]
17. Yilmaz, G.; Akcamete, A.; Demirors, O. A reference model for BIM capability assessments. *Autom. Constr.* **2019**, *101*, 245–263. [CrossRef]
18. Troiani, E.; Mahamadu, A.-M.; Manu, P.; Kissi, E.; Aigbavboa, C.; Oti, A. Macro-maturity factors and their influence on micro-level BIM implementation within design firms in Italy. *Archit. Eng. Des. Manag.* **2020**, *16*, 209–226. [CrossRef]
19. Ferraz, C.; Loures, E.R.; Deschamps, F. BIM maturity models evaluated by design principles. *Adv. Transdiscipl. Eng.* **2020**, *12*, 504–513. [CrossRef]
20. Olawumi, T.O.; Chan, D.W.M. Building information modelling and project information management framework for construction projects. *J. Civ. Eng. Manag.* **2019**, *25*, 53–75. [CrossRef]
21. Mahamadu, A.-M.; Manu, P.; Mahdjoubi, L.; Booth, C.; Aigbavboa, C.; Abanda, F.H. The importance of BIM capability assessment: An evaluation of post-selection performance of organisations on construction projects. *Eng. Constr. Archit. Manag.* **2020**, *27*, 24–48. [CrossRef]
22. CIC (Computer Integrated Construction Research Program). BIM Planning Guide for Facility Owners. 2013. Available online: https://www.bim.psu.edu/owners_guide/ (accessed on 12 July 2022).
23. Giel, B.; Issa, R.R.A. Framework for Evaluating the BIM Competencies of Facility Owners. *J. Manag. Eng.* **2016**, *32*, 4015024. [CrossRef]
24. Alankarage, S.; Chileshe, N.; Samaraweera, A.; Rameezdeen, R.; Edwards, D.J. Organisational BIM maturity models and their applications: A systematic literature review. *Archit. Eng. Des. Manag.* **2022**, 1–19. [CrossRef]
25. Vittal, S.A.; Parviz, F.R. Role of organizational project management maturity factors on project success. *Eng. Manag. J.* **2018**, *30*, 165–178. [CrossRef]
26. Viana, J.C.; Mota, C.M.M. Enhancing organizational project management maturity: A framework based on the value focused thinking model. *Producao* **2016**, *26*, 313–329. [CrossRef]
27. Zhang, L.; He, J.; Zhang, X. The project management maturity model and application based on PRINCE2. *Procedia Eng.* **2012**, *29*, 3691–3697. [CrossRef]
28. Zhang, Q.; Yang, S.; Liao, P.-C.; Chen, W. Influence mechanisms of factors on project management capability. *J. Manag. Eng.* **2020**, *36*, 0000812. [CrossRef]
29. Kerpedzhiev, G.D.; König, U.M.; Röglinger, M.; Rosemann, M. An Exploration into Future Business Process Management Capabilities in View of Digitalization. *Bus. Inf. Syst. Eng.* **2021**, *63*, 83–96. [CrossRef]
30. Wendler, R. The maturity of maturity model research: A systematic mapping study. *Inf. Softw. Technol.* **2012**, *54*, 1317–1339. [CrossRef]
31. Khoshgoftar, M.; Osman, O. Comparison of maturity models. In Proceedings of the 2009 2nd IEEE International Conference on Computer Science and Information Technology, Beijing, China, 8–11 August 2009; pp. 297–301. [CrossRef]
32. Kwak, Y.H.; Ibbs, C.W. Project management process maturity (PM)2 model. *J. Manag. Eng.* **2002**, *18*, 150–155. [CrossRef]
33. Permana, V.; Sucahyo, Y.G.; Gandhi, A. Measuring information technology project management maturity level: A case study from a project based organization in Indonesia. In Proceedings of the 2017 International Conference on Information Technology Systems and Innovation (ICITSI), Bandung, Indonesia, 23–24 October 2017; pp. 342–347. [CrossRef]
34. Crawford, J.K. The project management maturity model. *Inf. Syst. Manag.* **2006**, *23*, 50–58. [CrossRef]
35. Langston, C.; Ghanbaripour, A.N. A Management Maturity Model (MMM) for project-based organisational performance assessment. *Constr. Econ. Build.* **2016**, *16*, 68–85. [CrossRef]
36. Fabbro, E.; Tonchia, S. Project management maturity models: Literature review and new developments. *J. Mod. Proj. Manag.* **2021**, *8*, 31–45. [CrossRef]
37. Yeo, K.T.; Ren, Y.; Ren, Y. Risk management maturity in large complex rail projects: A case study. *Int. J. Proj. Organ. Manag.* **2016**, *8*, 301–323. [CrossRef]
38. Heravi, G.; Gholami, A. The Influence of Project Risk Management Maturity and Organizational Learning on the Success of Power Plant Construction Projects. *Proj. Manag. J.* **2018**, *49*, 22–37. [CrossRef]
39. Ma, L.; Fu, H. A Governance Framework for the Sustainable Delivery of Megaprojects: The Interaction of Megaproject Citizenship Behavior and Contracts. *J. Constr. Eng. Manag.* **2022**, *148*, 04022004. [CrossRef]

40. Seelhofer, D.; Graf, C.O. National project management maturity: A conceptual framework. *Cent. Eur. Bus. Rev.* **2018**, *7*, 1–20. [CrossRef]
41. Jugdev, K.; Thomas, J. Project management maturity models: The silver bullets of competitive advantage? *Proj. Manag. J.* **2002**, *33*, 4–14. [CrossRef]
42. Wu, C.; Xu, B.; Mao, C.; Li, X. Overview of BIM maturity measurement tools. *J. Inf. Technol. Constr.* **2017**, *22*, 34–62. Available online: https://www.scopus.com/inward/record.uri?eid=2-s2.0-85011655077&partnerID=40&md5=4c64b9ba28831a5c519542f9a5b147dc (accessed on 19 October 2022).
43. Sebastian, R.; Van Berlo, L. Tool for benchmarking BIM performance of design, engineering and construction firms in the Netherlands. *Archit. Eng. Des. Manag.* **2010**, *6*, 254–263. [CrossRef]
44. McCuen, T.L.; Suermann, P.C.; Krogulecki, M.J. Evaluating award-winning BIM projects using the National Building Information Model Standard Capability Maturity Model. *J. Manag. Eng.* **2012**, *28*, 224–230. [CrossRef]
45. Morlhon, R.; Pellerin, R.; Bourgault, M. Building information modeling implementation through maturity evaluation and critical success factors management. *Procedia Technol.* **2014**, *16*, 1126–1134. [CrossRef]
46. Sinoh, S.S.; Ibrahim, Z.; Othman, F.; Muhammad, N.L.N. Review of BIM literature and government initiatives to promote BIM in Malaysia. *IOP Conf. Ser. Mater. Sci. Eng.* **2020**, *943*, 012057. [CrossRef]
47. Kam, C.; Song, M.H.; Senaratna, D. VDC scorecard: Formulation, application, and validation. *J. Constr. Eng. Manag.* **2017**, *143*, 04016100. [CrossRef]
48. Indiana University. BIM Guidelines and Standards for Architects, Engineers, and Contractors. 2015. Available online: https://knowledge.autodesk.com/akn-aknsite-article-attachments/8a7f652b-edb7-4fb4-87a2-2eaec27ec6cc.pdf (accessed on 12 July 2022).
49. Succar, B. Building information modelling framework: A research and delivery foundation for industry stakeholders. *Autom. Constr.* **2009**, *18*, 357–375. [CrossRef]
50. Succar, B. Building information modelling maturity matrix. In *Handbook of Research on Building Information Modeling and Construction Informatics: Concepts and Technologies*; IGI Global: Hershey, PA, USA, 2010; pp. 65–103. [CrossRef]
51. Joblot, L.; Paviot, T.; Deneux, D.; Lamouri, S. Building Information Maturity Model specific to the renovation sector. *Autom. Constr.* **2019**, *101*, 140–159. [CrossRef]
52. Lu, W.; Chen, K.; Zetkulic, A.; Liang, C. Measuring building information modeling maturity: A Hong Kong case study. *Int. J. Constr. Manag.* **2021**, *21*, 299–311. [CrossRef]
53. Hu, W.; Li, D.; Hu, R. Three-dimensional complex construction project management maturity model: Case study of 2010 shanghai expo. *Appl. Mech. Mater.* **2012**, *209–211*, 1363–1369. [CrossRef]
54. ISACA 2012. COBIT 5: A Business Framework for the Governance and Management of Enterprise IT. ISACA, Rolling Meadows. Available online: https://www.isaca.org/resources/cobit/cobit-5#sort=relevancy (accessed on 12 July 2022).
55. PMI (Project Management Institute). *A Guide to the Project Management Body of Knowledge (PMBOK®Guide)*, 7th ed.; PMI: Newtown Square, PA, USA, 2001; Available online: https://www.pmi.org/pmbok-guide-standards/foundational/pmbok (accessed on 12 July 2022).
56. Kim, S.; Feldt, L.S. A Comparison of Tests for Equality of Two or More Independent Alpha Coefficients. *J. Educ. Meas.* **2008**, *45*, 179–193. [CrossRef]
57. Volk, R.; Stengel, J.; Schultmann, F. Building Information Modeling (BIM) for existing buildings -Literature review and future needs. *Autom. Constr.* **2014**, *38*, 109–127. [CrossRef]

 *sustainability*

Article

Workers' Unsafe Actions When Working at Heights: Detecting from Images

Qijun Hu [1,*], Yu Bai [1], Leping He [2], Jie Huang [3], Haoyu Wang [4] and Guangran Cheng [3]

1 School of Mechatronic Engineering, Southwest Petroleum University, Chengdu 610500, China; by_duo@163.com
2 School of Civil Engineering and Geomatics, Southwest Petroleum University, Chengdu 610500, China; 201231010028@swpu.edu.cn
3 CECEP Construction Engineering Design Institute Limited, Chengdu 610052, China; hi_huangjie@163.com (J.H.); gr-cheng@live.cn (G.C.)
4 College of Physics, Chongqing University, Chongqing 400044, China; wanghaoyu200007@163.com
* Correspondence: huqijunswpu@163.com; Tel.: +86-130-7283-2168

Abstract: Working at heights causes heavy casualties among workers during construction activities. Workers' unsafe action detection could play a vital role in strengthening the supervision of workers to avoid them falling from heights. Existing methods for managing workers' unsafe actions commonly rely on managers' observation, which consumes a lot of human resources and impossibly covers a whole construction site. In this research, we propose an automatic identification method for detecting workers' unsafe actions, considering a heights working environment, based on an improved Faster Regions with CNN features (Faster R-CNN) algorithm. We designed and carried out a series of experiments involving five types of unsafe actions to examine their efficiency and accuracy. The results illustrate and verify the method's feasibility for improving safety inspection and supervision, as well as its limitations.

Keywords: unsafe actions; working at heights; intelligent recognition; deep learning; construction site

Citation: Hu, Q.; Bai, Y.; He, L.; Huang, J.; Wang, H.; Cheng, G. Workers' Unsafe Actions When Working at Heights: Detecting from Images. *Sustainability* **2022**, *14*, 6126. https://doi.org/10.3390/su14106126

Academic Editors: Srinath Perera, Albert P. C. Chan, Xiaohua Jin, Dilanthi Amaratunga, Makarand Hastak, Patrizia Lombardi, Sepani Senaratne and Anil Sawhney

Received: 25 March 2022
Accepted: 11 May 2022
Published: 18 May 2022

Publisher's Note: MDPI stays neutral with regard to jurisdictional claims in published maps and institutional affiliations.

1. Introduction

Working at heights is associated with frequent injuries and the deaths of workers during construction activities [1]. Fatal injuries due to construction accidents exceed 60,000 injuries every year all over the world [2]. According to the Occupational Safety and Health Administration, a similar trend occurs in developed countries, even though infrastructure construction has almost been completed in these countries [3]. Statistics from the Ministry of Housing and Urban-Rural Development of China show that, from 2010 to 2019, there were an average of 603 production safety accidents per year, resulting in approximately 730 worker deaths per year [4]. Among these accidents, fall-from-height accidents accounted for at least 52.10%, followed by struck-by-object accidents (13.90%, Figure 1). Many researchers have noted that the root causes of safety accidents are workers' unsafe behaviors [5–7]. Heinrich's accident causation theory states that more than 80% of safety accidents are caused by workers' unsafe behaviors [8]. Therefore, management to minimize workers' unsafe behaviors is important to construction safety.

Behavior-based safety (BBS) plays an influential role in the supervision and management of workers' activities [5,9,10], in which workers' activities are recorded and their behaviors are analyzed through observation, interview, and survey. Most BBS studies have involved four necessary steps [11]: (1) create a list of workers' unsafe behaviors; (2) observe and record the frequency of unsafe behaviors; (3) educate and intervene in workers' behavior; (4) provide feedback and perform follow-up observations. BBS has gained its status in construction management because it has been more successful than other methods for solving problems caused by unsafe behaviors. Furthermore, researchers have recognized unsafe behaviors as the most important problem. The purpose of normal science is neither to discover new types of phenomena nor to invent new theories [12]. On the contrary, normal science research is to continuously improve phenomena and theories provided by existing paradigms, which are eternal challenges to researchers' skills and imagination [13]. BBS observation is a traditional form of worker behavior measurement, which has certain limitations in practical applications [14].

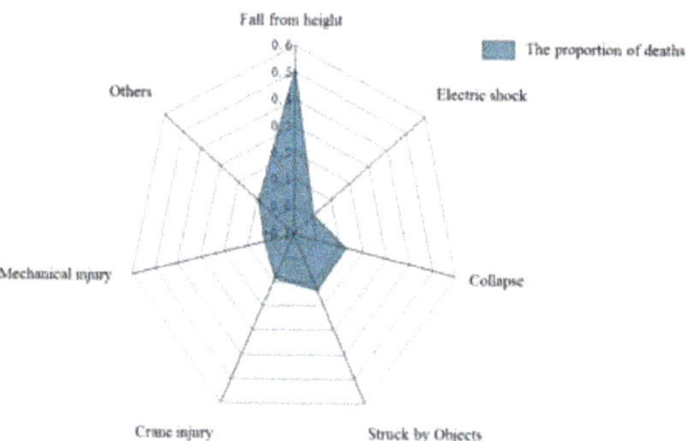

Figure 1. Distribution map of the causes of death during construction from 2010 to 2019 [4].

Observation—the second stage—is important because it can provide more data for the analysis of patterns. Ref. [15] puts forward a safety assessment method of leading indicators based on jobsite safety inspection (JSI) through a lot of accident data analyzing. Traditional unsafe behavior observation mainly relies on safety managers' manual observation and recording, which not only consumes a lot of time and cost, but it is also difficult to cover the whole construction site, or all workers. On the one hand, many human resources are needed for data acquisition due to large sample data requirements [16]. On the other hand, excessive reliance on workers' observations can easily cause personal impact since different people have different feelings about the same thing [17]. Therefore, an automated and reliable method that could efficiently measure unsafe behavior is needed to support BBS observation. Automation technology is already making its mark in the observations of workers' behaviors [18]. Proposed real-time positioning systems based on different types of sensors and the Internet of Things (IoT) have played considerable roles in workers' safety observations [19,20]. However, sensors can sometimes affect workers' normal work [21]. Computer vision technology can also be used for collecting and processing workers' safety information [22]. Its ability to provide a wide range of visual information at a low cost has attracted a lot of attention [23–26].

Construction workers' unsafe actions are a type of unsafe behavior that could be the main reason leading to construction accidents, primarily occurring when working at heights. Most unsafe actions are instantaneous, and therefore, it is difficult for safety supervisors to observe them in real time. Furthermore, detecting workers' unsafe actions is critical to the observation process of BBS. Computer vision technology for the automatic recognition and detection of workers' unsafe actions could tentatively replace manual observations of BBS. The gap in research could be significant for specific groups of construction workers. To date, there is no automatic method of detecting the unsafe behavior of workers in a high working environment. Therefore, the present study is to help improve the observation method of workers' unsafe behavior considering five unsafe actions that mostly appeared in the high working environment on construction sites. A series of experiments involving over 30 testees were implemented to verify the proposed method. This paper is structured as follows: First of all, the research method is presented, involving unsafe actions lists, dataset construction, and the Convolutional Neural Networks (CNN) model built. Following this, the results are presented and discussed. Finally, a conclusion is drawn.

2. Related Works

2.1. Safety Management in High Places

Falls from height in construction sites have earned science mapping research to reveal the existing research gaps [27]. The common causes that lead to falling accidents are defects in protective devices, poor work organization [28], and workers' unsafe actions, such as sleeping on the baseboard. Ref. [29] built a database that dissected the mechanics of workers falling off the baseboard. An IoT infrastructure, combined with the fuzzy markup language for falling objects on the construction site, could greatly help safety managers [30].

Deep learning enhances the automation capabilities of computer vision in safety monitoring [14,23,31]. CNN has shown exceptionally superior performance in high-dimensional data with intricate structures processing. Related research on using computer vision for worker safety management under working-at-height conditions can be described from the following three aspects:

Aspect 1: To automatically check a worker's safety equipment, such as their helmet and seat belt [31–33]. The detection of safety equipment originated from early feature engineering research, such as the histogram of oriented gradients (HOG) [34]. It has been proposed that increasing the setting of the color threshold of the helmet and the upper body detection could improve detection accuracy [35]. Deep learning has been used to develop multiple processing layers to extract unknown information, without the need to set the image features artificially [36]. In addition, a regional convolutional neural network (R-CNN) has been used to identify a helmet and has achieved good results [37].

Aspect 2: To automatically identify hazardous areas. This research is generally based on object recognition, including openings, rims, and groove edges in high places [38–40]. Computer vision has been used to detect whether workers pass through a support structure [39], since workers walking on support structures have a risk of falling.

Aspect 3: To monitor non-compliance of safety regulations [41], in particular, climbing scaffolding and carrying workers with a tower crane [42]. There has been a study on intelligent assessment for interactive work of workers and machines [24]. Research in this area needs to combine computer vision and safety assessment methods so that safety status can be explained using computer semantics [38,41].

2.2. Computer Vision in Construction

Scientists designed the CNN to describe the primary visual cortex (V1) of the brain by studying how neurons in a cat's brain responded to images projected precisely in front of the cat [43,44]. There are three fundamental properties to a CNN imitating the V1 [45]:

Property 1: The V1 can perform spatial mapping, and CNN describes this property through two-dimensional mapping.

Property 2: The V1 includes many simple cells, and the convolution kernel is used to simulate these simple cells' activities; that is, the linear function of the image in a particular receptive field.

Property 3: The V1 includes many complex cells, which inspire the pooling unit of CNN.

On all accounts, there are still two factors that determine the deep learning target detection model quality:

Factor 1: The dataset that is used to train the workers' unsafe actions recognition model should have unique unsafe action features that can be identified by a computer. There are many open-source datasets available for deep learning research, such as ImageNet [46] and COCO [47]; however, not all of them are suitable for object detection on construction sites. Many researchers have also established datasets related to construction engineering. For object recognition and detection on construction sites, there are datasets about workers [31]; construction machinery [31]; and on-site structures, such as railings [48]. For workers' activities on the construction sites, a dataset dedicated to steelworkers engaged in steel processing activities has been established [49]. Scholars have even enhanced datasets by preprocessing Red-Green-Blue (RGB) images to optical and gray images 49], which has provided novel ideas for dataset acquisition.

Factor 2: Deep learning algorithms and models as mathematical methods to find optimal solutions are important, as they affect the detection results. Le-Net is the foundation of deep learning models [50], which contains the basic modules of deep learning, convolutional layer, pooling layer, and fully connected layer. AlexNet came first in the 2012 ImageNet Large Scale Visual Recognition Challenge (ILSVRC2012) [36]. Since then, various neural network models have shown their accuracy and efficiency in feature extraction, including ZF-net [51], VGG-net [52], Res-net [53], Inception-net [54], etc. After obtaining the feature map, additional algorithms are needed to classify and locate the object. Faster R-CNN, YOLO, and SSD are deep learning algorithms that are widely used in many applications.

In this research, we propose an improved faster R-CNN utilizing a special dataset composed of unsafe actions when working at heights, which is based on ZF-net. This work's application is that it could provide construction managers with information on workers' unsafe actions, and therefore, assist with interventions. In addition, it could become a new procedure for BBS observation.

3. Methods

3.1. Research Framework

Feature engineering needs to set the transcendental extracted features artificially. Deep learning training is a process of finding the optimal solution based on a dataset using representation. Unlike feature engineering, representation learning is a group of machine learning methods that can automatically find useful input data features through a general-purpose learning procedure [55–57]. The representation learning is similar to a "black box", and therefore, it is difficult to understand how the internal nonlinear function works. As the dataset and the algorithm are the critical factors in applied research related to deep learning, the research process is designed by improving design science research [58]. In this study, this is more suitable for research on the automatic detection of workers' unsafe actions, as shown in Figure 2.

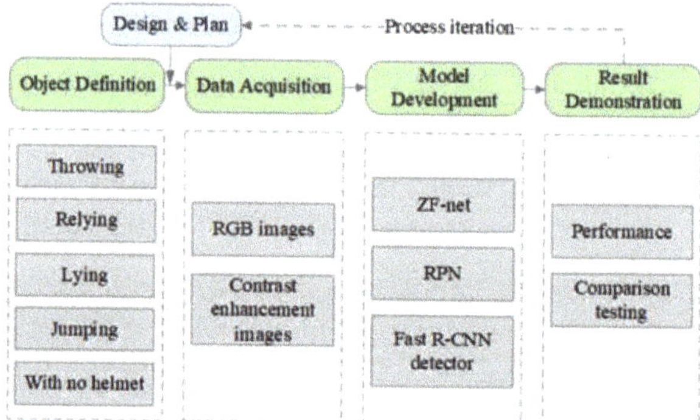

Figure 2. The research process of intelligent recognition for workers' unsafe action.

The method is structured as follows:

(1) Before building the automatic recognition model, the workers' unsafe actions that are to be detected should be defined. This research lists five kinds of worker action types that were likely to cause safety accidents when working at heights.
(2) The second stage is data acquisition, mainly to acquire images with features of workers' unsafe actions that could be used for deep learning training, validation, and testing. In this step, Red-Green-Blue (RGB) and contrast enhancement images are integrated to reinforce the dataset performance.
(3) Then, the model development stage involves deep learning training and testing.

Therefore, the test results demonstrate the model's performance and are the basis of the design and plan modifications.

3.2. Definition of Workers' Unsafe Actions

Fall-from-height and struck-by-object accidents are prone to happen when workers are working at heights; therefore, in this research, we analyzed workers' unsafe actions through investigating the regulations and reports on these two accident types [1,59,60]. Simultaneously, it should be considered whether the unsafe actions have features that could be identified through computer vision. The five main unsafe actions are summarized in Table 1, namely, throwing objects downwards, relying on railings, lying on scaffold boards and operating platforms, jumping up and down levels, and not wearing a helmet.

Table 1. Unsafe actions categories and descriptions.

No.	Categories	Unsafe Actions Descriptions
1	Throwing	1.1 Throw waste and leftover materials at will. 1.2 Throw fragments down when working on building exterior walls. 1.3 Throw tools and materials up and down. 1.4 Throw rubbish from windows. 1.5 Throw dismantled objects and remaining materials arbitrarily. 1.6 Throw broken glass downwards when installing skylights.
2	Relying	2.1 Relying on the protective railing. 2.2 Rely or ride on the window rails when painting windows.
3	Lying	3.1 Lie on scaffold boards and operating platforms.
4	Jumping	4.1 Jump up and down shelves.
5	With no helmet	5.1 Workers fail to use safety protection equipment correctly when entering a dangerous site of falling objects.

All of the unsafe actions are momentary actions, except for not wearing a helmet, and therefore, it is difficult for security managers to observe these actions instantly. Whether or not to wear a helmet is a relatively stable state of a person's action, which security managers can clearly observe and prevent in time. However, wearing a helmet is very important to the safety of workers, especially for working at heights; therefore, it is worthwhile including it in the research content.

3.3. Data Acquisition

The quality and quantity of the dataset is the decisive factor affecting detection accuracy. Many researchers have established datasets for construction safety management, such as datasets used to identify construction machines and workers and to identify the activities of steel processing workers. To date, there is no dataset for the characteristics of workers' unsafe actions when working at heights on construction sites; therefore, even if multiple publicly annotated open-source datasets could be used for deep learning detection model training and testing, none of them could be used as experimental data for this study directly. The dataset used in this research needs to include the five types of workers' unsafe actions that occur when working at heights, including throwing, relying, lying, jumping, and with no helmet. A total of 2200 original sample images were collected. The five unsafe actions sample distribution is shown in Table 2.

Table 2. Sample distribution.

Unsafe Actions	Distribution
Jumping	11.12%
Throwing	21.41%
Relying	22.12%
Lying	15.82%
Helmet	29.53%

3.4. Model Development

The detection algorithm is another decisive factor affecting the detection accuracy. Deep learning algorithms have received wide attention for their potential to improve construction safety and production efficiency. R-CNN and fast R-CNN have been proposed successively and have dramatically improved the accuracy of target recognition. Faster R-CNN, a deep learning algorithm with an "attention" mechanism, introduces the region proposal network (RPN), further shortening the model's training time and the detection network's running time. Based on the convolutional neural network framework of convolutional architecture for fast feature embedding (Caffe), a faster R-CNN is mainly composed of two modules. One of the modules is the RPN, which generates a more accurate, high-quality candidate frame position by premarking the targets' possible positions. The other module is the fast R-CNN detector, which aims to improve the target recognition area based on the RPN's candidate frame.

The entire training process works by alternate training of the RPN network and the fast R-CNN detector, and both use the Zeiler and Fergus network (ZF-net) [51]. Figure 3 shows the flowchart of the model training for workers' unsafe actions. The procedure to implement the workers' unsafe action detection model is described as follows:

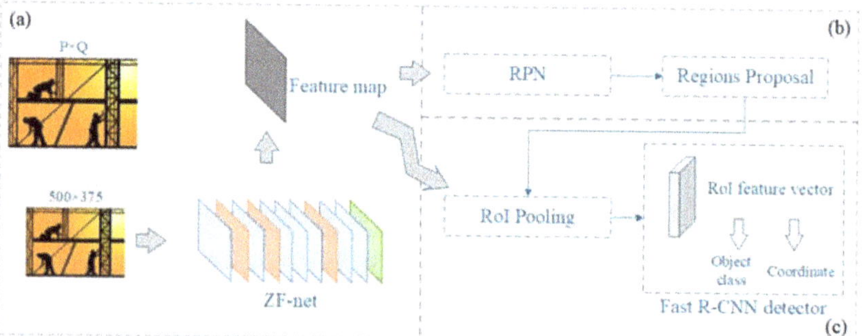

Figure 3. Flowchart of the model training. (**a**) Input the training samples to the ZF-net; (**b**) Generate region proposals using the RPN network; (**c**) Integrate feature maps and proposals and sent to the Faster R-CNN detector to determine the target category.

Stage 1: Input the training samples to the ZF-net for pretraining. Then, conduct alternate training of the model to acquire the first stage RPN and the fast R-CNN detector. Alternate training means to obtain the first stage ZF-net and RPN through the first round of training, and then the first round of the fast R-CNN detector training uses the training samples and the first stage RPN.

Stage 2: The second stage of training is almost the same as the first stage, except the input parameters are the results obtained in the first stage of training. After the second stage of training ends, the obtained ZF-net is saved as the final training network.

Stage 3: The validation sample is entered, and the final ZF-net is used to adjust and update the RPN network and the fast R-CNN. Finally, the proposed detection model is obtained.

4. Experiment and Results

4.1. Experiment

The data were mostly acquired from graduate students, while a small amount of data were acquired from workers on the construction site. College students and workers are adults, and their movements are similar. Since the dataset construction needed to consider image quality factors, such as illumination conditions and different shooting angles, using students as experimental subjects facilitated the collection of a large number of images. Therefore, we selected students' images as the training set for the model. There were

31 graduate students, with heights ranging from 158 cm to 181 cm, who participated in the experiments. The actions of throwing, relying, lying, jumping, and not wearing a helmet were examples of unsafe actions. The image sample collection followed 6 principles:

(1) Each student must perform the 5 actions, i.e., throwing, lying, relying, jumping, and without helmets, with their usual manner of behavior.

(2) Each type of unsafe action would be taken in different scenarios, with different shooting angles and lighting conditions.

(3) For each class of unsafe actions, 3–5 sequential images as a group were collected to reflect a continuously varying action.

(4) Images of poor quality were filtered out and deleted, such as indistinct images and targets with a small proportion of images.

(5) The originally collected images were preprocessed through contrast enhancement, and then included as part of the samples. Adding preprocessed images into the dataset increased the sample size of the dataset and improved the deep learning convolutional neural network model [31,49].

(6) Samples were reshaped in the dataset to a resolution of 375 × 500.

The data collected from construction workers were mainly about wearing a helmet. The total number of participants was approximately 50. Figure 4 shows examples of the 5 unsafe actions image samples. The samples were labeled using the sample labeling tool LabelImag, and the labeled annotation files were saved in the xml format file. The samples were finally divided into 3 groups, i.e., the training data, the validation data, and the testing data. Finally, the dataset was ready, which included 5 action types of images, annotation files, and image sets.

Figure 4. Sample examples.

At the implementation stage, the Faster-RCNN algorithm was trained using 20,000 iterations with a learning rate of 0.001. The dataset was the improved VOC 2007, which has been built in Section 3.3. The proposed method was implemented in MATLAB. For the hardware configuration, the model was tested on a computer with Intel(R) Core(TM) i7-6700 CPU @ 3.40 GHz, memory card of 16.0 GB, GPU of NVIDIA GeForce GTX 1080 Ti, and Windows 10 64-bit OS.

4.2. Results

After training, the ZF-net model file for detecting unsafe actions was obtained. The remaining samples were used to test the final model, and the testing results are shown in Figure 5. The average detection time of the sample test is 0.042 s.

Figure 5. Testing results.

Some concepts of target detection need to be explained as follows: True positive (TP), the input images are positive samples, and the detection results are also positive samples; false-positive (FP), the input images are negative samples, but the detection results are positive samples; true negative (TN), the input images are negative samples, and the detection results are negative samples; false-negative (FN), the input images are positive samples, but the detection results are negative. Take the action of relying on a railing as an example. If the testing result is "relying", it is recorded as TP; if the testing result is without relying, an FN is recorded. However, if no worker is relying on the test image's railing, but a "relying" is detected, FP is recorded. Figure 6 shows examples of TP, FN, and FP samples.

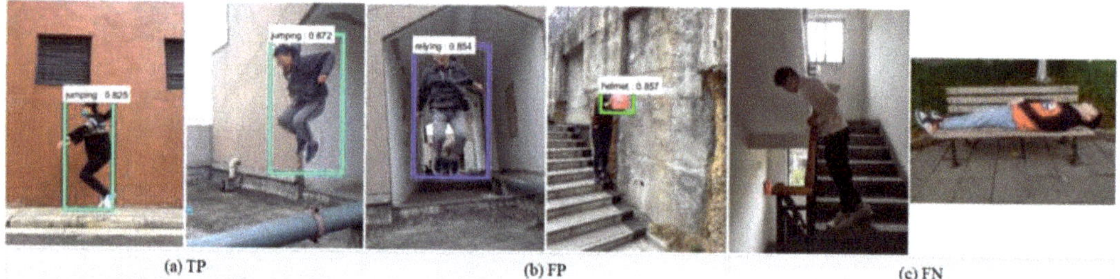

(a) TP   (b) FP   (c) FN

Figure 6. Examples of (**a**) TP, (**b**) FP, and (**c**) FN samples.

The analysis of TP, TN, FP, and FN's test results with the fourfold table are shown in Figure 7, which helps to understand the distribution of test results.

	Positive (actually)	Negative (actually)
Positive (detection)	93.5%	0.3%
Negative (detection)	6.2%	0

Figure 7. Test results with the fourfold table.

This experiment used four key performance indicators (KPIs) to assess the performance of the model for detecting unsafe actions: (1) accuracy, (2) precision, (3) recall, and (4) F1 measures.

Accuracy, i.e., the ratio of the TPs and TNs to the total detections, is generally used to evaluate the global accuracy of a training model. Precision is an indicator of the accuracy of prediction and recall is a measurement of the coverage area.

The F1 measure is the weighted harmonic average of precision and recall, which evaluates the model's quality. The higher the F1 measure, the more ideal the model.

The overall performance of the model is good. The test results were 93.46% for accuracy, 99.71% for precision, 93.72% for recall, and 96.63% for the F1 measure.

4.2.1. Sample Source Analysis

For the sample source analysis, we established a dataset of workers' unsafe actions and trained a CNN model for workers' unsafe action detection. The model's performance based on four indicators was good; the results for all of the indicators were above 90%. However, the model's performance indicators were based on the examinees added into the dataset. The images in the dataset had a fairly high degree of similarity. The results could be understood as the model mainly aimed at workers in the same project and construction scenes. However, unsafe actions outside of the dataset should be considered, for example, different scenarios and different examinees. Therefore, we tested the examinees outside the dataset as a comparison group to analyze whether the model applied to other types of scenarios and workers.

Similar to the previous dataset establishment method, the comparison group examinees were asked to perform the five unsafe actions according to their usual manner of behavior. The shooting scene of the comparison group was entirely different from the scene in the original dataset. The results are shown in Table 3 and Figure 8.

Table 3. Test results of the model's overall performance.

Indicators	Accuracy	Precision	Recall	F1-Measure
Original	93.46%	99.71%	93.72%	96.63%
Comparison	83.38%	87.79%	63.79%	73.89%

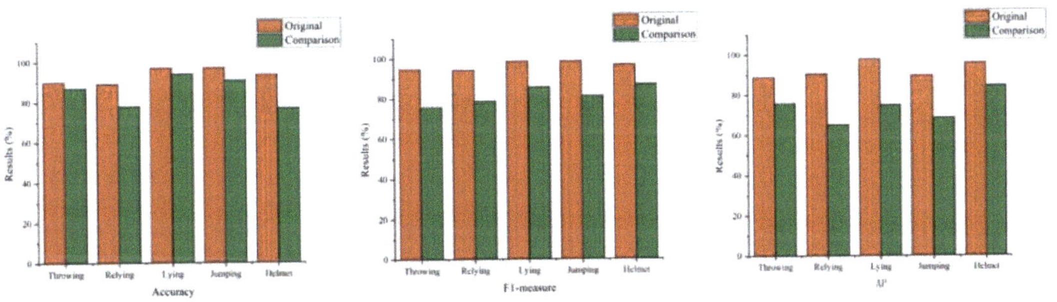

Figure 8. Test results of the model's performance on each unsafe action (between the original and comparison group).

We compared the testing results of the comparison group with the original dataset. The performance of the comparison group was worse than the original samples. Nevertheless, they were still more than 60 percent accurate. The reason could be attributed to the individual differences in the actions performed. There are significant numbers of construction sites, as well as numbers of workers. An unsafe action detection dataset built for a specific engineering construction site is more suitable for workers at a specific construction site for a long time period. To detect workers' unsafe actions at other engineering projects, rebuilding the unsafe action model dedicated to those projects would be more conducive to observing workers. The interference environment of images also affects the recognition accuracy, such as light, rain and fog.

4.2.2. FPs Analyzation

In the FP analysis, it was observed that FPs were usually due to the similar characteristics of two unsafe actions. For example, Figure 9 shows an image sequence of a throwing action. The testing result of the throwing action in Figure 9 was erroneously detected as relying on the railing. At the early stage of the throwing action, the detection result was relying (Figure 9a). As the image sequence changed, the detection was a result of both throwing and relying on the railing (Figure 9b). At the end of the throwing action, a throwing result was detected when the projectile was about to leave the hand (Figure 9c).

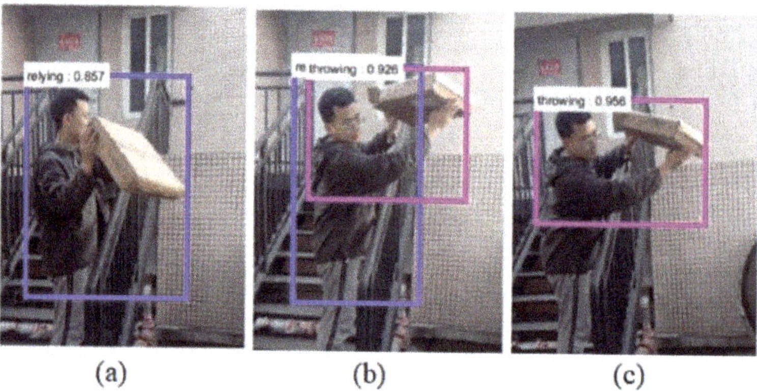

(a)   (b)   (c)

Figure 9. Testing results analysis for a FP sample. (**a**) At the early stage of the throwing action, the detection result was relying; (**b**) As the image sequence changed, the detection was a result of both throwing and relying on the railing; (**c**) At the end of the throwing action, a throwing result was detected when the projectile was about to leave the hand.

There are two explanations for the phenomenon that appeared above.

First, most of the image samples of the relying action in this study were relying on the railing. When performing sample labeling, the railing object was usually included in the sample labeling box of the relying action. The CNN learns all types of features of the images. These features include but are not limited to color, shape, and texture. Therefore, when the ZF-net was learning action features, it might mistake the railing for the feature of the relying action.

Secondly, the recognition method of the human skeleton may help to explain this problem. In previous studies, the posture of the human body could be distinguished by the distribution of human bone parameters [61], such as elbow angle and torso angle. The similarity of the human bone shape parameters in these two actions (throwing and relying) was very high, as shown in Figure 9. When a worker throws an object beside a railing, it is accompanied by a relying posture. Hence, it was not surprising that such a result was reached. In recent related research, Openpose has been used as an advanced algorithm that could achieve accurate human posture. The general process of CNN in learning image features has usually been divided into three layers [57], which could simply reveal the process of accepting and understanding the news for the V1. In the first layer, the learned features usually indicate whether there are edges in a specific direction and position in the image. In the second layer, the pattern is detected by discovering the specific arrangement of the edges, regardless of small changes in edge positions. In the third layer, the patterns are assembled into larger combinations of parts corresponding to familiar objects, and subsequent layers detect the object as a combination of these parts. This is a bottom-up strategy. In contrast to this learning strategy, part of the affinity field (PAF) is used in Openpose, which can improve the ability of computer vision in human pose estimation [62]. This provides a new research method in workers' unsafe actions management.

4.2.3. Helmet Detection Analysis

For the helmet detection analysis, the action detection of whether the worker wears a helmet is different from other unsafe action detection methods. Detecting only the helmet could not determine whether the worker is wearing a helmet. Figure 10 is an intuitive explanation of this conclusion. When a helmet is not worn, it could also be detected that the object is a helmet (Figure 10a). However, based on "people" detection, helmet detection could effectively avoid this problem. Therefore, on the one hand, for the helmet-wearing test, if there is a result that the person and the helmet are detected simultaneously, as shown in Figure 10b, it was defined as safe. On the other hand, if only one of them appears, it could not judge whether it was an unsafe action, Figure 10a.

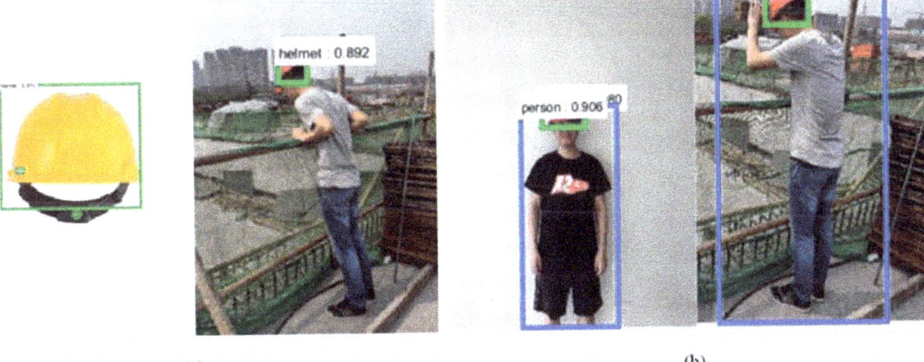

(a)   (b)

Figure 10. Helmet wearing test (**a**) Helmet; (**b**) Helmet & person.

5. Discussion

This research proposed an automatic identification method based on an improved Faster R-CNN algorithm to detect workers' unsafe actions considering a heights working environment. Based on the method, this research designs and carries out a series of experiments involving five types of unsafe actions to examine their efficiency and accuracy. The results illustrate and verify the method's feasibility for improving safety inspection and supervision, as well as its limitations. According to the experiment and results, it could be found that it is an excellent way to detect workers' unsafe behaviors through the proposed method.

Compared with previous studies that have utilized computer vision for construction management, this research has the following advantages. This work combines human knowledge with computer semantics that consider a high workplace. Thus, leading to better intervention in construction safety management when workers work at heights. For the observation and recording part of BBS, manually observing was a waste of human resources and could not capture the worker's action information comprehensively. Computer vision has an advantage in this aspect. Computer vision has been utilized to observe workers, including the efficiency of their activities [49], helmet-wearing [37], and even construction activities at night-time [63]. These works are significant to project management and achieved a good effect. However, accidents are most likely to occur when workers work at heights. There is a lack of research on observing workers' behavior in high working environments. The proposed method is for workers' observation in a high working place. There was a dataset of unsafe actions that commonly happen in high places on construction sites. It may provide a more reliable method for observing workers' behavior in a high places.

Three factors affect the robustness of the model:

(1) Whether a deep learning algorithm could fully extract image features.
(2) Whether the dataset could adequately represent the detection object. Deep learning is learning the features in a dataset. Therefore, the characteristics contained in the dataset determine the final effect of the model.
(3) The quality of the image used for recognition also affects the robustness. It includes the quality of data acquisition equipment, image acquisition angle and object obscured, and the influence of environmental factors, such as light, rain and fog on image quality.

This research also has application limitations. It lacks a large-scale dataset. There comprised approximately 2200 images utilized for the CNN model training, which was deemed to be small. The quality and quantity of images in the dataset will affect the effect of the model. Future studies will consider further dataset improvement to enhance the robustness of the model.

6. Conclusions

Most construction sites are equipped with cameras to observe their safety status. However, manual observation is laborious and may not accurately capture workers' unsafe behavior. Computer vision technology's time-sparing and intelligent advantages could help construction safety management when workers work at heights. This paper proposed a deep learning model that could be applied to detect unsafe actions when working in a high place automatically. The workers' unsafe actions worth observing and detecting were defined first to achieve this work. The dataset including five workers' unsafe actions for deep learning was built. Finally, an automatic recognition model was built for training, validation, and testing the unsafe actions model. The model's accuracy in detecting throwing, relying, lying, jumping actions, and helmet wearing were 90.14%, 89.19%, 97.18%, 97.22%, and 93.67%, respectively.

This work combines human knowledge with computer semantics, thus leading to a better intervention in construction safety management when workers work at heights. Its contribution is to enable computers to identify the unsafe behavior of workers in a high working environment. Its application is that it could provide workers' unsafe action information intelligently to safety managers and assist with intervention. Besides, an unsafe action detection dataset built for a specific engineering construction site is more suitable for workers who would engage with the specific construction site for a long time. It could become a new means for the BBS observation procedure. All of these would benefit workers, managers, and supervisors working in a hazardous construction environment.

Since the research mainly focuses on whether unsafe actions could be well-detected, all scenarios where workers made these actions were not fully considered in the dataset production process. According to the occurrence rules of the accident, an accident is often coupled with multiple factors. Although unsafe actions were an essential factor in accidents, they do not lead to accidents directly but in a specific scenario.

Author Contributions: Y.B. proposed the conceptualization and wrote this manuscript under the supervision of Q.H. L.H. provided the resources. J.H. and H.W. conducted the experiment planning and setup. G.C. provided valuable insight in preparing this manuscript. All authors have read and agreed to the published version of the manuscript.

Funding: This research was funded by the National Natural Science Foundation of China (52178357), the Sichuan Science and Technology Program (2021YFSY0307, 2021JDRC0076), and the Graduate Research and Innovation Fund Program of SWPU (2020cxyb009).

Data Availability Statement: All data generated or analyzed during the study are included in the submitted article.

Conflicts of Interest: The authors declare no conflict of interest.

References

1. Fang, W.; Ma, L.; Love, P.E.; Luo, H.; Ding, L.; Zhou, A. Knowledge graph for identifying hazards on construction sites: Integrating computer vision with ontology. *Autom. Constr.* **2020**, *119*, 103310. [CrossRef]
2. Lingard, H. Occupational health and safety in the construction industry. *Constr. Manag. Econ.* **2013**, *31*, 505–514. [CrossRef]
3. Available online: https://www.osha.gov/data/work (accessed on 16 February 2022).
4. Ministry of Housing and Urban-Rural Development of China (MHURDC), 2010–2019. Available online: https://www.mohurd.gov.cn/ (accessed on 16 February 2022).
5. Li, H.; Lu, M.; Hsu, S.-C.; Gray, M.; Huang, T. Proactive behavior-based safety management for construction safety improvement. *Saf. Sci.* **2015**, *75*, 107–117. [CrossRef]
6. Li, S.; Wu, X.; Wang, X.; Hu, S. Relationship between Social Capital, Safety Competency, and Safety Behaviors of Construction Workers. *J. Constr. Eng. Manag.* **2020**, *146*, 4020059. [CrossRef]
7. Guo, S.; Ding, L.; Luo, H.; Jiang, X. A Big-Data-based platform of workers' behavior: Observations from the field. *Accid. Anal. Prev.* **2016**, *93*, 299–309. [CrossRef]
8. Heinrich, H. *Industrial Accident Prevention: A Safety Management Approach*; Mcgraw-Hill Book Company: New York, NY, USA, 1980; Volume 468.
9. Chen, D.; Tian, H. Behavior Based Safety for Accidents Prevention and Positive Study in China Construction Project. In Proceedings of the International Symposium on Safety Science and Engineering in China, Beijing, China, 7–9 November 2012; Volume 43, pp. 528–534. [CrossRef]
10. Zhang, M.; Fang, D. A continuous Behavior-Based Safety strategy for persistent safety improvement in construction industry. *Autom. Constr.* **2013**, *34*, 101–107. [CrossRef]
11. Ismail, F.; Hashim, A.; Zuriea, W.; Ismail, W.; Kamarudin, H.; Baharom, Z. Behaviour Based Approach for Quality and Safety Environment Improvement: Malaysian Experience in the Oil and Gas Industry. *Procedia-Soc. Behav. Sci.* **2012**, *35*, 586–594. [CrossRef]
12. Barber, B. Resistance by Scientists to Scientific Discovery. *Science* **1961**, *134*, 596–602. [CrossRef]
13. Kuhn, T. *The Structure of Scientific Revolutions*; University of Chicago Press: Chicago, IL, USA, 1996.
14. Ding, L.; Fang, W.; Luo, H.; Love, P.; Zhong, B.T.; Ouyang, X. A deep hybrid learning model to detect unsafe behavior: Integrating convolution neural networks and long short-term memory. *Autom. Constr.* **2018**, *86*, 118–124. [CrossRef]
15. Bhagwat, K.; Kumar, V.S.; Nanthagopalan, P. Construction Safety Performance Measurement Using Leading Indicator-based Jobsite Safety Inspection Method—A Case Study of Building Construction Project. *Int. J. Occup. Saf. Ergon.* **2021**, 1–27. [CrossRef]
16. Khosrowpour, A.; Niebles, J.; Fard, M. Vision-based workface assessment using depth images for activity analysis of interior construction operations. *Autom. Constr.* **2014**, *48*, 74–87. [CrossRef]
17. Cheng, T.; Teizer, J.; Migliaccio, G.; Gatti, U. Automated task-level activity analysis through fusion of real time location sensors and worker's thoracic posture data. *Autom. Constr.* **2013**, *29*, 24–39. [CrossRef]
18. Guo, H.; Yu, Y.; Skitmore, M. Visualization technology-based construction safety management: A review. *Autom. Constr.* **2017**, *73*, 135–144. [CrossRef]
19. Li, H.; Li, X.; Luo, X.; Siebert, J. Investigation of the causality patterns of non-helmet use behavior of construction workers. *Autom. Constr.* **2017**, *80*, 95–103. [CrossRef]
20. Lingard, H.; Rowlinson, S. Behavior-based safety management in Hong Kong's construction industry. *J. Saf. Res.* **1997**, *28*, 243–256. [CrossRef]
21. Li, H.; Chan, G.; Wong, J.; Skitmore, M. Real-time locating systems applications in construction. *Autom. Constr.* **2016**, *63*, 37–47. [CrossRef]
22. Kim, H.; Kim, H.; Hong, Y.; Byun, Y. Detecting Construction Equipment Using a Region-Based Fully Convolutional Network and Transfer Learning. *J. Comput. Civ. Eng.* **2018**, *32*, 04017082. [CrossRef]
23. Angah, O.; Chen, A. Tracking multiple construction workers through deep learning and the gradient based method with re-matching based on multi-object tracking accuracy. *Autom. Constr.* **2020**, *119*, 103308. [CrossRef]
24. Hu, Q.; Bai, Y.; He, L.; Cai, Q. Intelligent Framework for Worker-Machine Safety Assessment. *J. Constr. Eng. Manag.* **2020**, *146*, 04020045. [CrossRef]
25. Hu, Q.; Ren, Y.; He, L.; Bai, Y. A Novel Vision Based Warning System For Proactive Prevention Of Accidenets Induced By Falling Objects Based On Building And Environmental Engineering Requirements. *Fresenius Environ. Bull.* **2020**, *29*, 7867–7876.
26. Liu, X.; Hu, Y.; Wang, F.; Liang, Y.; Liu, H. Design and realization of a video monitoring system based on the intelligent behavior identify technique. In Proceedings of the 9th International Congress on Image and Signal Processing, Biomedical Engineering and Informatics, Datong, China, 15–17 October 2016.
27. Vigneshkumar, C.; Salve, U.R. A scientometric analysis and review of fall from height research in construction. *Constr. Econ. Build.* **2020**, *20*, 17–35. [CrossRef]
28. Hoła, A.; Sawicki, M.; Szóstak, M. Methodology of Classifying the Causes of Occupational Accidents Involving Construction Scaffolding Using Pareto-Lorenz Analysis. *Appl. Sci.* **2018**, *8*, 48. [CrossRef]
29. Szóstak, M.; Hoła, B.; Bogusławski, P. Identification of accident scenarios involving scaffolding. *Autom. Constr.* **2021**, *126*, 103690. [CrossRef]

30. Martínez-Rojas, M.; Gacto, M.J.; Vitiello, A.; Acampora, G.; Soto-Hidalgo, J.M. An Internet of Things and Fuzzy Markup Language Based Approach to Prevent the Risk of Falling Object Accidents in the Execution Phase of Construction Projects. *Sensors* **2021**, *21*, 6461. [CrossRef] [PubMed]

31. Fang, W.; Ding, L.; Zhong, B.; Love, P.; Luo, H. Automated detection of workers and heavy equipment on construction sites: A convolutional neural network approach. *Adv. Eng. Inform.* **2018**, *37*, 139–149. [CrossRef]

32. Fan, H.; Su, H.; Guibas, L. A Point Set Generation Network for 3D Object Reconstruction from a Single Image. In Proceedings of the IEEE Conference on Computer Vision And Pattern Recognition (CVPR 2017), Honolulu, HI, USA, 21–26 July 2017; pp. 2463–2471. [CrossRef]

33. Zhu, J.; Wan, X.; Wu, C.; Xu, C. Object detection and localization in 3D environment by fusing raw fisheye image and attitude data. *J. Vis. Commun. Image Represent.* **2019**, *59*, 128–139. [CrossRef]

34. Dalal, N.; Triggs, B. Histograms of Oriented Gradients for Human Detection. In Proceedings of the Computer Vision and Pattern Recognition, San Diego, CA, USA, 20–26 June 2005. [CrossRef]

35. Rubaiyat, A.; Toma, T.; Masoumeh, K.; Rahman, S.; Chen, L.; Ye, Y.; Pan, C. Automatic Detection Of Helmet Uses For Construction Safety. In Proceedings of the IEEE/WIC/ACM International Conference on Web Intelligence Workshops, Leipzig, Germany, 23–26 August 2017.

36. Krizhevsky, A.; Sutskever, I.; Hinton, G. Imagenet Classification with Deep Convolutional Neural Networks. *Commun. ACM* **2012**, *60*, 84–90. [CrossRef]

37. Fang, Q.; Li, H.; Luo, X.; Ding, L.; Luo, H.; Rose, T.; An, W. Detecting non-hardhat-use by a deep learning method from far-field surveillance videos. *Autom. Constr.* **2018**, *85*, 1–9. [CrossRef]

38. Qiao, L.; Qie, Y.; Zhu, Z.; Zhu, Y.; Zaman, U.; Anwer, N. An ontology-based modelling and reasoning framework for assembly sequence planning. *Int. J. Adv. Manuf. Technol.* **2017**, *94*, 4187–4197. [CrossRef]

39. Fang, W.; Zhong, B.; Zhao, N.; Love, P.; Luo, H.; Xue, J.; Xu, S. A deep learning-based approach for mitigating falls from height with computer vision: Convolutional neural network. *Adv. Eng. Inform.* **2019**, *39*, 170–177. [CrossRef]

40. Fang, W.; Ding, L.; Luo, H.; Love, P. Falls from heights: A computer vision-based approach for safety harness detection. *Autom. Constr.* **2018**, *91*, 53–61. [CrossRef]

41. Lateef, F.; Ruichek, Y. Survey on semantic segmentation using deep learning techniques. *Neurocomputing* **2019**, *338*, 321–348. [CrossRef]

42. Zhu, J.; Zeng, H.; Liao, S.; Lei, Z.; Cai, C.; Zheng, L. Deep Hybrid Similarity Learning for Person Re-Identification. *IEEE Trans. Circuits Syst. Video Technol.* **2018**, *28*, 3183–3193. [CrossRef]

43. Hubel, D.; Wiesel, T. Receptive fields of single neurones in the cat's striate cortex. *J. Physiol.* **1959**, *148*, 574–591. [CrossRef] [PubMed]

44. Hubel, D.; Wiesel, T. Receptive fields, binocular interaction and functional architecture in the cat's visual cortex. *J. Physiol.* **1962**, *160*, 106–154. [CrossRef]

45. Goodfellow, I.; Bengio, Y.; Courville, A. *Deep Learning*; The MIT Press: Cambridge, MA, USA, 2016.

46. Russakovsky, O.; Deng, J.; Su, H.; Krause, J.; Satheesh, S.; Ma, S.; Huang, Z.; Karpathy, A.; Khosla, A.; Bernstein, M.; et al. ImageNet Large Scale Visual Recognition Challenge. *Int. J. Comput. Vis.* **2014**, *115*, 211–252. [CrossRef]

47. Lin, T.Y.; Maire, M.; Belongie, S.; Hays, J.; Zitnick, C.L. *Microsoft COCO: Common Objects in Context*; Springer International Publishing: Cham, Switzerland, 2014. [CrossRef]

48. Kolar, Z.; Chen, H.; Luo, X. Transfer learning and deep convolutional neural networks for safety guardrail detection in 2D images. *Autom. Constr.* **2018**, *89*, 58–70. [CrossRef]

49. Luo, H.; Xiong, C.; Fang, W.; Love, P.; Zhang, B.; Ouyang, X. Convolutional neural networks: Computer vision-based workforce activity assessment in construction. *Autom. Constr.* **2018**, *94*, 282–289. [CrossRef]

50. Lecun, Y.; Bottou, L.; Bengio, Y.; Haffner, P. Gradient-based learning applied to document recognition. *Proc. IEEE* **1998**, *86*, 2278–2324. [CrossRef]

51. Zeiler, M.; Fergus, R. Visualizing and Understanding Convolutional Networks. In Proceedings of the European Conference on Computer Vision, Zurich, Switzerland, 6–12 September 2014. [CrossRef]

52. Simonyan, K.; Zisserman, A. Very Deep Convolutional Networks for Large-Scale Image Recognition. *Comput. Sci.* **2014**. [CrossRef]

53. He, K.; Zhang, X.; Ren, S.; Sun, J. Deep residual learning for image recognition. In Proceedings of the IEEE Conference on Computer Vision and Pattern Recognition (CVPR), Las Vegas, NV, USA, 30 June 2016; pp. 770–778. [CrossRef]

54. Szegedy, C.; Vanhoucke, V.; Ioffe, S.; Shlens, J.; Wojna, Z. Rethinking the Inception Architecture for Computer Vision. In Proceedings of the IEEE Computer Society Conference on Computer Vision and Pattern Recognition, Boston, MA, USA, 7–12 June 2015; pp. 2818–2826.

55. Morteza, H. Learning representations from dendrograms. *Mach. Learn.* **2020**, *109*, 1779–1802. [CrossRef]

56. Glorot, X.; Bengio, Y. Understanding the difficulty of training deep feedforward neural networks. *J. Mach. Learn. Res.* **2010**, *9*, 249–256.

57. LeCun, Y.; Bengio, Y.; Hinton, G. Deep learning. *Nature* **2015**, *521*, 436–444. [CrossRef]

58. Chu, M.; Matthews, J.; Love, P. Integrating mobile Building Information Modelling and Augmented Reality systems: An experimental study. *Autom. Constr.* **2018**, *85*, 305–316. [CrossRef]

59. *GB 50870-2013*; Unified Code for Technique for Constructional Safety. China Planning Press: Beijing, China, 2013.

60. Wang, H.; Li, X. *Construction Safety Technical Manual*; China Building Materials Press: Beijing, China, 2008.
61. Yu, Y.; Guo, H.; Ding, Q.; Li, H.; Skitmore, M. An experimental study of real-time identification of construction workers' unsafe behaviors. *Autom. Constr.* **2017**, *82*, 193–206. [CrossRef]
62. Cao, Z.; Hidalgo, M.; Simon, T.; Wei, S.; Sheikh, Y. OpenPose: Realtime Multi-Person 2D Pose Estimation Using Part Affinity Fields. *IEEE Trans. Pattern Anal. Mach. Intell.* **2019**, *43*, 172–186. [CrossRef]
63. Xiao, B.; Lin, Q.; Chen, Y. A vision-based method for automatic tracking of construction machines at nighttime based on deep learning illumination enhancement. *Autom. Constr.* **2021**, *127*, 103721. [CrossRef]

MDPI
St. Alban-Anlage 66
4052 Basel
Switzerland
Tel. +41 61 683 77 34
Fax +41 61 302 89 18
www.mdpi.com

MDPI Books Editorial Office
E-mail: books@mdpi.com
www.mdpi.com/books